Texts in Computational Science and Engineering

22

Editors

More information about this series at http://www.springer.com/series/5151

Tom Lyche

Numerical Linear Algebra and Matrix Factorizations

Springer

Tom Lyche
Blindern
University of Oslo
Oslo, Norway

ISSN 1611-0994 ISSN 2197-179X (electronic)
Texts in Computational Science and Engineering
ISBN 978-3-030-36470-0 ISBN 978-3-030-36468-7 (eBook)
https://doi.org/10.1007/978-3-030-36468-7

Mathematics Subject Classification (2010): 15-XX, 65-XX

This Springer imprint is published by the registered company Springer Nature Switzerland AG.
The registered company address is: Gewerbestrasse 11, 6330 Cham, Switzerland

Foreword

It is a pleasure to write this foreword to the book "Numerical Linear Algebra and Matrix Factorizations" by Tom Lyche. I see this book project from three perspectives, corresponding to my three different roles: first, as a friend and close colleague of Tom for a number of years, secondly as the present department head, and, finally, as a researcher within the international linear algebra and matrix theory community. The book actually has a long history and started out as lecture notes that Tom wrote for a course in numerical linear algebra. For almost forty years this course has been an important and popular course for our students in mathematics, both in theoretical and more applied directions, as well as students in statistics, physics, mechanics and computer science. These notes have been revised multiple times during the years, and new topics have been added. I have had the pleasure to lecture the course myself, using Tom's lecture notes, and I believe that both the selection of topics and the combined approach of theory and algorithms is very appealing. This is also what our students point out when they have taken this course. As we know, the area presented in this book play a highly central role in many applications of mathematics and in scientific computing in general. Sometimes, in the international linear algebra and matrix theory community, one divides the area into numerical linear algebra, applied linear algebra and core (theoretical) linear algebra. This may serve some purpose, but often it is fruitful to have a more unified view on this, in order to see the interplay between theory, applications and algorithms. I think this view dominates this book, and that this makes the book interesting to a wide range of readers. Finally, I would like to thank Tom for his work with this book and the mentioned course, and for being a good colleague from whom I have learned a lot. I know that his international research community in spline theory also share this view. Most importantly, I hope that you, the reader, will enjoy the book!

Oslo, Norway
June 2019

Geir Dahl

Preface

This book, which has grown out of a one semester course at the University of Oslo, targets upper undergraduate and beginning graduate students in mathematics, statistics, computational physics and engineering who need a mathematical background in numerical linear algebra and related matrix factorizations.

Mastering the material in this book should enable a student to analyze computational problems and develop his or her own algorithms for solving problems of the following kind,

- *System of linear equations.* Given a (square) matrix $\boldsymbol{A}$ and a vector $\boldsymbol{b}$. Find a vector $\boldsymbol{x}$ such that $\boldsymbol{A}\boldsymbol{x} = \boldsymbol{b}$.
- *Least squares.* Given a (rectangular) matrix $\boldsymbol{A}$ and a vector $\boldsymbol{b}$. Find a vector $\boldsymbol{x}$ such that the sum of squares of the components of $\boldsymbol{b} - \boldsymbol{A}\boldsymbol{x}$ is as small as possible.
- *Eigenvalues and eigenvectors.* Given a (square) matrix $\boldsymbol{A}$. Find a number λ and/or a nonzero vector $\boldsymbol{x}$ such that $\boldsymbol{A}\boldsymbol{x} = \lambda\boldsymbol{x}$.

Such problems can be large and difficult to handle, so much can be gained by understanding and taking advantage of special structures. For this we need a good understanding of basic numerical linear algebra and matrix factorizations. Factoring a matrix into a product of simpler matrices is a crucial tool in numerical linear algebra for it allows one to tackle large problems through solving a sequence of easier ones.

The main characteristics of this book are as follows:

1. It is self-contained, only assuming first year calculus, an introductory course in linear algebra, and some experience in solving mathematical problems on a computer. A special feature of this book is the detailed proofs of practically all results. Parts of the book can be studied independently making it suitable for self study.
2. There are numerous exercises which can be found at the end of each chapter. In a separate book we offer solutions to all problems. Solutions of many exam problems given for this course at the University of Oslo are included in this separate volume.

3. The book, consisting of an introductory first chapter and 15 more chapters, naturally disaggregating into six thematically related parts. The chapters are designed to be suitable for a one week per chapter one semester course. Toward the goal of being self-contained, the first chapter contains a review of linear algebra, and is provided to the reader for convenient occasional reference.
4. Many of the chapters contain material beyond what might normally be covered in one week of lectures. A typical 15 week semester's curriculum could consist of the following curated material

LU and QR factorizations	2.4, 2.5, 3.2, 3.3, 3.5, 4.1, 4.2, 5.1 – 5.4, 5.6
SVD, norms and LSQ	6.1, 6.3, 7.1 – 7.4, 8.1 – 8.3, 9.1 – 9.3, 9.4.1
Kronecker products	10.1, 10.2, 10.3, 11.1, 11.2, 11.3
Iterative methods	12.1 – 12.4, 13.1 – 13.3, 13.5
Eigenpairs	14.1 – 14.5, 15.1 – 15.3

Chapters 2–4 give a rather complete treatment of various LU factorizations.

Chapters 5–9 cover QR and singular value factorizations, matrix norms, least squares methods and perturbation theory for linear equations and least squares problems.

Chapter 10 gives an introduction to Kronecker products. We illustrate their use by giving simple proofs of properties of the matrix arising from a discretization of the 2 dimensional Poison Equation. Also, we study fast methods based on eigenvector expansions and the Fast Fourier Transform in Chap. 11. Some background from Chaps. 2, 3 and 4 may be needed for Chaps. 10 and 11.

Iterative methods are studied in Chaps. 12 and 13. This includes the classical methods of Jacobi, Gauss Seidel Richardson and Successive Over Relaxation (SOR), as well as a derivation and convergence analysis of the methods of steepest descent and conjugate gradients. The preconditioned conjugate gradient method is introduced and applied to the Poisson problem with variable coefficients.

In Chap. 14 we consider perturbation theory for eigenvalues, the power method and its variants, and use the Inertia Theorem to find a single eigenvalue of a symmetric matrix. Chapter 15 gives a brief informal introduction to one of the most celebrated algorithms of the twentieth century, the QR method for finding all eigenvalues and eigenvectors of a matrix.
5. In this book we give many detailed numerical algorithms for solving linear algebra problems. We have written these algorithms as functions in MATLAB. A list of these functions and the page number where they can be found is included after the table of contents. Moreover, their listings can be found online at http://folk.uio.no/tom/numlinalg/code. Complexity is discussed briefly in Sect. 3.3.2. As for programming issues, we often vectorize the algorithms leading to shorter and more efficient programs. Stability is important both for the mathematical problems and for the numerical algorithms. Stability can be studied in terms of perturbation theory that leads to condition numbers, see Chaps. 8, 9 and 14. We

will often use phrases like "the algorithm is numerically stable" or "the algorithm is not numerically stable" without saying precisely what we mean by this. Loosely speaking, an algorithm is numerically stable if the solution, computed in floating point arithmetic, is the exact solution of a slightly perturbed problem. To determine upper bounds for these perturbations is the topic of *backward error analysis*. We refer to [7] and [17, 18] for an in-depths treatment.

A list of freely available software tools for solving linear algebra problems can be found at

www.netlib.org/utk/people/JackDongarra/la-sw.html

To supplement this volume the reader might consult Björck [2], Meyer [15] and Stewart [17, 18]. For matrix analysis the two volumes by Horn and Johnson [9, 10] contain considerable additional material.

Acknowledgments

I would like to thank my colleagues Elaine Cohen, Geir Dahl, Michael Floater, Knut Mørken, Richard Riesenfeld, Nils Henrik Risebro, Øyvind Ryan and Ragnar Winther for all the inspiring discussions we have had over the years. Earlier versions of this book were converted to LaTeX by Are Magnus Bruaset and Njål Foldnes with help for the final version from Øyvind Ryan. I thank Christian Schulz, Georg Muntingh and Øyvind Ryan who helped me with the exercise sessions and we have, in a separate volume, provided solutions to practically all problems in this book. I also thank an anonymous referee for useful suggestions. Finally, I would like to give a special thanks to Larry Schumaker for his enduring friendship and encouragement over the years.

Oslo, Norway
June 2019

Tom Lyche

Contents

Part V Iterative Methods for Large Linear Systems

List of Figures

List of Tables

Listings

Chapter 1
A Short Review of Linear Algebra

In this introductory chapter we give a compact introduction to linear algebra with emphasis on $\mathbb{R}^n$ and $\mathbb{C}^n$. For a more elementary introduction, see for example the book [13].

1.1 Notation

The following sets and notations will be used in this book.

1. The sets of natural numbers, integers, rational numbers, real numbers, and complex numbers are denoted by $\mathbb{N}$, $\mathbb{Z}$, $\mathbb{Q}$, $\mathbb{R}$, $\mathbb{C}$, respectively.
2. We use the "colon equal" symbol $v := e$ to indicate that the symbol v is defined by the expression e.
3. $\mathbb{R}^n$ is the set of n-tuples of real numbers which we will represent as bold face column vectors. Thus $\boldsymbol{x} \in \mathbb{R}^n$ means

$$\boldsymbol{x} = \begin{bmatrix} x_1 \\ x_2 \\ \vdots \\ x_n \end{bmatrix},$$

where $x_i \in \mathbb{R}$ for $i = 1, \ldots, n$. Row vectors are normally identified using the transpose operation. Thus if $\boldsymbol{x} \in \mathbb{R}^n$ then $\boldsymbol{x}$ is a column vector and $\boldsymbol{x}^T$ is a row vector.

T. Lyche, *Numerical Linear Algebra and Matrix Factorizations*,
Texts in Computational Science and Engineering 22,
https://doi.org/10.1007/978-3-030-36468-7_1

4. Addition and scalar multiplication are denoted and defined by

$$\boldsymbol{x}+\boldsymbol{y} := \begin{bmatrix} x_1+y_1 \\ \vdots \\ x_n+y_n \end{bmatrix}, \quad a\boldsymbol{x} := \begin{bmatrix} ax_1 \\ \vdots \\ ax_n \end{bmatrix}, \quad \boldsymbol{x}, \boldsymbol{y} \in \mathbb{R}^n, \quad a \in \mathbb{R}.$$

5. $\mathbb{R}^{m\times n}$ is the set of matrices $\boldsymbol{A}$ with real elements. The integers m and n are the number of rows and columns in the tableau

$$\boldsymbol{A} = \begin{bmatrix} a_{11} & a_{12} & \cdots & a_{1n} \\ a_{21} & a_{22} & \cdots & a_{2n} \\ \vdots & \vdots & \ddots & \vdots \\ a_{m1} & a_{m2} & \cdots & a_{mn} \end{bmatrix}.$$

The element in the ith row and jth column of $\boldsymbol{A}$ will be denoted by $a_{i,j}$, a_{ij}, $\boldsymbol{A}(i, j)$ or $(\boldsymbol{A})_{i,j}$. We use the notations

$$\boldsymbol{a}_{:j} := \begin{bmatrix} a_{1j} \\ a_{2j} \\ \vdots \\ a_{mj} \end{bmatrix}, \quad \boldsymbol{a}_{i:}^T := [a_{i1}, a_{i2}, \ldots, a_{in}], \quad \boldsymbol{A} = [\boldsymbol{a}_{:1}, \boldsymbol{a}_{:2}, \ldots \boldsymbol{a}_{:n}] = \begin{bmatrix} \boldsymbol{a}_{1:}^T \\ \boldsymbol{a}_{2:}^T \\ \vdots \\ \boldsymbol{a}_{m:}^T \end{bmatrix}$$

for the columns $\boldsymbol{a}_{:j}$ and rows $\boldsymbol{a}_{i:}^T$ of $\boldsymbol{A}$. We often drop the colon and write $\boldsymbol{a}_j$ and $\boldsymbol{a}_i^T$ with the risk of some confusion. If $m = 1$ then $\boldsymbol{A}$ is a row vector, if $n = 1$ then $\boldsymbol{A}$ is a column vector, while if $m = n$ then $\boldsymbol{A}$ is a square matrix. In this text we will denote matrices by boldface capital letters $\boldsymbol{A}, \boldsymbol{B}, \boldsymbol{C}, \cdots$ and vectors most often by boldface lower case letters $\boldsymbol{x}, \boldsymbol{y}, \boldsymbol{z}, \cdots$.

6. A **complex number** is a number written in the form $x = a + ib$, where a, b are real numbers and i, **the imaginary unit**, satisfies $i^2 = -1$. The set of all such numbers is denoted by $\mathbb{C}$. The numbers $a = \operatorname{Re} x$ and $b = \operatorname{Im} x$ are the **real and imaginary part** of x. The number $\overline{x} := a - ib$ is called the **complex conjugate** of $x = a + ib$, and $|x| := \sqrt{\overline{x}x} = \sqrt{a^2 + b^2}$ the **absolute value** or **modulus** of x. The **complex exponential function** can be defined by

$$e^x = e^{a+ib} := e^a(\cos b + i \sin b).$$

In particular,

$$e^{i\pi/2} = i, \quad e^{i\pi} = -1, \quad e^{2i\pi} = 1.$$

We have $e^{x+y} = e^x e^y$ for all $x, y \in \mathbb{C}$. The **polar form** of a complex number is

$$x = a + ib = re^{i\theta}, \quad r = |x| = \sqrt{a^2 + b^2}, \quad \cos\theta = \frac{a}{r}, \quad \sin\theta = \frac{b}{r}.$$

7. For matrices and vectors with complex elements we use the notation $\boldsymbol{A} \in \mathbb{C}^{m\times n}$ and $\boldsymbol{x} \in \mathbb{C}^n$. We define complex row vectors using either the transpose $\boldsymbol{x}^T$ or the conjugate transpose operation $\boldsymbol{x}^* := \overline{\boldsymbol{x}}^T = [\overline{x}_1, \ldots, \overline{x}_n]$. If $\boldsymbol{x} \in \mathbb{R}^n$ then $\boldsymbol{x}^* = \boldsymbol{x}^T$.
8. For $\boldsymbol{x}, \boldsymbol{y} \in \mathbb{C}^n$ and $a \in \mathbb{C}$ the operations of vector addition and scalar multiplication is defined by component operations as in the real case (cf. 4.).
9. The arithmetic operations on rectangular matrices are

- **matrix addition** $\boldsymbol{C} := \boldsymbol{A} + \boldsymbol{B}$ if $\boldsymbol{A}, \boldsymbol{B}, \boldsymbol{C}$ are matrices of the same size, i.e., with the same number of rows and columns, and $c_{ij} := a_{ij} + b_{ij}$ for all i, j.
- **multiplication by a scalar** $\boldsymbol{C} := \alpha\boldsymbol{A}$, where $c_{ij} := \alpha a_{ij}$ for all i, j.
- **matrix multiplication** $\boldsymbol{C} := \boldsymbol{A}\boldsymbol{B}$, $\boldsymbol{C} = \boldsymbol{A} \cdot \boldsymbol{B}$ or $\boldsymbol{C} = \boldsymbol{A} * \boldsymbol{B}$, where $\boldsymbol{A} \in \mathbb{C}^{m\times p}$, $\boldsymbol{B} \in \mathbb{C}^{p\times n}$, $\boldsymbol{C} \in \mathbb{C}^{m\times n}$, and $c_{ij} := \sum_{k=1}^{p} a_{ik}b_{kj}$ for $i = 1, \ldots, m$, $j = 1, \ldots, n$.
- **element-by-element matrix operations** $\boldsymbol{C} := \boldsymbol{A} \times \boldsymbol{B}$, $\boldsymbol{D} := \boldsymbol{A}/\boldsymbol{B}$, and $\boldsymbol{E} := \boldsymbol{A} \wedge r$ where all matrices are of the same size and $c_{ij} := a_{ij}b_{ij}$, $d_{ij} := a_{ij}/b_{ij}$ and $e_{ij} := a_{ij}^r$ for all i, j and suitable r. For the division $\boldsymbol{A}/\boldsymbol{B}$ we assume that all elements of $\boldsymbol{B}$ are nonzero. The element-by-element product $\boldsymbol{C} = \boldsymbol{A} \times \boldsymbol{B}$ is known as the **Schur product** and also the **Hadamard product**.

10. Let $\boldsymbol{A} \in \mathbb{R}^{m\times n}$ or $\boldsymbol{A} \in \mathbb{C}^{m\times n}$. The **transpose** $\boldsymbol{A}^T$ and **conjugate transpose** $\boldsymbol{A}^*$ are $n \times m$ matrices with elements $a_{ij}^T := a_{ji}$ and $a_{ij}^* := \overline{a}_{ji}$, respectively. If $\boldsymbol{B}$ is an n, p matrix then $(\boldsymbol{A}\boldsymbol{B})^T = \boldsymbol{B}^T\boldsymbol{A}^T$ and $(\boldsymbol{A}\boldsymbol{B})^* = \boldsymbol{B}^*\boldsymbol{A}^*$. A matrix $\boldsymbol{A} \in \mathbb{C}^{n\times n}$ is **symmetric** if $\boldsymbol{A}^T = \boldsymbol{A}$ and **Hermitian** if $\boldsymbol{A}^* = \boldsymbol{A}$.
11. The **unit vectors** in $\mathbb{R}^n$ and $\mathbb{C}^n$ are denoted by

$$\boldsymbol{e}_1 := \begin{bmatrix} 1 \\ 0 \\ 0 \\ \vdots \\ 0 \end{bmatrix}, \quad \boldsymbol{e}_2 := \begin{bmatrix} 0 \\ 1 \\ 0 \\ \vdots \\ 0 \end{bmatrix}, \quad \boldsymbol{e}_3 := \begin{bmatrix} 0 \\ 0 \\ 1 \\ \vdots \\ 0 \end{bmatrix}, \quad \ldots, \quad \boldsymbol{e}_n := \begin{bmatrix} 0 \\ 0 \\ 0 \\ \vdots \\ 1 \end{bmatrix},$$

while $\boldsymbol{I}_n = \boldsymbol{I} := [\delta_{ij}]_{i,j=1}^n$, where

$$\delta_{ij} := \begin{cases} 1 & \text{if } i = j, \\ 0 & \text{otherwise}, \end{cases} \tag{1.1}$$

is the **identity matrix** of order n. Both the columns and the transpose of the rows of $\boldsymbol{I}$ are the unit vectors $\boldsymbol{e}_1, \boldsymbol{e}_2, \ldots, \boldsymbol{e}_n$.

12. Some matrices with many zeros have names indicating their "shape". Suppose $\boldsymbol{A} \in \mathbb{R}^{n\times n}$ or $\boldsymbol{A} \in \mathbb{C}^{n\times n}$. Then $\boldsymbol{A}$ is
    - **diagonal** if $a_{ij} = 0$ for $i \neq j$.
    - **upper triangular** or **right triangular** if $a_{ij} = 0$ for $i > j$.
    - **lower triangular** or **left triangular** if $a_{ij} = 0$ for $i < j$.
    - **upper Hessenberg** if $a_{ij} = 0$ for $i > j + 1$.
    - **lower Hessenberg** if $a_{ij} = 0$ for $i < j + 1$.
    - **tridiagonal** if $a_{ij} = 0$ for $|i - j| > 1$.
    - d-**banded** if $a_{ij} = 0$ for $|i - j| > d$.
13. We use the following notations for diagonal- and tridiagonal $n \times n$ matrices

$$\operatorname{diag}(d_i) = \operatorname{diag}(d_1, \ldots, d_n) := \begin{bmatrix} d_1 & 0 & \cdots & 0 \\ 0 & d_2 & \cdots & 0 \\ \vdots & \vdots & \ddots & \vdots \\ 0 & 0 & \cdots & d_n \end{bmatrix} = \begin{bmatrix} d_1 & & \\ & \ddots & \\ & & d_n \end{bmatrix},$$

$$\boldsymbol{B} = \operatorname{tridiag}(a_i, d_i, c_i) = \operatorname{tridiag}(\boldsymbol{a}, \boldsymbol{d}, \boldsymbol{c}) := \begin{bmatrix} d_1 & c_1 & & & \\ a_1 & d_2 & c_2 & & \\ & \ddots & \ddots & \ddots & \\ & & a_{n-2} & d_{n-1} & c_{n-1} \\ & & & a_{n-1} & d_n \end{bmatrix}.$$

    Here $b_{ii} := d_i$ for $i = 1, \ldots, n$, $b_{i+1,i} := a_i$, $b_{i,i+1} := c_i$ for $i = 1, \ldots, n-1$, and $b_{ij} := 0$ otherwise.
14. Suppose $\boldsymbol{A} \in \mathbb{C}^{m\times n}$ and $1 \le i_1 < i_2 < \cdots < i_r \le m$, $1 \le j_1 < j_2 < \cdots < j_c \le n$. The matrix $\boldsymbol{A}(\boldsymbol{i}, \boldsymbol{j}) \in \mathbb{C}^{r\times c}$ is the submatrix of $\boldsymbol{A}$ consisting of rows $\boldsymbol{i} := [i_1, \ldots, i_r]$ and columns $\boldsymbol{j} := [j_1, \ldots, j_c]$

$$\boldsymbol{A}(\boldsymbol{i}, \boldsymbol{j}) := \boldsymbol{A}\begin{pmatrix} i_1 & i_2 & \cdots & i_r \\ j_1 & j_2 & \cdots & j_c \end{pmatrix} = \begin{bmatrix} a_{i_1,j_1} & a_{i_1,j_2} & \cdots & a_{i_1,j_c} \\ a_{i_2,j_1} & a_{i_2,j_2} & \cdots & a_{i_2,j_c} \\ \vdots & \vdots & \ddots & \vdots \\ a_{i_r,j_1} & a_{i_r,j_2} & \cdots & a_{i_r,j_c} \end{bmatrix}.$$

    For the special case of consecutive rows and columns we also use the notation

$$\boldsymbol{A}(r_1 : r_2, c_1 : c_2) := \begin{bmatrix} a_{r_1,c_1} & a_{r_1,c_1+1} & \cdots & a_{r_1,c_2} \\ a_{r_1+1,c_1} & a_{r_1+1,c_1+1} & \cdots & a_{r_1+1,c_2} \\ \vdots & & \vdots\ \ddots & \vdots \\ a_{r_2,c_1} & a_{r_2,c_1+1} & \cdots & a_{r_2,c_2} \end{bmatrix}.$$

1.2 Vector Spaces and Subspaces

Many mathematical systems have analogous properties to vectors in $\mathbb{R}^2$ or $\mathbb{R}^3$.

Definition 1.1 (Real Vector Space) A **real vector space** is a nonempty set $\mathcal{V}$, whose objects are called **vectors**, together with two operations $+ : \mathcal{V} \times \mathcal{V} \longrightarrow \mathcal{V}$ and $\cdot : \mathbb{R} \times \mathcal{V} \longrightarrow \mathcal{V}$, called **addition** and **scalar multiplication**, satisfying the following axioms for all vectors $\boldsymbol{u}$, $\boldsymbol{v}$, $\boldsymbol{w}$ in $\mathcal{V}$ and scalars c, d in $\mathbb{R}$.

(V1) The sum $\boldsymbol{u} + \boldsymbol{v}$ is in $\mathcal{V}$,
(V2) $\boldsymbol{u} + \boldsymbol{v} = \boldsymbol{v} + \boldsymbol{u}$,
(V3) $\boldsymbol{u} + (\boldsymbol{v} + \boldsymbol{w}) = (\boldsymbol{u} + \boldsymbol{v}) + \boldsymbol{w}$,
(V4) There is a **zero vector** $\mathbf{0}$ such that $\boldsymbol{u} + \mathbf{0} = \boldsymbol{u}$,
(V5) For each $\boldsymbol{u}$ in $\mathcal{V}$ there is a vector $-\boldsymbol{u}$ in $\mathcal{V}$ such that $\boldsymbol{u} + (-\boldsymbol{u}) = \mathbf{0}$,
(S1) The scalar multiple $c \cdot \boldsymbol{u}$ is in $\mathcal{V}$,
(S2) $c \cdot (\boldsymbol{u} + \boldsymbol{v}) = c \cdot \boldsymbol{u} + c \cdot \boldsymbol{v}$,
(S3) $(c + d) \cdot \boldsymbol{u} = c \cdot \boldsymbol{u} + d \cdot \boldsymbol{u}$,
(S4) $c \cdot (d \cdot \boldsymbol{u}) = (cd) \cdot \boldsymbol{u}$,
(S5) $1 \cdot \boldsymbol{u} = \boldsymbol{u}$.

The scalar multiplication symbol $\cdot$ is often omitted, writing $c\boldsymbol{v}$ instead of $c \cdot \boldsymbol{v}$. We define $\boldsymbol{u} - \boldsymbol{v} := \boldsymbol{u} + (-\boldsymbol{v})$. We call $\mathcal{V}$ a **complex vector space** if the scalars consist of all complex numbers $\mathbb{C}$. In this book a vector space is either real or complex.

From the axioms it follows that

1. The zero vector is unique.
2. For each $\boldsymbol{u} \in \mathcal{V}$ the **negative** $-\boldsymbol{u}$ of $\boldsymbol{u}$ is unique.
3. $0\boldsymbol{u} = \mathbf{0}$, $c\mathbf{0} = \mathbf{0}$, and $-\boldsymbol{u} = (-1)\boldsymbol{u}$.

Here are some examples

1. The spaces $\mathbb{R}^n$ and $\mathbb{C}^n$, where $n \in \mathbb{N}$, are real and complex vector spaces, respectively.
2. Let $\mathcal{D}$ be a subset of $\mathbb{R}$ and $d \in \mathbb{N}$. The set $\mathcal{V}$ of all functions $\boldsymbol{f}, \boldsymbol{g} : \mathcal{D} \to \mathbb{R}^d$ is a real vector space with

$$(\boldsymbol{f} + \boldsymbol{g})(t) := \boldsymbol{f}(t) + \boldsymbol{g}(t), \quad (c\boldsymbol{f})(t) := c\boldsymbol{f}(t), \quad t \in \mathcal{D}, \quad c \in \mathbb{R}.$$

   Two functions $\boldsymbol{f}, \boldsymbol{g}$ in $\mathcal{V}$ are equal if $\boldsymbol{f}(t) = \boldsymbol{g}(t)$ for all $t \in \mathcal{D}$. The zero element is the **zero function** given by $\boldsymbol{f}(t) = \mathbf{0}$ for all $t \in \mathcal{D}$ and the negative of $\boldsymbol{f}$ is given by $-\boldsymbol{f} = (-1)\boldsymbol{f}$. In the following we will use boldface letters for functions only if $d > 1$.
3. For $n \geq 0$ the space Π_n of polynomials of degree at most n consists of all polynomials $p : \mathbb{R} \to \mathbb{R}$, $p : \mathbb{R} \to \mathbb{C}$, or $p : \mathbb{C} \to \mathbb{C}$ of the form

$$p(t) := a_0 + a_1 t + a_2 t^2 + \cdots + a_n t^n, \tag{1.2}$$

where the coefficients $a_0, \ldots, a_n$ are real or complex numbers. p is called the **zero polynomial** if all coefficients are zero. All other polynomials are said to be **nontrivial**. The **degree** of a nontrivial polynomial p given by (1.2) is the smallest integer $0 \le k \le n$ such that $p(t) = a_0 + \cdots + a_k t^k$ with $a_k \ne 0$. The degree of the zero polynomial is not defined. Π_n is a vector space if we define addition and scalar multiplication as for functions.

Definition 1.2 (Linear Combination) For $n \ge 1$ let $\mathcal{X} := \{\boldsymbol{x}_1, \ldots, \boldsymbol{x}_n\}$ be a set of vectors in a vector space $\mathcal{V}$ and let $c_1, \ldots, c_n$ be scalars.

1. The sum $c_1\boldsymbol{x}_1 + \cdots + c_n\boldsymbol{x}_n$ is called a **linear combination** of $\boldsymbol{x}_1, \ldots, \boldsymbol{x}_n$.
2. The linear combination is **nontrivial** if $c_j\boldsymbol{x}_j \ne \boldsymbol{0}$ for at least one j.
3. The set of all linear combinations of elements in $\mathcal{X}$ is denoted span($\mathcal{X}$).
4. A vector space is **finite dimensional** if it has a finite spanning set; i.e., there exists $n \in \mathbb{N}$ and $\{\boldsymbol{x}_1, \ldots, \boldsymbol{x}_n\}$ in $\mathcal{V}$ such that $\mathcal{V} = \operatorname{span}(\{\boldsymbol{x}_1, \ldots, \boldsymbol{x}_n\})$.

Example 1.1 (Linear Combinations)

1. Any $\boldsymbol{x} = [x_1, \ldots, x_m]^T$ in $\mathbb{C}^m$ can be written as a linear combination of the unit vectors as $\boldsymbol{x} = x_1\boldsymbol{e}_1 + x_2\boldsymbol{e}_2 + \cdots + x_m\boldsymbol{e}_m$. Thus, $\mathbb{C}^m = \operatorname{span}(\{\boldsymbol{e}_1, \ldots, \boldsymbol{e}_m\})$ and $\mathbb{C}^m$ is finite dimensional. Similarly $\mathbb{R}^m$ is finite dimensional.
2. Let $\Pi = \cup_n \Pi_n$ be the space of all polynomials. Π is a vector space that is not finite dimensional. For suppose Π is finite dimensional. Then $\Pi = \operatorname{span}(\{p_1, \ldots, p_m\})$ for some polynomials $p_1, \ldots, p_m$. Let d be an integer such that the degree of p_j is less than d for $j = 1, \ldots, m$. A polynomial of degree d cannot be written as a linear combination of $p_1, \ldots, p_m$, a contradiction.

1.2.1 Linear Independence and Bases

Definition 1.3 (Linear Independence) A set $\mathcal{X} = \{\boldsymbol{x}_1, \ldots, \boldsymbol{x}_n\}$ of nonzero vectors in a vector space is **linearly dependent** if $\boldsymbol{0}$ can be written as a nontrivial linear combination of $\{\boldsymbol{x}_1, \ldots, \boldsymbol{x}_n\}$. Otherwise $\mathcal{X}$ is **linearly independent**.

A set of vectors $\mathcal{X} = \{\boldsymbol{x}_1, \ldots, \boldsymbol{x}_n\}$ is linearly independent if and only if

$$c_1\boldsymbol{x}_1 + \cdots + c_n\boldsymbol{x}_n = \boldsymbol{0} \quad \Longrightarrow \quad c_1 = \cdots = c_n = 0. \tag{1.3}$$

Suppose $\{\boldsymbol{x}_1, \ldots, \boldsymbol{x}_n\}$ is linearly independent. Then

1. If $\boldsymbol{x} \in \operatorname{span}(\mathcal{X})$ then the scalars $c_1, \ldots, c_n$ in the representation $\boldsymbol{x} = c_1\boldsymbol{x}_1 + \cdots + c_n\boldsymbol{x}_n$ are unique.
2. Any nontrivial linear combination of $\boldsymbol{x}_1, \ldots, \boldsymbol{x}_n$ is nonzero,

Lemma 1.1 (Linear Independence and Span) *Suppose $\boldsymbol{v}_1, \ldots, \boldsymbol{v}_n$ span a vector space $\mathcal{V}$ and that $\boldsymbol{w}_1, \ldots, \boldsymbol{w}_k$ are linearly independent vectors in $\mathcal{V}$. Then $k \le n$.*

Proof Suppose $k > n$. Write $\boldsymbol{w}_1$ as a linear combination of elements from the set $\mathcal{X}_0 := \{\boldsymbol{v}_1, \ldots, \boldsymbol{v}_n\}$, say $\boldsymbol{w}_1 = c_1\boldsymbol{v}_1 + \cdots + c_n\boldsymbol{v}_n$. Since $\boldsymbol{w}_1 \neq \boldsymbol{0}$ not all the c's are equal to zero. Pick a nonzero c, say c_{i_1}. Then $\boldsymbol{v}_{i_1}$ can be expressed as a linear combination of $\boldsymbol{w}_1$ and the remaining $\boldsymbol{v}$'s. So the set $\mathcal{X}_1 := \{\boldsymbol{w}_1, \boldsymbol{v}_1, \ldots, \boldsymbol{v}_{i_1-1}, \boldsymbol{v}_{i_1+1}, \ldots, \boldsymbol{v}_n\}$ must also be a spanning set for $\mathcal{V}$. We repeat this for $\boldsymbol{w}_2$ and $\mathcal{X}_1$. In the linear combination $\boldsymbol{w}_2 = d_{i_1}\boldsymbol{w}_1 + \sum_{j\neq i_1} d_j\boldsymbol{v}_j$, we must have $d_{i_2} \neq 0$ for some i_2 with $i_2 \neq i_1$. For otherwise $\boldsymbol{w}_2 = d_1\boldsymbol{w}_1$ contradicting the linear independence of the $\boldsymbol{w}$'s. So the set $\mathcal{X}_2$ consisting of the $\boldsymbol{v}$'s with $\boldsymbol{v}_{i_1}$ replaced by $\boldsymbol{w}_1$ and $\boldsymbol{v}_{i_2}$ replaced by $\boldsymbol{w}_2$ is again a spanning set for $\mathcal{V}$. Repeating this process $n-2$ more times we obtain a spanning set $\mathcal{X}_n$ where $\boldsymbol{v}_1, \ldots, \boldsymbol{v}_n$ have been replaced by $\boldsymbol{w}_1, \ldots, \boldsymbol{w}_n$. Since $k > n$ we can then write $\boldsymbol{w}_k$ as a linear combination of $\boldsymbol{w}_1, \ldots, \boldsymbol{w}_n$ contradicting the linear independence of the $\boldsymbol{w}$'s. We conclude that $k \leq n$. □

Definition 1.4 (Basis) A finite set of vectors $\{\boldsymbol{v}_1, \ldots, \boldsymbol{v}_n\}$ in a vector space $\mathcal{V}$ is a **basis** for $\mathcal{V}$ if

1. $\text{span}\{\boldsymbol{v}_1, \ldots, \boldsymbol{v}_n\} = \mathcal{V}$.
2. $\{\boldsymbol{v}_1, \ldots, \boldsymbol{v}_n\}$ is linearly independent.

Theorem 1.1 (Basis Subset of a Spanning Set) *Suppose $\mathcal{V}$ is a vector space and that $\{\boldsymbol{v}_1, \ldots, \boldsymbol{v}_n\}$ is a spanning set for $\mathcal{V}$. Then we can find a subset $\{\boldsymbol{v}_{i_1}, \ldots, \boldsymbol{v}_{i_k}\}$ that forms a basis for $\mathcal{V}$.*

Proof If $\{\boldsymbol{v}_1, \ldots, \boldsymbol{v}_n\}$ is linearly dependent we can express one of the $\boldsymbol{v}$'s as a nontrivial linear combination of the remaining $\boldsymbol{v}$'s and drop that $\boldsymbol{v}$ from the spanning set. Continue this process until the remaining $\boldsymbol{v}$'s are linearly independent. They still span the vector space and therefore form a basis. □

Corollary 1.1 (Existence of a Basis) *A vector space is finite dimensional (cf. Definition 1.2) if and only if it has a basis.*

Proof Let $\mathcal{V} = \text{span}\{\boldsymbol{v}_1, \ldots, \boldsymbol{v}_n\}$ be a finite dimensional vector space. By Theorem 1.1, $\mathcal{V}$ has a basis. Conversely, if $\mathcal{V} = \text{span}\{\boldsymbol{v}_1, \ldots, \boldsymbol{v}_n\}$ and $\{\boldsymbol{v}_1, \ldots, \boldsymbol{v}_n\}$ is a basis then it is by definition a finite spanning set. □

Theorem 1.2 (Dimension of a Vector Space) *Every basis for a vector space $\mathcal{V}$ has the same number of elements. This number is called the* ***dimension*** *of the vector space and denoted* $\dim \mathcal{V}$.

Proof Suppose $\mathcal{X} = \{\boldsymbol{v}_1, \ldots, \boldsymbol{v}_n\}$ and $\mathcal{Y} = \{\boldsymbol{w}_1, \ldots, \boldsymbol{w}_k\}$ are two bases for $\mathcal{V}$. By Lemma 1.1 we have $k \leq n$. Using the same Lemma with $\mathcal{X}$ and $\mathcal{Y}$ switched we obtain $n \leq k$. We conclude that $n = k$. □

The set of unit vectors $\{\boldsymbol{e}_1, \ldots, \boldsymbol{e}_n\}$ form a basis for both $\mathbb{R}^n$ and $\mathbb{C}^n$.

Theorem 1.3 (Enlarging Vectors to a Basis) *Every linearly independent set of vectors $\{\boldsymbol{v}_1, \ldots, \boldsymbol{v}_k\}$ in a finite dimensional vector space $\mathcal{V}$ can be enlarged to a basis for $\mathcal{V}$.*

Proof If $\{\boldsymbol{v}_1, \ldots, \boldsymbol{v}_k\}$ does not span $\mathcal{V}$ we can enlarge the set by one vector $\boldsymbol{v}_{k+1}$ which cannot be expressed as a linear combination of $\{\boldsymbol{v}_1, \ldots, \boldsymbol{v}_k\}$. The enlarged set is also linearly independent. Continue this process. Since the space is finite dimensional it must stop after a finite number of steps. □

1.2.2 Subspaces

Definition 1.5 (Subspace) A nonempty subset $\mathcal{S}$ of a real or complex vector space $\mathcal{V}$ is called a **subspace** of $\mathcal{V}$ if

(V1) The sum $\boldsymbol{u} + \boldsymbol{v}$ is in $\mathcal{S}$ for any $\boldsymbol{u}, \boldsymbol{v} \in \mathcal{S}$.
(S1) The scalar multiple $c\boldsymbol{u}$ is in $\mathcal{S}$ for any scalar c and any $\boldsymbol{u} \in \mathcal{S}$.

Using the operations in $\mathcal{V}$, any subspace $\mathcal{S}$ of $\mathcal{V}$ is a vector space, i.e., all 10 axioms $V1 - V5$ and $S1 - S5$ are satisfied for $\mathcal{S}$. In particular, $\mathcal{S}$ must contain the zero element in $\mathcal{V}$. This follows since the operations of vector addition and scalar multiplication are inherited from $\mathcal{V}$.

Example 1.2 (Examples of Subspaces)

1. $\{\boldsymbol{0}\}$, where $\boldsymbol{0}$ is the zero vector is a subspace, the **trivial subspace**. The dimension of the trivial subspace is defined to be zero. All other subspaces are **nontrivial**.
2. $\mathcal{V}$ is a subspace of itself.
3. $\operatorname{span}(\mathcal{X})$ is a subspace of $\mathcal{V}$ for any $\mathcal{X} = \{\boldsymbol{x}_1, \ldots, \boldsymbol{x}_n\} \subseteq \mathcal{V}$. Indeed, it is easy to see that **(V1)** and **(S1)** hold.
4. The **sum** of two subspaces $\mathcal{S}$ and $\mathcal{T}$ of a vector space $\mathcal{V}$ is defined by
$$\mathcal{S} + \mathcal{T} := \{\boldsymbol{s} + \boldsymbol{t} : \boldsymbol{s} \in \mathcal{S} \text{ and } \boldsymbol{t} \in \mathcal{T}\}. \tag{1.4}$$
Clearly **(V1)** and **(S1)** hold and it is a subspace of $\mathcal{V}$.
5. The **intersection** of two subspaces $\mathcal{S}$ and $\mathcal{T}$ of a vector space $\mathcal{V}$ is defined by
$$\mathcal{S} \cap \mathcal{T} := \{\boldsymbol{x} : \boldsymbol{x} \in \mathcal{S} \text{ and } \boldsymbol{x} \in \mathcal{T}\}. \tag{1.5}$$
It is a subspace of $\mathcal{V}$.
6. The **union** of two subspaces $\mathcal{S}$ and $\mathcal{T}$ of a vector space $\mathcal{V}$ is defined by
$$\mathcal{S} \cup \mathcal{T} := \{\boldsymbol{x} : \boldsymbol{x} \in \mathcal{S} \text{ or } \boldsymbol{x} \in \mathcal{T}\}. \tag{1.6}$$
In general it is not a subspace of $\mathcal{V}$.
7. A sum of two subspaces $\mathcal{S}$ and $\mathcal{T}$ of a vector space $\mathcal{V}$ is called a **direct sum** and denoted $\mathcal{S} \oplus \mathcal{T}$ if $\mathcal{S} \cap \mathcal{T} = \{\boldsymbol{0}\}$.

Theorem 1.4 (Dimension Formula for Sums of Subspaces) *Let $\mathcal{S}$ and $\mathcal{T}$ be two finite subspaces of a vector space $\mathcal{V}$. Then*

$$\dim(\mathcal{S}+\mathcal{T}) = \dim(\mathcal{S}) + \dim(\mathcal{T}) - \dim(\mathcal{S}\cap\mathcal{T}). \tag{1.7}$$

In particular, for a direct sum

$$\dim(\mathcal{S}\oplus\mathcal{T}) = \dim(\mathcal{S}) + \dim(\mathcal{T}). \tag{1.8}$$

Proof Let $\{\boldsymbol{u}_1,\dots,\boldsymbol{u}_p\}$ be a basis for $\mathcal{S}\cap\mathcal{T}$, where $\{\boldsymbol{u}_1,\dots,\boldsymbol{u}_p\}=\emptyset$, the empty set, in the case $\mathcal{S}\cap\mathcal{T}=\{\boldsymbol{0}\}$. We use Theorem 1.3 to extend $\{\boldsymbol{u}_1,\dots,\boldsymbol{u}_p\}$ to a basis $\{\boldsymbol{u}_1,\dots,\boldsymbol{u}_p,\boldsymbol{s}_1,\dots,\boldsymbol{s}_q\}$ for $\mathcal{S}$ and a basis $\{\boldsymbol{u}_1,\dots,\boldsymbol{u}_p,\boldsymbol{t}_1,\dots,\boldsymbol{t}_r\}$ for $\mathcal{T}$. Every $\boldsymbol{x}\in\mathcal{S}+\mathcal{T}$ can be written as a linear combination of

$$\{\boldsymbol{u}_1,\dots,\boldsymbol{u}_p,\boldsymbol{s}_1,\dots,\boldsymbol{s}_q,\boldsymbol{t}_1,\dots,\boldsymbol{t}_r\}$$

so these vectors span $\mathcal{S}+\mathcal{T}$. We show that they are linearly independent and hence a basis. Suppose $\boldsymbol{u}+\boldsymbol{s}+\boldsymbol{t}=\boldsymbol{0}$, where $\boldsymbol{u}:=\sum_{j=1}^p\alpha_j\boldsymbol{u}_j$, $\boldsymbol{s}:=\sum_{j=1}^q\rho_j\boldsymbol{s}_j$, and $\boldsymbol{t}:=\sum_{j=1}^r\sigma_j\boldsymbol{t}_j$. Now $\boldsymbol{s}=-(\boldsymbol{u}+\boldsymbol{t})$ belongs to both $\mathcal{S}$ and to $\mathcal{T}$ and hence $\boldsymbol{s}\in\mathcal{S}\cap\mathcal{T}$. Therefore $\boldsymbol{s}$ can be written as a linear combination of $\boldsymbol{u}_1,\dots,\boldsymbol{u}_p$ say $\boldsymbol{s}:=\sum_{j=1}^p\beta_j\boldsymbol{u}_j$. But then $\boldsymbol{0}=\sum_{j=1}^p\beta_j\boldsymbol{u}_j-\sum_{j=1}^q\rho_j\boldsymbol{s}_j$ and since

$$\{\boldsymbol{u}_1,\dots,\boldsymbol{u}_p,\boldsymbol{s}_1,\dots,\boldsymbol{s}_q\}$$

is linearly independent we must have $\beta_1=\cdots=\beta_p=\rho_1=\cdots=\rho_q=0$ and hence $\boldsymbol{s}=\boldsymbol{0}$. We then have $\boldsymbol{u}+\boldsymbol{t}=\boldsymbol{0}$ and by linear independence of $\{\boldsymbol{u}_1,\dots,\boldsymbol{u}_p,\boldsymbol{t}_1,\dots,\boldsymbol{t}_r\}$ we obtain $\alpha_1=\cdots=\alpha_p=\sigma_1=\cdots=\sigma_r=0$. We have shown that the vectors $\{\boldsymbol{u}_1,\dots,\boldsymbol{u}_p,\boldsymbol{s}_1,\dots,\boldsymbol{s}_q,\boldsymbol{t}_1,\dots,\boldsymbol{t}_r\}$ constitute a basis for $\mathcal{S}+\mathcal{T}$. But then

$$\dim(\mathcal{S}+\mathcal{T}) = p+q+r = (p+q)+(p+r)-p = \dim(\mathcal{S})+\dim(\mathcal{T})-\dim(\mathcal{S}\cap\mathcal{T})$$

and (1.7) follows. Equation (1.7) implies (1.8) since $\dim\{\boldsymbol{0}\}=0$. □

It is convenient to introduce a matrix transforming a basis in a subspace into a basis for the space itself.

Lemma 1.2 (Change of Basis Matrix) *Suppose $\mathcal{S}$ is a subspace of a finite dimensional vector space $\mathcal{V}$ and let $\{\boldsymbol{s}_1,\dots,\boldsymbol{s}_n\}$ be a basis for $\mathcal{S}$ and $\{\boldsymbol{v}_1,\dots,\boldsymbol{v}_m\}$ a basis for $\mathcal{V}$. Then each $\boldsymbol{s}_j$ can be expressed as a linear combination of $\boldsymbol{v}_1,\dots,\boldsymbol{v}_m$, say*

$$\boldsymbol{s}_j=\sum_{i=1}^m a_{ij}\boldsymbol{v}_i \text{ for } j=1,\dots,n. \tag{1.9}$$

If $\boldsymbol{x} \in \mathcal{S}$ *then* $\boldsymbol{x} = \sum_{j=1}^{n} c_j \boldsymbol{s}_j = \sum_{i=1}^{m} b_i \boldsymbol{v}_i$ *for some coefficients* $\boldsymbol{b} := [b_1, \ldots, b_m]^T$, $\boldsymbol{c} := [c_1, \ldots, c_n]^T$. *Moreover* $\boldsymbol{b} = \boldsymbol{A}\boldsymbol{c}$, *where* $\boldsymbol{A} = [a_{ij}] \in \mathbb{C}^{m\times n}$ *is given by* (1.9). *The matrix* $\boldsymbol{A}$ *has linearly independent columns.*

Proof Equation (1.9) holds for some a_{ij} since $\boldsymbol{s}_j \in \mathcal{V}$ and $\{\boldsymbol{v}_1, \ldots, \boldsymbol{v}_m\}$ spans $\mathcal{V}$. Since $\{\boldsymbol{s}_1, \ldots, \boldsymbol{s}_n\}$ is a basis for $\mathcal{S}$ and $\{\boldsymbol{v}_1, \ldots, \boldsymbol{v}_m\}$ a basis for $\mathcal{V}$, every $\boldsymbol{x} \in \mathcal{S}$ can be written $\boldsymbol{x} = \sum_{j=1}^{n} c_j \boldsymbol{s}_j = \sum_{i=1}^{m} b_i \boldsymbol{v}_i$ for some scalars (c_j) and (b_i). But then

$$\sum_{i=1}^{m} b_i \boldsymbol{v}_i = \boldsymbol{x} = \sum_{j=1}^{n} c_j \boldsymbol{s}_j \overset{(1.9)}{=} \sum_{j=1}^{n} c_j \Big(\sum_{i=1}^{m} a_{ij} \boldsymbol{v}_i\Big) = \sum_{i=1}^{m} \Big(\sum_{j=1}^{n} a_{ij} c_j\Big) \boldsymbol{v}_i.$$

Since $\{\boldsymbol{v}_1, \ldots, \boldsymbol{v}_m\}$ is linearly independent it follows that $b_i = \sum_{j=1}^{n} a_{ij} c_j$ for $i = 1, \ldots, m$ or $\boldsymbol{b} = \boldsymbol{A}\boldsymbol{c}$. Finally, to show that $\boldsymbol{A}$ has linearly independent columns suppose $\boldsymbol{b} := \boldsymbol{A}\boldsymbol{c} = \boldsymbol{0}$ for some $\boldsymbol{c} = [c_1, \ldots, c_n]^T$. Define $\boldsymbol{x} := \sum_{j=1}^{n} c_j \boldsymbol{s}_j$. Then $\boldsymbol{x} = \sum_{i=1}^{m} b_i \boldsymbol{v}_i$ and since $\boldsymbol{b} = \boldsymbol{0}$ we have $\boldsymbol{x} = \boldsymbol{0}$. But since $\{\boldsymbol{s}_1, \ldots, \boldsymbol{s}_n\}$ is linearly independent it follows that $\boldsymbol{c} = \boldsymbol{0}$. □

The matrix $\boldsymbol{A}$ in Lemma 1.2 is called a **change of basis matrix**.

1.2.3 The Vector Spaces $\mathbb{R}^n$ *and* $\mathbb{C}^n$

When $\mathcal{V} = \mathbb{R}^m$ or $\mathbb{C}^m$ we can think of n vectors in $\mathcal{V}$, say $\boldsymbol{x}_1, \ldots, \boldsymbol{x}_n$, as a set $\mathcal{X} := \{\boldsymbol{x}_1, \ldots, \boldsymbol{x}_n\}$ or as the columns of an $m \times n$ matrix $\boldsymbol{X} = [\boldsymbol{x}_1, \ldots, \boldsymbol{x}_n]$. A linear combination can then be written as a matrix times vector $\boldsymbol{X}\boldsymbol{c}$, where $\boldsymbol{c} = [c_1, \ldots, c_n]^T$ is the vector of scalars. Thus

$$\mathcal{R}(\boldsymbol{X}) := \{\boldsymbol{X}\boldsymbol{c} : \boldsymbol{c} \in \mathbb{R}^n\} = \operatorname{span}(\mathcal{X}).$$

Definition 1.6 (Column Space, Null Space, Inner Product and Norm) Associated with an $m \times n$ matrix $\boldsymbol{X} = [\boldsymbol{x}_1, \ldots, \boldsymbol{x}_n]$, where $\boldsymbol{x}_j \in \mathcal{V}$, $j = 1, \ldots, n$ are the following subspaces of $\mathcal{V}$.

1. The subspace $\mathcal{R}(\boldsymbol{X})$ is called the **column space** of $\boldsymbol{X}$. It is the smallest subspace containing $\mathcal{X} = \{\boldsymbol{x}_1, \ldots, \boldsymbol{x}_n\}$. The dimension of $\mathcal{R}(\boldsymbol{X})$ is called the **rank** of $\boldsymbol{X}$. The matrix $\boldsymbol{X}$ has rank n if and only if it has linearly independent columns.
2. $\mathcal{R}(\boldsymbol{X}^T)$ is called the **row space** of $\boldsymbol{X}$. It is generated by the rows of $\boldsymbol{X}$ written as column vectors.
3. The subspace $\mathcal{N}(\boldsymbol{X}) := \{\boldsymbol{y} \in \mathbb{R}^n : \boldsymbol{X}\boldsymbol{y} = \boldsymbol{0}\}$ is called the **null space** or **kernel space** of $\boldsymbol{X}$. The dimension of $\mathcal{N}(\boldsymbol{X})$ is called the **nullity** of $\boldsymbol{X}$ and denoted $\operatorname{null}(\boldsymbol{X})$.

4. The **standard inner product** is

$$\langle \boldsymbol{x}, \boldsymbol{y} \rangle := \boldsymbol{y}^* \boldsymbol{x} = \boldsymbol{x}^T \overline{\boldsymbol{y}} = \sum_{j=1}^{n} x_j \overline{y_j}. \tag{1.10}$$

5. The **Euclidian norm** is defined by

$$\|\boldsymbol{x}\|_2 := \Big(\sum_{j=1}^{n} |x_j|^2 \Big)^{1/2} = \sqrt{\boldsymbol{x}^* \boldsymbol{x}}. \tag{1.11}$$

Clearly $\mathcal{N}(\boldsymbol{X})$ is nontrivial if and only if $\boldsymbol{X}$ has linearly dependent columns. Inner products and norms are treated in more generality in Chaps. 5 and 8.

The following Theorem is shown in any basic course in linear algebra. See Exercise 7.10 for a simple proof using the singular value decomposition.

Theorem 1.5 (Counting Dimensions of Fundamental Subspaces) *Suppose* $\boldsymbol{X} \in \mathbb{C}^{m \times n}$. *Then*

1. $\operatorname{rank}(\boldsymbol{X}) = \operatorname{rank}(\boldsymbol{X}^*)$.
2. $\operatorname{rank}(\boldsymbol{X}) + \operatorname{null}(\boldsymbol{X}) = n$,
3. $\operatorname{rank}(\boldsymbol{X}) + \operatorname{null}(\boldsymbol{X}^*) = m$,

1.3 Linear Systems

Consider a linear system

$$\begin{array}{ccccccccc} a_{11}x_1 & + & a_{12}x_2 & + & \cdots & + & a_{1n}x_n & = & b_1 \\ a_{21}x_1 & + & a_{22}x_2 & + & \cdots & + & a_{2n}x_n & = & b_2 \\ \vdots & & \vdots & & & & \vdots & & \vdots \\ a_{m1}x_1 & + & a_{m2}x_2 & + & \cdots & + & a_{mn}x_n & = & b_m \end{array}$$

of m equations in n unknowns. Here for all i, j, the coefficients a_{ij}, the unknowns x_j, and the components b_i of the right hand side are real or complex numbers. The system can be written as a vector equation

$$x_1 \boldsymbol{a}_1 + x_2 \boldsymbol{a}_2 + \cdots + x_n \boldsymbol{a}_n = \boldsymbol{b},$$

where $\boldsymbol{a}_j = [a_{1j}, \ldots, a_{mj}]^T \in \mathbb{C}^m$ for $j = 1, \ldots, n$ and $\boldsymbol{b} = [b_1, \ldots, b_m]^T \in \mathbb{C}^m$. It can also be written as a matrix equation

$$\boldsymbol{A}\boldsymbol{x} = \begin{bmatrix} a_{11} & a_{12} & \cdots & a_{1n} \\ a_{21} & a_{22} & \cdots & a_{2n} \\ \vdots & \vdots & \ddots & \vdots \\ a_{m1} & a_{m2} & \cdots & a_{mn} \end{bmatrix} \begin{bmatrix} x_1 \\ x_2 \\ \vdots \\ x_n \end{bmatrix} = \begin{bmatrix} b_1 \\ b_2 \\ \vdots \\ b_m \end{bmatrix} = \boldsymbol{b}.$$

The system is **homogeneous** if $\boldsymbol{b} = \boldsymbol{0}$ and it is said to be **underdetermined**, **square**, or **overdetermined** if $m < n$, $m = n$, or $m > n$, respectively.

1.3.1 Basic Properties

A linear system has a unique solution, infinitely many solutions, or no solution. To discuss this we first consider the real case, and a homogeneous underdetermined system.

Lemma 1.3 (Underdetermined System) *Suppose $\boldsymbol{A} \in \mathbb{R}^{m \times n}$ with $m < n$. Then there is a nonzero $\boldsymbol{x} \in \mathbb{R}^n$ such that $\boldsymbol{A}\boldsymbol{x} = \boldsymbol{0}$.*

Proof Suppose $\boldsymbol{A} \in \mathbb{R}^{m \times n}$ with $m < n$. The n columns of $\boldsymbol{A}$ span a subspace of $\mathbb{R}^m$. Since $\mathbb{R}^m$ has dimension m the dimension of this subspace is at most m. By Lemma 1.1 the columns of $\boldsymbol{A}$ must be linearly dependent. It follows that there is a nonzero $\boldsymbol{x} \in \mathbb{R}^n$ such that $\boldsymbol{A}\boldsymbol{x} = \boldsymbol{0}$. □

A square matrix is either **nonsingular** or **singular**.

Definition 1.7 (Real Nonsingular or Singular Matrix) A square matrix $\boldsymbol{A} \in \mathbb{R}^{n \times n}$ is said to be **nonsingular** if the only real solution of the homogeneous system $\boldsymbol{A}\boldsymbol{x} = \boldsymbol{0}$ is $\boldsymbol{x} = \boldsymbol{0}$. The matrix is **singular** if there is a nonzero $\boldsymbol{x} \in \mathbb{R}^n$ such that $\boldsymbol{A}\boldsymbol{x} = \boldsymbol{0}$.

Theorem 1.6 (Linear Systems; Existence and Uniqueness) *Suppose $\boldsymbol{A} \in \mathbb{R}^{n \times n}$. The linear system $\boldsymbol{A}\boldsymbol{x} = \boldsymbol{b}$ has a unique solution $\boldsymbol{x} \in \mathbb{R}^n$ for any $\boldsymbol{b} \in \mathbb{R}^n$ if and only if the matrix $\boldsymbol{A}$ is nonsingular.*

Proof Suppose $\boldsymbol{A}$ is nonsingular. We define $\boldsymbol{B} = [\boldsymbol{A}\ \boldsymbol{b}] \in \mathbb{R}^{n \times (n+1)}$ by adding a column to $\boldsymbol{A}$. By Lemma 1.3 there is a nonzero $\boldsymbol{z} \in \mathbb{R}^{n+1}$ such that $\boldsymbol{B}\boldsymbol{z} = \boldsymbol{0}$. If we write $\boldsymbol{z} = \begin{bmatrix} \tilde{\boldsymbol{z}} \\ z_{n+1} \end{bmatrix}$ where $\tilde{\boldsymbol{z}} = [z_1, \ldots, z_n]^T \in \mathbb{R}^n$ and $z_{n+1} \in \mathbb{R}$, then

$$\boldsymbol{B}\boldsymbol{z} = [\boldsymbol{A}\ \boldsymbol{b}] \begin{bmatrix} \tilde{\boldsymbol{z}} \\ z_{n+1} \end{bmatrix} = \boldsymbol{A}\tilde{\boldsymbol{z}} + z_{n+1}\boldsymbol{b} = \boldsymbol{0}.$$

We cannot have $z_{n+1} = 0$ for then $\boldsymbol{A}\tilde{\boldsymbol{z}} = \boldsymbol{0}$ for a nonzero $\tilde{\boldsymbol{z}}$, contradicting the nonsingularity of $\boldsymbol{A}$. Define $\boldsymbol{x} := -\tilde{\boldsymbol{z}}/z_{n+1}$. Then

$$\boldsymbol{A}\boldsymbol{x} = -\boldsymbol{A}\Big(\frac{\tilde{\boldsymbol{z}}}{z_{n+1}}\Big) = -\frac{1}{z_{n+1}}\boldsymbol{A}\tilde{\boldsymbol{z}} = -\frac{1}{z_{n+1}}\big(-z_{n+1}\boldsymbol{b}\big) = \boldsymbol{b},$$

so $\boldsymbol{x}$ is a solution.

Suppose $\boldsymbol{A}\boldsymbol{x} = \boldsymbol{b}$ and $\boldsymbol{A}\boldsymbol{y} = \boldsymbol{b}$ for $\boldsymbol{x}, \boldsymbol{y} \in \mathbb{R}^n$. Then $\boldsymbol{A}(\boldsymbol{x} - \boldsymbol{y}) = \boldsymbol{0}$ and since $\boldsymbol{A}$ is nonsingular we conclude that $\boldsymbol{x} - \boldsymbol{y} = \boldsymbol{0}$ or $\boldsymbol{x} = \boldsymbol{y}$. Thus the solution is unique.

Conversely, if $\boldsymbol{A}\boldsymbol{x} = \boldsymbol{b}$ has a unique solution for any $\boldsymbol{b} \in \mathbb{R}^n$ then $\boldsymbol{A}\boldsymbol{x} = \boldsymbol{0}$ has a unique solution which must be $\boldsymbol{x} = \boldsymbol{0}$. Thus $\boldsymbol{A}$ is nonsingular. □

For the complex case we have

Lemma 1.4 (Complex Underdetermined System) *Suppose $\boldsymbol{A} \in \mathbb{C}^{m\times n}$ with $m < n$. Then there is a nonzero $\boldsymbol{x} \in \mathbb{C}^n$ such that $\boldsymbol{A}\boldsymbol{x} = \boldsymbol{0}$.*

Definition 1.8 (Complex Nonsingular Matrix) A square matrix $\boldsymbol{A} \in \mathbb{C}^{n\times n}$ is said to be **nonsingular** if the only complex solution of the homogeneous system $\boldsymbol{A}\boldsymbol{x} = \boldsymbol{0}$ is $\boldsymbol{x} = \boldsymbol{0}$. The matrix is **singular** if it is not nonsingular.

Theorem 1.7 (Complex Linear System; Existence and Uniqueness) *Suppose $\boldsymbol{A} \in \mathbb{C}^{n\times n}$. The linear system $\boldsymbol{A}\boldsymbol{x} = \boldsymbol{b}$ has a unique solution $\boldsymbol{x} \in \mathbb{C}^n$ for any $\boldsymbol{b} \in \mathbb{C}^n$ if and only if the matrix $\boldsymbol{A}$ is nonsingular.*

1.3.2 The Inverse Matrix

Suppose $\boldsymbol{A} \in \mathbb{C}^{n\times n}$ is a square matrix. A matrix $\boldsymbol{B} \in \mathbb{C}^{n\times n}$ is called a **right inverse** of $\boldsymbol{A}$ if $\boldsymbol{A}\boldsymbol{B} = \boldsymbol{I}$. A matrix $\boldsymbol{C} \in \mathbb{C}^{n\times n}$ is said to be a **left inverse** of $\boldsymbol{A}$ if $\boldsymbol{C}\boldsymbol{A} = \boldsymbol{I}$. We say that $\boldsymbol{A}$ is **invertible** if it has both a left- and a right inverse. If $\boldsymbol{A}$ has a right inverse $\boldsymbol{B}$ and a left inverse $\boldsymbol{C}$ then

$$\boldsymbol{C} = \boldsymbol{C}\boldsymbol{I} = \boldsymbol{C}(\boldsymbol{A}\boldsymbol{B}) = (\boldsymbol{C}\boldsymbol{A})\boldsymbol{B} = \boldsymbol{I}\boldsymbol{B} = \boldsymbol{B}$$

and this common inverse is called the **inverse** of $\boldsymbol{A}$ and denoted by $\boldsymbol{A}^{-1}$. Thus the inverse satisfies $\boldsymbol{A}^{-1}\boldsymbol{A} = \boldsymbol{A}\boldsymbol{A}^{-1} = \boldsymbol{I}$.

We want to characterize the class of invertible matrices and start with a lemma.

Theorem 1.8 (Product of Nonsingular Matrices) *If $\boldsymbol{A}, \boldsymbol{B}, \boldsymbol{C} \in \mathbb{C}^{n\times n}$ with $\boldsymbol{A}\boldsymbol{B} = \boldsymbol{C}$ then $\boldsymbol{C}$ is nonsingular if and only if both $\boldsymbol{A}$ and $\boldsymbol{B}$ are nonsingular. In particular, if either $\boldsymbol{A}\boldsymbol{B} = \boldsymbol{I}$ or $\boldsymbol{B}\boldsymbol{A} = \boldsymbol{I}$ then $\boldsymbol{A}$ is nonsingular and $\boldsymbol{A}^{-1} = \boldsymbol{B}$.*

Proof Suppose both $\boldsymbol{A}$ and $\boldsymbol{B}$ are nonsingular and let $\boldsymbol{C}\boldsymbol{x} = \boldsymbol{0}$. Then $\boldsymbol{A}\boldsymbol{B}\boldsymbol{x} = \boldsymbol{0}$ and since $\boldsymbol{A}$ is nonsingular we see that $\boldsymbol{B}\boldsymbol{x} = \boldsymbol{0}$. Since $\boldsymbol{B}$ is nonsingular we have $\boldsymbol{x} = \boldsymbol{0}$. We conclude that $\boldsymbol{C}$ is nonsingular.

For the converse suppose first that $\boldsymbol{B}$ is singular and let $\boldsymbol{x} \in \mathbb{C}^n$ be a nonzero vector so that $\boldsymbol{Bx} = \boldsymbol{0}$. But then $\boldsymbol{Cx} = (\boldsymbol{AB})\boldsymbol{x} = \boldsymbol{A}(\boldsymbol{Bx}) = \boldsymbol{A0} = \boldsymbol{0}$ so $\boldsymbol{C}$ is singular. Finally suppose $\boldsymbol{B}$ is nonsingular, but $\boldsymbol{A}$ is singular. Let $\tilde{\boldsymbol{x}}$ be a nonzero vector such that $\boldsymbol{A}\tilde{\boldsymbol{x}} = \boldsymbol{0}$. By Theorem 1.7 there is a vector $\boldsymbol{x}$ such that $\boldsymbol{Bx} = \tilde{\boldsymbol{x}}$ and $\boldsymbol{x}$ is nonzero since $\tilde{\boldsymbol{x}}$ is nonzero. But then $\boldsymbol{Cx} = (\boldsymbol{AB})\boldsymbol{x} = \boldsymbol{A}(\boldsymbol{Bx}) = \boldsymbol{A}\tilde{\boldsymbol{x}} = \boldsymbol{0}$ for a nonzero vector $\boldsymbol{x}$ and $\boldsymbol{C}$ is singular. □

Theorem 1.9 (When Is a Square Matrix Invertible?) *A square matrix is invertible if and only if it is nonsingular.*

Proof Suppose first $\boldsymbol{A}$ is a nonsingular matrix. By Theorem 1.7 each of the linear systems $\boldsymbol{Ab}_i = \boldsymbol{e}_i$ has a unique solution $\boldsymbol{b}_i$ for $i = 1, \dots, n$. Let $\boldsymbol{B} = \left[\boldsymbol{b}_1, \dots, \boldsymbol{b}_n\right]$. Then $\boldsymbol{AB} = \left[\boldsymbol{Ab}_1, \dots, \boldsymbol{Ab}_n\right] = \left[\boldsymbol{e}_1, \dots, \boldsymbol{e}_n\right] = \boldsymbol{I}$ so that $\boldsymbol{A}$ has a right inverse $\boldsymbol{B}$. By Theorem 1.8 $\boldsymbol{B}$ is nonsingular since $\boldsymbol{I}$ is nonsingular and $\boldsymbol{AB} = \boldsymbol{I}$. Since $\boldsymbol{B}$ is nonsingular we can use what we have shown for $\boldsymbol{A}$ to conclude that $\boldsymbol{B}$ has a right inverse $\boldsymbol{C}$, i.e. $\boldsymbol{BC} = \boldsymbol{I}$. But then $\boldsymbol{AB} = \boldsymbol{BC} = \boldsymbol{I}$ so $\boldsymbol{B}$ has both a right inverse and a left inverse which must be equal so $\boldsymbol{A} = \boldsymbol{C}$. Since $\boldsymbol{BC} = \boldsymbol{I}$ we have $\boldsymbol{BA} = \boldsymbol{I}$, so $\boldsymbol{B}$ is also a left inverse of $\boldsymbol{A}$ and $\boldsymbol{A}$ is invertible.

Conversely, if $\boldsymbol{A}$ is invertible then it has a right inverse $\boldsymbol{B}$. Since $\boldsymbol{AB} = \boldsymbol{I}$ and $\boldsymbol{I}$ is nonsingular, we again use Theorem 1.8 to conclude that $\boldsymbol{A}$ is nonsingular. □

To verify that some matrix $\boldsymbol{B}$ is an inverse of another matrix $\boldsymbol{A}$ it is enough to show that $\boldsymbol{B}$ is either a left inverse or a right inverse of $\boldsymbol{A}$. This calculation also proves that $\boldsymbol{A}$ is nonsingular. We use this observation to give simple proofs of the following results.

Corollary 1.2 (Basic Properties of the Inverse Matrix) *Suppose $\boldsymbol{A}, \boldsymbol{B} \in \mathbb{C}^{n\times n}$ are nonsingular and c is a nonzero constant.*

1. *$\boldsymbol{A}^{-1}$ is nonsingular and $(\boldsymbol{A}^{-1})^{-1} = \boldsymbol{A}$.*
2. *$\boldsymbol{C} = \boldsymbol{AB}$ is nonsingular and $\boldsymbol{C}^{-1} = \boldsymbol{B}^{-1}\boldsymbol{A}^{-1}$.*
3. *$\boldsymbol{A}^T$ is nonsingular and $(\boldsymbol{A}^T)^{-1} = (\boldsymbol{A}^{-1})^T =: \boldsymbol{A}^{-T}$.*
4. *$\boldsymbol{A}^*$ is nonsingular and $(\boldsymbol{A}^*)^{-1} = (\boldsymbol{A}^{-1})^* =: \boldsymbol{A}^{-*}$.*
5. *$c\boldsymbol{A}$ is nonsingular and $(c\boldsymbol{A})^{-1} = \frac{1}{c}\boldsymbol{A}^{-1}$.*

Proof

1. Since $\boldsymbol{A}^{-1}\boldsymbol{A} = \boldsymbol{I}$ the matrix $\boldsymbol{A}$ is a right inverse of $\boldsymbol{A}^{-1}$. Thus $\boldsymbol{A}^{-1}$ is nonsingular and $(\boldsymbol{A}^{-1})^{-1} = \boldsymbol{A}$.
2. We note that $(\boldsymbol{B}^{-1}\boldsymbol{A}^{-1})(\boldsymbol{AB}) = \boldsymbol{B}^{-1}(\boldsymbol{A}^{-1}\boldsymbol{A})\boldsymbol{B} = \boldsymbol{B}^{-1}\boldsymbol{B} = \boldsymbol{I}$. Thus $\boldsymbol{AB}$ is invertible with the indicated inverse since it has a left inverse.
3. Now $\boldsymbol{I} = \boldsymbol{I}^T = (\boldsymbol{A}^{-1}\boldsymbol{A})^T = \boldsymbol{A}^T(\boldsymbol{A}^{-1})^T$ showing that $(\boldsymbol{A}^{-1})^T$ is a right inverse of $\boldsymbol{A}^T$. The proof of part 4 is similar.
4. The matrix $\frac{1}{c}\boldsymbol{A}^{-1}$ is a one sided inverse of $c\boldsymbol{A}$.

□

1.4 Determinants

The first systematic treatment of determinants was given by Cauchy in 1812. He adopted the word "determinant". The first use of determinants was made by Leibniz in 1693 in a letter to De L'Hôspital. By the beginning of the twentieth century the theory of determinants filled four volumes of almost 2000 pages (Muir, 1906–1923. Historic references can be found in this work). The main use of determinants in this text will be to study the characteristic polynomial of a matrix and to show that a matrix is nonsingular.

For any $\boldsymbol{A} \in \mathbb{C}^{n\times n}$ the determinant of $\boldsymbol{A}$ is defined by the number

$$\det(\boldsymbol{A}) = \sum_{\sigma\in S_n} \operatorname{sign}(\sigma) a_{\sigma(1),1} a_{\sigma(2),2} \cdots a_{\sigma(n),n}. \tag{1.12}$$

This sum ranges of all $n!$ permutations of $\{1, 2, \ldots, n\}$. Moreover, $\operatorname{sign}(\sigma)$ equals the number of times a bigger integer precedes a smaller one in σ. We also denote the determinant by (Cayley, 1841)

$$\begin{vmatrix} a_{11} & a_{12} & \cdots & a_{1n} \\ a_{21} & a_{22} & \cdots & a_{2n} \\ \vdots & \vdots & & \vdots \\ a_{n1} & a_{n2} & \cdots & a_{nn} \end{vmatrix}.$$

From the definition we have

$$\begin{vmatrix} a_{11} & a_{12} \\ a_{21} & a_{22} \end{vmatrix} = a_{11}a_{22} - a_{21}a_{12}.$$

The first term on the right corresponds to the identity permutation ϵ given by $\epsilon(i) = i$, $i = 1, 2$. The second term comes from the permutation $\sigma = \{2, 1\}$. For $n = 3$ there are six permutations of $\{1, 2, 3\}$. Then

$$\begin{vmatrix} a_{11} & a_{12} & a_{13} \\ a_{21} & a_{22} & a_{23} \\ a_{31} & a_{32} & a_{33} \end{vmatrix} = a_{11}a_{22}a_{33} - a_{11}a_{32}a_{23} - a_{21}a_{12}a_{33} + a_{21}a_{32}a_{13} + a_{31}a_{12}a_{23} - a_{31}a_{22}a_{13}.$$

This follows since $\operatorname{sign}(\{1, 2, 3\}) = \operatorname{sign}(\{2, 3, 1\}) = \operatorname{sign}(\{3, 1, 2\}) = 1$, and noting that interchanging two numbers in a permutation reverses it sign we find $\operatorname{sign}(\{2, 1, 3\}) = \operatorname{sign}(\{3, 2, 1\}) = \operatorname{sign}(\{1, 3, 2\}) = -1$.

To compute the value of a determinant from the definition can be a trying experience. It is often better to use elementary operations on rows or columns to reduce it to a simpler form. For example, if $\boldsymbol{A}$ is triangular then $\det(\boldsymbol{A}) =$

$a_{11}a_{22}\cdots a_{nn}$, the product of the diagonal elements. In particular, for the identity matrix $\det(\boldsymbol{I}) = 1$. The elementary operations using either rows or columns are

1. Interchanging two rows(columns): $\det(\boldsymbol{B}) = -\det(\boldsymbol{A})$,
2. Multiply a row(column) by a scalar: α, $\det(\boldsymbol{B}) = \alpha\det(\boldsymbol{A})$,
3. Add a constant multiple of one row(column) to another row(column): $\det(\boldsymbol{B}) = \det(\boldsymbol{A})$.

where $\boldsymbol{B}$ is the result of performing the indicated operation on $\boldsymbol{A}$.

If only a few elements in a row or column are nonzero then a **cofactor expansion** can be used. These expansions take the form

$$\det(\boldsymbol{A}) = \sum_{j=1}^{n}(-1)^{i+j}a_{ij}\det(\boldsymbol{A}_{ij}) \text{ for } i = 1,\ldots,n, \text{ row} \tag{1.13}$$

$$\det(\boldsymbol{A}) = \sum_{i=1}^{n}(-1)^{i+j}a_{ij}\det(\boldsymbol{A}_{ij}) \text{ for } j = 1,\ldots,n, \text{ column.} \tag{1.14}$$

Here $\boldsymbol{A}_{i,j}$ denotes the submatrix of $\boldsymbol{A}$ obtained by deleting the ith row and jth column of $\boldsymbol{A}$. For $\boldsymbol{A} \in \mathbb{C}^{n\times n}$ and $1 \le i, j \le n$ the determinant $\det(\boldsymbol{A}_{ij})$ is called the **cofactor** of a_{ij}.

Example 1.3 (Determinant Equation for a Straight Line) The equation for a straight line through two points (x_1, y_1) and (x_2, y_2) in the plane can be written as the equation

$$\det(\boldsymbol{A}) := \begin{vmatrix} 1 & x & y \\ 1 & x_1 & y_1 \\ 1 & x_2 & y_2 \end{vmatrix} = 0$$

involving a determinant of order 3. We can compute this determinant using row operations of type 3. Subtracting row 2 from row 3 and then row 1 from row 2, and then using a cofactor expansion on the first column we obtain

$$\begin{vmatrix} 1 & x & y \\ 1 & x_1 & y_1 \\ 1 & x_2 & y_2 \end{vmatrix} = \begin{vmatrix} 1 & x & y \\ 0 & x_1 - x & y_1 - y \\ 0 & x_2 - x_1 & y_2 - y_1 \end{vmatrix}$$

$$= \begin{vmatrix} x_1 - x & y_1 - y \\ x_2 - x_1 & y_2 - y_1 \end{vmatrix} = (x_1 - x)(y_2 - y_1) - (y_1 - y)(x_2 - x_1).$$

Rearranging the equation $\det(\boldsymbol{A}) = 0$ we obtain

$$y - y_1 = \frac{y_2 - y_1}{x_2 - x_1}(x - x_1)$$

which is the slope form of the equation of a straight line.

We will freely use, without proofs, the following properties of determinants. If $\boldsymbol{A}, \boldsymbol{B}$ are square matrices of order n with real or complex elements, then

1. $\det(\boldsymbol{AB}) = \det(\boldsymbol{A})\det(\boldsymbol{B})$.
2. $\det(\boldsymbol{A}^T) = \det(\boldsymbol{A})$, and $\det(\boldsymbol{A}^*) = \overline{\det(\boldsymbol{A})}$, (complex conjugate).
3. $\det(a\boldsymbol{A}) = a^n \det(\boldsymbol{A})$, for $a \in \mathbb{C}$.
4. $\boldsymbol{A}$ is singular if and only if $\det(\boldsymbol{A}) = 0$.
5. If $\boldsymbol{A} = \begin{bmatrix} \boldsymbol{C} & \boldsymbol{D} \\ \boldsymbol{0} & \boldsymbol{E} \end{bmatrix}$ for some square matrices $\boldsymbol{C}, \boldsymbol{E}$ then $\det(\boldsymbol{A}) = \det(\boldsymbol{C})\det(\boldsymbol{E})$.
6. **Cramer's rule** Suppose $\boldsymbol{A} \in \mathbb{C}^{n\times n}$ is nonsingular and $\boldsymbol{b} \in \mathbb{C}^n$. Let $\boldsymbol{x} = [x_1, x_2, \dots, x_n]^T$ be the unique solution of $\boldsymbol{Ax} = \boldsymbol{b}$. Then

$$x_j = \frac{\det(\boldsymbol{A}_j(\boldsymbol{b}))}{\det(\boldsymbol{A})}, \quad j = 1, 2, \dots, n,$$

   where $\boldsymbol{A}_j(\boldsymbol{b})$ denote the matrix obtained from $\boldsymbol{A}$ by replacing the jth column of $\boldsymbol{A}$ by $\boldsymbol{b}$.
7. **Adjoint formula for the inverse**. If $\boldsymbol{A} \in \mathbb{C}^{n\times n}$ is nonsingular then

$$\boldsymbol{A}^{-1} = \frac{1}{\det(\boldsymbol{A})} \operatorname{adj}(\boldsymbol{A}),$$

   where the matrix $\operatorname{adj}(\boldsymbol{A}) \in \mathbb{C}^{n\times n}$ with elements $\operatorname{adj}(\boldsymbol{A})_{i,j} = (-1)^{i+j}\det(\boldsymbol{A}_{j,i})$ is called the **adjoint** of $\boldsymbol{A}$. Moreover, $\boldsymbol{A}_{j,i}$ denotes the submatrix of $\boldsymbol{A}$ obtained by deleting the jth row and ith column of $\boldsymbol{A}$.
8. **Cauchy-Binet formula**: Let $\boldsymbol{A} \in \mathbb{C}^{m\times p}$, $\boldsymbol{B} \in \mathbb{C}^{p\times n}$ and $\boldsymbol{C} = \boldsymbol{AB}$. Suppose $1 \le r \le \min\{m, n, p\}$ and let $\boldsymbol{i} = \{i_1, \dots, i_r\}$ and $\boldsymbol{j} = \{j_1, \dots, j_r\}$ be integers with $1 \le i_1 < i_2 < \cdots < i_r \le m$ and $1 \le j_1 < j_2 < \cdots < j_r \le n$. Then

$$\begin{bmatrix} c_{i_1,j_1} & \cdots & c_{i_1,j_r} \\ \vdots & & \vdots \\ c_{i_r,j_1} & \cdots & c_{i_r,j_r} \end{bmatrix} = \sum_{\boldsymbol{k}} \begin{bmatrix} a_{i_1,k_1} & \cdots & a_{i_1,k_r} \\ \vdots & & \vdots \\ a_{i_r,k_1} & \cdots & a_{i_r,k_r} \end{bmatrix} \begin{bmatrix} b_{k_1,j_1} & \cdots & b_{k_1,j_r} \\ \vdots & & \vdots \\ b_{k_r,j_1} & \cdots & b_{k_r,j_r} \end{bmatrix},$$

   where we sum over all $\boldsymbol{k} = \{k_1, \dots, k_r\}$ with $1 \le k_1 < k_2 < \cdots < k_r \le p$. More compactly,

$$\det\big(\boldsymbol{C}(\boldsymbol{i}, \boldsymbol{j})\big) = \sum_{\boldsymbol{k}} \det\big(\boldsymbol{A}(\boldsymbol{i}, \boldsymbol{k})\big)\det\big(\boldsymbol{B}(\boldsymbol{k}, \boldsymbol{j})\big), \tag{1.15}$$

   Note the resemblance to the formula for matrix multiplication.

1.5 Eigenvalues, Eigenvectors and Eigenpairs

Suppose $\boldsymbol{A} \in \mathbb{C}^{n\times n}$ is a square matrix, $\lambda \in \mathbb{C}$ and $\boldsymbol{x} \in \mathbb{C}^n$. We say that $(\lambda, \boldsymbol{x})$ is an **eigenpair** for $\boldsymbol{A}$ if $\boldsymbol{A}\boldsymbol{x} = \lambda\boldsymbol{x}$ and $\boldsymbol{x}$ is nonzero. The scalar λ is called an **eigenvalue** and $\boldsymbol{x}$ is said to be an **eigenvector**.[1] The set of eigenvalues is called the **spectrum** of $\boldsymbol{A}$ and is denoted by $\sigma(\boldsymbol{A})$. For example, $\sigma(\boldsymbol{I}) = \{1, \dots, 1\} = \{1\}$.

Eigenvalues are the roots of the characteristic polynomial.

Lemma 1.5 (Characteristic Equation) *For any* $\boldsymbol{A} \in \mathbb{C}^{n\times n}$ *we have* $\lambda \in \sigma(\boldsymbol{A}) \iff \det(\boldsymbol{A} - \lambda\boldsymbol{I}) = 0$.

Proof Suppose $(\lambda, \boldsymbol{x})$ is an eigenpair for $\boldsymbol{A}$. The equation $\boldsymbol{A}\boldsymbol{x} = \lambda\boldsymbol{x}$ can be written $(\boldsymbol{A} - \lambda\boldsymbol{I})\boldsymbol{x} = \boldsymbol{0}$. Since $\boldsymbol{x}$ is nonzero the matrix $\boldsymbol{A} - \lambda\boldsymbol{I}$ must be singular with a zero determinant. Conversely, if $\det(\boldsymbol{A} - \lambda\boldsymbol{I}) = 0$ then $\boldsymbol{A} - \lambda\boldsymbol{I}$ is singular and $(\boldsymbol{A} - \lambda\boldsymbol{I})\boldsymbol{x} = \boldsymbol{0}$ for some nonzero $\boldsymbol{x} \in \mathbb{C}^n$. Thus $\boldsymbol{A}\boldsymbol{x} = \lambda\boldsymbol{x}$ and $(\lambda, \boldsymbol{x})$ is an eigenpair for $\boldsymbol{A}$. □

The expression $\det(\boldsymbol{A} - \lambda\boldsymbol{I})$ is a polynomial of exact degree n in λ. For $n = 3$ we have

$$\det(\boldsymbol{A} - \lambda\boldsymbol{I}) = \begin{vmatrix} a_{11} - \lambda & a_{12} & a_{13} \\ a_{21} & a_{22} - \lambda & a_{23} \\ a_{31} & a_{32} & a_{33} - \lambda \end{vmatrix}.$$

Expanding this determinant by the first column we find

$$\det(\boldsymbol{A} - \lambda\boldsymbol{I}) = (a_{11} - \lambda)\begin{vmatrix} a_{22} - \lambda & a_{23} \\ a_{32} & a_{33} - \lambda \end{vmatrix} - a_{21}\begin{vmatrix} a_{12} & a_{13} \\ a_{32} & a_{33} - \lambda \end{vmatrix}$$
$$+ a_{31}\begin{vmatrix} a_{12} & a_{13} \\ a_{22} - \lambda & a_{23} \end{vmatrix} = (a_{11} - \lambda)(a_{22} - \lambda)(a_{33} - \lambda) + r(\lambda)$$

for some polynomial r of degree at most one. In general

$$\det(\boldsymbol{A} - \lambda\boldsymbol{I}) = (a_{11} - \lambda)(a_{22} - \lambda)\cdots(a_{nn} - \lambda) + r(\lambda), \tag{1.16}$$

where each term in $r(\lambda)$ has at most $n - 2$ factors containing λ. It follows that r is a polynomial of degree at most $n - 2$, $\det(\boldsymbol{A} - \lambda\boldsymbol{I})$ is a polynomial of exact degree n in λ and the eigenvalues are the roots of this polynomial.

We observe that $\det(\boldsymbol{A} - \lambda\boldsymbol{I}) = (-1)^n \det(\lambda\boldsymbol{I} - \boldsymbol{A})$ so $\det(\boldsymbol{A} - \lambda\boldsymbol{I}) = 0$ if and only if $\det(\lambda\boldsymbol{I} - \boldsymbol{A}) = 0$.

[1]The word "eigen" is derived from German and means "own".

Definition 1.9 (Characteristic Polynomial of a Matrix) The function $\pi_{\boldsymbol{A}}: \mathbb{C} \to \mathbb{C}$ given by $\pi_{\boldsymbol{A}}(\lambda) = \det(\boldsymbol{A} - \lambda \boldsymbol{I})$ is called the **characteristic polynomial** of $\boldsymbol{A}$. The equation $\det(\boldsymbol{A} - \lambda \boldsymbol{I}) = 0$ is called the **characteristic equation** of $\boldsymbol{A}$.

By the fundamental theorem of algebra an $n \times n$ matrix has, counting multiplicities, precisely n eigenvalues $\lambda_1, \ldots, \lambda_n$ some of which might be complex even if $\boldsymbol{A}$ is real. The complex eigenpairs of a real matrix occur in complex conjugate pairs. Indeed, taking the complex conjugate on both sides of the equation $\boldsymbol{A}\boldsymbol{x} = \lambda \boldsymbol{x}$ with $\boldsymbol{A}$ real gives $\boldsymbol{A}\overline{\boldsymbol{x}} = \overline{\lambda}\overline{\boldsymbol{x}}$.

Theorem 1.10 (Sums and Products of Eigenvalues; Trace) *For any $\boldsymbol{A} \in \mathbb{C}^{n\times n}$*

$$\operatorname{trace}(\boldsymbol{A}) = \lambda_1 + \lambda_2 + \cdots + \lambda_n, \quad \det(\boldsymbol{A}) = \lambda_1\lambda_2\cdots\lambda_n, \tag{1.17}$$

*where the **trace** of $\boldsymbol{A} \in \mathbb{C}^{n\times n}$ is the sum of its diagonal elements*

$$\operatorname{trace}(\boldsymbol{A}) := a_{11} + a_{22} + \cdots + a_{nn}. \tag{1.18}$$

Proof We compare two different expansions of $\pi_{\boldsymbol{A}}$. On the one hand from (1.16) we find

$$\pi_{\boldsymbol{A}}(\lambda) = (-1)^n\lambda^n + c_{n-1}\lambda^{n-1} + \cdots + c_0,$$

where $c_{n-1} = (-1)^{n-1}\operatorname{trace}(\boldsymbol{A})$ and $c_0 = \pi_{\boldsymbol{A}}(0) = \det(\boldsymbol{A})$. On the other hand

$$\pi_{\boldsymbol{A}}(\lambda) = (\lambda_1 - \lambda)\cdots(\lambda_n - \lambda) = (-1)^n\lambda^n + d_{n-1}\lambda^{n-1} + \cdots + d_0,$$

where $d_{n-1} = (-1)^{n-1}(\lambda_1 + \cdots + \lambda_n)$ and $d_0 = \lambda_1 \cdots \lambda_n$. Since $c_j = d_j$ for all j we obtain (1.17). □

For a 2×2 matrix the characteristic equation takes the convenient form

$$\lambda^2 - \operatorname{trace}(\boldsymbol{A})\lambda + \det(\boldsymbol{A}) = 0. \tag{1.19}$$

Thus, if $\boldsymbol{A} = \begin{bmatrix} 2 & 1 \\ 1 & 2 \end{bmatrix}$ then $\operatorname{trace}(\boldsymbol{A}) = 4$, $\det(\boldsymbol{A}) = 3$ so that $\pi_{\boldsymbol{A}}(\lambda) = \lambda^2 - 4\lambda + 3$.

Since $\boldsymbol{A}$ is singular $\iff$ $\boldsymbol{A}\boldsymbol{x} = \boldsymbol{0}$, some $\boldsymbol{x} \neq 0$ $\iff$ $\boldsymbol{A}\boldsymbol{x} = \boldsymbol{0}\boldsymbol{x}$, some $\boldsymbol{x} \neq 0$ $\iff$ zero is an eigenvalue of $\boldsymbol{A}$, we obtain

Theorem 1.11 (Zero Eigenvalue) *The matrix $\boldsymbol{A} \in \mathbb{C}^{n\times n}$ is singular if and only if zero is an eigenvalue.*

Since the determinant of a triangular matrix is equal to the product of the diagonal elements the eigenvalues of a triangular matrix are found on the diagonal. In general it is not easy to find all eigenvalues of a matrix. However, sometimes the dimension of the problem can be reduced. Since the determinant of a block triangular matrix is equal to the product of the determinants of the diagonal blocks we obtain

Theorem 1.12 (Eigenvalues of a Block Triangular Matrix) *If* $A = \begin{bmatrix} B & D \\ 0 & C \end{bmatrix}$ *is block triangular then* $\pi_A = \pi_B \cdot \pi_C$.

1.6 Exercises Chap. 1

1.6.1 Exercises Sect. 1.1

Exercise 1.1 (Strassen Multiplication (Exam Exercise 2017-1)) (By arithmetic operations we mean additions, subtractions, multiplications and divisions.)

Let $\boldsymbol{A}$ and $\boldsymbol{B}$ be $n \times n$ real matrices.

a) With $\boldsymbol{A}, \boldsymbol{B} \in \mathbb{R}^{n \times n}$, how many arithmetic operations are required to form the product $\boldsymbol{AB}$?

b) Consider the $2n \times 2n$ block matrix

$$\begin{bmatrix} W & X \\ Y & Z \end{bmatrix} = \begin{bmatrix} A & B \\ C & D \end{bmatrix} \begin{bmatrix} E & F \\ G & H \end{bmatrix},$$

where all matrices $\boldsymbol{A}, \ldots, \boldsymbol{Z}$ are in $\mathbb{R}^{n \times n}$. How many operations does it take to compute $\boldsymbol{W}, \boldsymbol{X}, \boldsymbol{Y}$ and $\boldsymbol{Z}$ by the obvious algorithm?

c) An alternative method to compute $\boldsymbol{W}, \boldsymbol{X}, \boldsymbol{Y}$ and $\boldsymbol{Z}$ is to use Strassen's formulas:

$$\begin{aligned}
\mathbf{P}_1 &= (\boldsymbol{A} + \boldsymbol{D})(\boldsymbol{E} + \boldsymbol{H}), & & \\
\mathbf{P}_2 &= (\boldsymbol{C} + \boldsymbol{D})\boldsymbol{E}, & \mathbf{P}_5 &= (\boldsymbol{A} + \boldsymbol{B})\boldsymbol{H}, \\
\mathbf{P}_3 &= \boldsymbol{A}(\boldsymbol{F} - \boldsymbol{H}), & \mathbf{P}_6 &= (\boldsymbol{C} - \boldsymbol{A})(\boldsymbol{E} + \boldsymbol{F}), \\
\mathbf{P}_4 &= \boldsymbol{D}(\boldsymbol{G} - \boldsymbol{E}), & \mathbf{P}_7 &= (\boldsymbol{B} - \boldsymbol{D})(\boldsymbol{G} + \boldsymbol{H}), \\
\boldsymbol{W} &= \mathbf{P}_1 + \mathbf{P}_4 - \mathbf{P}_5 + \mathbf{P}_7, & \boldsymbol{X} &= \mathbf{P}_3 + \mathbf{P}_5, \\
\boldsymbol{Y} &= \mathbf{P}_2 + \mathbf{P}_4, & \boldsymbol{Z} &= \mathbf{P}_1 + \mathbf{P}_3 - \mathbf{P}_2 + \mathbf{P}_6.
\end{aligned}$$

You do not have to verify these formulas. What is the operation count for this method?

d) Describe a recursive algorithm, based on Strassen's formulas, which given two matrices $\boldsymbol{A}$ and $\boldsymbol{B}$ of size $m \times m$, with $m = 2^k$ for some $k \geq 0$, calculates the product $\boldsymbol{AB}$.

e) Show that the operation count of the recursive algorithm is $\mathcal{O}\left(m^{\log_2(7)}\right)$. Note that $\log_2(7) \approx 2.8 < 3$, so this is less costly than straightforward matrix multiplication.

1.6.2 Exercises Sect. 1.3

Exercise 1.2 (The Inverse of a General 2×2 Matrix) Show that

$$\begin{bmatrix} a & b \\ c & d \end{bmatrix}^{-1} = \alpha \begin{bmatrix} d & -b \\ -c & a \end{bmatrix}, \quad \alpha = \frac{1}{ad - bc},$$

for any a, b, c, d such that $ad - bc \neq 0$.

Exercise 1.3 (The Inverse of a Special 2×2 Matrix) Find the inverse of

$$A = \begin{bmatrix} \cos\theta & -\sin\theta \\ \sin\theta & \cos\theta \end{bmatrix}.$$

Exercise 1.4 (Sherman-Morrison Formula) Suppose $\boldsymbol{A} \in \mathbb{C}^{n\times n}$, and $\boldsymbol{B}, \boldsymbol{C} \in \mathbb{R}^{n\times m}$ for some $n, m \in \mathbb{N}$. If $(\boldsymbol{I} + \boldsymbol{C}^T\boldsymbol{A}^{-1}\boldsymbol{B})^{-1}$ exists then

$$(\boldsymbol{A} + \boldsymbol{B}\boldsymbol{C}^T)^{-1} = \boldsymbol{A}^{-1} - \boldsymbol{A}^{-1}\boldsymbol{B}(\boldsymbol{I} + \boldsymbol{C}^T\boldsymbol{A}^{-1}\boldsymbol{B})^{-1}\boldsymbol{C}^T\boldsymbol{A}^{-1}.$$

Exercise 1.5 (Inverse Update (Exam Exercise 1977-1))

a) Let $\boldsymbol{u}, \boldsymbol{v} \in \mathbb{R}^n$ and suppose $\boldsymbol{v}^T\boldsymbol{u} \neq 1$. Show that $\boldsymbol{I} - \boldsymbol{u}\boldsymbol{v}^T$ has an inverse given by $\boldsymbol{I} - \tau\boldsymbol{u}\boldsymbol{v}^T$, where $\tau = 1/(\boldsymbol{v}^T\boldsymbol{u} - 1)$.

b) Let $\boldsymbol{A} \in \mathbb{R}^{n\times n}$ be nonsingular with inverse $\boldsymbol{C} := \boldsymbol{A}^{-1}$, and let $\boldsymbol{a} \in \mathbb{R}^n$. Let $\overline{\boldsymbol{A}}$ be the matrix which differs from $\boldsymbol{A}$ by exchanging the ith row of $\boldsymbol{A}$ with $\boldsymbol{a}^T$, i.e., $\overline{\boldsymbol{A}} = \boldsymbol{A} - \boldsymbol{e}_i(\boldsymbol{e}_i^T\boldsymbol{A} - \boldsymbol{a}^T)$, where $\boldsymbol{e}_i$ is the ith column in the identity matrix $\boldsymbol{I}$. Show that if

$$\lambda := \boldsymbol{a}^T\boldsymbol{C}\boldsymbol{e}_i \neq 0, \tag{1.20}$$

then $\overline{\boldsymbol{A}}$ has an inverse $\overline{\boldsymbol{C}} = \overline{\boldsymbol{A}}^{-1}$ given by

$$\overline{\boldsymbol{C}} = \boldsymbol{C}\big(\boldsymbol{I} + \frac{1}{\lambda}\boldsymbol{e}_i(\boldsymbol{e}_i^T - \boldsymbol{a}^T\boldsymbol{C})\big) \tag{1.21}$$

c) Write an algorithm which to given $\boldsymbol{C}$ and $\boldsymbol{a}$ checks if (1.20) holds and computes $\overline{\boldsymbol{C}}$ provided $\lambda \neq 0$. (hint: Use (1.21) to find formulas for computing each column in $\overline{\boldsymbol{C}}$.)

Exercise 1.6 (Matrix Products (Exam Exercise 2009-1)) Let $\boldsymbol{A}, \boldsymbol{B}, \boldsymbol{C}, \boldsymbol{E} \in \mathbb{R}^{n\times n}$ be matrices where $\boldsymbol{A}^T = \boldsymbol{A}$. In this problem an (arithmetic) operation is an addition or a multiplication. We ask about exact numbers of operations.

a) How many operations are required to compute the matrix product $\boldsymbol{B}\boldsymbol{C}$? How many operations are required if $\boldsymbol{B}$ is lower triangular?

b) Show that there exists a lower triangular matrix $\boldsymbol{L} \in \mathbb{R}^{n\times n}$ such that $\boldsymbol{A} = \boldsymbol{L}+\boldsymbol{L}^T$.

c) We have $\boldsymbol{E}^T\boldsymbol{A}\boldsymbol{E} = \boldsymbol{S} + \boldsymbol{S}^T$ where $\boldsymbol{S} = \boldsymbol{E}^T\boldsymbol{L}\boldsymbol{E}$. How many operations are required to compute $\boldsymbol{E}^T\boldsymbol{A}\boldsymbol{E}$ in this way?

1.6.3 Exercises Sect. 1.4

Exercise 1.7 (Cramer's Rule; Special Case) Solve the following system by Cramer's rule:

$$\begin{bmatrix} 1 & 2 \\ 2 & 1 \end{bmatrix} \begin{bmatrix} x_1 \\ x_2 \end{bmatrix} = \begin{bmatrix} 3 \\ 6 \end{bmatrix}$$

Exercise 1.8 (Adjoint Matrix; Special Case) Show that if

$$\boldsymbol{A} = \begin{bmatrix} 2 & -6 & 3 \\ 3 & -2 & -6 \\ 6 & 3 & 2 \end{bmatrix},$$

then

$$\text{adj}(\boldsymbol{A}) = \begin{bmatrix} 14 & 21 & 42 \\ -42 & -14 & 21 \\ 21 & -42 & 14 \end{bmatrix}.$$

Moreover,

$$\text{adj}(\boldsymbol{A})\boldsymbol{A} = \begin{bmatrix} 343 & 0 & 0 \\ 0 & 343 & 0 \\ 0 & 0 & 343 \end{bmatrix} = \det(\boldsymbol{A})\boldsymbol{I}.$$

Exercise 1.9 (Determinant Equation for a Plane) Show that

$$\begin{vmatrix} x & y & z & 1 \\ x_1 & y_1 & z_1 & 1 \\ x_2 & y_2 & z_2 & 1 \\ x_3 & y_3 & z_3 & 1 \end{vmatrix} = 0.$$

is the equation for a plane through three points (x_1, y_1, z_1), (x_2, y_2, z_2), (x_3, y_3, z_3) in space.

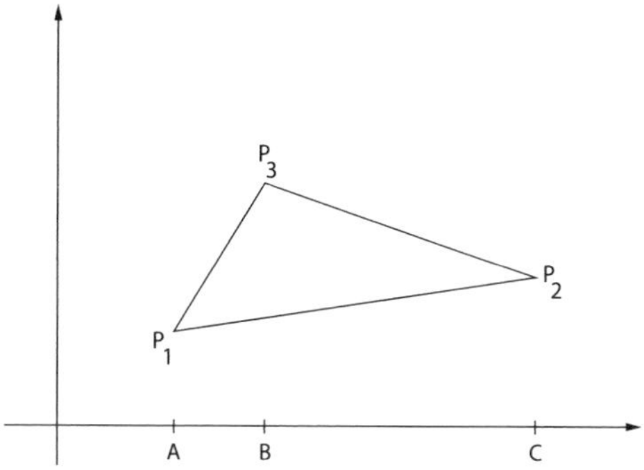

Fig. 1.1 The triangle T defined by the three points P_1, P_2 and P_3

Exercise 1.10 (Signed Area of a Triangle) Let $P_i = (x_i, y_i)$, $i = 1, 2, 3$, be three points in the plane defining a triangle T. Show that the area of T is[2]

$$A(T) = \frac{1}{2} \begin{vmatrix} 1 & 1 & 1 \\ x_1 & x_2 & x_3 \\ y_1 & y_2 & y_3 \end{vmatrix}.$$

The area is positive if we traverse the vertices in counterclockwise order.

Exercise 1.11 (Vandermonde Matrix) Show that

$$\begin{vmatrix} 1 & x_1 & x_1^2 & \cdots & x_1^{n-1} \\ 1 & x_2 & x_2^2 & \cdots & x_2^{n-1} \\ \vdots & \vdots & \vdots & & \vdots \\ 1 & x_n & x_n^2 & \cdots & x_n^{n-1} \end{vmatrix} = \prod_{i>j} (x_i - x_j),$$

where $\prod_{i>j}(x_i - x_j) = \prod_{i=2}^{n}(x_i - x_1)(x_i - x_2)\cdots(x_i - x_{i-1})$. This determinant is called the **Vandermonde determinant**.[3]

Exercise 1.12 (Cauchy Determinant (1842)) Let $\boldsymbol{\alpha} = [\alpha_1, \ldots, \alpha_n]^T$, $\boldsymbol{\beta} = [\beta_1, \ldots, \beta_n]^T$ be in $\mathbb{R}^n$.

a) Consider the matrix $\boldsymbol{A} \in \mathbb{R}^{n\times n}$ with elements $a_{i,j} = 1/(\alpha_i + \beta_j)$, $i, j = 1, 2, \ldots, n$. Show that

$$\det(\boldsymbol{A}) = P g(\boldsymbol{\alpha}) g(\boldsymbol{\beta})$$

[2]Hint: $A(T) = A(ABP_3P_1) + A(P_3BCP_2) - A(P_1ACP_2)$, c.f. Fig. 1.1.

[3]Hint: subtract x_n^k times column k from column $k+1$ for $k = n-1, n-2, \ldots, 1$.

where $P = \prod_{i=1}^{n}\prod_{j=1}^{n} a_{ij}$, and for $\boldsymbol{\gamma} = [\gamma_1, \dots, \gamma_n]^T$

$$g(\boldsymbol{\gamma}) = \prod_{i=2}^{n}(\gamma_i - \gamma_1)(\gamma_i - \gamma_2)\cdots(\gamma_i - \gamma_{i-1})$$

Hint: Multiply the ith row of $\boldsymbol{A}$ by $\prod_{j=1}^{n}(\alpha_i + \beta_j)$ for $i = 1, 2, \dots, n$. Call the resulting matrix $\boldsymbol{C}$. Each element of $\boldsymbol{C}$ is a product of n–1 factors $\alpha_r + \beta_s$. Hence $\det(\boldsymbol{C})$ is a sum of terms where each term contain precisely $n(n-1)$ factors $\alpha_r + \beta_s$. Thus $\det(\boldsymbol{C}) = q(\alpha, \beta)$ where q is a polynomial of degree at most $n(n-1)$ in α_i and β_j. Since $\det(\boldsymbol{A})$ and therefore $\det(\boldsymbol{C})$ vanishes if $\alpha_i = \alpha_j$ for some $i \neq j$ or $\beta_r = \beta_s$ for some $r \neq s$, we have that $q(\boldsymbol{\alpha}, \boldsymbol{\beta})$ must be divisible by each factor in $g(\boldsymbol{\alpha})$ and $g(\boldsymbol{\beta})$. Since $g(\boldsymbol{\alpha})$ and $g(\boldsymbol{\beta})$ is a polynomial of degree $n(n-1)$, we have

$$q(\boldsymbol{\alpha}, \boldsymbol{\beta}) = k g(\boldsymbol{\alpha}) g(\boldsymbol{\beta})$$

for some constant k independent of $\boldsymbol{\alpha}$ and $\boldsymbol{\beta}$. Show that $k = 1$ by choosing $\beta_i + \alpha_i = 0, i = 1, 2, \dots, n$.

b) Notice that the cofactor of any element in the above matrix $\boldsymbol{A}$ is the determinant of a matrix of similar form. Use the cofactor and determinant of $\boldsymbol{A}$ to represent the elements of $\boldsymbol{A}^{-1} = (b_{j,k})$. Answer:

$$b_{j,k} = (\alpha_k + \beta_j) A_k(-\beta_j) B_j(-\alpha_k),$$

where

$$A_k(x) = \prod_{s \neq k}\left(\frac{\alpha_s - x}{\alpha_s - \alpha_k}\right), \qquad B_k(x) = \prod_{s \neq k}\left(\frac{\beta_s - x}{\beta_s - \beta_k}\right).$$

Exercise 1.13 (Inverse of the Hilbert Matrix) Let $\boldsymbol{H}_n = (h_{i,j})$ be the $n \times n$ matrix with elements $h_{i,j} = 1/(i+j-1)$. Use Exercise 1.12 to show that the elements $t_{i,j}^n$ in $\boldsymbol{T}_n = \boldsymbol{H}_n^{-1}$ are given by

$$t_{i,j}^n = \frac{f(i) f(j)}{i + j - 1},$$

where

$$f(i+1) = \left(\frac{i^2 - n^2}{i^2}\right) f(i), \quad i = 1, 2, \dots, \quad f(1) = -n.$$

Part I
LU and QR Factorizations

The first three chapters in this part consider ways of factoring a matrix $\boldsymbol{A}$ into a lower triangular matrix $\boldsymbol{L}$ and an upper triangular matrix $\boldsymbol{U}$ resulting in the product $\boldsymbol{A} = \boldsymbol{L}\boldsymbol{U}$. We also consider the factorization $\boldsymbol{A} = \boldsymbol{L}\boldsymbol{D}\boldsymbol{U}$, where $\boldsymbol{L}$ is lower triangular, $\boldsymbol{D}$ is diagonal and $\boldsymbol{U}$ is upper triangular. Moreover, $\boldsymbol{L}$ and $\boldsymbol{U}$ have ones on their respective diagonals.

Three simple introductory problems and related LU factorizations are considered in Chap. 2. We also consider some basic properties of triangular matrices and the powerful tool of block multiplication. We consider Gaussian elimination, it's relation to LU factorization, and the general theory of LU factorizations in Chap. 3. Symmetric positive definite matrices, where LU factorizations play an important role, are considered in Chap. 4.

There exists problems where Gaussian elimination leads to inaccurate results. Such problems can to a large extent be avoided by using an alternative method based on QR factorization. Here $\boldsymbol{A} = \boldsymbol{Q}\boldsymbol{R}$, where $\boldsymbol{Q}$ is unitary, i.e., $\boldsymbol{Q}^*\boldsymbol{Q} = \boldsymbol{I}$, and $\boldsymbol{R}$ is upper triangular. QR factorization is related to Gram-Schmidt orthogonalization of a basis in a vector space. The QR factorization plays an important role in computational least squares and eigenvalue problems.

Chapter 2
Diagonally Dominant Tridiagonal Matrices; Three Examples

In this chapter we consider three problems originating from:

- cubic spline interpolation,
- a two point boundary value problem,
- an eigenvalue problem for a two point boundary value problem.

Each of these problems leads to a linear algebra problem with a matrix which is diagonally dominant and tridiagonal. Taking advantage of structure we can show existence, uniqueness and characterization of a solution, and derive efficient and stable algorithms based on LU factorization to compute a numerical solution.

For a particular tridiagonal test matrix we determine all its eigenvectors and eigenvalues. We will need these later when studying more complex problems.

We end the chapter with an introduction to block multiplication, a powerful tool in matrix analysis and numerical linear algebra. Block multiplication is applied to derive some basic facts about triangular matrices.

2.1 Cubic Spline Interpolation

We consider the following interpolation problem.

Given an interval $[a, b]$, $n + 1 \geq 2$ equidistant sites in $[a, b]$

$$x_i = a + \frac{i-1}{n}(b-a), \quad i = 1, 2, \ldots, n+1 \tag{2.1}$$

and y values $\boldsymbol{y} := [y_1, \ldots, y_{n+1}]^T \in \mathbb{R}^{n+1}$. We seek a function $g : [a, b] \to \mathbb{R}$ such that

$$g(x_i) = y_i, \text{ for } i = 1, \ldots, n+1. \tag{2.2}$$

T. Lyche, *Numerical Linear Algebra and Matrix Factorizations*, Texts in Computational Science and Engineering 22,
https://doi.org/10.1007/978-3-030-36468-7_2

For simplicity we only consider equidistant sites. More generally they could be any $a \le x_1 < x_2 < \cdots < x_{n+1} \le b$.

2.1.1 Polynomial Interpolation

Since there are $n+1$ interpolation conditions in (2.2) a natural choice for a function g is a polynomial of degree n. As shown in most books on numerical methods such a g is uniquely defined and there are good algorithms for computing it. Evidently, when $n = 1$, g is the straight line

$$g(x) = y_1 + \frac{y_2 - y_1}{x_2 - x_1}(x - x_1), \tag{2.3}$$

known as the **linear interpolation polynomial**.

Polynomial interpolation is an important technique which often gives good results, but the interpolant g can have undesirable oscillations when n is large. As an example, consider the function given by

$$f(x) = \arctan(10x) + \pi/2, \quad x \in [-1, 1].$$

The function f and the polynomial g of degree at most 13 satisfying (2.2) with $[a, b] = [-1, 1]$ and $y_i = f(x_i)$, $i = 1, \ldots, 14$ is shown in Fig. 2.1. The interpolant has large oscillations near the end of the range. This is an example of the **Runge phenomenon**. Using larger n will only make the oscillations bigger.[1]

2.1.2 Piecewise Linear and Cubic Spline Interpolation

To avoid oscillations like the one in Fig. 2.1 piecewise linear interpolation can be used. An example is shown in Fig. 2.2. The interpolant g approximates the original function quite well, and for some applications, like plotting, the linear interpolant using many points is what is used. Note that g is a piecewise polynomial of the form

$$g(x) := \begin{cases} p_1(x), & \text{if } x_1 \le x < x_2, \\ p_2(x), & \text{if } x_2 \le x < x_3, \\ \vdots & \\ p_{n-1}(x), & \text{if } x_{n-1} \le x < x_n, \\ p_n(x), & \text{if } x_n \le x \le x_{n+1}, \end{cases} \tag{2.4}$$

[1]This is due to the fact that the sites are uniformly spaced. High degree interpolation converges uniformly to the function being interpolated when a sequence consisting of the extrema of the Chebyshev polynomial of increasing degree is used as sites. This is not true for any continuous function (the Faber theorem), but holds if the function is Lipschitz continuous.

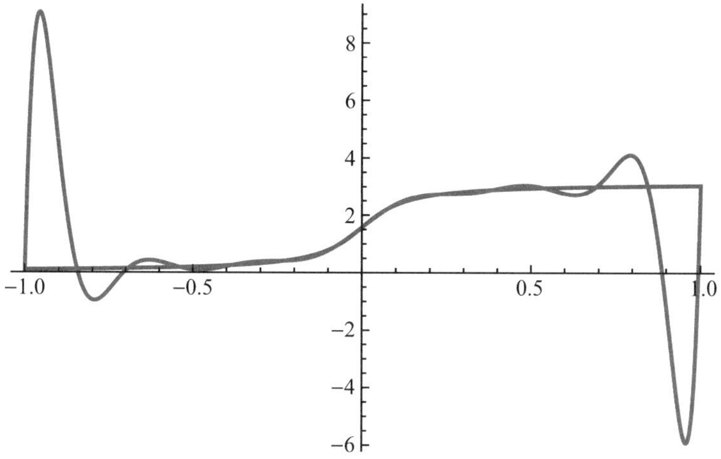

Fig. 2.1 The polynomial of degree 13 interpolating $f(x) = \arctan(10x) + \pi/2$ on $[-1, 1]$. See text

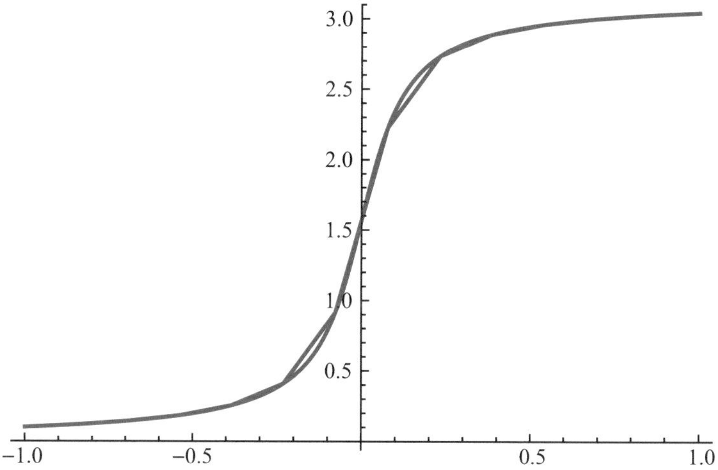

Fig. 2.2 The piecewise linear polynomial interpolating $f(x) = \arctan(10x) + \pi/2$ at $n = 14$ uniform points on $[-1, 1]$

where each p_i is a polynomial of degree ≤ 1. In particular, p_1 is given in (2.3) and the other polynomials p_i are given by similar expressions.

The piecewise linear interpolant is continuous, but the first derivative will usually have jumps at the interior sites. We can obtain a smoother approximation by letting g be a piecewise polynomial of higher degree. With degree 3 (cubic) we obtain continuous derivatives of order ≤ 2 (C^2). We consider here the following functions giving examples of C^2 **cubic spline interpolants**.

Definition 2.1 (The D_2-Spline Problem) Given $n \in \mathbb{N}$, an interval $[a, b]$, $\boldsymbol{y} \in \mathbb{R}^{n+1}$, knots (sites) $x_1, \ldots, x_{n+1}$ given by (2.1) and numbers μ_1, μ_{n+1}. The problem is to find a function $g : [a, b] \to \mathbb{R}$ such that

- **piecewise cubic polynomial**: g is of the form (2.4) with each p_i a cubic polynomial, ,
- **smoothness**: $g \in C^2[a, b]$, i.e., derivatives of order ≤ 2 are continuous on $\mathbb{R}$,
- **interpolation**: $g(x_i) = y_i, \quad i = 1, 2, \ldots, n+1$,
- **D_2 boundary conditions**: $g''(a) = \mu_1, \quad g''(b) = \mu_{n+1}$.

We call g a **D_2-spline**. It is called an **N-spline** or **natural spline** if $\mu_1 = \mu_{n+1} = 0$.

Example 2.1 (A D_2-Spline) Suppose we choose $n = 2$ and sample data from the function $f : [0, 2] \to \mathbb{R}$ given by $f(x) = x^4$. Thus we consider the D_2-spline problem with $[a, b] = [0, 2]$, $\boldsymbol{y} := [0, 1, 16]^T$ and $\mu_1 = g''(0) = 0$, $\mu_3 = g''(2) = 48$. The knots are $x_1 = 0$, $x_2 = 1$ and $x_3 = 2$. The function g given by

$$g(x) := \begin{cases} p_1(x) = -\frac{1}{2}x + \frac{3}{2}x^3, & \text{if } 0 \leq x < 1, \\ p_2(x) = 1 + 4(x-1) + \frac{9}{2}(x-1)^2 + \frac{13}{2}(x-1)^3, & \text{if } 1 \leq x \leq 2, \end{cases} \tag{2.5}$$

is a D_2-spline solving this problem. Indeed, p_1 and p_2 are cubic polynomials. For smoothness we find $p_1(1) = p_2(1) = 1$, $p_1'(1) = p_2'(1) = 4$, $p_1''(1) = p_2''(1) = 9$ which implies that $g \in C^2[0, 2]$. Finally we check that the interpolation and boundary conditions hold. Indeed, $g(0) = p_1(0) = 0$, $g(1) = p_2(1) = 1$, $g(2) = p_2(2) = 16$, $g''(0) = p_1''(0) = 0$ and $g''(2) = p_2''(2) = 48$. Note that $p_1'''(x) = 9 \neq 39 = p_2'''(x)$ showing that the third derivative of g is piecewise constant with a jump discontinuity at the interior knot. A plot of f and g is shown in Fig. 2.3. It is hard to distinguish one from the other.

We note that

- The C^2 condition is equivalent to

$$p_{i-1}^{(j)}(x_i) = p_i^{(j)}(x_i), \quad j = 0, 1, 2, \quad i = 2, \ldots, n.$$

- The extra boundary conditions D_2 or N are introduced to obtain a unique interpolant. Indeed counting requirements we have $3(n-1)$ C^2 conditions, $n+1$ conditions (2.2), and two boundary conditions, adding to $4n$. Since a cubic polynomial has four coefficients, this number is equal to the number of coefficients of the n polynomials $p_1, \ldots, p_n$ and give hope for uniqueness of the interpolant.

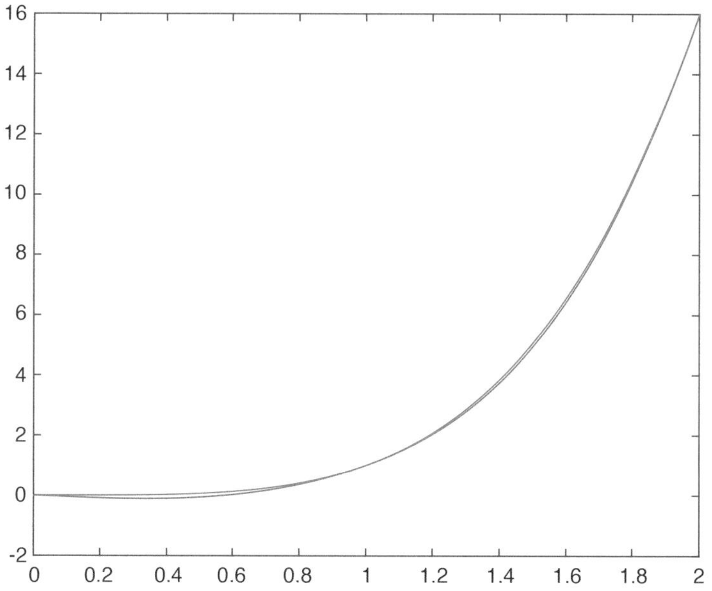

Fig. 2.3 A cubic spline with one knot interpolating $f(x) = x^4$ on $[0, 2]$

2.1.3 *Give Me a Moment*

Existence and uniqueness of a solution of the D_2-spline problem hinges on the nonsingularity of a linear system of equations that we now derive. The unknowns are derivatives at the knots. Here we use second derivatives which are sometimes called **moments**. We start with the following lemma.

Lemma 2.1 (Representing Each p_i Using $(0, 2)$ Interpolation) *Given $a < b$, $h = (b-a)/n$ with $n \geq 2$, $x_i = a+(i-1)h$, and numbers y_i, μ_i for $i = 1, \ldots, n+1$. For $i = 1, \ldots, n$ there are unique cubic polynomials p_i such that*

$$p_i(x_i) = y_i, \ p_i(x_{i+1}) = y_{i+1}, \quad p_i''(x_i) = \mu_i, \ p_i''(x_{i+1}) = \mu_{i+1}. \tag{2.6}$$

Moreover,

$$p_i(x) = c_{i,1} + c_{i,2}(x - x_i) + c_{i,3}(x - x_i)^2 + c_{i,4}(x - x_i)^3 \quad i = 1, \ldots, n, \tag{2.7}$$

where

$$c_{i1} = y_i, \ c_{i2} = \frac{y_{i+1} - y_i}{h} - \frac{h}{3}\mu_i - \frac{h}{6}\mu_{i+1}, \ c_{i,3} = \frac{\mu_i}{2}, \ c_{i,4} = \frac{\mu_{i+1} - \mu_i}{6h}. \tag{2.8}$$

Proof Consider p_i in the form (2.7) for some $1 \le i \le n$. Evoking (2.6) we find $p_i(x_i) = c_{i,1} = y_i$. Since $p_i''(x) = 2c_{i,3}+6c_{i,4}(x-x_i)$ we obtain $c_{i,3}$ from $p_i''(x_i) = 2c_{i,3} = \mu_i$ (a moment), and then $c_{i,4}$ from $p_i''(x_{i+1}) = \mu_i + 6hc_{i,4} = \mu_{i+1}$. Finally we find $c_{i,2}$ by solving $p_i(x_{i+1}) = y_i + c_{i,2}h + \frac{\mu_i}{2}h^2 + \frac{\mu_{i+1}-\mu_i}{6h}h^3 = y_{i+1}$. For $j = 0, 1, 2, 3$ the shifted powers $(x - x_i)^j$ constitute a basis for cubic polynomials and the formulas (2.8) are unique by construction. It follows that p_i is unique. □

Theorem 2.1 (Constructing a D_2-Spline) *Suppose for some moments $\mu_1, \dots, \mu_{n+1}$ that each p_i is given as in Lemma 2.1 for $i = 1, \dots, n$. If in addition*

$$\mu_{i-1} + 4\mu_i + \mu_{i+1} = \frac{6}{h^2}(y_{i+1} - 2y_i + y_{i-1}), \quad i = 2, \dots, n, \tag{2.9}$$

then the function g given by (2.4) *solves a D_2-spline problem.*

Proof Suppose for $1 \le i \le n$ that p_i is given as in Lemma 2.1 for some $\mu_1, \dots, \mu_{n+1}$. Consider the C^2 requirement. Since $p_{i-1}(x_i) = p_i(x_i) = y_i$ and $p_{i-1}''(x_i) = p_i''(x_i) = \mu_i$ for $i = 2, \dots, n$ it follows that $g \in C^2$ if and only if $p_{i-1}'(x_i) = p_i'(x_i)$ for $i = 2, \dots, n$. By (2.7)

$$\begin{aligned} p_{i-1}'(x_i) &= c_{i-1,2} + 2hc_{i-1,3} + 3h^2c_{i-1,4} \\ &= \frac{y_i - y_{i-1}}{h} - \frac{h}{3}\mu_{i-1} - \frac{h}{6}\mu_i + 2h\frac{\mu_{i-1}}{2} + 3h^2\frac{\mu_i - \mu_{i-1}}{6h} \\ &= \frac{y_i - y_{i-1}}{h} + \frac{h}{6}\mu_{i-1} + \frac{h}{3}\mu_i \\ p_i'(x_i) &= c_{i2} = \frac{y_{i+1} - y_i}{h} - \frac{h}{3}\mu_i - \frac{h}{6}\mu_{i+1}. \end{aligned} \tag{2.10}$$

A simple calculation shows that $p_{i-1}'(x_i) = p_i'(x_i)$ if and only if (2.9) holds.

Finally consider the function g given by (2.4). If (2.9) holds then $g \in C^2[a, b]$. By construction $g(x_i) = y_i$, $i = 1, \dots, n+1$, $g''(a) = p_1''(x_1) = \mu_1$ and $g''(b) = p_n''(x_{n+1}) = \mu_{n+1}$. It follows that g solves the D_2-spline problem. □

In order for the D_2-spline to exist we need to show that $\mu_2, \dots, \mu_n$ always can be determined from (2.9). For $n \ge 3$ and with μ_1 and μ_{n+1} given (2.9) can be written in the form

$$\begin{bmatrix} 4 & 1 & & & \\ 1 & 4 & 1 & & \\ & \ddots & \ddots & \ddots & \\ & & 1 & 4 & 1 \\ & & & 1 & 4 \end{bmatrix} \begin{bmatrix} \mu_2 \\ \mu_3 \\ \vdots \\ \mu_{n-1} \\ \mu_n \end{bmatrix} = \frac{6}{h^2} \begin{bmatrix} \delta^2 y_2 - \mu_1 \\ \delta^2 y_3 \\ \vdots \\ \delta^2 y_{n-1} \\ \delta^2 y_n - \mu_{n+1} \end{bmatrix}, \quad \delta^2 y_i := y_{i+1} - 2y_i + y_{i-1}. \tag{2.11}$$

This is a square linear system of equations. We recall (see Theorem 1.6) that a square system $\boldsymbol{Ax} = \boldsymbol{b}$ has a solution for all right hand sides $\boldsymbol{b}$ if and only if the coefficient matrix $\boldsymbol{A}$ is nonsingular, i.e., the homogeneous system $\boldsymbol{Ax} = \boldsymbol{0}$ only has the solution $\boldsymbol{x} = \boldsymbol{0}$. Moreover, the solution is unique. We need to show that the coefficient matrix in (2.11) is nonsingular.

We observe that the matrix in (2.11) is strictly diagonally dominant in accordance with the following definition.

Definition 2.2 (Strict Diagonal Dominance) The matrix $\boldsymbol{A} = [a_{ij}] \in \mathbb{C}^{n\times n}$ is **strictly diagonally dominant** if

$$|a_{ii}| > \sum_{j\neq i} |a_{ij}|, \; i = 1, \ldots, n. \tag{2.12}$$

Theorem 2.2 (Strict Diagonal Dominance) *A strictly diagonally dominant matrix is nonsingular. Moreover, if* $\mathbf{A} \in \mathbb{C}^{n\times n}$ *is strictly diagonally dominant then the solution* $\mathbf{x}$ *of* $\mathbf{Ax} = \mathbf{b}$ *is bounded as follows:*

$$\max_{1\le i\le n} |x_i| \le \max_{1\le i\le n} \left(\frac{|b_i|}{\sigma_i}\right), \; \textit{where } \sigma_i := |a_{ii}| - \sum_{j\neq i} |a_{ij}|. \tag{2.13}$$

Proof We first show that the bound (2.13) holds for any solution $\boldsymbol{x}$. Choose k so that $|x_k| = \max_i |x_i|$. Then

$$|b_k| = |a_{kk}x_k + \sum_{j\neq k} a_{kj}x_j| \ge |a_{kk}||x_k| - \sum_{j\neq k} |a_{kj}||x_j| \ge |x_k|\big(|a_{kk}| - \sum_{j\neq k} |a_{kj}|\big),$$

and this implies $\max_{1\le i\le n}|x_i| = |x_k| \le \frac{|b_k|}{\sigma_k} \le \max_{1\le i\le n}\left(\frac{|b_i|}{\sigma_i}\right)$. For the nonsingularity, if $\boldsymbol{Ax} = \boldsymbol{0}$, then $\max_{1\le i\le n}|x_i| \le 0$ by (2.13), and so $\boldsymbol{x} = \boldsymbol{0}$. □

For an alternative simple proof of the nonsingularity based on Gershgorin circle theorem see Exercise 14.3.

Theorem 2.2 implies that the system (2.11) has a unique solution giving rise to a function g detailed in Lemma 2.1 and solving the D_2-spline problem. For uniqueness suppose g_1 and g_2 are two D_2-splines interpolating the same data. Then $g := g_1 - g_2$ is an N-spline satisfying (2.2) with $\boldsymbol{y} = \boldsymbol{0}$. The solution $[\mu_2, \ldots, \mu_n]^T$ of (2.11) and also $\mu_1 = \mu_{n+1}$ are zero. It follows from (2.8) that all coefficients $c_{i,j}$ are zero. We conclude that $g = 0$ and $g_1 = g_2$.

Example 2.2 (Cubic B-Spline) For the N-spline with $[a, b] = [0, 4]$ and $\boldsymbol{y} = [0, \frac{1}{6}, \frac{2}{3}, \frac{1}{6}, 0]$ the linear system (2.9) takes the form

$$\begin{bmatrix} 4 & 1 & 0 \\ 1 & 4 & 1 \\ 0 & 1 & 4 \end{bmatrix} \begin{bmatrix} \mu_2 \\ \mu_3 \\ \mu_4 \end{bmatrix} = \begin{bmatrix} 2 \\ -6 \\ 2 \end{bmatrix}.$$

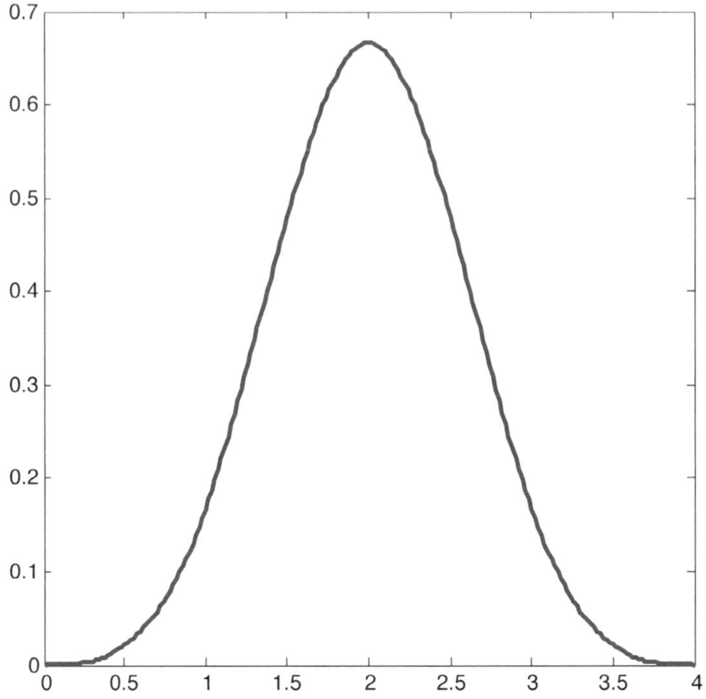

Fig. 2.4 A cubic B-spline

The solution is $\mu_2 = \mu_4 = 1$, $\mu_3 = -2$. The knotset is $\{0, 1, 2, 3, 4\}$. Using (2.8) (cf. Exercise 2.6) we find

$$g(x) := \begin{cases} p_1(x) = \frac{1}{6}x^3, & \text{if } 0 \le x < 1, \\ p_2(x) = \frac{1}{6} + \frac{1}{2}(x-1) + \frac{1}{2}(x-1)^2 - \frac{1}{2}(x-1)^3, & \text{if } 1 \le x < 2, \\ p_3(x) = \frac{2}{3} - (x-2)^2 + \frac{1}{2}(x-2)^3, & \text{if } 2 \le x < 3, \\ p_4(x) = \frac{1}{6} - \frac{1}{2}(x-3) + \frac{1}{2}(x-3)^2 - \frac{1}{6}(x-3)^3, & \text{if } 3 \le x \le 4, \end{cases} \tag{2.14}$$

A plot of this spline is shown in Fig. 2.4. On $(0, 4)$ the function g equals the nonzero part of a function known as a C^2 cubic B-spline.

2.1.4 LU Factorization of a Tridiagonal System

To find the D^2-spline g we have to solve the triangular system (2.11). Consider solving a general tridiagonal linear system $\boldsymbol{A}\boldsymbol{x} = \boldsymbol{b}$ where $\boldsymbol{A} = \text{tridiag}(a_i, d_i, c_i) \in$

$\mathbb{C}^{n\times n}$. Instead of using Gaussian elimination directly, we can construct two matrices $\boldsymbol{L}$ and $\boldsymbol{U}$ such that $\boldsymbol{A} = \boldsymbol{LU}$. Since $\boldsymbol{Ax} = \boldsymbol{LUx} = \boldsymbol{b}$ we can find $\boldsymbol{x}$ by solving two systems $\boldsymbol{Lz} = \boldsymbol{b}$ and $\boldsymbol{Ux} = \boldsymbol{z}$. Moreover $\boldsymbol{L}$ and $\boldsymbol{U}$ are both triangular and bidiagonal, and if in addition they are nonsingular the two systems can be solved easily without using elimination.

In our case we write the product $\boldsymbol{A} = \boldsymbol{LU}$ in the form

$$\begin{bmatrix} d_1 & c_1 & & & \\ a_1 & d_2 & c_2 & & \\ & \ddots & \ddots & \ddots & \\ & & a_{n-2} & d_{n-1} & c_{n-1} \\ & & & a_{n-1} & d_n \end{bmatrix} = \begin{bmatrix} 1 & & & \\ l_1 & 1 & & \\ & \ddots & \ddots & \\ & & l_{n-1} & 1 \end{bmatrix} \begin{bmatrix} u_1 & c_1 & & \\ & \ddots & \ddots & \\ & & u_{n-1} & c_{n-1} \\ & & & u_n \end{bmatrix}. \tag{2.15}$$

To find $\boldsymbol{L}$ and $\boldsymbol{U}$ we first consider the case $n = 3$. Equation (2.15) takes the form

$$\begin{bmatrix} d_1 & c_1 & 0 \\ a_1 & d_2 & c_2 \\ 0 & a_2 & d_3 \end{bmatrix} = \begin{bmatrix} 1 & 0 & 0 \\ l_1 & 1 & 0 \\ 0 & l_2 & 1 \end{bmatrix} \begin{bmatrix} u_1 & c_1 & 0 \\ 0 & u_2 & c_2 \\ 0 & 0 & u_3 \end{bmatrix} = \begin{bmatrix} u_1 & c_1 & 0 \\ l_1u_1 & l_1c_1 + u_2 & c_2 \\ 0 & l_2u_2 & l_2c_2 + u_3 \end{bmatrix},$$

and the systems $\boldsymbol{Lz} = \boldsymbol{b}$ and $\boldsymbol{Ux} = \boldsymbol{z}$ can be written

$$\begin{bmatrix} 1 & 0 & 0 \\ l_1 & 1 & 0 \\ 0 & l_2 & 1 \end{bmatrix} \begin{bmatrix} z_1 \\ z_2 \\ z_3 \end{bmatrix} = \begin{bmatrix} b_1 \\ b_2 \\ b_3 \end{bmatrix}, \quad \begin{bmatrix} u_1 & c_1 & 0 \\ 0 & u_2 & c_2 \\ 0 & 0 & u_3 \end{bmatrix} \begin{bmatrix} x_1 \\ x_2 \\ x_3 \end{bmatrix} = \begin{bmatrix} z_1 \\ z_2 \\ z_3 \end{bmatrix}.$$

Comparing elements we find

$$\begin{aligned} &u_1 = d_1, \quad l_1 = a_1/u_1, \quad u_2 = d_2 - l_1c_1, \quad l_2 = a_2/u_2, \quad u_3 = d_3 - l_2c_2, \\ &z_1 = b_1, \quad z_2 = b_2 - l_1z_1, \quad z_3 = b_3 - l_2z_2, \\ &x_3 = z_3/u_3, \quad x_2 = (z_2 - c_2x_3)/u_2, \quad x_1 = (z_1 - c_1x_2)/u_1. \end{aligned}$$

In general, if

$$u_1 = d_1, \quad l_k = a_k/u_k, \quad u_{k+1} = d_{k+1} - l_kc_k, \quad k = 1, 2, \ldots, n-1, \tag{2.16}$$

then $\boldsymbol{A} = \boldsymbol{LU}$. If $u_1, u_2, \ldots, u_{n-1}$ are nonzero then (2.16) is well defined. If in addition $u_n \neq 0$ then we can solve $\boldsymbol{Lz} = \boldsymbol{b}$ and $\boldsymbol{Ux} = \boldsymbol{z}$ for $\boldsymbol{z}$ and $\boldsymbol{x}$. We formulate this as two algorithms. In `trifactor`, vectors $\boldsymbol{l} \in \mathbb{C}^{n-1}$, $\boldsymbol{u} \in \mathbb{C}^n$ are computed from $\boldsymbol{a}, \boldsymbol{c} \in \mathbb{C}^{n-1}$, $\boldsymbol{d} \in \mathbb{C}^n$. This implements the LU factorization of a tridiagonal matrix:

```
function [l,u]=trifactor(a,d,c)
% [l,u]=trifactor(a,d,c)
u=d; l=a;
for k =1:length(a)
    l(k)=a(k)/u(k);
    u(k+1)=d(k+1)-l(k)*c(k);
end
```

Listing 2.1 trifactor

In `trisolve`, the solution $\boldsymbol{x}$ of a tridiagonal system with r right hand sides is computed from a previous call to `trifactor`. Here $\boldsymbol{l}, \in \mathbb{C}^{n-1}$ and $\boldsymbol{u} \in \mathbb{C}^n$ are output from `trifactor` and $\boldsymbol{b} \in \mathbb{C}^{n,r}$ for some $r \in \mathbb{N}$:

```
function x = trisolve (l,u,c,b)
% x = trisolve (l,u,c,b)
x=b;
n= size(b,1);
for k =2:n
    x(k,:)=b(k,:)-l(k-1)*x(k-1,:);
end
x(n,:)=x(n,:)/u(n);
for k=(n-1):-1:1
    x(k,:)=(x(k,:)-c(k)*x(k+1,:))/u(k);
end
```

Listing 2.2 trisolve

Since division by zero can occur, the algorithms will not work in general, but for tridiagonal strictly diagonally dominant systems we have

Theorem 2.3 (LU of a Tridiagonal Strictly Dominant System) *A strictly diagonally dominant tridiagonal matrix has a unique LU factorization of the form* (2.15).

Proof We show that the u_k's in (2.16) are nonzero for $k = 1, \ldots, n$. For this it is sufficient to show by induction that

$$|u_k| \ge \sigma_k + |c_k|, \quad \text{where, } \sigma_k := |d_k| - |a_{k-1}| - |c_k| > 0, \; k = 1, \ldots, n, \tag{2.17}$$

and where $a_0 := c_n := 0$. By assumption $|u_1| = |d_1| = \sigma_1 + |c_1|$. Suppose $|u_k| \ge \sigma_k + |c_k|$ for some $1 \le k \le n-1$. Then $|c_k|/|u_k| < 1$ and by (2.16) and strict diagonal dominance

$$\begin{aligned} |u_{k+1}| = |d_{k+1} - l_k c_k| = |d_{k+1} - \frac{a_k c_k}{u_k}| &\ge |d_{k+1}| - \frac{|a_k||c_k|}{|u_k|} \\ &\ge |d_{k+1}| - |a_k| = \sigma_{k+1} + |c_{k+1}|. \end{aligned} \tag{2.18}$$

□

Corollary 2.1 (Stability of the LU Factorization) *Suppose $A \in \mathbb{C}^{n\times n}$ is tridiagonal and strictly diagonally dominant with computed elements in the LU factorization given by* (2.16)*. Then* (2.17) *holds, $u_1 = d_1$ and*

$$|l_k| = \frac{|a_k|}{|u_k|} \le \frac{|a_k|}{|d_k| - |a_{k-1}|}, \quad |u_{k+1}| \le |d_{k+1}| + \frac{|a_k||c_k|}{|d_k| - |a_{k-1}|}, \quad k = 1, \ldots, n-1. \tag{2.19}$$

Proof Using (2.16) and (2.17) for $1 \le k \le n-1$ we find

$$|l_k| = \frac{|a_k|}{|u_k|} \le \frac{|a_k|}{|d_k| - |a_{k-1}|}, \quad |u_{k+1}| \le |d_{k+1}| + |l_k||c_k| \le |d_{k+1}| + \frac{|a_k||c_k|}{|d_k| - |a_{k-1}|}.$$

□

- For a strictly diagonally dominant tridiagonal matrix it follows from Corollary 2.1 that the LU factorization algorithm `trifactor` is stable meaning that we cannot have severe growth in the computed elements u_k and l_k.
- The number of arithmetic operations to compute the LU factorization of a tridiagonal matrix of order n using (2.16) is $3n-3$, while the number of arithmetic operations for Algorithm `trisolve` is $r(5n-4)$, where r is the number of right-hand sides. This means that the complexity to solve a tridiagonal system is $O(n)$, or more precisely $8n-7$ when $r = 1$, and this number only grows linearly[2] with n.

2.2 A Two Point Boundary Value Problem

Consider the simple **two point boundary value problem**

$$-u''(x) = f(x), \quad x \in [0, 1], \quad u(0) = 0, \; u(1) = 0, \tag{2.20}$$

where f is a given continuous function on $[0, 1]$ and u is an unknown function. This problem is also known as the **one-dimensional (1D) Poisson problem**. In principle it is easy to solve (2.20) exactly. We just integrate f twice and determine the two integration constants so that the homogeneous boundary conditions $u(0) = u(1) = 0$ are satisfied. For example, if $f(x) = 1$ then $u(x) = x(x-1)/2$ is the solution.

Suppose f cannot be integrated exactly. Problem (2.20) can then be solved approximately using the **finite difference method**. We need a difference approximation to the second derivative. If g is a function differentiable at x then

$$g'(x) = \lim_{h\to 0} \frac{g(x+\frac{h}{2}) - g(x-\frac{h}{2})}{h}$$

[2]We show in Sect. 3.3.2 that Gaussian elimination on a full $n \times n$ system is an $O(n^3)$ process.

and applying this to a function u that is twice differentiable at x

$$u''(x) = \lim_{h\to 0} \frac{u'(x+\frac{h}{2}) - u'(x-\frac{h}{2})}{h} = \lim_{h\to 0} \frac{\frac{u(x+h)-u(x)}{h} - \frac{u(x)-u(x-h)}{h}}{h}$$
$$= \lim_{h\to 0} \frac{u(x+h) - 2u(x) + u(x-h)}{h^2}.$$

To define the points where this difference approximation is used we choose a positive integer m, let $h := 1/(m+1)$ be the discretization parameter, and replace the interval $[0, 1]$ by grid points $x_j := jh$ for $j = 0, 1, \dots, m+1$. We then obtain approximations v_j to the exact solution $u(x_j)$ for $j = 1, \dots, m$ by replacing the differential equation by the difference equation

$$\frac{-v_{j-1} + 2v_j - v_{j+1}}{h^2} = f(jh), \quad j = 1, \dots, m, \quad v_0 = v_{m+1} = 0.$$

Moving the h^2 factor to the right hand side this can be written as an $m \times m$ linear system

$$\boldsymbol{T}\boldsymbol{v} = \begin{bmatrix} 2 & -1 & 0 & & & \\ -1 & 2 & -1 & & & \\ 0 & \ddots & \ddots & \ddots & & \\ & & & & & 0 \\ & & & -1 & 2 & -1 \\ & & & 0 & -1 & 2 \end{bmatrix} \begin{bmatrix} v_1 \\ v_2 \\ \vdots \\ v_{m-1} \\ v_m \end{bmatrix} = h^2 \begin{bmatrix} f(h) \\ f(2h) \\ \vdots \\ f\big((m-1)h\big) \\ f(mh) \end{bmatrix} =: \boldsymbol{b}. \tag{2.21}$$

The matrix $\boldsymbol{T}$ is called the **second derivative matrix** and will occur frequently in this book. It is our second example of a tridiagonal matrix, $\boldsymbol{T} = \text{tridiag}(a_i, d_i, c_i) \in \mathbb{R}^{m\times m}$, where in this case $a_i = c_i = -1$ and $d_i = 2$ for all i.

2.2.1 *Diagonal Dominance*

We want to show that (2.21) has a unique solution. Note that $\boldsymbol{T}$ is not strictly diagonally dominant. However, $\boldsymbol{T}$ is weakly diagonally dominant in accordance with the following definition.

Definition 2.3 (Diagonal Dominance) The matrix $\boldsymbol{A} = [a_{ij}] \in \mathbb{C}^{n\times n}$ is **weakly diagonally dominant** if

$$|a_{ii}| \ge \sum_{j\ne i} |a_{ij}|, \; i = 1, \dots, n. \tag{2.22}$$

We showed in Theorem 2.2 that a strictly diagonally dominant matrix is nonsingular. This is in general not true in the weakly diagonally dominant case. Consider the 3 matrices

$$\boldsymbol{A}_1 = \begin{bmatrix} 1 & 1 & 0 \\ 1 & 2 & 1 \\ 0 & 1 & 1 \end{bmatrix}, \quad \boldsymbol{A}_2 = \begin{bmatrix} 1 & 0 & 0 \\ 0 & 0 & 0 \\ 0 & 0 & 1 \end{bmatrix}, \quad \boldsymbol{A}_3 = \begin{bmatrix} 2 & -1 & 0 \\ -1 & 2 & -1 \\ 0 & -1 & 2 \end{bmatrix}.$$

They are all weakly diagonally dominant, but $\boldsymbol{A}_1$ and $\boldsymbol{A}_2$ are singular, while $\boldsymbol{A}_3$ is nonsingular. Indeed, for $\boldsymbol{A}_1$ column two is the sum of columns one and three, $\boldsymbol{A}_2$ has a zero row, and $\det(\boldsymbol{A}_3) = 4 \neq 0$. It follows that for the nonsingularity and existence of an LU factorization of a weakly diagonally dominant matrix we need some additional conditions. Here are some sufficient conditions.

Theorem 2.4 (Weak Diagonal Dominance) *Suppose* $\boldsymbol{A} = tridiag(a_i, d_i, c_i) \in \mathbb{C}^{n\times n}$ *is tridiagonal and weakly diagonally dominant. If in addition* $|d_1| > |c_1|$ *and* $a_i \neq 0$ *for* $i = 1, \ldots, n-2$, *then* $\boldsymbol{A}$ *has a unique LU factorization* (2.15). *If in addition* $d_n \neq 0$, *then* $\boldsymbol{A}$ *is nonsingular.*

Proof The proof is similar to the proof of Theorem 2.2. The matrix $\boldsymbol{A}$ has an LU factorization if the u_k's in (2.16) are nonzero for $k = 1, \ldots, n-1$. For this it is sufficient to show by induction that $|u_k| > |c_k|$ for $k = 1, \ldots, n-1$. By assumption $|u_1| = |d_1| > |c_1|$. Suppose $|u_k| > |c_k|$ for some $1 \leq k \leq n-2$. Then $|c_k|/|u_k| < 1$ and by (2.16) and since $a_k \neq 0$

$$|u_{k+1}| = |d_{k+1} - l_k c_k| = |d_{k+1} - \frac{a_k c_k}{u_k}| \geq |d_{k+1}| - \frac{|a_k||c_k|}{|u_k|} > |d_{k+1}| - |a_k|. \tag{2.23}$$

This also holds for $k = n-1$ if $a_{n-1} \neq 0$. By (2.23) and weak diagonal dominance $|u_{k+1}| > |d_{k+1}| - |a_k| \geq |c_{k+1}|$ and it follows by induction that an LU factorization exists. It is unique since any LU factorization must satisfy (2.16). For the nonsingularity we need to show that $u_n \neq 0$. For then by Lemma 2.5, both $\boldsymbol{L}$ and $\boldsymbol{U}$ are nonsingular, and this is equivalent to $\boldsymbol{A} = \boldsymbol{LU}$ being nonsingular. If $a_{n-1} \neq 0$ then by (2.16) $|u_n| > |d_n| - |a_{n-1}| \geq 0$ by weak diagonal dominance, while if $a_{n-1} = 0$ then again by (2.23) $|u_n| \geq |d_n| > 0$. □

Consider now the special system $\boldsymbol{T}\boldsymbol{v} = \boldsymbol{b}$ given by (2.21). The matrix $\boldsymbol{T}$ is weakly diagonally dominant and satisfies the additional conditions in Theorem 2.4. Thus it is nonsingular and we can solve the system in $O(n)$ arithmetic operations using the algorithms `trifactor` and `trisolve`.

We could use the explicit inverse of $\boldsymbol{T}$, given in Exercise 2.15, to compute the solution of $\boldsymbol{T}\boldsymbol{v} = \boldsymbol{b}$ as $\boldsymbol{v} = \boldsymbol{T}^{-1}\boldsymbol{b}$. However, this is not a good idea. In fact, all elements in $\boldsymbol{T}^{-1}$ are nonzero and the calculation of $\boldsymbol{T}^{-1}\boldsymbol{b}$ requires $O(n^2)$ operations.

2.3 An Eigenvalue Problem

Recall that if $\boldsymbol{A} \in \mathbb{C}^{n\times n}$ is a square matrix and $\boldsymbol{A}\boldsymbol{x} = \lambda\boldsymbol{x}$ for some nonzero $\boldsymbol{x} \in \mathbb{C}^n$, then $\lambda \in \mathbb{C}$ is called an **eigenvalue** and $\boldsymbol{x}$ an **eigenvector**. We call $(\lambda, \boldsymbol{x})$ an **eigenpair** of $\boldsymbol{A}$.

2.3.1 The Buckling of a Beam

Consider a horizontal beam of length L located between 0 and L on the x-axis of the plane. We assume that the beam is fixed at $x = 0$ and $x = L$ and that a force F is applied at $(L, 0)$ in the direction towards the origin. This situation can be modeled by the boundary value problem

$$Ry''(x) = -Fy(x), \quad y(0) = y(L) = 0, \tag{2.24}$$

where $y(x)$ is the vertical displacement of the beam at x, and R is a constant defined by the rigidity of the beam. We can transform the problem to the unit interval $[0, 1]$ by considering the function $u : [0, 1] \to \mathbb{R}$ given by $u(t) := y(tL)$. Since $u''(t) = L^2y''(tL)$, the problem (2.24) then becomes

$$u''(t) = -Ku(t), \quad u(0) = u(1) = 0, \quad K := \frac{FL^2}{R}. \tag{2.25}$$

Clearly $u = 0$ is a solution, but we can have nonzero solutions corresponding to certain values of the K known as eigenvalues. The corresponding function u is called an eigenfunction. If $F = 0$ then $K = 0$ and $u = 0$ is the only solution, but if the force is increased it will reach a critical value where the beam will buckle and maybe break. This critical value corresponds to the smallest eigenvalue of (2.25). With $u(t) = \sin(\pi t)$ we find $u''(t) = -\pi^2u(t)$ and this u is a solution if $K = \pi^2$. It can be shown that this is the smallest eigenvalue of (2.25) and solving for F we find $F = \frac{\pi^2 R}{L^2}$.

We can approximate this eigenvalue numerically. Choosing $m \in \mathbb{N}$, $h := 1/(m+1)$ and using for the second derivative the approximation

$$u''(jh) \approx \frac{u((j+1)h) - 2u(jh) + u((j-1)h)}{h^2}, \quad j = 1, \ldots, m,$$

(this is the same finite difference approximation as in Sect. 2.2) we obtain

$$\frac{-v_{j-1} + 2v_j - v_{j+1}}{h^2} = Kv_j, \quad j = 1, \ldots, m, \; h = \frac{1}{m+1}, \quad v_0 = v_{m+1} = 0,$$

where $v_j \approx u(jh)$ for $j = 0, \ldots, m+1$. If we define $\lambda := h^2 K$ then we obtain the equation

$$\boldsymbol{T}\boldsymbol{v} = \lambda \boldsymbol{v}, \text{ with } \boldsymbol{v} = [v_1, \ldots, v_m]^T, \tag{2.26}$$

and

$$\boldsymbol{T} = \boldsymbol{T}_m := \text{tridiag}_m(-1, 2, -1) - = \begin{bmatrix} 2 & -1 & 0 & & & \\ -1 & 2 & -1 & & & \\ 0 & \ddots & \ddots & \ddots & & \\ & & & & & 0 \\ & & & -1 & 2 & -1 \\ & & & 0 & -1 & 2 \end{bmatrix} \in \mathbb{R}^{m \times m}. \tag{2.27}$$

The problem now is to determine the eigenvalues of $\boldsymbol{T}$. Normally we would need a numerical method to determine the eigenvalues of a matrix, but for this simple problem the eigenvalues can be determined exactly. We show in the next subsection that the smallest eigenvalue of (2.26) is given by $\lambda = 4\sin^2(\pi h/2)$. Since $\lambda = h^2 K = \frac{h^2 F L^2}{R}$ we can solve for F to obtain

$$F = \frac{4\sin^2(\pi h/2) R}{h^2 L^2}.$$

For small h this is a good approximation to the value $\frac{\pi^2 R}{L^2}$ we computed above.

2.4 The Eigenpairs of the 1D Test Matrix

The second derivative matrix $\boldsymbol{T} = \text{tridiag}(-1, 2, -1)$ is a special case of the tridiagonal matrix

$$\boldsymbol{T}_1 := \text{tridiag}(a, d, a) \tag{2.28}$$

where $a, d \in \mathbb{R}$. We call this the **1D test matrix**. It is symmetric and strictly diagonally dominant if $|d| > 2|a|$.

We show that the eigenvectors are the columns of the **sine matrix** defined by

$$\boldsymbol{S} = \left[\sin \frac{jk\pi}{m+1} \right]_{j,k=1}^{m} \in \mathbb{R}^{m \times m}. \tag{2.29}$$

For $m = 3$,

$$\boldsymbol{S} = [\boldsymbol{s}_1, \boldsymbol{s}_2, \boldsymbol{s}_3] = \begin{bmatrix} \sin\frac{\pi}{4} & \sin\frac{2\pi}{4} & \sin\frac{3\pi}{4} \\ \sin\frac{2\pi}{4} & \sin\frac{4\pi}{4} & \sin\frac{6\pi}{4} \\ \sin\frac{3\pi}{4} & \sin\frac{6\pi}{4} & \sin\frac{9\pi}{4} \end{bmatrix} = \begin{bmatrix} t & 1 & t \\ 1 & 0 & -1 \\ t & -1 & t \end{bmatrix}, \quad t := \frac{1}{\sqrt{2}}.$$

Lemma 2.2 (Eigenpairs of 1D Test Matrix) *Suppose* $\boldsymbol{T}_1 = (t_{kj})_{k,j} = \operatorname{tridiag}(a, d, a) \in \mathbb{R}^{m\times m}$ *with* $m \geq 2$, $a, d \in \mathbb{R}$, *and let* $h = 1/(m+1)$.

1. *We have* $\boldsymbol{T}_1\boldsymbol{s}_j = \lambda_j\boldsymbol{s}_j$ *for* $j = 1, \ldots, m$, *where*

$$\boldsymbol{s}_j = [\sin(j\pi h), \sin(2j\pi h), \ldots, \sin(mj\pi h)]^T, \tag{2.30}$$

$$\lambda_j = d + 2a\cos(j\pi h). \tag{2.31}$$

2. *The eigenvalues are distinct and the eigenvectors are orthogonal*

$$\boldsymbol{s}_j^T\boldsymbol{s}_k = \frac{m+1}{2}\delta_{j,k} = \frac{1}{2h}\delta_{j,k}, \quad j, k = 1, \ldots, m. \tag{2.32}$$

Proof We find for $1 < k < m$

$$\begin{aligned}(\boldsymbol{T}_1\boldsymbol{s}_j)_k &= \sum_{l=1}^{m} t_{k,l}\sin(lj\pi h) \\ &= a\big[\sin\big((k-1)j\pi h\big) + \sin\big((k+1)j\pi h\big)\big] + d\sin(kj\pi h) \\ &= 2a\cos(j\pi h)\sin(kj\pi h) + d\sin(kj\pi h) = \lambda_j s_{k,j}.\end{aligned}$$

This also holds for $k = 1, m$, and part 1 follows. Since $j\pi h = j\pi/(m+1) \in (0, \pi)$ for $j = 1, \ldots, m$ and the cosine function is strictly monotone decreasing on $(0, \pi)$ the eigenvalues are distinct, and since $\boldsymbol{T}_1$ is symmetric it follows from Lemma 2.3 below that the eigenvectors $\boldsymbol{s}_j$ are orthogonal. To finish the proof of (2.32) we compute

$$\begin{aligned}\boldsymbol{s}_j^T\boldsymbol{s}_j &= \sum_{k=1}^{m}\sin^2(kj\pi h) = \sum_{k=0}^{m}\sin^2(kj\pi h) = \frac{1}{2}\sum_{k=0}^{m}\big(1 - \cos(2kj\pi h)\big) \\ &= \frac{m+1}{2} - \frac{1}{2}\sum_{k=0}^{m}\cos(2kj\pi h) = \frac{m+1}{2},\end{aligned}$$

since the last cosine sum is zero. We show this by summing a geometric series of complex exponentials. With $i = \sqrt{-1}$ we find

$$\sum_{k=0}^{m} \cos(2kj\pi h) + i \sum_{k=0}^{m} \sin(2kj\pi h) = \sum_{k=0}^{m} e^{2ikj\pi h} = \frac{e^{2i(m+1)j\pi h} - 1}{e^{2ij\pi h} - 1} = 0,$$

and (2.32) follows. □

Recall that the conjugate transpose of a matrix is defined by $\boldsymbol{A}^* := \overline{\boldsymbol{A}}^T$, where $\overline{\boldsymbol{A}}$ is obtained from $\boldsymbol{A}$ by taking the complex conjugate of all elements. A matrix $\boldsymbol{A} \in \mathbb{C}^{n \times n}$ is **Hermitian** if $\boldsymbol{A}^* = \boldsymbol{A}$. A real symmetric matrix is Hermitian.

Lemma 2.3 (Eigenpairs of a Hermitian Matrix) *The eigenvalues of a Hermitian matrix are real. Moreover, eigenvectors corresponding to distinct eigenvalues are orthogonal.*

Proof Suppose $\boldsymbol{A}^* = \boldsymbol{A}$ and $\boldsymbol{A}\boldsymbol{x} = \lambda\boldsymbol{x}$ with $\boldsymbol{x} \neq 0$. We multiply both sides of $\boldsymbol{A}\boldsymbol{x} = \lambda\boldsymbol{x}$ from the left by $\boldsymbol{x}^*$ and divide by $\boldsymbol{x}^*\boldsymbol{x}$ to obtain $\lambda = \frac{\boldsymbol{x}^*\boldsymbol{A}\boldsymbol{x}}{\boldsymbol{x}^*\boldsymbol{x}}$. Taking complex conjugates we find $\overline{\lambda} = \lambda^* = \frac{(\boldsymbol{x}^*\boldsymbol{A}\boldsymbol{x})^*}{(\boldsymbol{x}^*\boldsymbol{x})^*} = \frac{\boldsymbol{x}^*\boldsymbol{A}^*\boldsymbol{x}}{\boldsymbol{x}^*\boldsymbol{x}} = \frac{\boldsymbol{x}^*\boldsymbol{A}\boldsymbol{x}}{\boldsymbol{x}^*\boldsymbol{x}} = \lambda$, and λ is real. Suppose that $(\lambda, \boldsymbol{x})$ and $(\mu, \boldsymbol{y})$ are two eigenpairs for $\boldsymbol{A}$ with $\mu \neq \lambda$. Multiplying $\boldsymbol{A}\boldsymbol{x} = \lambda\boldsymbol{x}$ by $\boldsymbol{y}^*$ gives

$$\lambda\boldsymbol{y}^*\boldsymbol{x} = \boldsymbol{y}^*\boldsymbol{A}\boldsymbol{x} = (\boldsymbol{x}^*\boldsymbol{A}^*\boldsymbol{y})^* = (\boldsymbol{x}^*\boldsymbol{A}\boldsymbol{y})^* = (\mu\boldsymbol{x}^*\boldsymbol{y})^* = \mu\boldsymbol{y}^*\boldsymbol{x},$$

using that μ is real. Since $\lambda \neq \mu$ it follows that $\boldsymbol{y}^*\boldsymbol{x} = 0$, which means that $\boldsymbol{x}$ and $\boldsymbol{y}$ are orthogonal. □

2.5 Block Multiplication and Triangular Matrices

Block multiplication is a powerful and essential tool for dealing with matrices. It will be used extensively in this book. We will also need some basic facts about triangular matrices.

2.5.1 Block Multiplication

A rectangular matrix $\boldsymbol{A}$ can be partitioned into submatrices by drawing horizontal lines between selected rows and vertical lines between selected columns. For example, the matrix

$$\boldsymbol{A} = \begin{bmatrix} 1 & 2 & 3 \\ 4 & 5 & 6 \\ 7 & 8 & 9 \end{bmatrix}$$

can be partitioned as

$$(i)\ \begin{bmatrix} A_{11} & A_{12} \\ A_{21} & A_{22} \end{bmatrix} = \left[\begin{array}{c|cc} 1 & 2 & 3 \\ \hline 4 & 5 & 6 \\ 7 & 8 & 9 \end{array}\right], \quad (ii)\ \begin{bmatrix} a_{:1}, a_{:2}, a_{:3} \end{bmatrix} = \left[\begin{array}{c|c|c} 1 & 2 & 3 \\ 4 & 5 & 6 \\ 7 & 8 & 9 \end{array}\right],$$

$$(iii)\ \begin{bmatrix} a_{1:}^T \\ a_{2:}^T \\ a_{3:}^T \end{bmatrix} = \left[\begin{array}{ccc} 1 & 2 & 3 \\ \hline 4 & 5 & 6 \\ \hline 7 & 8 & 9 \end{array}\right], \quad (iv)\ \begin{bmatrix} A_{11}, A_{12} \end{bmatrix} = \left[\begin{array}{c|cc} 1 & 2 & 3 \\ 4 & 5 & 6 \\ 7 & 8 & 9 \end{array}\right].$$

In (i) the matrix A is divided into four submatrices

$$A_{11} = \begin{bmatrix} 1 \end{bmatrix}, \ A_{12} = \begin{bmatrix} 2, 3 \end{bmatrix}, \ A_{21} = \begin{bmatrix} 4 \\ 7 \end{bmatrix}, \text{ and } A_{22} = \begin{bmatrix} 5 & 6 \\ 8 & 9 \end{bmatrix},$$

while in (ii) and (iii) A has been partitioned into columns and rows, respectively. The submatrices in a partition are often referred to as **blocks** and a partitioned matrix is sometimes called a **block matrix**.

In the following we assume that $A \in \mathbb{C}^{m \times p}$ and $B \in \mathbb{C}^{p \times n}$. Here are some rules and observations for block multiplication.

1. If $B = \begin{bmatrix} b_{:1}, \ldots, b_{:n} \end{bmatrix}$ is partitioned into columns then the partition of the product AB into columns is

$$AB = \begin{bmatrix} Ab_{:1}, Ab_{:2}, \ldots, Ab_{:n} \end{bmatrix}.$$

   In particular, if I is the identity matrix of order p then

$$A = AI = A\begin{bmatrix} e_1, e_2, \ldots, e_p \end{bmatrix} = \begin{bmatrix} Ae_1, Ae_2, \ldots, Ae_p \end{bmatrix}$$

   and we see that column j of A can be written Ae_j for $j = 1, \ldots, p$.
2. Similarly, if A is partitioned into rows then

$$AB = \begin{bmatrix} a_{1:}^T \\ a_{2:}^T \\ \vdots \\ a_{m:}^T \end{bmatrix} B = \begin{bmatrix} a_{1:}^T B \\ a_{2:}^T B \\ \vdots \\ a_{m:}^T B \end{bmatrix},$$

   and taking $A = I$ it follows that row i of B can be written $e_i^T B$ for $i = 1, \ldots, m$.
3. It is often useful to write the matrix-vector product Ax as a linear combination of the columns of A

$$Ax = x_1 a_{:1} + x_2 a_{:2} + \cdots + x_p a_{:p}.$$

4. If $\boldsymbol{B} = [\boldsymbol{B}_1, \boldsymbol{B}_2]$, where $\boldsymbol{B}_1 \in \mathbb{C}^{p\times r}$ and $\boldsymbol{B}_2 \in \mathbb{C}^{p\times(n-r)}$ then

$$\boldsymbol{A}[\boldsymbol{B}_1, \boldsymbol{B}_2] = [\boldsymbol{A}\boldsymbol{B}_1, \boldsymbol{A}\boldsymbol{B}_2].$$

This follows from Rule 1. by an appropriate grouping of columns.

5. If $\boldsymbol{A} = \begin{bmatrix} \boldsymbol{A}_1 \\ \boldsymbol{A}_2 \end{bmatrix}$, where $\boldsymbol{A}_1 \in \mathbb{C}^{k\times p}$ and $\boldsymbol{A}_2 \in \mathbb{C}^{(m-k)\times p}$ then

$$\begin{bmatrix} \boldsymbol{A}_1 \\ \boldsymbol{A}_2 \end{bmatrix} \boldsymbol{B} = \begin{bmatrix} \boldsymbol{A}_1\boldsymbol{B} \\ \boldsymbol{A}_2\boldsymbol{B} \end{bmatrix}.$$

This follows from Rule 2. by a grouping of rows.

6. If $\boldsymbol{A} = [\boldsymbol{A}_1, \boldsymbol{A}_2]$ and $\boldsymbol{B} = \begin{bmatrix} \boldsymbol{B}_1 \\ \boldsymbol{B}_2 \end{bmatrix}$, where $\boldsymbol{A}_1 \in \mathbb{C}^{m\times s}$, $\boldsymbol{A}_2 \in \mathbb{C}^{m\times(p-s)}$, $\boldsymbol{B}_1 \in \mathbb{C}^{s\times n}$ and $\boldsymbol{B}_2 \in \mathbb{C}^{(p-s)\times n}$ then

$$[\boldsymbol{A}_1, \boldsymbol{A}_2] \begin{bmatrix} \boldsymbol{B}_1 \\ \boldsymbol{B}_2 \end{bmatrix} = [\boldsymbol{A}_1\boldsymbol{B}_1 + \boldsymbol{A}_2\boldsymbol{B}_2].$$

Indeed,

$$\begin{aligned}(\boldsymbol{A}\boldsymbol{B})_{ij} &= \sum_{k=1}^{p} a_{ik}b_{kj} = \sum_{k=1}^{s} a_{ik}b_{kj} + \sum_{k=s+1}^{p} a_{ik}b_{kj} \\ &= (\boldsymbol{A}_1\boldsymbol{B}_1)_{ij} + (\boldsymbol{A}_2\boldsymbol{B}_2)_{ij} = (\boldsymbol{A}_1\boldsymbol{B}_1 + \boldsymbol{A}_2\boldsymbol{B}_2)_{ij}.\end{aligned}$$

7. If $\boldsymbol{A} = \begin{bmatrix} \boldsymbol{A}_{11} & \boldsymbol{A}_{12} \\ \boldsymbol{A}_{21} & \boldsymbol{A}_{22} \end{bmatrix}$ and $\boldsymbol{B} = \begin{bmatrix} \boldsymbol{B}_{11} & \boldsymbol{B}_{12} \\ \boldsymbol{B}_{21} & \boldsymbol{B}_{22} \end{bmatrix}$ then

$$\begin{bmatrix} \boldsymbol{A}_{11} & \boldsymbol{A}_{12} \\ \boldsymbol{A}_{21} & \boldsymbol{A}_{22} \end{bmatrix} \begin{bmatrix} \boldsymbol{B}_{11} & \boldsymbol{B}_{12} \\ \boldsymbol{B}_{21} & \boldsymbol{B}_{22} \end{bmatrix} = \begin{bmatrix} \boldsymbol{A}_{11}\boldsymbol{B}_{11} + \boldsymbol{A}_{12}\boldsymbol{B}_{21} & \boldsymbol{A}_{11}\boldsymbol{B}_{12} + \boldsymbol{A}_{12}\boldsymbol{B}_{22} \\ \boldsymbol{A}_{21}\boldsymbol{B}_{11} + \boldsymbol{A}_{22}\boldsymbol{B}_{21} & \boldsymbol{A}_{21}\boldsymbol{B}_{12} + \boldsymbol{A}_{22}\boldsymbol{B}_{22} \end{bmatrix},$$

provided the vertical partition in $\boldsymbol{A}$ matches the horizontal one in $\boldsymbol{B}$, i.e. the number of columns in $\boldsymbol{A}_{11}$ and $\boldsymbol{A}_{21}$ equals the number of rows in $\boldsymbol{B}_{11}$ and $\boldsymbol{B}_{12}$ and the number of columns in $\boldsymbol{A}$ equals the number of rows in $\boldsymbol{B}$. To show this we use Rule 4. to obtain

$$\boldsymbol{A}\boldsymbol{B} = \left[\begin{bmatrix} \boldsymbol{A}_{11} & \boldsymbol{A}_{12} \\ \boldsymbol{A}_{21} & \boldsymbol{A}_{22} \end{bmatrix} \begin{bmatrix} \boldsymbol{B}_{11} \\ \boldsymbol{B}_{21} \end{bmatrix}, \begin{bmatrix} \boldsymbol{A}_{11} & \boldsymbol{A}_{12} \\ \boldsymbol{A}_{21} & \boldsymbol{A}_{22} \end{bmatrix} \begin{bmatrix} \boldsymbol{B}_{12} \\ \boldsymbol{B}_{22} \end{bmatrix}\right].$$

We complete the proof using Rules 5. and 6.

8. Consider finally the general case. If all the matrix products $\boldsymbol{A}_{ik}\boldsymbol{B}_{kj}$ in

$$\boldsymbol{C}_{ij} = \sum_{k=1}^{s} \boldsymbol{A}_{ik}\boldsymbol{B}_{kj}, \quad i = 1, \ldots, p, \ j = 1, \ldots, q$$

are well defined then

$$\begin{bmatrix} \boldsymbol{A}_{11} & \cdots & \boldsymbol{A}_{1s} \\ \vdots & & \vdots \\ \boldsymbol{A}_{p1} & \cdots & \boldsymbol{A}_{ps} \end{bmatrix} \begin{bmatrix} \boldsymbol{B}_{11} & \cdots & \boldsymbol{B}_{1q} \\ \vdots & & \vdots \\ \boldsymbol{B}_{s1} & \cdots & \boldsymbol{B}_{sq} \end{bmatrix} = \begin{bmatrix} \boldsymbol{C}_{11} & \cdots & \boldsymbol{C}_{1q} \\ \vdots & & \vdots \\ \boldsymbol{C}_{p1} & \cdots & \boldsymbol{C}_{pq} \end{bmatrix}.$$

The requirements are that

- the number of columns in $\boldsymbol{A}$ is equal to the number of rows in $\boldsymbol{B}$.
- the position of the vertical partition lines in $\boldsymbol{A}$ has to mach the position of the horizontal partition lines in $\boldsymbol{B}$. The horizontal lines in $\boldsymbol{A}$ and the vertical lines in $\boldsymbol{B}$ can be anywhere.

2.5.2 *Triangular Matrices*

We need some basic facts about triangular matrices and we start with

Lemma 2.4 (Inverse of a Block Triangular Matrix) *Suppose*

$$\boldsymbol{A} = \begin{bmatrix} \boldsymbol{A}_{11} & \boldsymbol{A}_{12} \\ \boldsymbol{0} & \boldsymbol{A}_{22} \end{bmatrix}$$

where $\boldsymbol{A}$, $\boldsymbol{A}_{11}$ and $\boldsymbol{A}_{22}$ are square matrices. Then $\boldsymbol{A}$ is nonsingular if and only if both $\boldsymbol{A}_{11}$ and $\boldsymbol{A}_{22}$ are nonsingular. In that case

$$\boldsymbol{A}^{-1} = \begin{bmatrix} \boldsymbol{A}_{11}^{-1} & \boldsymbol{C} \\ \boldsymbol{0} & \boldsymbol{A}_{22}^{-1} \end{bmatrix}, \tag{2.33}$$

for some matrix $\boldsymbol{C}$.

Proof Suppose $\boldsymbol{A}$ is nonsingular. We partition $\boldsymbol{B} := \boldsymbol{A}^{-1}$ conformally with $\boldsymbol{A}$ and have

$$\boldsymbol{B}\boldsymbol{A} = \begin{bmatrix} \boldsymbol{B}_{11} & \boldsymbol{B}_{12} \\ \boldsymbol{B}_{21} & \boldsymbol{B}_{22} \end{bmatrix} \begin{bmatrix} \boldsymbol{A}_{11} & \boldsymbol{A}_{12} \\ \boldsymbol{0} & \boldsymbol{A}_{22} \end{bmatrix} = \begin{bmatrix} \boldsymbol{I} & \boldsymbol{0} \\ \boldsymbol{0} & \boldsymbol{I} \end{bmatrix} = \boldsymbol{I}$$

Using block-multiplication we find

$$\boldsymbol{B}_{11}\boldsymbol{A}_{11} = \boldsymbol{I}, \ \boldsymbol{B}_{21}\boldsymbol{A}_{11} = \boldsymbol{0}, \ \boldsymbol{B}_{21}\boldsymbol{A}_{12} + \boldsymbol{B}_{22}\boldsymbol{A}_{22} = \boldsymbol{I}, \ \boldsymbol{B}_{11}\boldsymbol{A}_{12} + \boldsymbol{B}_{12}\boldsymbol{A}_{22} = \boldsymbol{0}.$$

The first equation implies that $\boldsymbol{A}_{11}$ is nonsingular, this in turn implies that $\boldsymbol{B}_{21} = \boldsymbol{0}\boldsymbol{A}_{11}^{-1} = \boldsymbol{0}$ in the second equation, and then the third equation simplifies to $\boldsymbol{B}_{22}\boldsymbol{A}_{22} = \boldsymbol{I}$. We conclude that also $\boldsymbol{A}_{22}$ is nonsingular. From the fourth equation we find

$$\boldsymbol{B}_{12} = \boldsymbol{C} = -\boldsymbol{A}_{11}^{-1}\boldsymbol{A}_{12}\boldsymbol{A}_{22}^{-1}.$$

Conversely, if $\boldsymbol{A}_{11}$ and $\boldsymbol{A}_{22}$ are nonsingular then

$$\begin{bmatrix} \boldsymbol{A}_{11}^{-1} & -\boldsymbol{A}_{11}^{-1}\boldsymbol{A}_{12}\boldsymbol{A}_{22}^{-1} \\ \boldsymbol{0} & \boldsymbol{A}_{22}^{-1} \end{bmatrix} \begin{bmatrix} \boldsymbol{A}_{11} & \boldsymbol{A}_{12} \\ \boldsymbol{0} & \boldsymbol{A}_{22} \end{bmatrix} = \begin{bmatrix} \boldsymbol{I} & \boldsymbol{0} \\ \boldsymbol{0} & \boldsymbol{I} \end{bmatrix} = \boldsymbol{I}$$

and $\boldsymbol{A}$ is nonsingular with the indicated inverse. □

Consider now a triangular matrix.

Lemma 2.5 (Inverse of a Triangular Matrix) *An upper (lower) triangular matrix $\boldsymbol{A} = [a_{ij}] \in \mathbb{C}^{n\times n}$ is nonsingular if and only if the diagonal elements a_{ii}, $i = 1, \dots, n$ are nonzero. In that case the inverse is upper (lower) triangular with diagonal elements a_{ii}^{-1}, $i = 1, \dots, n$.*

Proof We use induction on n. The result holds for $n = 1$. The 1-by-1 matrix $\boldsymbol{A} = [a_{11}]$ is nonsingular if and only if $a_{11} \neq \boldsymbol{0}$ and in that case $\boldsymbol{A}^{-1} = [a_{11}^{-1}]$. Suppose the result holds for $n = k$ and let $\boldsymbol{A} \in \mathbb{C}^{(k+1)\times(k+1)}$ be upper triangular. We partition $\boldsymbol{A}$ in the form

$$\boldsymbol{A} = \begin{bmatrix} \boldsymbol{A}_k & \boldsymbol{a}_k \\ \boldsymbol{0} & a_{k+1,k+1} \end{bmatrix}$$

and note that $\boldsymbol{A}_k \in \mathbb{C}^{k\times k}$ is upper triangular. By Lemma 2.4 $\boldsymbol{A}$ is nonsingular if and only if $\boldsymbol{A}_k$ and $(a_{k+1,k+1})$ are nonsingular and in that case

$$\boldsymbol{A}^{-1} = \begin{bmatrix} \boldsymbol{A}_k^{-1} & \boldsymbol{c} \\ \boldsymbol{0} & a_{k+1,k+1}^{-1} \end{bmatrix},$$

for some $\boldsymbol{c} \in \mathbb{C}^n$. By the induction hypothesis $\boldsymbol{A}_k$ is nonsingular if and only if the diagonal elements $a_{11}, \dots, a_{kk}$ of $\boldsymbol{A}_k$ are nonzero and in that case $\boldsymbol{A}_k^{-1}$ is upper triangular with diagonal elements a_{ii}^{-1}, $i = 1, \dots, k$. The result for $\boldsymbol{A}$ follows. □

Lemma 2.6 (Product of Triangular Matrices) *The product $\boldsymbol{C} = \boldsymbol{A}\boldsymbol{B} = (c_{ij})$ of two upper (lower) triangular matrices $\boldsymbol{A} = (a_{ij})$ and $\boldsymbol{B} = (b_{ij})$ is upper (lower) triangular with diagonal elements $c_{ii} = a_{ii}b_{ii}$ for all i.*

Proof Exercise. □

A matrix is called **unit triangular** if it is triangular with 1's on the diagonal.

Lemma 2.7 (Unit Triangular Matrices) *For a unit upper (lower) triangular matrix $\mathbf{A} \in \mathbb{C}^{n \times n}$:*

1. *$\mathbf{A}$ is nonsingular and the inverse is unit upper(lower) triangular.*
2. *The product of two unit upper (lower) triangular matrices is unit upper (lower) triangular.*

Proof 1. follows from Lemma 2.5, while Lemma 2.6 implies 2. □

2.6 Exercises Chap. 2

2.6.1 Exercises Sect. 2.1

Exercise 2.1 (The Shifted Power Basis Is a Basis) Show that the polynomials $\{(x - x_i)^j\}_{0 \le j \le n}$ is a basis for polynomials of degree n.[3]

Exercise 2.2 (The Natural Spline, $n = 1$) How can one define an N-spline when $n = 1$?

Exercise 2.3 (Bounding the Moments) Show that for the N-spline the solution of the linear system (2.11) is bounded as follows[4]:

$$\max_{2 \le j \le n} |\mu_j| \le \frac{3}{h^2} \max_{2 \le i \le n} |y_{i+1} - 2y_i + y_{i-1}|.$$

Exercise 2.4 (Moment Equations for 1. Derivative Boundary Conditions) Suppose instead of the D_2 boundary conditions we use D_1 conditions given by $g'(a) = s_1$ and $g'(b) = s_{n+1}$ for some given numbers s_1 and s_{n+1}. Show that the linear system for the moments of a D_1-spline can be written

$$\begin{bmatrix} 2 & 1 & & & \\ 1 & 4 & 1 & & \\ & \ddots & \ddots & \ddots & \\ & & 1 & 4 & 1 \\ & & & 1 & 2 \end{bmatrix} \begin{bmatrix} \mu_1 \\ \mu_2 \\ \vdots \\ \mu_n \\ \mu_{n+1} \end{bmatrix} = \frac{6}{h^2} \begin{bmatrix} y_2 - y_1 - hs_1 \\ \delta^2 y_2 \\ \delta^2 y_3 \\ \vdots \\ \delta^2 y_{n-1} \\ \delta^2 y_n \\ hs_{n+1} - y_{n+1} + y_n \end{bmatrix}, \tag{2.34}$$

where $\delta^2 y_i := y_{i+1} - 2y_i + y_{i-1}$, $i = 2, \ldots, n$. Hint: Use (2.10) to compute $g'(x_1)$ and $g'(x_{n+1})$, Is g unique?

---

[3]Hint: consider an arbitrary polynomial of degree n and expanded it in Taylor series around x_i.

[4]Hint, use Theorem 2.2.

Exercise 2.5 (Minimal Norm Property of the Natural Spline) Study proof of the following theorem.[5]

Theorem 2.5 (Minimal Norm Property of a Cubic Spline) *Suppose g is an N-spline. Then*

$$\int_a^b \left(g''(x)\right)^2 dx \le \int_a^b \left(h''(x)\right)^2 dx$$

for all $h \in C^2[a,b]$ such that $h(x_i) = g(x_i)$, $i = 1, \ldots, n+1$.

Proof Let h be any interpolant as in the theorem. We first show the orthogonality condition

$$\int_a^b g''e'' = 0, \quad e := h - g. \tag{2.35}$$

Integration by parts gives $\int_a^b g''e'' = \left[g''e'\right]_a^b - \int_a^b g'''e'$. The first term is zero since g'' is continuous and $g''(b) = g''(a) = 0$. For the second term, since g''' is equal to a constant v_i on each subinterval (x_i, x_{i+1}) and $e(x_i) = 0$, for $i = 1, \ldots, n+1$

$$\int_a^b g'''e' = \sum_{i=1}^n \int_{x_i}^{x_{i+1}} g'''e' = \sum_{i=1}^n v_i \int_{x_i}^{x_{i+1}} e' = \sum_{i=1}^n v_i \big(e(x_{i+1}) - e(x_i)\big) = 0.$$

Writing $h = g + e$ and using (2.35)

$$\begin{aligned}\int_a^b (h'')^2 &= \int_a^b (g'' + e'')^2 \\ &= \int_a^b (g'')^2 + \int_a^b (e'')^2 + 2\int_a^b g''e'' \\ &= \int_a^b (g'')^2 + \int_a^b (e'')^2 \ge \int_a^b (g'')^2\end{aligned}$$

and the proof is complete. □

Exercise 2.6 (Computing the D_2-Spline) Let g be the D_2-spline corresponding to an interval $[a,b]$, a vector $\boldsymbol{y} \in \mathbb{R}^{n+1}$ and μ_1, μ_{n+1}. The vector $\boldsymbol{x} = [x_1, \ldots, x_n]$ and

[5]The name spline is inherited from a"physical analogue", an elastic ruler that is used to draw smooth curves. Heavy weights, called **ducks**, are used to force the ruler to pass through, or near given locations. The ruler will take a shape that minimizes its potential energy. Since the potential energy is proportional to the integral of the square of the curvature, and the curvature can be approximated by the second derivative it follows from Theorem 2.5 that the mathematical N-spline approximately models the physical spline.

the coefficient matrix $\boldsymbol{C} \in \mathbb{R}^{n\times 4}$ in (2.7) are returned in the following algorithm. It uses Algorithms 2.1 and 2.2 to solve the tridiagonal linear system.

```
function [x,C]=splineint(a,b,y,mu1,munp1)
% [x,C]=splineint(a,b,y,mu1,munp1)
y=y(:); n=length(y)-1;
h=(b-a)/n; x=a:h:b-h; c=ones(n-2,1);
[l,u]= trifactor(c,4*ones(n-1,1),c);
b1=6/h^2*(y(3:n+1)-2*y(2:n)+y(1:n-1));
b1(1)=b1(1)-mu1; b1(n-1)=b1(n-1)-munp1;
mu= [mu1;trisolve(l,u,c,b1);munp1];
C=zeros(4*n,1);
C(1:4:4*n-3)=y(1:n);
C(2:4:4*n-2)=(y(2:n+1)-y(1:n))/h...
   -h*mu(1:n)/3-h*mu(2:n+1)/6;
C(3:4:4*n-1)=mu(1:n)/2;
C(4:4:4*n)=(mu(2:n+1)-mu(1:n))/(6*h);
C=reshape(C,4,n)';
end
```

Listing 2.3 splineint

Use the algorithm to compute the $c_{i,j}$ in Example 2.2.

Exercise 2.7 (Spline Evaluation) To plot a piecewise polynomial g in the form (2.4) we need to compute values $g(r_j)$ at a number of sites $\boldsymbol{r} = [r_1, \ldots, r_m] \in \mathbb{R}^m$ for some reasonably large integer m. To determine $g(r_j)$ for some j we need to find an integer i_j so that $g(r_j) = p_{i_j}(r_j)$.

Given $k \in \mathbb{N}$, $\boldsymbol{t} = [t_1, \ldots, t_k]$ and a real number x. We consider the problem of computing an integer i so that $i = 1$ if $x < t_2$, $i = k$ if $x \geq t_k$, and $t_i \leq x < t_{i+1}$ otherwise. If $\boldsymbol{x} \in \mathbb{R}^m$ is a vector then an m-vector $\boldsymbol{i}$ should be computed, such that the jth component of $\boldsymbol{i}$ gives the location of the jth component of $\boldsymbol{x}$. The following MATLAB function determines $\boldsymbol{i} = [i_1, \ldots, i_m]$. It uses the built in MATLAB functions `length, min, sort, find`.

```
function i = findsubintervals(t,x)
%i = findsubintervals(t,x)
k=length(t); m=length(x);
 if k<2
     i=ones(m,1);
 else
   t(1)=min(x(1),t(1))-1;
   [~,j]=sort([t(:)',x(:)']);
   i=(find(j>k)-(1:m))';
 end
```

Listing 2.4 findsubintervals

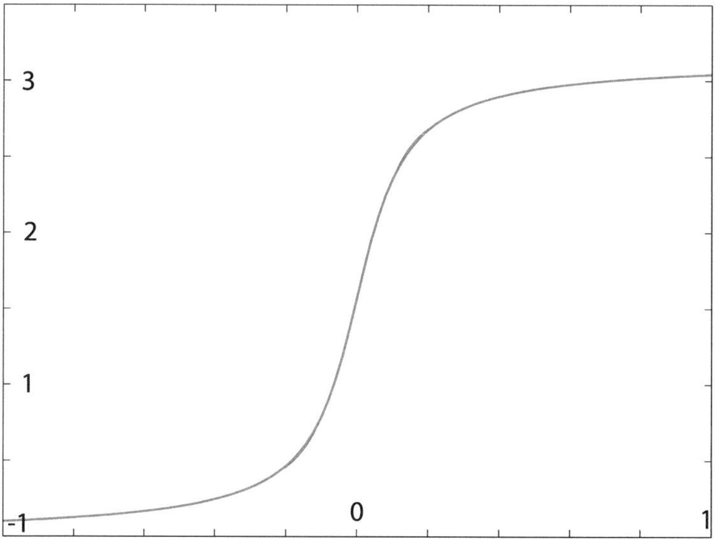

Fig. 2.5 The cubic spline interpolating $f(x) = \arctan(10x) + \pi/2$ at 14 equidistant sites on $[-1, 1]$. The exact function is also shown

Use `findsubintervals` and the algorithm `splineval` below to make the plots in Fig. 2.5.

```
function [X,G]=splineval(x,C,X)
% [X,G]=splineval(x,C,X)
m=length(X);
i=findsubintervals(x,X);
G=zeros(m,1);
for j=1:m
    k=i(j);
    t=X(j)-x(k);
   G(j)=[1,t,t^2,t^3]*C(k,:)';
end
```

Listing 2.5 splineval

Given output $\boldsymbol{x}$, $\boldsymbol{C}$ of `splineint`, defining a cubic spline g, and a vector $\boldsymbol{X}$, `splineval`ăcomputes the vector $\boldsymbol{G} = g(\boldsymbol{X})$.

2.6.2 Exercises Sect. 2.2

Exercise 2.8 (Central Difference Approximation of 2. Derivative) Consider

$$\delta^2 f(x) := \frac{f(x+h) - 2f(x) + f(x-h)}{h^2}, \quad h > 0, \quad f : [x-h, x+h] \to \mathbb{R}.$$

a) Show using Taylor expansion that if $f \in C^2[x-h, x+h]$ then for some η_2

$$\delta^2 f(x) = f''(\eta_2), \quad x - h < \eta_2 < x + h.$$

b) Show that if $f \in C^4[x-h, x+h]$ then for some η_4

$$\delta^2 f(x) = f''(x) + \frac{h^2}{12} f^{(4)}(\eta_4), \quad x - h < \eta_4 < x + h.$$

$\delta^2 f(x)$ is known as the **central difference approximation** to the second derivative at x.

Exercise 2.9 (Two Point Boundary Value Problem) We consider a finite difference method for the two point boundary value problem

$$\begin{aligned} -u''(x) + r(x)u'(x) + q(x)u(x) &= f(x), \text{ for } x \in [a, b], \\ u(a) = g_0, \quad u(b) &= g_1. \end{aligned} \tag{2.36}$$

We assume that the given functions f, q and r are continuous on $[a, b]$ and that $q(x) \geq 0$ for $x \in [a, b]$. It can then be shown that (2.36) has a unique solution u.

To solve (2.36) numerically we choose $m \in \mathbb{N}$, $h = (b-a)/(m+1)$, $x_j = a + jh$ for $j = 0, 1, \ldots, m+1$ and solve the difference equation

$$\frac{-v_{j-1} + 2v_j - v_{j+1}}{h^2} + r(x_j)\frac{v_{j+1} - v_{j-1}}{2h} + q(x_j)v_j = f(x_j), \quad j = 1, \ldots, m, \tag{2.37}$$

with $v_0 = g_0$ and $v_{m+1} = g_1$.

a) Show that (2.37) leads to a tridiagonal linear system $\boldsymbol{A}\boldsymbol{v} = \boldsymbol{b}$, where $\boldsymbol{A} = \operatorname{tridiag}(a_j, d_j.c_j) \in \mathbb{R}^{m \times m}$ has elements

$$a_j = -1 - \frac{h}{2}r(x_j), \; c_j = -1 + \frac{h}{2}r(x_j), \; d_j = 2 + h^2 q(x_j),$$

and

$$b_j = \begin{cases} h^2 f(x_1) - a_1 g_0, & \text{if } j = 1, \\ h^2 f(x_j), & \text{if } 2 \le j \le m-1, \\ h^2 f(x_m) - c_m g_1, & \text{if } j = m. \end{cases}$$

b) Show that the linear system satisfies the conditions in Theorem 2.4 if the spacing h is so small that $\frac{h}{2}|r(x)| < 1$ for all $x \in [a, b]$.

(c) Propose a method to find $v_1, \ldots, v_m$.

Exercise 2.10 (Two Point Boundary Value Problem; Computation)

a) Consider the problem (2.36) with $r = 0$, $f = q = 1$ and boundary conditions $u(0) = 1$, $u(1) = 0$. The exact solution is $u(x) = 1 - \sinh x / \sinh 1$. Write a computer program to solve (2.37) for $h = 0.1, 0.05, 0.025, 0.0125$, and compute the "error" $\max_{1 \le j \le m} |u(x_j) - v_j|$ for each h.

b) Make a combined plot of the solution u and the computed points v_j, $j = 0, \ldots, m+1$ for $h = 0.1$.

c) One can show that the error is proportional to h^p for some integer p. Estimate p based on the error for $h = 0.1, 0.05, 0.025, 0.0125$.

2.6.3 Exercises Sect. 2.3

Exercise 2.11 (Approximate Force) Show that

$$F = \frac{4\sin^2(\pi h/2) R}{h^2 L^2} = \frac{\pi^2 R}{L^2} + O(h^2).$$

Exercise 2.12 (Symmetrize Matrix (Exam Exercise 1977-3)) Let $\boldsymbol{A} \in \mathbb{R}^{n \times n}$ be tridiagonal and suppose $a_{i,i+1} a_{i+1,i} > 0$ for $i = 1, \ldots, n-1$. Show that there exists a diagonal matrix $\boldsymbol{D} = \operatorname{diag}(d_1, \ldots, d_n)$ with $d_i > 0$ for all i such that $\boldsymbol{B} := \boldsymbol{D}\boldsymbol{A}\boldsymbol{D}^{-1}$ is symmetric.

2.6.4 Exercises Sect. 2.4

Exercise 2.13 (Eigenpairs $\boldsymbol{T}$ of Order 2) Compute directly the eigenvalues and eigenvectors for $\boldsymbol{T}$ when $n = 2$ and thus verify Lemma 2.2 in this case.

Exercise 2.14 (LU Factorization of 2. Derivative Matrix) Show that $\boldsymbol{T} = \boldsymbol{L}\boldsymbol{U}$, where

$$\boldsymbol{L} = \begin{bmatrix} 1 & 0 & \cdots & \cdots & 0 \\ -\frac{1}{2} & 1 & \ddots & & \vdots \\ 0 & -\frac{2}{3} & 1 & \ddots & \vdots \\ \vdots & \ddots & \ddots & \ddots & 0 \\ 0 & \cdots & 0 & -\frac{m-1}{m} & 1 \end{bmatrix}, \quad \boldsymbol{U} = \begin{bmatrix} 2 & -1 & 0 & \cdots & 0 \\ 0 & \frac{3}{2} & -1 & \ddots & \vdots \\ \vdots & \ddots & \ddots & \ddots & 0 \\ \vdots & & \ddots & \frac{m}{m-1} & -1 \\ 0 & \cdots & \cdots & 0 & \frac{m+1}{m} \end{bmatrix}. \tag{2.38}$$

This is the LU factorization of $\boldsymbol{T}$.

Exercise 2.15 (Inverse of the 2. Derivative Matrix) Let $\boldsymbol{S} \in \mathbb{R}^{m \times m}$ have elements s_{ij} given by

$$s_{i,j} = s_{j,i} = \frac{1}{m+1} j\big(m + 1 - i\big), \quad 1 \le j \le i \le m. \tag{2.39}$$

Show that $\boldsymbol{S}\boldsymbol{T} = \boldsymbol{I}$ and conclude that $\boldsymbol{T}^{-1} = \boldsymbol{S}$.

2.6.5 Exercises Sect. 2.5

Exercise 2.16 (Matrix Element as a Quadratic Form) For any matrix $\boldsymbol{A}$ show that $a_{ij} = \boldsymbol{e}_i^T \boldsymbol{A} \boldsymbol{e}_j$ for all i, j.

Exercise 2.17 (Outer Product Expansion of a Matrix) For any matrix $\boldsymbol{A} \in \mathbb{C}^{m \times n}$ show that $\boldsymbol{A} = \sum_{i=1}^m \sum_{j=1}^n a_{ij} \boldsymbol{e}_i \boldsymbol{e}_j^T$.

Exercise 2.18 (The Product $\boldsymbol{A}^T\boldsymbol{A}$) Let $\boldsymbol{B} = \boldsymbol{A}^T\boldsymbol{A}$. Explain why this product is defined for any matrix $\boldsymbol{A}$. Show that $b_{ij} = \boldsymbol{a}_{:i}^T \boldsymbol{a}_{:j}$ for all i, j.

Exercise 2.19 (Outer Product Expansion) For $\boldsymbol{A} \in \mathbb{R}^{m \times n}$ and $\boldsymbol{B} \in \mathbb{R}^{p \times n}$ show that

$$\boldsymbol{A}\boldsymbol{B}^T = \boldsymbol{a}_{:1}\boldsymbol{b}_{:1}^T + \boldsymbol{a}_{:2}\boldsymbol{b}_{:2}^T + \cdots + \boldsymbol{a}_{:n}\boldsymbol{b}_{:n}^T.$$

This is called the **outer product expansion** of the columns of $\boldsymbol{A}$ and $\boldsymbol{B}$.

Exercise 2.20 (System with Many Right Hand Sides; Compact Form) Suppose $\boldsymbol{A} \in \mathbb{R}^{m \times n}$, $\boldsymbol{B} \in \mathbb{R}^{m \times p}$, and $\boldsymbol{X} \in \mathbb{R}^{n \times p}$. Show that

$$\boldsymbol{A}\boldsymbol{X} = \boldsymbol{B} \quad \iff \quad \boldsymbol{A}\boldsymbol{x}_{:j} = \boldsymbol{b}_{:j}, \; j = 1, \ldots, p.$$

Exercise 2.21 (Block Multiplication Example) Suppose $\boldsymbol{A} = [\boldsymbol{A}_1, \boldsymbol{A}_2]$ and $\boldsymbol{B} = \begin{bmatrix} \boldsymbol{B}_1 \\ \boldsymbol{0} \end{bmatrix}$. When is $\boldsymbol{A}\boldsymbol{B} = \boldsymbol{A}_1\boldsymbol{B}_1$?

Exercise 2.22 (Another Block Multiplication Example) Suppose $\boldsymbol{A}, \boldsymbol{B}, \boldsymbol{C} \in \mathbb{R}^{n\times n}$ are given in block form by

$$\boldsymbol{A} := \begin{bmatrix} \lambda & \boldsymbol{a}^T \\ \boldsymbol{0} & \boldsymbol{A}_1 \end{bmatrix}, \quad \boldsymbol{B} := \begin{bmatrix} 1 & \boldsymbol{0}^T \\ \boldsymbol{0} & \boldsymbol{B}_1 \end{bmatrix}, \quad \boldsymbol{C} := \begin{bmatrix} 1 & \boldsymbol{0}^T \\ \boldsymbol{0} & \boldsymbol{C}_1 \end{bmatrix},$$

where $\boldsymbol{A}_1, \boldsymbol{B}_1, \boldsymbol{C}_1 \in \mathbb{R}^{(n-1)\times(n-1)}$. Show that

$$\boldsymbol{C}\boldsymbol{A}\boldsymbol{B} = \begin{bmatrix} \lambda & \boldsymbol{a}^T\boldsymbol{B}_1 \\ \boldsymbol{0} & \boldsymbol{C}_1\boldsymbol{A}_1\boldsymbol{B}_1 \end{bmatrix}.$$

2.7 Review Questions

2.7.1 How do we define nonsingularity of a matrix?

2.7.2 Define the second derivative matrix $\boldsymbol{T}$. How did we show that it is nonsingular?

2.7.3 Why do we not use the explicit inverse of $\boldsymbol{T}$ to solve the linear system $\boldsymbol{T}\boldsymbol{x} = \boldsymbol{b}$?

2.7.4 What are the eigenpairs of the matrix $\boldsymbol{T}$?

2.7.5 Why are the diagonal elements of a Hermitian matrix real?

2.7.6 Is the matrix $\begin{bmatrix} 1 & 1+i \\ 1+i & 2 \end{bmatrix}$ Hermitian? Symmetric?

2.7.7 Is a weakly diagonally dominant matrix nonsingular?

2.7.8 Is a strictly diagonally dominant matrix always nonsingular?

2.7.9 Does a tridiagonal matrix always have an LU factorization?

Chapter 3
Gaussian Elimination and LU Factorizations

In this chapter we first review Gaussian elimination. Gaussian elimination leads to an LU factorization of the coefficient matrix or more generally to a PLU factorization, if row interchanges are introduced. Here $\boldsymbol{P}$ is a permutation matrix, $\boldsymbol{L}$ is lower triangular and $\boldsymbol{U}$ is upper triangular.

We also consider in great detail the general theory of LU factorizations.

3.1 3 by 3 Example

Gaussian elimination with row interchanges is the classical method for solving n linear equations in n unknowns.[1] We first recall how it works on a 3×3 system.

Example 3.1 (Gaussian Elimination on a 3×3 System) Consider a nonsingular system of three equations in three unknowns:

$$\begin{aligned} a_{11}^{(1)}x_1 + a_{12}^{(1)}x_2 + a_{13}^{(1)}x_3 &= b_1^{(1)}, \quad \text{I} \\ a_{21}^{(1)}x_1 + a_{22}^{(1)}x_2 + a_{23}^{(1)}x_3 &= b_2^{(1)}, \quad \text{II} \\ a_{31}^{(1)}x_1 + a_{32}^{(1)}x_2 + a_{33}^{(1)}x_3 &= b_3^{(1)}. \quad \text{III.} \end{aligned}$$

[1]The method was known long before Gauss used it in 1809. It was further developed by Doolittle in 1881, see [4].

T. Lyche, *Numerical Linear Algebra and Matrix Factorizations*,
Texts in Computational Science and Engineering 22,
https://doi.org/10.1007/978-3-030-36468-7_3

To solve this system by Gaussian elimination suppose $a_{11}^{(1)} \neq 0$. We subtract $l_{21}^{(1)} := a_{21}^{(1)}/a_{11}^{(1)}$ times equation I from equation II and $l_{31}^{(1)} := a_{31}^{(1)}/a_{11}^{(1)}$ times equation I from equation III. The result is

$$\begin{aligned} a_{11}^{(1)}x_1 + a_{12}^{(1)}x_2 + a_{13}^{(1)}x_3 &= b_1^{(1)}, \quad \text{I} \\ a_{22}^{(2)}x_2 + a_{23}^{(2)}x_3 &= b_2^{(2)}, \quad \text{II}' \\ a_{32}^{(2)}x_2 + a_{33}^{(2)}x_3 &= b_3^{(2)}, \quad \text{III}', \end{aligned}$$

where $b_i^{(2)} = b_i^{(1)} - l_{i1}^{(1)}b_i^{(1)}$ for $i = 2, 3$ and $a_{ij}^{(2)} = a_{ij}^{(1)} - l_{i,1}^{(1)}a_{1j}^{(1)}$ for $i, j = 2, 3$. If $a_{11}^{(1)} = 0$ and $a_{21}^{(1)} \neq 0$ we first interchange equation I and equation II. If $a_{11}^{(1)} = a_{21}^{(1)} = 0$ we interchange equation I and III. Since the system is nonsingular the first column cannot be zero and an interchange is always possible.

If $a_{22}^{(2)} \neq 0$ we subtract $l_{32}^{(2)} := a_{32}^{(2)}/a_{22}^{(2)}$ times equation II′ from equation III′ to obtain

$$\begin{aligned} a_{11}^{(1)}x_1 + a_{12}^{(1)}x_2 + a_{13}^{(1)}x_3 &= b_1^{(1)}, \quad \text{I} \\ a_{22}^{(2)}x_2 + a_{23}^{(2)}x_3 &= b_2^{(2)}, \quad \text{II}' \\ a_{33}^{(3)}x_3 &= b_3^{(3)}, \quad \text{III}'', \end{aligned}$$

where $a_{33}^{(3)} = a_{33}^{(2)} - l_{32}^{(2)}a_{23}^{(2)}$ and $b_3^{(3)} = b_3^{(2)} - l_{32}^{(2)}b_2^{(2)}$. If $a_{22}^{(2)} = 0$ then $a_{32}^{(2)} \neq 0$ (cf. Sect. 3.4) and we first interchange equation II′ and equation III′. The reduced system is easy to solve since it is upper triangular. Starting from the bottom and moving upwards we find

$$\begin{aligned} x_3 &= b_3^{(3)}/a_{33}^{(3)} \\ x_2 &= (b_2^{(2)} - a_{23}^{(2)}x_3)/a_{22}^{(2)} \\ x_1 &= (b_1^{(1)} - a_{12}^{(1)}x_2 - a_{13}^{(1)}x_3)/a_{11}^{(1)}. \end{aligned}$$

This is known as **back substitution**. Gauss elimination leads to an LU factorization. Indeed, if $a_{kk}^{(k)} \neq 0$, $k = 1, 2$ then

$$\boldsymbol{LU} := \begin{bmatrix} 1 & 0 & 0 \\ l_{21}^{(1)} & 1 & 0 \\ l_{31}^{(1)} & l_{32}^{(2)} & 1 \end{bmatrix} \begin{bmatrix} a_{11}^{(1)} & a_{12}^{(1)} & a_{13}^{(1)} \\ 0 & a_{22}^{(2)} & a_{23}^{(2)} \\ 0 & 0 & a_{33}^{(3)} \end{bmatrix}$$

$$= \begin{bmatrix} a_{11}^{(1)} & a_{12}^{(1)} & a_{13}^{(1)} \\ l_{21}^{(1)}a_{11}^{(1)} & l_{21}^{(1)}a_{12}^{(1)} + a_{22}^{(2)} & l_{21}^{(1)}a_{13}^{(1)} + a_{23}^{(2)} \\ l_{31}^{(1)}a_{11}^{(1)} & l_{31}^{(1)}a_{12}^{(1)} + l_{32}^{(2)}a_{22}^{(2)} & l_{31}^{(1)}a_{13}^{(1)} + l_{32}^{(2)}a_{23}^{(2)} + a_{33}^{(3)} \end{bmatrix} = \begin{bmatrix} a_{11}^{(1)} & a_{12}^{(1)} & a_{13}^{(1)} \\ a_{21}^{(1)} & a_{22}^{(1)} & a_{23}^{(1)} \\ a_{31}^{(1)} & a_{32}^{(1)} & a_{33}^{(1)} \end{bmatrix} = \boldsymbol{A}.$$

Thus Gaussian elimination leads to an LU factorization of the coefficient matrix $\boldsymbol{A}^{(1)}$ (cf. the proof of Theorem 3.2).

3.2 Gauss and LU

In **Gaussian elimination** without row interchanges we start with a linear system $\boldsymbol{A}\boldsymbol{x} = \boldsymbol{b}$ and generate a sequence of equivalent systems $\boldsymbol{A}^{(k)}\boldsymbol{x} = \boldsymbol{b}^{(k)}$ for $k = 1, \ldots, n$, where $\boldsymbol{A}^{(1)} = \boldsymbol{A}$, $\boldsymbol{b}^{(1)} = \boldsymbol{b}$, and $\boldsymbol{A}^{(k)}$ has zeros under the diagonal in its first $k-1$ columns. Thus $\boldsymbol{A}^{(n)}$ is upper triangular and the system $\boldsymbol{A}^{(n)}\boldsymbol{x} = \boldsymbol{b}^{(n)}$ is easy to solve. The process is illustrated in Fig. 3.1.

The matrix $\boldsymbol{A}^{(k)}$ takes the form

$$\boldsymbol{A}^{(k)} = \left[\begin{array}{ccc|ccccc} a_{1,1}^{(1)} & \cdots & a_{1,k-1}^{(1)} & a_{1,k}^{(1)} & \cdots & a_{1,j}^{(1)} & \cdots & a_{1,n}^{(1)} \\ & \ddots & \vdots & \vdots & & \vdots & & \vdots \\ & & a_{k-1,k-1}^{(k-1)} & a_{k-1,k}^{(k-1)} & \cdots & a_{k-1,j}^{(k-1)} & \cdots & a_{k-1,n}^{(k-1)} \\ \hline & & & a_{k,k}^{(k)} & \cdots & a_{k,j}^{(k)} & \cdots & a_{k,n}^{(k)} \\ & & & \vdots & & \vdots & & \vdots \\ & & & a_{i,k}^{(k)} & \cdots & a_{i,j}^{(k)} & \cdots & a_{i,n}^{(k)} \\ & & & \vdots & & \vdots & & \vdots \\ & & & a_{n,k}^{(k)} & \cdots & a_{n,j}^{(k)} & \cdots & a_{n,n}^{(k)} \end{array}\right]. \tag{3.1}$$

The process transforming $\boldsymbol{A}^{(k)}$ into $\boldsymbol{A}^{(k+1)}$ for $k = 1, \ldots, n-1$ can be described as follows.

$$\boxed{\begin{array}{l} \text{for } i = k+1 : n \\ \quad l_{ik}^{(k)} = a_{ik}^{(k)} / a_{kk}^{(k)} \\ \quad \text{for } j = k : n \\ \qquad a_{ij}^{(k+1)} = a_{ij}^{(k)} - l_{ik}^{(k)} a_{kj}^{(k)} \end{array}} \tag{3.2}$$

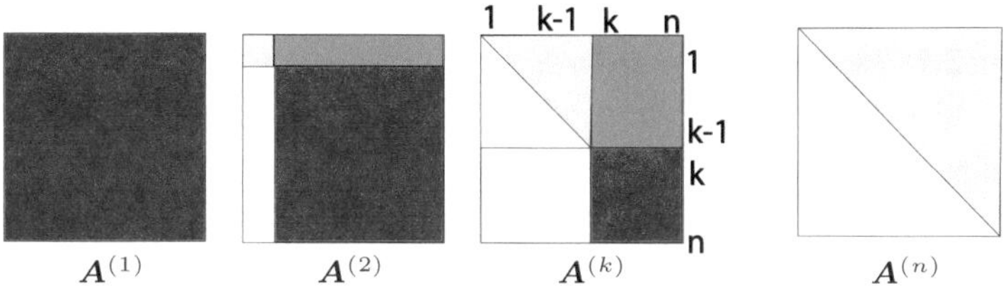

Fig. 3.1 Gaussian elimination

For $j = k$ it follows from (3.2) that $a_{ik}^{(k+1)} = a_{ik}^{(k)} - \frac{a_{ik}^{(k)}}{a_{kk}^{(k)}} a_{kk}^{(k)} = 0$ for $i = k+1, \ldots, n$. Thus $\boldsymbol{A}^{(k+1)}$ will have zeros under the diagonal in its first k columns and the elimination is carried one step further. The numbers $l_{ik}^{(k)}$ in (3.2) are called **multipliers**.

Gaussian elimination with no row interchanges is valid if and only if the **pivots** $a_{kk}^{(k)}$ are nonzero for $k = 1, \ldots, n-1$. This depends on certain submatrices of $\boldsymbol{A}$ known as **principal submatrices**.

Definition 3.1 (Principal Submatrix) For $k = 1, \ldots, n$ the matrices $\boldsymbol{A}_{[k]} \in \mathbb{C}^{k \times k}$ given by

$$\boldsymbol{A}_{[k]} := \boldsymbol{A}(1:k, 1:k) = \begin{bmatrix} a_{11} & \cdots & a_{k1} \\ \vdots & & \vdots \\ a_{k1} & \cdots & a_{kk} \end{bmatrix}$$

are called the **leading principal submatrices** of $\boldsymbol{A} \in \mathbb{C}^{n \times n}$. More generally, a matrix $\boldsymbol{B} \in \mathbb{C}^{k \times k}$ is called a **principal submatrix** of $\boldsymbol{A}$ if $\boldsymbol{B} = \boldsymbol{A}(\boldsymbol{r}, \boldsymbol{r})$, where $\boldsymbol{r} = [r_1, \ldots, r_k]$ for some $1 \le r_1 < \cdots < r_k \le n$. Thus,

$$b_{i,j} = a_{r_i, r_j}, \quad i, j = 1, \ldots, k.$$

The determinant of a (leading) principal submatrix is called a **(leading) principal minor**.

A principal submatrix is leading if $r_j = j$ for $j = 1, \ldots, k$. Also a principal submatrix is special in that it uses the same rows and columns of $\boldsymbol{A}$. For $k = 1$ The only principal submatrices of order $k = 1$ are the diagonal elements of $\boldsymbol{A}$.

Example 3.2 (Principal Submatrices) The principal submatrices of $\boldsymbol{A} = \begin{bmatrix} 1 & 2 & 3 \\ 4 & 5 & 6 \\ 7 & 8 & 9 \end{bmatrix}$ are

$$[1],\ [5],\ [9],\ \begin{bmatrix} 1 & 2 \\ 4 & 5 \end{bmatrix},\ \begin{bmatrix} 1 & 3 \\ 7 & 9 \end{bmatrix},\ \begin{bmatrix} 5 & 6 \\ 8 & 9 \end{bmatrix},\ \boldsymbol{A}.$$

The leading principal submatrices are

$$[1],\ \begin{bmatrix} 1 & 2 \\ 4 & 5 \end{bmatrix},\ \boldsymbol{A}.$$

Theorem 3.1 *We have $a_{k,k}^{(k)} \neq 0$ for $k = 1, \ldots, n-1$ if and only if the leading principal submatrices $\boldsymbol{A}_{[k]}$ of $\boldsymbol{A}$ are nonsingular for $k = 1, \ldots, n-1$. Moreover*

$$\det(\boldsymbol{A}_{[k]}) = a_{11}^{(1)} a_{22}^{(2)} \cdots a_{kk}^{(k)}, \quad k = 1, \ldots, n. \tag{3.3}$$

Proof Let $\boldsymbol{B}_k = \boldsymbol{A}_{k-1}^{(k)}$ be the upper left $k-1$ corner of $\boldsymbol{A}^{(k)}$ given by (3.1). Observe that the elements of $\boldsymbol{B}_k$ are computed from $\boldsymbol{A}$ by using only elements from $\boldsymbol{A}_{[k-1]}$. Since the determinant of a matrix does not change under the operation of subtracting a multiple of one row from another row the determinant of $\boldsymbol{A}_{[k]}$ equals the product of diagonal elements of $\boldsymbol{B}_{k+1}$ and (3.3) follows. But then $a_{11}^{(1)} \cdots a_{kk}^{(k)} \neq 0$ for $k = 1, \ldots, n-1$ if and only if $\det(\boldsymbol{A}_{[k]}) \neq 0$ for $k = 1, \ldots, n-1$, or equivalently $\boldsymbol{A}_{[k]}$ is nonsingular for $k = 1, \ldots, n-1$. □

Gaussian elimination is a way to compute the LU factorization of the coefficient matrix.

Theorem 3.2 *Suppose $\boldsymbol{A} \in \mathbb{C}^{n\times n}$ and that the leading principal submatrices $\boldsymbol{A}_{[k]}$ are nonsingular for $k = 1, \ldots, n-1$. Then Gaussian elimination with no row interchanges results in an LU factorization of $\boldsymbol{A}$. In particular $\boldsymbol{A} = \boldsymbol{LU}$, where*

$$\boldsymbol{L} = \begin{bmatrix} 1 & & & \\ l_{21}^{(1)} & 1 & & \\ \vdots & & \ddots & \\ l_{n1}^{(1)} & l_{n2}^{(2)} & \cdots & 1 \end{bmatrix}, \quad \boldsymbol{U} = \begin{bmatrix} a_{11}^{(1)} & \cdots & a_{1n}^{(1)} \\ & \ddots & \vdots \\ & & a_{nn}^{(n)} \end{bmatrix}, \tag{3.4}$$

where the $l_{ij}^{(j)}$ and $a_{ij}^{(i)}$ are given by (3.2).

Proof From (3.2) we have for all i, j

$$l_{ik}^{(k)} a_{kj}^{(k)} = a_{ij}^{(k)} - a_{ij}^{(k+1)} \text{ for } k < \min(i, j), \text{ and } l_{ij}^{(k)} a_{jj}^{(j)} = a_{ij}^{(j)} \text{ for } i > j.$$

Thus for $i \le j$ we find

$$(\boldsymbol{LU})_{ij} = \sum_{k=1}^{i-1} l_{ik}^{(k)} a_{kj}^{(k)} + a_{ij}^{(i)} = \sum_{k=1}^{i-1} \left(a_{ij}^{(k)} - a_{ij}^{(k+1)}\right) + a_{ij}^{(i)} = a_{ij}^{(1)} = a_{ij}, \tag{3.5}$$

while for $i > j$

$$(\boldsymbol{LU})_{ij} = \sum_{k=1}^{j-1} l_{ik}^{(k)} a_{kj}^{(k)} + l_{ij} a_{jj}^{(j)} = \sum_{k=1}^{j-1} \left(a_{ij}^{(k)} - a_{ij}^{(k+1)}\right) + a_{ij}^{(j)} = a_{ij}. \tag{3.6}$$

□

Note that this Theorem holds even if $\boldsymbol{A}$ is singular. Since $\boldsymbol{L}$ is nonsingular the matrix $\boldsymbol{U}$ is then singular, and we must have $a_{nn}^{(n)} = 0$ when $\boldsymbol{A}$ is singular.

3.3 Banded Triangular Systems

Once we know an LU factorization of $\boldsymbol{A}$ the system $\boldsymbol{Ax} = \boldsymbol{b}$ is solved in two steps. Since $\boldsymbol{LUx} = \boldsymbol{b}$ we have $\boldsymbol{Ly} = \boldsymbol{b}$, where $\boldsymbol{y} := \boldsymbol{Ux}$. We first solve $\boldsymbol{Ly} = \boldsymbol{b}$, for $\boldsymbol{y}$ and then $\boldsymbol{Ux} = \boldsymbol{y}$ for $\boldsymbol{x}$.

3.3.1 Algorithms for Triangular Systems

A nonsingular triangular linear system $\boldsymbol{Ax} = \boldsymbol{b}$ is easy to solve. By Lemma 2.5 $\boldsymbol{A}$ has nonzero diagonal elements. Consider first the lower triangular case. For $n = 3$ the system is

$$\begin{bmatrix} a_{11} & 0 & 0 \\ a_{21} & a_{22} & 0 \\ a_{31} & a_{32} & a_{33} \end{bmatrix} \begin{bmatrix} x_1 \\ x_2 \\ x_3 \end{bmatrix} = \begin{bmatrix} b_1 \\ b_2 \\ b_3 \end{bmatrix}.$$

From the first equation we find $x_1 = b_1/a_{11}$. Solving the second equation for x_2 we obtain $x_2 = (b_2 - a_{21}x_1)/a_{22}$. Finally the third equation gives $x_3 = (b_3 - a_{31}x_1 - a_{32}x_2)/a_{33}$. This process is known as forward substitution. In general

$$x_k = \big(b_k - \sum_{j=1}^{k-1} a_{k,j}x_j\big)/a_{kk}, \quad k = 1, 2, \dots, n. \tag{3.7}$$

When $\boldsymbol{A}$ is a lower triangular band matrix the number of arithmetic operations necessary to find $\boldsymbol{x}$ can be reduced. Suppose $\boldsymbol{A}$ is a lower triangular d-banded, so that $a_{k,j} = 0$ for $j \notin \{l_k, l_k + 1, \dots, k$ for $k = 1, 2, \dots, n$, and where $l_k := \max(1, k - d)$, see Fig. 3.2. For a lower triangular d-band matrix the calculation in (3.7) can be simplified as follows

$$x_k = \big(b_k - \sum_{j=l_k}^{k-1} a_{k,j}x_j\big)/a_{kk}, \quad k = 1, 2, \dots, n. \tag{3.8}$$

Note that (3.8) reduces to (3.7) if $d = n$. Letting $A(k, l_k : (k-1)) * x(l_k : (k-1))$ denote the sum $\sum_{j=l_k}^{k-1} a_{kj}x_j$ we arrive at the following algorithm, where the initial

$$\begin{bmatrix} a_{11} & 0 & 0 & 0 & 0 \\ a_{21} & a_{22} & 0 & 0 & 0 \\ 0 & a_{32} & a_{33} & 0 & 0 \\ 0 & 0 & a_{43} & a_{44} & 0 \\ 0 & 0 & 0 & a_{54} & a_{55} \end{bmatrix}, \quad \begin{bmatrix} a_{11} & 0 & 0 & 0 & 0 \\ a_{21} & a_{22} & 0 & 0 & 0 \\ a_{31} & a_{32} & a_{33} & 0 & 0 \\ 0 & a_{42} & a_{43} & a_{44} & 0 \\ 0 & 0 & a_{53} & a_{54} & a_{55} \end{bmatrix}$$

Fig. 3.2 Lower triangular 5×5 band matrices: $d = 1$ (left) and $d = 2$ right

"r" in the name signals that this algorithm is row oriented. The algorithm takes a nonsingular lower triangular d-banded matrix $\boldsymbol{A} \in \mathbb{C}^{n\times n}$, and $\boldsymbol{b} \in \mathbb{C}^n$, as input, and returns an $\boldsymbol{x} \in \mathbb{C}^n$ so that $\boldsymbol{Ax} = \boldsymbol{b}$. For each k we take the inner product of a part of a row with the already computed unknowns.

```
function x=rforwardsolve(A,b,d)
% x=rforwardsolve(A,b,d)
n=length(b); x=b;
x(1)=b(1)/A(1,1);
for k=2:n
    lk=max(1,k-d);
    x(k)=(b(k)-A(k,lk:(k-1))*x(lk:(k-1)))/A(k,k);
end
```

Listing 3.1 rforwardsolve

A system $\boldsymbol{Ax} = \boldsymbol{b}$, where $\boldsymbol{A}$ is upper triangular must be solved by back substitution or 'bottom-up'. We first find x_n from the last equation and then move upwards for the remaining unknowns. For an upper triangular d-banded matrix this leads to the following algorithm, which takes a nonsingular upper triangular d-banded matrix $\boldsymbol{A} \in \mathbb{C}^{n\times n}$, $\boldsymbol{b} \in \mathbb{C}^n$ and d, as input, and returns an $\boldsymbol{x} \in \mathbb{C}^n$ so that $\boldsymbol{Ax} = \boldsymbol{b}$.

```
function x=rbacksolve(A,b,d)
% x=rbacksolve(A,b,d)
n=length(b); x=b;
x(n)=b(n)/A(n,n);
for k=n-1:-1:1
    uk=min(n,k+d);
    x(k)=(b(k)-A(k,(k+1):uk)*x((k+1):uk))/A(k,k);
end
```

Listing 3.2 rbacksolve

Example 3.3 (Column Oriented Forwardsolve) In this example we develop a column oriented vectorized version of forward substitution. For a backward substitution see Exercise 3.1. Consider the system $\boldsymbol{Ax} = \boldsymbol{b}$, where $\boldsymbol{A} \in \mathbb{C}^{n\times n}$ is lower triangular. Suppose after $k-1$ steps of the algorithm we have a reduced system in the form

$$\begin{bmatrix} a_{k,k} & 0 & \cdots & 0 \\ a_{k+1,k} & a_{k+1,k+1} & \cdots & 0 \\ \vdots & & \ddots & \vdots \\ a_{n,k} & & \cdots & a_{n\times n} \end{bmatrix} \begin{bmatrix} x_k \\ x_{k+1} \\ \vdots \\ x_n \end{bmatrix} = \begin{bmatrix} b_k \\ b_{k+1} \\ \vdots \\ b_n \end{bmatrix}.$$

This system is of order $n-k+1$. The unknowns are $x_k, \ldots, x_n$.

We see that $x_k = b_k/a_{k,k}$ and eliminating x_k from the remaining equations we obtain a system of order $n-k$ with unknowns $x_{k+1},\ldots,x_n$

$$\begin{bmatrix} a_{k+1,k+1} & 0 & \cdots & 0 \\ a_{k+2,k+1} & a_{k+2,k+2} & \cdots & 0 \\ \vdots & & \ddots & \vdots \\ a_{n,k+1} & & \cdots & a_{n,n} \end{bmatrix} \begin{bmatrix} x_{k+1} \\ \vdots \\ x_n \end{bmatrix} = \begin{bmatrix} b_{k+1} \\ \vdots \\ b_n \end{bmatrix} - x_k \begin{bmatrix} a_{k+1,k} \\ \vdots \\ a_{n,k} \end{bmatrix}.$$

Thus at the kth step, $k = 1, 2, \ldots n$ we set $x_k = b_k/A(k,k)$ and update b as follows:

$$b((k+1):n) = b((k+1):n) - x(k) * A((k+1):n, k).$$

This leads to the following algorithm for column oriented forward solve, which takes a nonsingular lower triangular d-banded matrix $\boldsymbol{A} \in \mathbb{C}^{n\times n}$, $\boldsymbol{b} \in \mathbb{C}^n$, and d as input, and returns an $\boldsymbol{x} \in \mathbb{C}^n$ so that $\boldsymbol{A}\boldsymbol{x} = \boldsymbol{b}$.

```
function x=cforwardsolve(A,b,d)
%x=cforwardsolve(A,b,d)
x=b; n=length(b);
for k=1:n-1
  x(k)=b(k)/A(k,k); uk=min(n,k+d);
  b((k+1):uk)=b((k+1):uk)-A((k+1):uk,k)*x(k);
end
x(n)=b(n)/A(n,n);
end
```

Listing 3.3 cforwardsolve

3.3.2 *Counting Operations*

It is useful to have a number which indicates the amount of work an algorithm requires. In this book we measure this by estimating the total number of (complex) arithmetic operations. We count both additions, subtractions, multiplications and divisions, but not work on indices. As an example we show that the LU factorization of a full matrix of order n using Gaussian elimination requires exactly

$$N_{LU} := \frac{2}{3}n^3 - \frac{1}{2}n^2 - \frac{1}{6}n \tag{3.9}$$

operations. Let M, D, A, S be the number of (complex) multiplications, divisions, additions, and subtractions. In (3.2) the multiplications and subtractions occur in the calculation of $a_{ij}^{k+1} = a_{ij}^{(k)} - l_{ik}^{(k)} a_{kj}^{(k)}$ which is carried out $(n-k)^2$ times. Moreover, each calculation involves one subtraction and one multiplication. Thus we find $M + S = 2\sum_{k=1}^{n-1}(n-k)^2 = 2\sum_{m=1}^{n-1} m^2 = \frac{2}{3}n(n-1)(n-\frac{1}{2})$. For each k there are $n-k$ divisions giving a sum of $\sum_{k=1}^{n-1}(n-k) = \frac{1}{2}n(n-1)$. Since there are no additions we obtain the total

$$M + D + A + S = \frac{2}{3}n(n-1)(n-\frac{1}{2}) + \frac{1}{2}n(n-1) = N_{LU}$$

given by (3.9).

We are only interested in N_{LU} when n is large and for such n the term $\frac{2}{3}n^3$ dominates. We therefore regularly ignore lower order terms and use **number of operations** both for the exact count and for the highest order term. We also say more loosely that the number of operations is $O(n^3)$. We will use the number of operations counted in one of these ways as a measure of the **complexity of an algorithm** and say that the complexity of LU factorization of a full matrix is $O(n^3)$ or more precisely $\frac{2}{3}n^3$.

We will compare the number of arithmetic operations of many algorithms with the number of arithmetic operations of Gaussian elimination and define for $n \in \mathbb{N}$ the number G_n as follows:

Definition 3.2 ($G_n := \frac{2}{3}n^3$) We define $G_n := \frac{2}{3}n^3$.

There is a quick way to arrive at the leading term $2n^3/3$. We only consider the operations contributing to this term. In (3.2) the leading term comes from the inner loop contributing to $M+S$. Then we replace sums by integrals letting the summation indices be continuous variables and adjust limits of integration in an insightful way to simplify the calculation. Thus,

$$M + S = 2\sum_{k=1}^{n-1}(n-k)^2 \approx 2\int_1^{n-1}(n-k)^2 dk \approx 2\int_0^n (n-k)^2 dk = \frac{2}{3}n^3$$

and this is the correct leading term.

Consider next N_S, the number of forward plus backward substitutions. By (3.7) we obtain

$$N_S = 2\sum_{k=1}^{n}(2k-1) \approx 2\int_1^n (2k-1)dk \approx 4\int_0^n kdk = 2n^2.$$

The last integral actually give the exact value for the sum in this case (cf. (3.26)).

We see that LU factorization is an $O(n^3)$ process while solving a triangular system requires $O(n^2)$ arithmetic operations. Thus, if $n = 10^6$ and one arithmetic operation requires $c = 10^{-14}$ seconds of computing time then $cn^3 = 10^4$ seconds ≈ 3 hours and $cn^2 = 0.01$ second, giving dramatic differences in computing time.

3.4 The PLU Factorization

Theorem 3.1 shows that Gaussian elimination without row interchanges can fail on a nonsingular system. A simple example is $\begin{bmatrix} 0 & 1 \\ 1 & 1 \end{bmatrix}\begin{bmatrix} x_1 \\ x_2 \end{bmatrix} = \begin{bmatrix} 1 \\ 1 \end{bmatrix}$. We show here that any nonsingular linear system can be solved by Gaussian elimination if we incorporate row interchanges.

3.4.1 *Pivoting*

Interchanging two rows (and/or two columns) during Gaussian elimination is known as **pivoting**. The element which is moved to the diagonal position (k, k) is called the **pivot element** or **pivot** for short, and the row containing the pivot is called the **pivot row**. Gaussian elimination with row pivoting can be described as follows.

1. **Choose** $r_k \geq k$ so that $a_{r_k,k}^{(k)} \neq 0$.
2. **Interchange** rows r_k and k of $\boldsymbol{A}^{(k)}$.
3. **Eliminate** by computing $l_{ik}^{(k)}$ and $a_{ij}^{(k+1)}$ using (3.2).

To show that Gaussian elimination can always be carried to completion by using suitable row interchanges suppose by induction on k that $\boldsymbol{A}^{(k)}$ is nonsingular. Since $\boldsymbol{A}^{(1)} = \boldsymbol{A}$ this holds for $k = 1$. By Lemma 2.4 the lower right diagonal block in $\boldsymbol{A}^{(k)}$ is nonsingular. But then at least one element in the first column of that block must be nonzero and it follows that r_k exists so that $a_{r_k,k}^{(k)} \neq 0$. But then $\boldsymbol{A}^{(k+1)}$ is nonsingular since it is computed from $\boldsymbol{A}^{(k)}$ using row operations preserving the nonsingularity. We conclude that $\boldsymbol{A}^{(k)}$ is nonsingular for $k = 1, \ldots, n$.

3.4.2 *Permutation Matrices*

Row interchanges can be described in terms of permutation matrices.

Definition 3.3 A **permutation matrix** is a matrix of the form

$$\boldsymbol{P} = \boldsymbol{I}(:, \boldsymbol{p}) = [\boldsymbol{e}_{i_1}, \boldsymbol{e}_{i_2}, \ldots, \boldsymbol{e}_{i_n}] \in \mathbb{R}^{n \times n},$$

where $\boldsymbol{e}_{i_1}, \ldots, \boldsymbol{e}_{i_n}$ is a permutation of the unit vectors $\boldsymbol{e}_1, \ldots, \boldsymbol{e}_n \in \mathbb{R}^n$.

Every permutation $\boldsymbol{p} = [i_1, \ldots, i_n]^T$ of the integers $1, 2, \ldots, n$ gives rise to a permutation matrix and vice versa. Post-multiplying a matrix $\boldsymbol{A}$ by a permutation matrix results in a permutation of the columns, while pre-multiplying by a permutation matrix gives a permutation of the rows. In symbols

$$\boldsymbol{A}\boldsymbol{P} = \boldsymbol{A}(:, \boldsymbol{p}), \quad \boldsymbol{P}^T\boldsymbol{A} = \boldsymbol{A}(\boldsymbol{p}, :). \tag{3.10}$$

Indeed, $\boldsymbol{AP} = (\boldsymbol{Ae}_{i_1}, \ldots, \boldsymbol{Ae}_{i_n}) = \boldsymbol{A}(:, \boldsymbol{p})$ and $\boldsymbol{P}^T\boldsymbol{A} = (\boldsymbol{A}^T\boldsymbol{P})^T = (\boldsymbol{A}^T(:, \boldsymbol{p}))^T = \boldsymbol{A}(\boldsymbol{p}, :)$.

Since $\boldsymbol{P}^T\boldsymbol{P} = \boldsymbol{I}$ the inverse of $\boldsymbol{P}$ is equal to its transpose, $\boldsymbol{P}^{-1} = \boldsymbol{P}^T$ and $\boldsymbol{P}\boldsymbol{P}^T = \boldsymbol{I}$ as well. We will use a particularly simple permutation matrix.

Definition 3.4 We define a (j, k)-**Interchange matrix** $\boldsymbol{I}_{jk}$ by interchanging column j and k of the identity matrix.

Since $\boldsymbol{I}_{jk} = \boldsymbol{I}_{kj}$, and we obtain the identity by applying $\boldsymbol{I}_{jk}$ twice, we see that $\boldsymbol{I}_{jk}^2 = \boldsymbol{I}$ and an interchange matrix is symmetric and equal to its own inverse. Premultiplying a matrix by an interchange matrix interchanges two rows of the matrix, while post-multiplication interchanges two columns.

We can keep track of the row interchanges using **pivot vectors** $\boldsymbol{p}_k$. We define

$$\boldsymbol{p} := \boldsymbol{p}_n, \text{ where } \boldsymbol{p}_1 := [1, 2, \ldots, n]^T, \text{ and } \boldsymbol{p}_{k+1} := \boldsymbol{I}_{r_k,k}\boldsymbol{p}_k \text{ for } k = 1, \ldots, n-1. \tag{3.11}$$

We obtain $\boldsymbol{p}_{k+1}$ from $\boldsymbol{p}_k$ by interchanging the entries r_k and k in $\boldsymbol{p}_k$. In particular, since $r_k \geq k$, the first $k-1$ components in $\boldsymbol{p}_k$ and $\boldsymbol{p}_{k+1}$ are the same.

There is a close relation between the pivot vectors $\boldsymbol{p}_k$ and the corresponding interchange matrices $\boldsymbol{P}_k := \boldsymbol{I}_{r_k,k}$. Since $\boldsymbol{P}_k\boldsymbol{I}(\boldsymbol{p}_k, :) = \boldsymbol{I}(\boldsymbol{P}_k\boldsymbol{p}_k, :) = \boldsymbol{I}(\boldsymbol{p}_{k+1}, :)$ we obtain

$$\boldsymbol{P}^T = \boldsymbol{P}_{n-1}\cdots\boldsymbol{P}_1 = \boldsymbol{I}(\boldsymbol{p}, :), \quad \boldsymbol{P} = \boldsymbol{P}_1\boldsymbol{P}_2\cdots\boldsymbol{P}_{n-1} = \boldsymbol{I}(:, \boldsymbol{p}). \tag{3.12}$$

Instead of interchanging the rows of $\boldsymbol{A}$ during elimination we can keep track of the ordering of the rows using the pivot vectors $\boldsymbol{p}_k$. Gaussian elimination with row pivoting starting with $a_{ij}^{(1)} = a_{ij}$ can be described as follows:

$$\begin{array}{|l|}\hline \boldsymbol{p} = [1, \ldots, n]^T; \\ \text{for } k = 1 : n-1 \\ \quad \text{choose } r_k \geq k \text{ so that } a_{p_{r_k},k}^{(k)} \neq 0. \\ \quad \boldsymbol{p} = I_{r_k,k}\boldsymbol{p} \\ \quad \text{for } i = k+1 : n \\ \quad\quad a_{p_i,k}^{(k)} = a_{p_i,k}^{(k)}/a_{p_k,k}^{(k)} \\ \quad\quad \text{for } j = k : n \\ \quad\quad\quad a_{p_i,j}^{(k+1)} = a_{p_i,j}^{(k)} - a_{p_i,k}^{(k)}a_{p_k,j}^{(k)} \\ \hline \end{array} \tag{3.13}$$

This leads to the following factorization:

Theorem 3.3 *Gaussian elimination with row pivoting on a nonsingular matrix* $\boldsymbol{A} \in \mathbb{C}^{n\times n}$ *leads to the factorization* $\boldsymbol{A} = \boldsymbol{PLU}$, *where* $\boldsymbol{P}$ *is a permutation matrix,* $\boldsymbol{L}$ *is lower triangular with ones on the diagonal, and* $\boldsymbol{U}$ *is upper triangular. More explicitly,* $\boldsymbol{P} = \boldsymbol{I}(:,\boldsymbol{p})$, *where* $\boldsymbol{p} = \boldsymbol{I}_{r_{n-1},n-1}\cdots\boldsymbol{I}_{r_1,1}[1,\ldots,n]^T$, *and*

$$\boldsymbol{L} = \begin{bmatrix} 1 & & & \\ a^{(1)}_{p_2,1} & 1 & & \\ \vdots & & \ddots & \\ a^{(1)}_{p_n,1} & a^{(2)}_{p_n,2} & \cdots & 1 \end{bmatrix}, \quad \boldsymbol{U} = \begin{bmatrix} a^{(1)}_{p_1,1} & \cdots & a^{(1)}_{p_1,n} \\ & \ddots & \vdots \\ & & a^{(n)}_{p_n,n} \end{bmatrix}. \tag{3.14}$$

Proof The proof is analogous to the proof for LU factorization without pivoting. From (3.13) we have for all i, j

$$a^{(k)}_{p_i,k}a^{(k)}_{p_k,j} = a^{(k)}_{p_i,j} - a^{(k+1)}_{p_i,j} \text{ for } k < \min(i,j), \text{ and } a^{(k)}_{p_i,j}a^{(j)}_{p_j,j} = a^{(j)}_{p_i,j} \text{ for } i > j.$$

Thus for $i \le j$ we find

$$\begin{aligned}(\boldsymbol{LU})_{ij} &= \sum_{k=1}^{n} l_{i,k}u_{kj} = \sum_{k=1}^{i-1} a^{(k)}_{p_i,k}a^{(k)}_{p_k,j} + a^{(i)}_{p_i,j} \\ &= \sum_{k=1}^{i-1}\left(a^{(k)}_{p_i,j} - a^{(k+1)}_{p_i,j}\right) + a^{(i)}_{p_i,j} = a^{(1)}_{p_i,j} = a_{p_i,j} = \left(\boldsymbol{P}^T\boldsymbol{A}\right)_{ij},\end{aligned}$$

while for $i > j$

$$\begin{aligned}(\boldsymbol{LU})_{ij} &= \sum_{k=1}^{n} l^{(k)}_{ik}u_{kj} = \sum_{k=1}^{j-1} a^{(k)}_{p_i,k}a^{(k)}_{p_k,j} + a^{(k)}_{p_i,j}a^{(j)}_{p_j,j} \\ &= \sum_{k=1}^{j-1}\left(a^{(k)}_{p_i,j} - a^{(k+1)}_{p_i,j}\right) + a^{(j)}_{p_i,j} = a^{(1)}_{p_i,j} = a_{p_i,j} = \left(\boldsymbol{P}^T\boldsymbol{A}\right)_{ij}.\end{aligned}$$

□

The PLU factorization can also be written $\boldsymbol{P}^T\boldsymbol{A} = \boldsymbol{LU}$. This shows that for a nonsingular matrix there is a permutation of the rows of $\boldsymbol{A}$ so that the permuted matrix has an LU factorization.

3.4.3 Pivot Strategies

The choice of pivot element in (3.13) is not unique. In **partial pivoting** we select the largest element

$$|a_{r_k,k}^{(k)}| := \max\{|a_{i,k}^{(k)}| : k \le i \le n\}$$

with r_k the smallest such index in case of a tie. The following example illustrating that small pivots should be avoided.

Example 3.4 Applying Gaussian elimination without row interchanges to the linear system

$$\begin{aligned} 10^{-4}x_1 + 2x_2 &= 4 \\ x_1 + x_2 &= 3 \end{aligned}$$

we obtain the upper triangular system

$$\begin{aligned} 10^{-4}x_1 + 2x_2 &= 4 \\ (1 - 2 \times 10^4)x_2 &= 3 - 4 \times 10^4 \end{aligned}$$

The exact solution is

$$x_2 = \frac{-39997}{-19999} \approx 2, \quad x_1 = \frac{4 - 2x_2}{10^{-4}} = \frac{20000}{19999} \approx 1.$$

Suppose we round the result of each arithmetic operation to three digits. The solutions $\mathrm{fl}(x_1)$ and $\mathrm{fl}(x_2)$ computed in this way is

$$\mathrm{fl}(x_2) = 2, \quad \mathrm{fl}(x_1) = 0.$$

The computed value 0 of x_1 is completely wrong. Suppose instead we apply Gaussian elimination to the same system, but where we have interchanged the equations. The system is

$$\begin{aligned} x_1 + x_2 &= 3 \\ 10^{-4}x_1 + 2x_2 &= 4 \end{aligned}$$

and we obtain the upper triangular system

$$\begin{aligned} x_1 + x_2 &= 3 \\ (2 - 10^{-4})x_2 &= 4 - 3 \times 10^{-4} \end{aligned}$$

Now the solution is computed as follows

$$x_2 = \frac{3.9997}{1.9999} \approx 2, \quad x_1 = 3 - x_2 \approx 1.$$

In this case rounding each calculation to three digits produces $\text{fl}(x_1) = 1$ and $\text{fl}(x_2) = 2$ which is quite satisfactory since it is the exact solution rounded to three digits.

Related to partial pivoting is **scaled partial pivoting**. Here r_k is the smallest index such that

$$\frac{|a_{r_k,k}^{(k)}|}{s_k} := \max\left\{\frac{|a_{i,k}^{(k)}|}{s_k} : k \le i \le n\right\}, \quad s_k := \max_{1\le j\le n} |a_{kj}|.$$

This can sometimes give more accurate results if the coefficient matrix have coefficients of wildly different sizes. Note that the scaling factors s_k are computed using the initial matrix.

It also is possible to interchange both rows and columns. The choice

$$a_{r_k,s_k}^{(k)} := \max\{|a_{i,j}^{(k)}| : k \le i, j \le n\}$$

with r_k, s_k the smallest such indices in case of a tie, is known as **complete pivoting**. Complete pivoting is known to be more numerically stable than partial pivoting, but requires a lot of search and is seldom used in practice.

3.5 The LU and LDU Factorizations

Gaussian elimination without row interchanges is one way of computing an LU factorization of a matrix. There are other ways that can be advantageous for certain kind of problems. Here we consider the general theory of LU factorizations. Recall that $\boldsymbol{A} = \boldsymbol{L}\boldsymbol{U}$ is an **LU factorization** of $\boldsymbol{A} \in \mathbb{C}^{n\times n}$ if $\boldsymbol{L} \in \mathbb{C}^{n\times n}$ is lower triangular and $\boldsymbol{U} \in \mathbb{C}^{n\times n}$ is upper triangular , i.e.,

$$\boldsymbol{L} = \begin{bmatrix} l_{1,1} & \cdots & 0 \\ \vdots & \ddots & \vdots \\ l_{n,1} & \cdots & l_{n,n} \end{bmatrix}, \quad \boldsymbol{U} = \begin{bmatrix} u_{1,1} & \cdots & u_{1,n} \\ \vdots & \ddots & \vdots \\ 0 & \cdots & u_{n,n} \end{bmatrix}.$$

To find an LU factorization there is one equation for each of the n^2 elements in $\boldsymbol{A}$, and $\boldsymbol{L}$ and $\boldsymbol{U}$ contain a total of $n^2 + n$ unknown elements. There are several ways to restrict the number of unknowns to n^2.

L1U: $l_{ii} = 1$ all i,
LU1: $u_{ii} = 1$ all i,
LDU: $\boldsymbol{A} = \boldsymbol{LDU}$, $l_{ii} = u_{ii} = 1$ all i, $\boldsymbol{D} = \operatorname{diag}(d_{11}, \ldots, d_{nn})$.

3.5.1 Existence and Uniqueness

Consider the L1U factorization. Three things can happen. An L1U factorization exists and is unique, it exists, but it is not unique, or it does not exist. The 2×2 case illustrates this.

Example 3.5 (L1U of 2×2 *Matrix)* Let $a, b, c, d \in \mathbb{C}$. An L1U factorization of $\boldsymbol{A} = \left[\begin{smallmatrix} a & b \\ c & d \end{smallmatrix}\right]$ must satisfy the equations

$$\begin{bmatrix} a & b \\ c & d \end{bmatrix} = \begin{bmatrix} 1 & 0 \\ l_1 & 1 \end{bmatrix} \begin{bmatrix} u_1 & u_2 \\ 0 & u_3 \end{bmatrix} = \begin{bmatrix} u_1 & u_2 \\ u_1 l_1 & u_2 l_1 + u_3 \end{bmatrix}$$

for the unknowns l_1 in $\boldsymbol{L}$ and u_1, u_2, u_3 in $\boldsymbol{U}$. The equations are

$$u_1 = a, \quad u_2 = b, \quad a l_1 = c, \quad b l_1 + u_3 = d. \tag{3.15}$$

These equations do not always have a solution. Indeed, the main problem is the equation $a l_1 = c$. There are essentially three cases

1. $a \neq 0$: The matrix has a unique L1U factorization.
2. $a = c = 0$: The L1U factorization exists, but it is not unique. Any value for l_1 can be used.
3. $a = 0$, $c \neq 0$: No L1U factorization exists.

Consider the four matrices

$$\boldsymbol{A}_1 := \begin{bmatrix} 2 & -1 \\ -1 & 2 \end{bmatrix}, \quad \boldsymbol{A}_2 := \begin{bmatrix} 0 & 1 \\ 1 & 1 \end{bmatrix}, \quad \boldsymbol{A}_3 := \begin{bmatrix} 0 & 1 \\ 0 & 2 \end{bmatrix}, \quad \boldsymbol{A}_4 := \begin{bmatrix} 1 & 1 \\ 1 & 1 \end{bmatrix}.$$

From the previous discussion it follows that $\boldsymbol{A}_1$ has a unique L1U factorization, $\boldsymbol{A}_2$ has no L1U factorization, $\boldsymbol{A}_3$ has an L1U factorization but it is not unique, and $\boldsymbol{A}_4$ has a unique L1U factorization even if it is singular.

In preparation for the main theorem about LU factorization we prove a simple lemma. Recall that

$$\boldsymbol{A}_{[k]} := \begin{bmatrix} a_{11} & \cdots & a_{k1} \\ \vdots & & \vdots \\ a_{k1} & \cdots & a_{kk} \end{bmatrix}$$

is called a leading principal submatrix of $\boldsymbol{A}$.

Lemma 3.1 (L1U of Leading Principal Submatrices) *Suppose* $\boldsymbol{A} = \boldsymbol{LU}$ *is an L1U factorization of* $\boldsymbol{A} \in \mathbb{C}^{n\times n}$. *For* $k = 1, \ldots, n$ *let* $\boldsymbol{A}_{[k]}, \boldsymbol{L}_{[k]}, \boldsymbol{U}_{[k]}$ *be the leading principal submatrices of* $\boldsymbol{A}, \boldsymbol{L}, \boldsymbol{U}$, *respectively. Then* $\boldsymbol{A}_{[k]} = \boldsymbol{L}_{[k]}\boldsymbol{U}_{[k]}$ *is an L1U factorization of* $\boldsymbol{A}_{[k]}$ *for* $k = 1, \ldots, n$.

Proof For $k = 1, \ldots, n-1$ we partition $\boldsymbol{A} = \boldsymbol{LU}$ as follows:

$$\begin{bmatrix} \boldsymbol{A}_{[k]} & \boldsymbol{B}_k \\ \boldsymbol{C}_k & \boldsymbol{F}_k \end{bmatrix} = \begin{bmatrix} \boldsymbol{L}_{[k]} & \boldsymbol{0} \\ \boldsymbol{M}_k & \boldsymbol{N}_k \end{bmatrix} \begin{bmatrix} \boldsymbol{U}_{[k]} & \boldsymbol{S}_k \\ \boldsymbol{0} & \boldsymbol{T}_k \end{bmatrix} = \begin{bmatrix} \boldsymbol{L}_{[k]}\boldsymbol{U}_{[k]} & \boldsymbol{L}_{[k]}\boldsymbol{S}_k \\ \boldsymbol{M}_k\boldsymbol{U}_{[k]} & \boldsymbol{M}_k\boldsymbol{S}_k + \boldsymbol{N}_k\boldsymbol{T}_k \end{bmatrix}, \tag{3.16}$$

where $\boldsymbol{F}_k, \boldsymbol{N}_k, \boldsymbol{T}_k \in \mathbb{C}^{n-k,n-k}$. Comparing blocks we find $\boldsymbol{A}_{[k]} = \boldsymbol{L}_{[k]}\boldsymbol{U}_{[k]}$. Since $\boldsymbol{L}_{[k]}$ is unit lower triangular and $\boldsymbol{U}_{[k]}$ is upper triangular this is an L1U factorization of $\boldsymbol{A}_{[k]}$. □

The following theorem gives a necessary and sufficient condition for existence of a unique LU factorization. The conditions are the same for the three factorizations L1U, LU1 and LDU.

Theorem 3.4 (LU Theorem) *A square matrix* $\boldsymbol{A} \in \mathbb{C}^{n\times n}$ *has a unique L1U (LU1, LDU) factorization if and only if the leading principal submatrices* $\boldsymbol{A}_{[k]}$ *of* $\boldsymbol{A}$ *are nonsingular for* $k = 1, \ldots, n-1$.

Proof Suppose $\boldsymbol{A}_{[k]}$ is nonsingular for $k = 1, \ldots, n-1$. Under these conditions Gaussian elimination gives an L1U factorization (cf. Theorem 3.2). We give another proof here that in addition to showing uniqueness also gives alternative ways to compute the L1U factorization. The proofs for the LU1 and LDU factorizations are similar and left as exercises.

We use induction on n to show that $\boldsymbol{A}$ has a unique L1U factorization. The result is clearly true for $n = 1$, since the unique L1U factorization of a 1×1 matrix is $[a_{11}] = [1][a_{11}]$. Suppose that $\boldsymbol{A}_{[n-1]}$ has a unique L1U factorization $\boldsymbol{A}_{[n-1]} = \boldsymbol{L}_{n-1}\boldsymbol{U}_{n-1}$, and that $\boldsymbol{A}_{[1]}, \ldots, \boldsymbol{A}_{[n-1]}$ are nonsingular. By block multiplication

$$\boldsymbol{A} = \begin{bmatrix} \boldsymbol{A}_{[n-1]} & \boldsymbol{c}_n \\ \boldsymbol{r}_n^T & a_{nn} \end{bmatrix} = \begin{bmatrix} \boldsymbol{L}_{n-1} & \boldsymbol{0} \\ \boldsymbol{l}_n^T & 1 \end{bmatrix} \begin{bmatrix} \boldsymbol{U}_{n-1} & \boldsymbol{u}_n \\ 0 & u_{nn} \end{bmatrix} = \begin{bmatrix} \boldsymbol{L}_{n-1}\boldsymbol{U}_{n-1} & \boldsymbol{L}_{n-1}\boldsymbol{u}_n \\ \boldsymbol{l}_n^T\boldsymbol{U}_{n-1} & \boldsymbol{l}_n^T\boldsymbol{u}_n + u_{nn} \end{bmatrix}, \tag{3.17}$$

if and only if $\boldsymbol{A}_{[n-1]} = \boldsymbol{L}_{n-1}\boldsymbol{U}_{n-1}$ and $\boldsymbol{l}_n, \boldsymbol{u}_n \in \mathbb{C}^{n-1}$ and $u_{nn} \in \mathbb{C}$ are determined from

$$\boldsymbol{U}_{n-1}^T\boldsymbol{l}_n = \boldsymbol{r}_n, \quad \boldsymbol{L}_{n-1}\boldsymbol{u}_n = \boldsymbol{c}_n, \quad u_{nn} = a_{nn} - \boldsymbol{l}_n^T\boldsymbol{u}_n. \tag{3.18}$$

Since $\boldsymbol{A}_{[n-1]}$ is nonsingular it follows that $\boldsymbol{L}_{n-1}$ and $\boldsymbol{U}_{n-1}$ are nonsingular and therefore $\boldsymbol{l}_n, \boldsymbol{u}_n$, and u_{nn} are uniquely given. Thus (3.17) gives a unique L1U factorization of $\boldsymbol{A}$.

Conversely, suppose $\boldsymbol{A}$ has a unique L1U factorization $\boldsymbol{A} = \boldsymbol{LU}$. By Lemma 3.1 $\boldsymbol{A}_{[k]} = \boldsymbol{L}_{[k]}\boldsymbol{U}_{[k]}$ is an L1U factorization of $\boldsymbol{A}_{[k]}$ for $k = 1, \ldots, n-1$. Suppose

$A_{[k]}$ is singular for some $k \le n-1$. We will show that this leads to a contradiction. Let k be the smallest integer so that $A_{[k]}$ is singular. Since $A_{[j]}$ is nonsingular for $j \le k-1$ it follows from what we have already shown that $A_{[k]} = L_{[k]} U_{[k]}$ is the unique L1U factorization of $A_{[k]}$. The matrix $U_{[k]}$ is singular since $A_{[k]}$ is singular and $L_{[k]}$ is nonsingular. By (3.16) we have $U_{[k]}^T M_k^T = C_k^T$. This can be written as $n-k$ linear systems for the columns of M_k^T. By assumption M_k^T exists, but since $U_{[k]}^T$ is singular M_k is not unique, a contradiction. □

By combining the last two equations in (3.18) we obtain with $k = n$

$$U_{k-1}^T l_k = r_k, \quad \begin{bmatrix} L_{k-1} & 0 \\ l_k^T & 1 \end{bmatrix} \begin{bmatrix} u_k \\ u_{kk} \end{bmatrix} = \begin{bmatrix} c_k \\ a_{kk} \end{bmatrix}.$$

This can be used in an algorithm to compute the L1U factorization. Moreover, if A is d-banded then the first $k-d$ components in r_k and c_k are zero so both L and U will be d-banded. Thus we can use the banded `rforwardsolve` Algorithm 3.1 to solve the lower triangular system $U_{k-1}^T l_k = r_k$ for the kth row l_k^T in L and the kth column $\left[\begin{smallmatrix} u_k \\ u_{kk} \end{smallmatrix}\right]$ in U for $k = 2, \ldots, n$. This leads to the following algorithm to compute the L1U factorization of a d-banded matrix A with $d \ge 1$. The algorithm will fail if the conditions in the LU theorem are not satisfied.

```
function [L,U]=L1U(A,d)
% [L,U]=L1U(A,d)
n=length(A);
L=eye(n,n); U=zeros(n,n);U(1,1)=A(1,1);
for k=2:n
  km=max(1,k-d);
  L(k,km:(k-1))=rforwardsolve(U(km:(k-1) ...
  ,km:(k-1))',A(k,km:(k-1))',d)';
  U(km:k,k)=rforwardsolve(L(km:k,km:k),A(km:k,k),d);
end
```

Listing 3.4 L1U

For each k we essentially solve a lower triangular linear system of order d. Thus the number of arithmetic operation for this algorithm is $O(d^2 n)$.

Remark 3.1 (LU of Upper Triangular Matrix) A matrix $A \in \mathbb{C}^{n \times n}$ can have an LU factorization even if $A_{[k]}$ is singular for some $k < n$. By Theorem 4.1 such an LU factorization cannot be unique. An L1U factorization of an upper triangular matrix A is $A = IA$ so it always exists even if A has zeros somewhere on the diagonal. By Lemma 2.5, if some a_{kk} is zero then $A_{[k]}$ is singular and the L1U factorization is not unique. In particular, for the zero matrix any unit lower triangular matrix can be used as L in an L1U factorization.

3.6 Block LU Factorization

Suppose $\boldsymbol{A} \in \mathbb{C}^{n\times n}$ is a block matrix of the form

$$\boldsymbol{A} := \begin{bmatrix} \boldsymbol{A}_{11} & \cdots & \boldsymbol{A}_{1m} \\ \vdots & & \vdots \\ \boldsymbol{A}_{m1} & \cdots & \boldsymbol{A}_{mm} \end{bmatrix}, \tag{3.19}$$

where each diagonal block $\boldsymbol{A}_{ii}$ is square. We call the factorization

$$\boldsymbol{A} = \boldsymbol{L}\boldsymbol{U} = \begin{bmatrix} \boldsymbol{I} & & & \\ \boldsymbol{L}_{21} & \boldsymbol{I} & & \\ \vdots & & \ddots & \\ \boldsymbol{L}_{m1} & \cdots & \boldsymbol{L}_{m,m-1} & \boldsymbol{I} \end{bmatrix} \begin{bmatrix} \boldsymbol{U}_{11} & & \cdots & \boldsymbol{U}_{1m} \\ & \boldsymbol{U}_{22} & \cdots & \boldsymbol{U}_{2m} \\ & & \ddots & \vdots \\ & & & \boldsymbol{U}_{mm} \end{bmatrix} \tag{3.20}$$

a **block L1U factorization of** A. Here the ith diagonal blocks $\boldsymbol{I}$ and $\boldsymbol{U}_{ii}$ in $\boldsymbol{L}$ and $\boldsymbol{U}$ have the same size as $\boldsymbol{A}_{ii}$, the ith diagonal block in $\boldsymbol{A}$. Moreover, the $\boldsymbol{U}_{ii}$ are not necessarily upper triangular. Block LU1 and block LDU factorizations are defined similarly.

The results for element-wise LU factorization carry over to block LU factorization as follows.

Theorem 3.5 (Block LU Theorem) *Suppose $\boldsymbol{A} \in \mathbb{C}^{n\times n}$ is a block matrix of the form* (3.19). *Then $\boldsymbol{A}$ has a unique block LU factorization* (3.20) *if and only if the* **leading principal block submatrices**

$$\boldsymbol{A}_{\{k\}} := \begin{bmatrix} \boldsymbol{A}_{11} & \cdots & \boldsymbol{A}_{1k} \\ \vdots & & \vdots \\ \boldsymbol{A}_{k1} & \cdots & \boldsymbol{A}_{kk} \end{bmatrix}$$

are nonsingular for $k = 1, \ldots, m-1$.

Proof Suppose $\boldsymbol{A}_{\{k\}}$ is nonsingular for $k = 1, \ldots, m-1$. Following the proof in Theorem 3.4 suppose $\boldsymbol{A}_{\{m-1\}}$ has a unique block LU factorization $\boldsymbol{A}_{\{m-1\}} = \boldsymbol{L}_{\{m-1\}}\boldsymbol{U}_{\{m-1\}}$, and that $\boldsymbol{A}_{\{1\}}, \ldots, \boldsymbol{A}_{\{m-1\}}$ are nonsingular. Then $\boldsymbol{L}_{\{m-1\}}$ and $\boldsymbol{U}_{\{m-1\}}$ are nonsingular and

$$\begin{aligned} \boldsymbol{A} &= \begin{bmatrix} \boldsymbol{A}_{\{m-1\}} & \boldsymbol{B} \\ \boldsymbol{C}^T & \boldsymbol{A}_{mm} \end{bmatrix} \\ &= \begin{bmatrix} \boldsymbol{L}_{\{m-1\}} & \boldsymbol{0} \\ \boldsymbol{C}^T\boldsymbol{U}_{\{m-1\}}^{-1} & \boldsymbol{I} \end{bmatrix} \begin{bmatrix} \boldsymbol{U}_{\{m-1\}} & \boldsymbol{L}_{\{m-1\}}^{-1}\boldsymbol{B} \\ 0 & \boldsymbol{A}_{mm} - \boldsymbol{C}^T\boldsymbol{U}_{\{m-1\}}^{-1}\boldsymbol{L}_{\{m-1\}}^{-1}\boldsymbol{B} \end{bmatrix}, \end{aligned} \tag{3.21}$$

is a block LU factorization of $\boldsymbol{A}$. It is unique by derivation. Conversely, suppose $\boldsymbol{A}$ has a unique block LU factorization $\boldsymbol{A} = \boldsymbol{L}\boldsymbol{U}$. Then as in Lemma 3.1 it is easily seen that $\boldsymbol{A}_{\{k\}} = \boldsymbol{L}_{\{k\}}\boldsymbol{U}_{\{k\}}$ is the unique block LU factorization of $\boldsymbol{A}_{[k]}$ for $k = 1, \ldots, m$. The rest of the proof is similar to the proof of Theorem 3.4. □

Remark 3.2 (Comparing LU and Block LU) The number of arithmetic operations for the block LU factorization is the same as for the ordinary LU factorization. An advantage of the block method is that it combines many of the operations into matrix operations.

Remark 3.3 (A Block LU Is Not an LU) Note that (3.20) is not an LU factorization of $\boldsymbol{A}$ since the $\boldsymbol{U}_{ii}$'s are not upper triangular in general. To relate the block LU factorization to the usual LU factorization we assume that each $\boldsymbol{U}_{ii}$ has an LU factorization $\boldsymbol{U}_{ii} = \tilde{\boldsymbol{L}}_{ii}\tilde{\boldsymbol{U}}_{ii}$. Then $\boldsymbol{A} = \hat{\boldsymbol{L}}\hat{\boldsymbol{U}}$, where $\hat{\boldsymbol{L}} := \boldsymbol{L}\,\mathrm{diag}(\tilde{\boldsymbol{L}}_{ii})$ and $\hat{\boldsymbol{U}} := \mathrm{diag}(\tilde{\boldsymbol{L}}_{ii}^{-1})\boldsymbol{U}$, and this is an ordinary LU factorization of $\boldsymbol{A}$.

3.7 Exercises Chap. 3

3.7.1 Exercises Sect. 3.3

Exercise 3.1 (Column Oriented Backsolve) Suppose $\boldsymbol{A} \in \mathbb{C}^{n\times n}$ is nonsingular, upper triangular, d-banded, and $\boldsymbol{b} \in \mathbb{C}^n$. Justify the following column oriented vectorized algorithm for solving $\boldsymbol{A}\boldsymbol{x} = \boldsymbol{b}$.

```
function x=cbacksolve(A,b,d)
% x=cbacksolve(A,b,d)
x=b; n=length(b);
for k=n:-1:2
   x(k)=b(k)/A(k,k); lk=max(1,k-d);
   b(lk:(k-1))=b(lk:(k-1))-A(lk:(k-1),k)*x(k);
end
x(1)=b(1)/A(1,1);
end
```

Listing 3.5 cbacksolve

Exercise 3.2 (Computing the Inverse of a Triangular Matrix) Suppose $\boldsymbol{A} \in \mathbb{C}^{n\times n}$ is a nonsingular lower triangular matrix. By Lemma 2.5 the inverse $\boldsymbol{B} = [\boldsymbol{b}_1, \ldots, \boldsymbol{b}_n]$ is also lower triangular. The kth column $\boldsymbol{b}_k$ of $\boldsymbol{B}$ is the solution of the linear systems $\boldsymbol{A}\boldsymbol{b}_k = \boldsymbol{e}_k$. Show that $b_k(k) = 1/a(k,k)$ for $k = 1, \ldots, n$, and explain why we can find $\boldsymbol{b}_k$ by solving the linear systems

$$\boldsymbol{A}((k+1):n, (k+1):n)\boldsymbol{b}_k((k+1):n) = -\boldsymbol{A}((k+1):n, k)b_k(k), \quad k = 1, \ldots, n-1. \tag{3.22}$$

Is it possible to store the interesting part of $\boldsymbol{b}_k$ in $\boldsymbol{A}$ as soon as it is computed?

When $\boldsymbol{A}$ instead is upper triangular, show also that we can find $\boldsymbol{b}_k$ by solving the linear systems

$$\boldsymbol{A}(1{:}k, 1{:}k)\boldsymbol{b}_k(1{:}k) = \boldsymbol{I}(1{:}k, k), \quad k = n, n-1, \ldots, 1, \tag{3.23}$$

for $k = n, n-1, \ldots, 1$.

Exercise 3.3 (Finite Sums of Integers) Use induction on m, or some other method, to show that

$$1 + 2 + \cdots + m = \frac{1}{2}m(m+1), \tag{3.24}$$

$$1^2 + 2^2 + \cdots + m^2 = \frac{1}{3}m(m+\frac{1}{2})(m+1), \tag{3.25}$$

$$1 + 3 + 5 + \cdots + 2m - 1 = m^2, \tag{3.26}$$

$$1 * 2 + 2 * 3 + 3 * 4 + \cdots + (m-1)m = \frac{1}{3}(m-1)m(m+1). \tag{3.27}$$

Exercise 3.4 (Multiplying Triangular Matrices) Show that the matrix multiplication $\boldsymbol{AB}$ can be done in $\frac{1}{3}n(2n^2+1) \approx G_n$ arithmetic operations when $\boldsymbol{A} \in \mathbb{R}^{n\times n}$ is lower triangular and $\boldsymbol{B} \in \mathbb{R}^{n\times n}$ is upper triangular. What about $\boldsymbol{BA}$?

3.7.2 Exercises Sect. 3.4

Exercise 3.5 (Using PLU for A^*) Suppose we know the PLU factors $\boldsymbol{P}, \boldsymbol{L}, \boldsymbol{U}$ in a PLU factorization $\boldsymbol{A} = \boldsymbol{PLU}$ of $\boldsymbol{A} \in \mathbb{C}^{n\times n}$. Explain how we can solve the system $\boldsymbol{A}^*\boldsymbol{x} = \boldsymbol{b}$ economically.

Exercise 3.6 (Using PLU for Determinant) Suppose we know the PLU factors $\boldsymbol{P}, \boldsymbol{L}, \boldsymbol{U}$ in a PLU factorization $\boldsymbol{A} = \boldsymbol{PLU}$ of $\boldsymbol{A} \in \mathbb{C}^{n\times n}$. Explain how we can use this to compute the determinant of $\boldsymbol{A}$.

Exercise 3.7 (Using PLU for A^{-1}) Suppose the factors $\boldsymbol{P}, \boldsymbol{L}, \boldsymbol{U}$ in a PLU factorization of $\boldsymbol{A} \in \mathbb{C}^{n\times n}$ are known. Use Exercise 3.4 to show that it takes approximately $2G_n$ arithmetic operations to compute $\boldsymbol{A}^{-1} = \boldsymbol{U}^{-1}\boldsymbol{L}^{-1}\boldsymbol{P}^T$. Here we have not counted the final multiplication with $\boldsymbol{P}^T$ which amounts to n row interchanges.

Exercise 3.8 (Upper Hessenberg System (Exam Exercise (1994-2)) Gaussian elimination with row pivoting can be written in the following form if for each k we exchange rows k and $k+1$

Algorithm 1

1. for $k = 1, 2, \ldots, n-1$
   (a) exchange $a_{k,j}$ and $a_{k+1,j}$ for $j = k, k+1, \ldots, n$
   (b) for $i = k+1, k+2, \ldots, n$
      i. $a_{i,k} = m_{i,k} = a_{i,k}/a_{k,k}$
      ii. $a_{i,j} = a_{i,j} - m_{i,k}a_{k,j}$ for $j = k+1, k+2, \ldots, n$

To solve the set of equations $\boldsymbol{Ax} = \boldsymbol{b}$ we have the following algorithm:

Algorithm 2

1. for $k = 1, 2, \ldots, n-1$
   (a) exchange b_k and b_{k+1}
   (b) $b_i = b_i - a_{i,k}b_k$ for $i = k+1, k+2, \ldots, n$
2. $x_n = b_n/a_{n,n}$
3. for $k = n-1, n-2, \ldots, 1$
   (a) $sum = 0$
   (b) $sum = sum + a_{k,j}x_j$ for $j = k+1, k+2, \ldots, n$
   (c) $x_k = (b_k - sum)/a_{k,k}$

We say that $\boldsymbol{H} \in \mathbb{R}^{n\times n}$ is **unreduced upper Hessenberg** if it is upper Hessenberg and the subdiagonal elements $h_{i,i-1} \neq 0$ for $i = 2, \ldots, n$.

a) Let $\boldsymbol{H} \in \mathbb{R}^{n\times n}$ be unreduced upper Hessenberg. Give an $O(n^2)$ algorithm for solving the linear system $\boldsymbol{Hx} = \boldsymbol{b}$ using suitable specializations of Algorithms 1 and 2.

b) Find the number of multiplications/divisions in the algorithm you developed in exercise a). Is division by zero possible?

c) Let $\boldsymbol{U} \in \mathbb{R}^{n\times n}$ be upper triangular and nonsingular. We define

$$\boldsymbol{C} := \boldsymbol{U} + \boldsymbol{v}\boldsymbol{e}_1^T, \tag{3.28}$$

where $\boldsymbol{v} \in \mathbb{R}^n$ and $\boldsymbol{e}_1$ is the first unit vector in $\mathbb{R}^n$. We also let

$$\boldsymbol{P} := \boldsymbol{I}_{1,2}\boldsymbol{I}_{2,3}\cdots\boldsymbol{I}_{n-1,n}, \tag{3.29}$$

where the $\boldsymbol{I}_{i,j}$ are obtained from the identity matrix by interchanging rows i and j. Explain why the matrix $\boldsymbol{E} := \boldsymbol{CP}$ is unreduced upper Hessenberg.

d) Let $\boldsymbol{A} \in \mathbb{R}^{n\times n}$ be nonsingular. We assume that $\boldsymbol{A}$ has a unique L1U factorization $\boldsymbol{A} = \boldsymbol{L}\boldsymbol{U}$. To a given $\boldsymbol{W} \in \mathbb{R}^n$ we define a rank one modification of $\boldsymbol{A}$ by

$$\boldsymbol{B} := \boldsymbol{A} + \boldsymbol{w}\boldsymbol{e}_1^T. \tag{3.30}$$

Show that $\boldsymbol{B}$ has the factorization $\boldsymbol{B} = \boldsymbol{L}\boldsymbol{H}\boldsymbol{P}^T$, where $\boldsymbol{L}$ is unit lower triangular, $\boldsymbol{P}$ is given by (3.29) and $\boldsymbol{H}$ is unreduced upper Hessenberg.

e) Use the results above to sketch an $O(n^2)$ algorithm for solving the linear system $\boldsymbol{B}\boldsymbol{x} = \boldsymbol{b}$, where $\boldsymbol{B}$ is given by (3.30). We assume that the matrices $\boldsymbol{L}$ and $\boldsymbol{U}$ in the L1U factorization of $\boldsymbol{A}$ have already been computed.

3.7.3 Exercises Sect. 3.5

Exercise 3.9 (# Operations for Banded Triangular Systems) Show that for $1 \le d \le n$ Algorithm 3.4, with $A(k,k) = 1$ for $k = 1, \ldots, n$ in Algorithm 3.1, requires exactly $N_{LU}(n,d) := (2d^2 + d)n - (d^2 + d)(8d + 1)/6 = O(d^2 n)$ operations.[2] In particular, for a full matrix $d = n - 1$ and we find $N_{LU}(n,n) = \frac{2}{3}n^3 - \frac{1}{2}n^2 - \frac{1}{6}n \approx G_n$ in agreement with the exact count (3.9) for Gaussian elimination, while for a tridiagonal matrix $N_{LU}(n,1) = 3n - 3 = O(n)$.

Exercise 3.10 (L1U and LU1) Show that the matrix $\boldsymbol{A}_3$ in Example 3.5 has no LU1 or LDU factorization. Give an example of a matrix that has an LU1 factorization, but no LDU or L1U factorization.

Exercise 3.11 (LU of Nonsingular Matrix) Show that the following are equivalent for a nonsingular matrix $\boldsymbol{A} \in \mathbb{C}^{n\times n}$.

1. $\boldsymbol{A}$ has an LDU factorization.
2. $\boldsymbol{A}$ has an L1U factorization.
3. $\boldsymbol{A}$ has an LU1 factorization.

Exercise 3.12 (Row Interchange) Show that $\boldsymbol{A} = \left[\begin{smallmatrix}1 & 1\\ 0 & 1\end{smallmatrix}\right]$ has a unique L1U factorization. Note that we have only interchanged rows in Example 3.5.

Exercise 3.13 (LU and Determinant) Suppose $\boldsymbol{A}$ has an L1U factorization $\boldsymbol{A} = \boldsymbol{L}\boldsymbol{U}$. Show that

$$\det(\boldsymbol{A}_{[k]}) = u_{11}u_{22}\cdots u_{kk} \text{ for } k = 1, \ldots, n.$$

[2] Hint: Consider the cases $2 \le k \le d$ and $d + 1 \le k \le n$ separately.

Exercise 3.14 (Diagonal Elements in U) Suppose $A \in \mathbb{C}^{n\times n}$ and $A_{[k]}$ is nonsingular for $k = 1, \ldots, n-1$. Use Exercise 3.13 to show that the diagonal elements u_{kk} in the L1U factorization are

$$u_{11} = a_{11}, \quad u_{kk} = \frac{\det(A_{[k]})}{\det(A_{[k-1]})}, \text{ for } k = 2, \ldots, n. \tag{3.31}$$

Exercise 3.15 (Proof of LDU Theorem) Give a proof of the LU theorem for the LDU case.

Exercise 3.16 (Proof of LU1 Theorem) Give a proof of the LU theorem for the LU1 case.

Exercise 3.17 (Computing the Inverse (Exam Exercise 1978-1)) Let $A \in \mathbb{R}^{n\times n}$ be nonsingular and with a unique L1U factorization $A = LU$. We partition L and U as follows

$$L = \begin{bmatrix} 1 & 0 \\ \boldsymbol{\ell}_1 & L_{2,2} \end{bmatrix}, \quad U = \begin{bmatrix} u_{1,1} & \boldsymbol{u}_1^T \\ 0 & U_{2,2} \end{bmatrix}, \tag{3.32}$$

where $L_{2,2}, U_{2,2} \in \mathbb{R}^{(n-1)\times(n-1)}$. Define $A_{2,2} := L_{2,2}U_{2,2}$ and $B_{2,2} := A_{2,2}^{-1}$.

a) Show that $A^{-1} = B$, where

$$B := \begin{bmatrix} (1 + \boldsymbol{u}_1^T B_{2,2}\boldsymbol{\ell}_1)/u_{1,1} & -\boldsymbol{u}_1^T B_{2,2}/u_{1,1} \\ -B_{2,2}\boldsymbol{\ell}_1 & B_{2,2} \end{bmatrix}. \tag{3.33}$$

b) Suppose that the elements $l_{i,j}$, $i > j$ in L and $u_{i,j}$, $j \ge i$ in U are stored in A with elements $a_{i,j}$. Write an algorithm that overwrites the elements in A with ones in A^{-1}. Only one extra vector $\boldsymbol{s} \in \mathbb{R}^n$ should be used.

Exercise 3.18 (Solving $THx = b$ (Exam Exercise 1981-3)) In this exercise we consider nonsingular matrices $T, H, S \in \mathbb{R}^{n\times n}$ with $T = (t_{ij})$ upper triangular, $H = (h_{ij})$ upper Hessenberg and $S := TH$. We assume that H has a unique LU1 factorization $H = LU$ with $\|L\|_\infty \|U\|_\infty \le K\|H\|_\infty$ for a constant K not too large. In this exercise the number of operations is the highest order term in the number of multiplications and divisions.

a) Give an algorithm which computes S from T and H without using the lower parts ($t_{ij}, i > j$) of T and ($h_{ij}, i > j+1$) of H. In what order should the elements in S be computed if S overwrites the elements in H? What is the number of operations of the algorithm?

b) Show that L is upper Hessenberg.

c) Give a detailed algorithm for finding the LU1-factorization of H stored in H. Determine the number of operations in the algorithm.

d) Given $\boldsymbol{b} \in \mathbb{R}^n$ and $\boldsymbol{T}, \boldsymbol{H}$ as before. Suppose $\boldsymbol{S}$ and the LU1-factorization are not computed. We want to find $\boldsymbol{x} \in \mathbb{R}^n$ such that $\boldsymbol{S}\boldsymbol{x} = \boldsymbol{b}$. We have the 2 following methods

Method 1:

1. $\boldsymbol{S} = \boldsymbol{T}\boldsymbol{H}$
2. Solve $\boldsymbol{S}\boldsymbol{x} = \boldsymbol{b}$

Method 2:

1. Solve $\boldsymbol{T}\boldsymbol{z} = \boldsymbol{b}$
2. Solve $\boldsymbol{H}\boldsymbol{x} = \boldsymbol{z}$

What method would you prefer? Give reasons for your answer.

Exercise 3.19 (L1U Factorization Update (Exam Exercise 1983-1)) Let $\boldsymbol{A} \in \mathbb{R}^{n \times n}$ be nonsingular with columns $\boldsymbol{a}_1, \boldsymbol{a}_2, \ldots, \boldsymbol{a}_n$. We assume that $\boldsymbol{A}$ has a unique L1U factorization $\boldsymbol{A} = \boldsymbol{L}\boldsymbol{U}$.

For a positive integer $p \le n$ and $\boldsymbol{b} \in \mathbb{R}^n$ we define

$$\boldsymbol{B} := [\boldsymbol{a}_1, \ldots, \boldsymbol{a}_{p-1}, \boldsymbol{a}_{p+1}, \ldots, \boldsymbol{a}_n, \boldsymbol{b}] \in \mathbb{R}^{n \times n}.$$

a) Show that $\boldsymbol{H} := \boldsymbol{L}^{-1}\boldsymbol{B}$ is upper Hessenberg. We assume that $\boldsymbol{H}$ has a unique L1U factorization. $\boldsymbol{H} = \boldsymbol{L}_H \boldsymbol{U}_H$.

b) Describe briefly how many multiplications/divisions are required to find the L1U factorization of $\boldsymbol{H}$?

c) Suppose we have found the L1U factorization $\boldsymbol{H} := \boldsymbol{L}_H \boldsymbol{U}_H$ of $\boldsymbol{H}$. Explain how we can find the L1U factorization of $\boldsymbol{B}$ from $\boldsymbol{L}_H$ and $\boldsymbol{U}_H$.

Exercise 3.20 (U1L Factorization (Exam Exercise 1990-1)) We say that $\boldsymbol{A} \in \mathbb{R}^{n \times n}$ has a U1L factorization if $\boldsymbol{A} = \boldsymbol{U}\boldsymbol{L}$ for an upper triangular matrix $\boldsymbol{U} \in \mathbb{R}^{n \times n}$ with ones on the diagonal and a lower triangular $\boldsymbol{L} \in \mathbb{R}^{n \times n}$. A UL and the more common LU factorization are analogous, but normally not the same.

a) Find a U1L factorization of the matrix

$$\boldsymbol{A} := \begin{bmatrix} -3 & -2 \\ 4 & 2 \end{bmatrix}.$$

b) Let the columns of $\boldsymbol{P} \in \mathbb{R}^{n \times n}$ be the unit vectors in reverse order, i.e.,

$$\boldsymbol{P} := [\boldsymbol{e}_n, \boldsymbol{e}_{n-1}, \ldots, \boldsymbol{e}_1].$$

Show that $\boldsymbol{P}^T = \boldsymbol{P}$ and $\boldsymbol{P}^2 = \boldsymbol{I}$. What is the connection between the elements in $\boldsymbol{A}$ and $\boldsymbol{PA}$?

c) Let $\boldsymbol{B} := \boldsymbol{PAP}$. Find integers r, s, depending on i, j, n, such that $b_{i,j} = a_{r,s}$.

d) Make a detailed algorithm which to given $\boldsymbol{A} \in \mathbb{R}^{n\times n}$ determines $\boldsymbol{B} := \boldsymbol{PAP}$. The elements $b_{i,j}$ in $\boldsymbol{B}$ should be stored in position i, j in $\boldsymbol{A}$. You should not use other matrices than $\boldsymbol{A}$ and a scalar $w \in \mathbb{R}$.

e) Let $\boldsymbol{PAP} = \boldsymbol{MR}$ be an L1U factorization of $\boldsymbol{PAP}$, i.e., $\boldsymbol{M}$ is lower triangular with ones on the diagonal and $\boldsymbol{R}$ is upper triangular. Express the matrices $\boldsymbol{U}$ and $\boldsymbol{L}$ in a U1L factorization of $\boldsymbol{A}$ in terms of $\boldsymbol{M}$, $\boldsymbol{R}$ and $\boldsymbol{P}$.

f) Give necessary and sufficient conditions for a matrix to have a unique U1L factorization.

3.7.4 Exercises Sect. 3.6

Exercise 3.21 (Making Block LU into LU) Show that $\hat{\boldsymbol{L}}$ is unit lower triangular and $\hat{\boldsymbol{U}}$ is upper triangular.

3.8 Review Questions

3.8.1 When is a triangular matrix nonsingular?

3.8.2 What is the general condition for Gaussian elimination without row interchanges to be well defined?

3.8.3 What is the content of the LU theorem?

3.8.4 Approximately how many arithmetic operations are needed for

- the multiplication of two square matrices?
- The LU factorization of a matrix?
- the solution of $\boldsymbol{Ax} = \boldsymbol{b}$, when $\boldsymbol{A}$ is triangular?

3.8.5 What is a PLU factorization? When does it exist?

3.8.6 What is complete pivoting?

Chapter 4
LDL* Factorization and Positive Definite Matrices

In this chapter we consider LU factorizations of Hermitian and positive definite matrices. Recall that a matrix $\boldsymbol{A} \in \mathbb{C}^{n \times n}$ is **Hermitian** if $\boldsymbol{A}^* = \boldsymbol{A}$, i.e., $a_{ji} = \overline{a}_{ij}$ for all i, j. A real Hermitian matrix is symmetric. Since $a_{ii} = \overline{a}_{ii}$ the diagonal elements of a Hermitian matrix must be real.

4.1 The LDL* Factorization

There are special versions of the LU factorization for Hermitian and positive definite matrices which takes advantage of the special properties of such matrices. The most important ones are

1. the LDL* factorization which is an LDU factorization with $\boldsymbol{U} = \boldsymbol{L}^*$ and $\boldsymbol{D}$ a diagonal matrix with real diagonal elements
2. the LL* factorization which is an LU factorization with $\boldsymbol{U} = \boldsymbol{L}^*$ and $l_{ii} > 0$ all i.

A matrix $\boldsymbol{A}$ having an LDL* factorization must be Hermitian since $\boldsymbol{D}$ is real so that $\boldsymbol{A}^* = (\boldsymbol{LDL}^*)^* = \boldsymbol{LD}^*\boldsymbol{L}^* = \boldsymbol{A}$. The LL* factorization is called a **Cholesky factorization** .

Example 4.1 (LDL of* 2×2 *Hermitian Matrix)* Let $a, d \in \mathbb{R}$ and $b \in \mathbb{C}$. An LDL* factorization of a 2×2 Hermitian matrix must satisfy the equations

$$\begin{bmatrix} a & \overline{b} \\ b & d \end{bmatrix} = \begin{bmatrix} 1 & 0 \\ l_1 & 1 \end{bmatrix} \begin{bmatrix} d_1 & 0 \\ 0 & d_2 \end{bmatrix} \begin{bmatrix} 1 & \overline{l}_1 \\ 0 & 1 \end{bmatrix} = \begin{bmatrix} d_1 & d_1\overline{l}_1 \\ d_1 l_1 & d_1|l_1|^2 + d_2 \end{bmatrix}$$

T. Lyche, *Numerical Linear Algebra and Matrix Factorizations*,
Texts in Computational Science and Engineering 22,
https://doi.org/10.1007/978-3-030-36468-7_4

for the unknowns l_1 in $\boldsymbol{L}$ and d_1, d_2 in $\boldsymbol{D}$. They are determined from

$$d_1 = a. \quad al_1 = b, \quad d_2 = d - a|l_1|^2. \tag{4.1}$$

There are essentially three cases

1. $a \neq 0$: The matrix has a unique LDL* factorization. Note that d_1 and d_2 are real.
2. $a = b = 0$: The LDL* factorization exists, but it is not unique. Any value for l_1 can be used.
3. $a = 0, b \neq 0$: No LDL* factorization exists.

Lemma 3.1 carries over to the Hermitian case.

Lemma 4.1 (LDL* of Leading Principal Sub Matrices) *Suppose* $\boldsymbol{A} = \boldsymbol{L}\boldsymbol{D}\boldsymbol{L}^*$ *is an LDL* factorization of* $\boldsymbol{A} \in \mathbb{C}^{n\times n}$. *For* $k = 1, \ldots, n$ *let* $\boldsymbol{A}_{[k]}$, $\boldsymbol{L}_{[k]}$ *and* $\boldsymbol{D}_{[k]}$ *be the leading principal submatrices of* $\boldsymbol{A}, \boldsymbol{L}$ *and* $\boldsymbol{D}$, *respectively. Then* $\boldsymbol{A}_{[k]} = \boldsymbol{L}_{[k]}\boldsymbol{D}_{[k]}\boldsymbol{L}_{[k]}^*$ *is an LDL* factorization of* $\boldsymbol{A}_{[k]}$ *for* $k = 1, \ldots, n$.

Proof For $k = 1, \ldots, n-1$ we partition $\boldsymbol{A} = \boldsymbol{L}\boldsymbol{D}\boldsymbol{L}^*$ as follows:

$$\boldsymbol{A} = \begin{bmatrix} \boldsymbol{A}_{[k]} & \boldsymbol{B}_k^* \\ \boldsymbol{B}_k & \boldsymbol{F}_k \end{bmatrix} = \begin{bmatrix} \boldsymbol{L}_{[k]} & \boldsymbol{0} \\ \boldsymbol{M}_k & \boldsymbol{N}_k \end{bmatrix} \begin{bmatrix} \boldsymbol{D}_{[k]} & \boldsymbol{0} \\ \boldsymbol{0} & \boldsymbol{E}_k \end{bmatrix} \begin{bmatrix} \boldsymbol{L}_{[k]}^* & \boldsymbol{M}_k^* \\ \boldsymbol{0} & \boldsymbol{N}_k^* \end{bmatrix} = \boldsymbol{L}\boldsymbol{D}\boldsymbol{U}, \tag{4.2}$$

where $\boldsymbol{F}_k, \boldsymbol{N}_k, \boldsymbol{E}_k \in \mathbb{C}^{n-k,n-k}$. Block multiplication gives $\boldsymbol{A}_{[k]} = \boldsymbol{L}_{[k]}\boldsymbol{D}_{[k]}\boldsymbol{L}_{[k]}^*$. Since $\boldsymbol{L}_{[k]}$ is unit lower triangular and $\boldsymbol{D}_{[k]}$ is real and diagonal this is an LDL* factorization of $\boldsymbol{A}_{[k]}$. □

Theorem 4.1 (LDL* Theorem) *The matrix* $\boldsymbol{A} \in \mathbb{C}^{n\times n}$ *has a unique LDL* factorization if and only if* $\boldsymbol{A} = \boldsymbol{A}^*$ *and* $\boldsymbol{A}_{[k]}$ *is nonsingular for* $k = 1, \ldots, n-1$.

Proof We essentially repeat the proof of Theorem 3.4 incorporating the necessary changes. Suppose $\boldsymbol{A}^* = \boldsymbol{A}$ and that $\boldsymbol{A}_{[k]}$ is nonsingular for $k = 1, \ldots, n-1$. Note that $\boldsymbol{A}_{[k]}^* = \boldsymbol{A}_{[k]}$ for $k = 1, \ldots, n$. We use induction on n to show that $\boldsymbol{A}$ has a unique LDL* factorization. The result is clearly true for $n = 1$, since the unique LDL* factorization of a 1-by-1 matrix is $[a_{11}] = [1][a_{11}][1]$ and a_{11} is real since $\boldsymbol{A}^* = \boldsymbol{A}$. Suppose that $\boldsymbol{A}_{[n-1]}$ has a unique LDL* factorization $\boldsymbol{A}_{[n-1]} = \boldsymbol{L}_{n-1}\boldsymbol{D}_{n-1}\boldsymbol{L}_{n-1}^*$, and that $\boldsymbol{A}_{[1]}, \ldots, \boldsymbol{A}_{[n-1]}$ are nonsingular. By definition $\boldsymbol{D}_{n-1}$ is real. Using block multiplication

$$\begin{aligned} \boldsymbol{A} = \begin{bmatrix} \boldsymbol{A}_{[n-1]} & \boldsymbol{a}_n \\ \boldsymbol{a}_n^* & a_{nn} \end{bmatrix} &= \begin{bmatrix} \boldsymbol{L}_{n-1} & \boldsymbol{0} \\ \boldsymbol{l}_n^* & 1 \end{bmatrix} \begin{bmatrix} \boldsymbol{D}_{n-1} & \boldsymbol{0} \\ 0 & d_{nn} \end{bmatrix} \begin{bmatrix} \boldsymbol{L}_{n-1}^* & \boldsymbol{l}_n \\ \boldsymbol{0}^* & 1 \end{bmatrix} \\ &= \begin{bmatrix} \boldsymbol{L}_{n-1}\boldsymbol{D}_{n-1}\boldsymbol{L}_{n-1}^* & \boldsymbol{L}_{n-1}\boldsymbol{D}_{n-1}\boldsymbol{l}_n \\ \boldsymbol{l}_n^*\boldsymbol{D}_{n-1}\boldsymbol{L}_{n-1}^* & \boldsymbol{l}_n^*\boldsymbol{D}_{n-1}\boldsymbol{l}_n + d_{nn} \end{bmatrix} \end{aligned} \tag{4.3}$$

if and only if $\boldsymbol{A}_{[n-1]} = \boldsymbol{L}_{n-1}\boldsymbol{D}_{n-1}\boldsymbol{L}_{n-1}^*$, and

$$\boldsymbol{a}_n = \boldsymbol{L}_{n-1}\boldsymbol{D}_{n-1}\boldsymbol{l}_n, \quad a_{nn} = \boldsymbol{l}_n^*\boldsymbol{D}_{n-1}\boldsymbol{l}_n + d_{nn}. \tag{4.4}$$

Thus we obtain an LDL* factorization of $\boldsymbol{A}$ that is unique since $\boldsymbol{L}_{n-1}$ and $\boldsymbol{D}_{n-1}$ are nonsingular. Also d_{nn} is real since a_{nn} and $\boldsymbol{D}_{n-1}$ are real.

For the converse we use Lemma 4.1 in the same way as Lemma 3.1 was used to prove Theorem 3.4. □

Here is an analog of Algorithm 3.4 that tries to compute the LDL* factorization of a d-banded matrix $\boldsymbol{A}$ with $d \geq 1$. It uses the upper part of the matrix.

```
function [L,dg]=LDLs(A,d)
% [L,dg]=LDLs(A,d)
n=length(A);
L=eye(n,n); dg=zeros(n,1);dg(1)=A(1,1);
for k=2:n
    m=rforwardsolve(L(1:k-1,1:k-1),A(1:k-1,k),d);
    L(k,1:k-1)=m./dg(1:k-1);
    dg(k)=A(k,k)-L(k,1:k-1)*m;
end
```

Listing 4.1 LDLs

The number of arithmetic operations for the LDL* factorization is approximately $\frac{1}{2}G_n$, half the number of operations needed for the LU factorization. Indeed, in the L1U factorization we needed to solve two triangular systems to find the vectors $\boldsymbol{s}$ and $\boldsymbol{m}$, while only one such system is needed to find $\boldsymbol{m}$ in the Hermitian case (4.3). The work to find d_{nn} is $O(n)$ and does not contribute to the highest order term.

Example 4.2 (A Factorization) Is the factorization

$$\begin{bmatrix} 3 & 1 \\ 1 & 3 \end{bmatrix} = \begin{bmatrix} 1 & 0 \\ 1/3 & 1 \end{bmatrix} \begin{bmatrix} 3 & 0 \\ 0 & 8/3 \end{bmatrix} \begin{bmatrix} 1 & 1/3 \\ 0 & 1 \end{bmatrix}$$

an LDL* factorization?

4.2 Positive Definite and Semidefinite Matrices

Given $\boldsymbol{A} \in \mathbb{C}^{n\times n}$. The function $f : \mathbb{C}^n \to \mathbb{R}$ given by

$$f(\boldsymbol{x}) = \boldsymbol{x}^*\boldsymbol{A}\boldsymbol{x} = \sum_{i=1}^{n}\sum_{j=1}^{n} a_{ij}\overline{x}_i x_j$$

is called a **quadratic form**. Note that f is real valued if A is Hermitian. Indeed, $\overline{f(x)} = \overline{x^* A x} = (x^* A x)^* = x^* A^* x = f(x)$.

Definition 4.1 (Positive Definite Matrix) We say that a matrix $A \in \mathbb{C}^{n\times n}$ is

(i) **positive definite** if $A^* = A$ and $x^* A x > 0$ for all nonzero $x \in \mathbb{C}^n$;
(ii) **positive semidefinite** if $A^* = A$ and $x^* A x \geq 0$ for all $x \in \mathbb{C}^n$;
(iii) **negative (semi)definite** if $-A$ is positive (semi)definite.

We observe that

1. The zero-matrix is positive semidefinite, while the unit matrix is positive definite.
2. The matrix A is positive definite if and only if it is positive semidefinite and $x^* A x = 0 \implies x = \mathbf{0}$.
3. A positive definite matrix A is nonsingular. For if $Ax = \mathbf{0}$ then $x^* A x = 0$ and this implies that $x = \mathbf{0}$.
4. It follows from Lemma 4.6 that a nonsingular positive semidefinite matrix is positive definite.
5. If A is real then it is enough to show definiteness for real vectors only. Indeed, if $A \in \mathbb{R}^{n\times n}$, $A^T = A$ and $x^T A x > 0$ for all nonzero $x \in \mathbb{R}^n$ then $z^* A z > 0$ for all nonzero $z \in \mathbb{C}^n$. For if $z = x + i y \neq \mathbf{0}$ with $x, y \in \mathbb{R}^n$ then

$$\begin{aligned} z^* A z &= (x - i y)^T A (x + i y) = x^T A x - i y^T A x + i x^T A y - i^2 y^T A y \\ &= x^T A x + y^T A y, \end{aligned}$$

and this is positive since at least one of the real vectors x, y is nonzero.

Example 4.3 (Gradient and Hessian) Symmetric positive definite matrices is important in nonlinear optimization. Consider (cf. (16.1)) the gradient ∇f and hessian Hf of a function $f : \Omega \subset \mathbb{R}^n \to \mathbb{R}$

$$\nabla f(x) = \begin{bmatrix} \frac{\partial f(x)}{\partial x_1} \\ \vdots \\ \frac{\partial f(x)}{\partial x_n} \end{bmatrix} \in \mathbb{R}^n, \quad Hf(x) = \begin{bmatrix} \frac{\partial^2 f(x)}{\partial x_1 \partial x_1} & \cdots & \frac{\partial^2 f(x)}{\partial x_1 \partial x_n} \\ \vdots & & \vdots \\ \frac{\partial^2 f(x)}{\partial x_n \partial x_1} & \cdots & \frac{\partial^2 f(x)}{\partial x_n \partial x_n} \end{bmatrix} \in \mathbb{R}^{n\times n}.$$

We assume that f has continuous first and second order partial derivatives on Ω.

Under suitable conditions on the domain Ω it is shown in advanced calculus texts that if $\nabla f(x) = \mathbf{0}$ and $Hf(x)$ is positive definite then x is a local minimum for f. This can be shown using the second-order Taylor expansion (16.2). Moreover, x is a local maximum if $\nabla f(x) = \mathbf{0}$ and $Hf(x)$ is negative definite.

Lemma 4.2 (The Matrix A^*A) *The matrix A^*A is positive semidefinite for any $m, n \in \mathbb{N}$ and $A \in \mathbb{C}^{m\times n}$. It is positive definite if and only if A has linearly independent columns or equivalently rank n.*

Proof Clearly $\boldsymbol{A}^*\boldsymbol{A}$ is Hermitian. Let $\boldsymbol{x} \in \mathbb{C}^n$ and set $\boldsymbol{z} := \boldsymbol{A}\boldsymbol{x}$. By the definition (1.11) of the Euclidean norm we have $\boldsymbol{x}^*\boldsymbol{A}^*\boldsymbol{A}\boldsymbol{x} = \boldsymbol{z}^*\boldsymbol{z} = \|\boldsymbol{z}\|_2^2 = \|\boldsymbol{A}\boldsymbol{x}\|_2^2 \geq 0$ with equality if and only if $\boldsymbol{A}\boldsymbol{x} = \boldsymbol{0}$. It follows that $\boldsymbol{A}^*\boldsymbol{A}$ is positive semidefinite and positive definite if and only if $\boldsymbol{A}$ has linearly independent columns. But this is equivalent to $\boldsymbol{A}$ having rank n (cf. Definition 1.6). □

Lemma 4.3 ($\boldsymbol{T}$ Is Positive Definite) *The second derivative matrix* $\boldsymbol{T} = \text{tridiag}(-1, 2, -1) \in \mathbb{R}^{n\times n}$ *is positive definite.*

Proof Clearly $\boldsymbol{T}$ is symmetric. For any $\boldsymbol{x} \in \mathbb{R}^n$

$$\begin{aligned}
\boldsymbol{x}^T\boldsymbol{T}\boldsymbol{x} &= 2\sum_{i=1}^{n} x_i^2 - \sum_{i=1}^{n-1} x_i x_{i+1} - \sum_{i=2}^{n} x_{i-1}x_i \\
&= \sum_{i=1}^{n-1} x_i^2 - 2\sum_{i=1}^{n-1} x_i x_{i+1} + \sum_{i=1}^{n-1} x_{i+1}^2 + x_1^2 + x_n^2 \\
&= x_1^2 + x_n^2 + \sum_{i=1}^{n-1}(x_{i+1} - x_i)^2.
\end{aligned}$$

Thus $\boldsymbol{x}^T\boldsymbol{T}\boldsymbol{x} \geq 0$ and if $\boldsymbol{x}^T\boldsymbol{T}\boldsymbol{x} = 0$ then $x_1 = x_n = 0$ and $x_i = x_{i+1}$ for $i = 1, \ldots, n-1$ which implies that $\boldsymbol{x} = 0$. Hence $\boldsymbol{T}$ is positive definite. □

4.2.1 The Cholesky Factorization

Recall that a **principal submatrix** $\boldsymbol{B} = \boldsymbol{A}(\boldsymbol{r}, \boldsymbol{r}) \in \mathbb{C}^{k\times k}$ of a matrix $\boldsymbol{A} \in \mathbb{C}^{n\times n}$ has elements $b_{i,j} = a_{r_i,r_j}$ for $i, j = 1, \ldots, k$, where $1 \leq r_1 < \cdots < r_k \leq n$. It is a **leading principal submatrix**, denoted $\boldsymbol{A}_{[k]}$ if $\boldsymbol{r} = [1, 2, \ldots, k]^T$. We have

$$\boldsymbol{A}(\boldsymbol{r}, \boldsymbol{r}) = \boldsymbol{X}^*\boldsymbol{A}\boldsymbol{X}, \quad \boldsymbol{X} := [\boldsymbol{e}_{r_1}, \ldots, \boldsymbol{e}_{r_k}] \in \mathbb{C}^{n\times k}. \tag{4.5}$$

Lemma 4.4 (Submatrices) *Any principal submatrix of a positive (semi)definite matrix is positive (semi)definite.*

Proof Let $\boldsymbol{X}$ and $\boldsymbol{B} := \boldsymbol{A}(\boldsymbol{r}, \boldsymbol{r})$ be given by (4.5). If $\boldsymbol{A}$ is positive semidefinite then $\boldsymbol{B}$ is positive semidefinite since

$$\boldsymbol{y}^*\boldsymbol{B}\boldsymbol{y} = \boldsymbol{y}^*\boldsymbol{X}^*\boldsymbol{A}\boldsymbol{X}\boldsymbol{y} = \boldsymbol{x}^*\boldsymbol{A}\boldsymbol{x} \geq 0, \quad \boldsymbol{y} \in \mathbb{C}^k, \quad \boldsymbol{x} := \boldsymbol{X}\boldsymbol{y}. \tag{4.6}$$

Suppose $\boldsymbol{A}$ is positive definite and $\boldsymbol{y}^*\boldsymbol{B}\boldsymbol{y} = 0$. By (4.6) we have $\boldsymbol{x} = \boldsymbol{0}$ and since $\boldsymbol{X}$ has linearly independent columns it follows that $\boldsymbol{y} = \boldsymbol{0}$. We conclude that $\boldsymbol{B}$ is positive definite. □

Theorem 4.2 (LDL* and LL*) *The following is equivalent for a matrix* $\boldsymbol{A} \in \mathbb{C}^{n\times n}$.

1. $\boldsymbol{A}$ *is positive definite,*
2. $\boldsymbol{A}$ *has an LDL* factorization with positive diagonal elements in* $\boldsymbol{D}$,
3. $\boldsymbol{A}$ *has a Cholesky factorization.*

If the Cholesky factorization exists it is unique.

Proof Recall that $\boldsymbol{A}^{-*} := (\boldsymbol{A}^{-1})^* = (\boldsymbol{A}^*)^{-1}$.
We show that $1 \implies 2 \implies 3 \implies 1$.

$1 \implies 2$: Suppose $\boldsymbol{A}$ is positive definite. By Lemma 4.4 the leading principal submatrices $\boldsymbol{A}_{[k]} \in \mathbb{C}^{k\times k}$ are positive definite and therefore nonsingular for $k = 1,\ldots,n-1$. Since $\boldsymbol{A}$ is Hermitian it has by Theorem 4.1 a unique LDL* factorization $\boldsymbol{A} = \boldsymbol{LDL}^*$. To show that the ith diagonal element in $\boldsymbol{D}$ is positive we note that $\boldsymbol{x}_i := \boldsymbol{L}^{-*}\boldsymbol{e}_i$ is nonzero since $\boldsymbol{L}^{-*}$ is nonsingular. But then $d_{ii} = \boldsymbol{e}_i^*\boldsymbol{D}\boldsymbol{e}_i = \boldsymbol{e}_i^*\boldsymbol{L}^{-1}\boldsymbol{A}\boldsymbol{L}^{-*}\boldsymbol{e}_i = \boldsymbol{x}_i^*\boldsymbol{A}\boldsymbol{x}_i > 0$ since $\boldsymbol{A}$ is positive definite.

$2 \implies 3$: Suppose $\boldsymbol{A}$ has an LDL* factorization $\boldsymbol{A} = \boldsymbol{LDL}^*$ with positive diagonal elements d_{ii} in $\boldsymbol{D}$. Then $\boldsymbol{A} = \boldsymbol{SS}^*$, where $\boldsymbol{S} := \boldsymbol{LD}^{1/2}$ and $\boldsymbol{D}^{1/2} := \operatorname{diag}(\sqrt{d_{11}},\ldots,\sqrt{d_{nn}})$, and this is a Cholesky factorization of $\boldsymbol{A}$.

$3 \implies 1$: Suppose $\boldsymbol{A}$ has a Cholesky factorization $\boldsymbol{A} = \boldsymbol{LL}^*$. Clearly $\boldsymbol{A}^* = \boldsymbol{A}$. Since $\boldsymbol{L}$ has positive diagonal elements it is nonsingular and $\boldsymbol{A}$ is positive definite by Lemma 4.2.

For uniqueness suppose $\boldsymbol{LL}^* = \boldsymbol{SS}^*$ are two Cholesky factorizations of the positive definite matrix $\boldsymbol{A}$. Since $\boldsymbol{A}$ is nonsingular both $\boldsymbol{L}$ and $\boldsymbol{S}$ are nonsingular. Then $\boldsymbol{S}^{-1}\boldsymbol{L} = \boldsymbol{S}^*\boldsymbol{L}^{-*}$, where by Lemma 2.5 $\boldsymbol{S}^{-1}\boldsymbol{L}$ is lower triangular and $\boldsymbol{S}^*\boldsymbol{L}^{-*}$ is upper triangular, with diagonal elements ℓ_{ii}/s_{ii} and s_{ii}/ℓ_{ii}, respectively. But then both matrices must be equal to the same diagonal matrix and $\ell_{ii}^2 = s_{ii}^2$. By positivity $\ell_{ii} = s_{ii}$ and we conclude that $\boldsymbol{S}^{-1}\boldsymbol{L} = \boldsymbol{I} = \boldsymbol{S}^*\boldsymbol{L}^{-*}$ which means that $\boldsymbol{L} = \boldsymbol{S}$. □

A Cholesky factorization can also be written in the equivalent form $\boldsymbol{A} = \boldsymbol{R}^*\boldsymbol{R}$, where $\boldsymbol{R} = \boldsymbol{L}^*$ is upper triangular with positive diagonal elements.

Example 4.4 (2 × 2) The matrix $\boldsymbol{A} = \begin{bmatrix} 2 & -1 \\ -1 & 2 \end{bmatrix}$ has an LDL* and a Cholesky-factorization given by

$$\begin{bmatrix} 2 & -1 \\ -1 & 2 \end{bmatrix} = \begin{bmatrix} 1 & 0 \\ -\frac{1}{2} & 1 \end{bmatrix}\begin{bmatrix} 2 & 0 \\ 0 & \frac{3}{2} \end{bmatrix}\begin{bmatrix} 1 & -\frac{1}{2} \\ 0 & 1 \end{bmatrix} = \begin{bmatrix} \sqrt{2} & 0 \\ -1/\sqrt{2} & \sqrt{3/2} \end{bmatrix}\begin{bmatrix} \sqrt{2} & -1/\sqrt{2} \\ 0 & \sqrt{3/2} \end{bmatrix}.$$

There are many good algorithms for finding the Cholesky factorization of a matrix, see [3]. The following version for finding the factorization of a matrix $\boldsymbol{A}$ with bandwidth $d \geq 1$ uses the LDL* factorization Algorithm 4.1. Only the upper part of $\boldsymbol{A}$ is used. The algorithm uses the MATLAB command `diag`.

```
function L=bandcholesky(A,d)
%L=bandcholesky(A,d)
[L,dg]=LDL(A,d);
L=L*diag(sqrt(dg));
end
```

Listing 4.2 bandcholesky

As for the LDL* factorization the leading term in an operation count for a band matrix is $O(d^2n)$. When d is small this is a considerable saving compared to the count $\frac{1}{2}G_n = n^3/3$ for a full matrix.

4.2.2 Positive Definite and Positive Semidefinite Criteria

Not all Hermitian matrices are positive definite, and sometimes we can tell just by glancing at the matrix that it cannot be positive definite. Here are some necessary conditions.

Theorem 4.3 (Necessary Conditions for Positive (Semi)Definiteness) *If $\boldsymbol{A} \in \mathbb{C}^{n\times n}$ is positive (semi)definite then for all i, j with $i \neq j$*

1. $a_{ii} > 0$, $(a_{ii} \geq 0)$,
2. $|Re\,(a_{ij})| < (a_{ii} + a_{jj})/2$, $(|Re\,(a_{ij})| \leq (a_{ii} + a_{jj})/2)$,
3. $|a_{ij}| < \sqrt{a_{ii}a_{jj}}$, $(|a_{ij}| \leq \sqrt{a_{ii}a_{jj}})$,
4. *If $\boldsymbol{A}$ is positive semidefinite and $a_{ii} = 0$ for some i then $a_{ij} = a_{ji} = 0$ for $j = 1, \ldots, n$.*

Proof Clearly $a_{ii} = \boldsymbol{e}_i^T \boldsymbol{A}\boldsymbol{e}_i > (\geq)0$ and Part 1 follows. If $\alpha, \beta \in \mathbb{C}$ and $\alpha\boldsymbol{e}_i + \beta\boldsymbol{e}_j \neq 0$ then

$$0 < (\leq)(\alpha\boldsymbol{e}_i + \beta\boldsymbol{e}_j)^*\boldsymbol{A}(\alpha\boldsymbol{e}_i + \beta\boldsymbol{e}_j) = |\alpha|^2 a_{ii} + |\beta|^2 a_{jj} + 2\mathrm{Re}\,(\overline{\alpha}\beta a_{ij}). \tag{4.7}$$

Taking $\alpha = 1$, $\beta = \pm 1$ we obtain $a_{ii} + a_{jj} \pm 2\mathrm{Re}\, a_{ij} > 0$ and this implies Part 2. We first show 3. when $\boldsymbol{A}$ is positive definite. Taking $\alpha = -a_{ij}$, $\beta = a_{ii}$ in (4.7) we find

$$0 < |a_{ij}|^2 a_{ii} + a_{ii}^2 a_{jj} - 2|a_{ij}|^2 a_{ii} = a_{ii}(a_{ii}a_{jj} - |a_{ij}|^2).$$

Since $a_{ii} > 0$ Part 3 follows in the positive definite case.

Suppose now $\boldsymbol{A}$ is positive semidefinite. For $\varepsilon > 0$ we define $\boldsymbol{B} := \boldsymbol{A} + \varepsilon\boldsymbol{I}$. The matrix $\boldsymbol{B}$ is positive definite since it is Hermitian and $\boldsymbol{x}^*\boldsymbol{B}\boldsymbol{x} \geq \varepsilon\|\boldsymbol{x}\|_2^2 > 0$ for any nonzero $\boldsymbol{x} \in \mathbb{C}^n$. From what we have shown

$$|a_{ij}| = |b_{ij}| < \sqrt{b_{ii}b_{jj}} = \sqrt{(a_{ii} + \varepsilon)(a_{jj} + \varepsilon)}, \quad i \neq j.$$

Since $\varepsilon > 0$ is arbitrary Part 3 follows in the semidefinite case. Since $\boldsymbol{A}$ is Hermitian Part 3 implies Part 4. □

Example 4.5 (Not Positive Definite) Consider the matrices

$$\boldsymbol{A}_1 = \begin{bmatrix} 0 & 1 \\ 1 & 1 \end{bmatrix}, \quad \boldsymbol{A}_2 = \begin{bmatrix} 1 & 2 \\ 2 & 2 \end{bmatrix}, \quad \boldsymbol{A}_3 = \begin{bmatrix} -2 & 1 \\ 1 & 2 \end{bmatrix}.$$

Here $\boldsymbol{A}_1$ and $\boldsymbol{A}_3$ are not positive definite, since a diagonal element is not positive. $\boldsymbol{A}_2$ is not positive definite since neither Part 2 nor Part 3 in Theorem 4.3 are satisfied.

The matrix $\left[\begin{smallmatrix} 2 & 1 \\ 1 & 2 \end{smallmatrix}\right]$ enjoys all the necessary conditions in Theorem 4.3. But to decide if it is positive definite it is nice to have sufficient conditions as well.

We start by considering eigenvalues of a positive (semi)definite matrix.

Lemma 4.5 (Positive Eigenvalues) *A matrix is positive (semi)definite if and only if it is Hermitian and all its eigenvalues are positive (nonnegative).*

Proof Suppose $\boldsymbol{A}$ is positive (semi)definite. Then $\boldsymbol{A}$ is Hermitian by definition, and if $\boldsymbol{A}\boldsymbol{x} = \lambda\boldsymbol{x}$ and $\boldsymbol{x}$ is nonzero, then $\boldsymbol{x}^*\boldsymbol{A}\boldsymbol{x} = \lambda\boldsymbol{x}^*\boldsymbol{x}$. This implies that $\lambda > 0 (\geq 0)$ since $\boldsymbol{A}$ is positive (semi)definite and $\boldsymbol{x}^*\boldsymbol{x} = \|\boldsymbol{x}\|_2^2 > 0$. Conversely, suppose $\boldsymbol{A} \in \mathbb{C}^{n\times n}$ is Hermitian with positive (nonnegative) eigenvalues $\lambda_1, \ldots, \lambda_n$. By Theorem 6.9 (the spectral theorem) there is a matrix $\boldsymbol{U} \in \mathbb{C}^{n\times n}$ with $\boldsymbol{U}^*\boldsymbol{U} = \boldsymbol{U}\boldsymbol{U}^* = \boldsymbol{I}$ such that $\boldsymbol{U}^*\boldsymbol{A}\boldsymbol{U} = \operatorname{diag}(\lambda_1, \ldots, \lambda_n)$. Let $\boldsymbol{x} \in \mathbb{C}^n$ and define $\boldsymbol{z} := \boldsymbol{U}^*\boldsymbol{x} = [z_1, \ldots, z_n]^T \in \mathbb{C}^n$. Then $\boldsymbol{x} = \boldsymbol{U}\boldsymbol{U}^*\boldsymbol{x} = \boldsymbol{U}\boldsymbol{z}$ and by the spectral theorem

$$\boldsymbol{x}^*\boldsymbol{A}\boldsymbol{x} = \boldsymbol{z}^*\boldsymbol{U}^*\boldsymbol{A}\boldsymbol{U}\boldsymbol{z} = \boldsymbol{z}^* \operatorname{diag}(\lambda_1, \ldots, \lambda_n)\boldsymbol{z} = \sum_{j=1}^{n} \lambda_j |z_j|^2 \geq 0.$$

It follows that $\boldsymbol{A}$ is positive semidefinite. Since $\boldsymbol{U}^*$ is nonsingular we see that $\boldsymbol{z} = \boldsymbol{U}^*\boldsymbol{x}$ is nonzero if $\boldsymbol{x}$ is nonzero, and therefore $\boldsymbol{A}$ is positive definite. □

Lemma 4.6 (Positive Semidefinite and Nonsingular) *A matrix is positive definite if and only if it is positive semidefinite and nonsingular.*

Proof If $\boldsymbol{A}$ is positive definite then it is positive semidefinite and if $\boldsymbol{A}\boldsymbol{x} = \boldsymbol{0}$ then $\boldsymbol{x}^*\boldsymbol{A}\boldsymbol{x} = 0$ which implies that $\boldsymbol{x} = \boldsymbol{0}$. Conversely, if $\boldsymbol{A}$ is positive semidefinite then it is Hermitian with nonnegative eigenvalues (cf. Lemma 4.5). If it is nonsingular all eigenvalues are positive (cf. Theorem 1.11), and it follows from Lemma 4.5 that $\boldsymbol{A}$ is positive definite. □

The following necessary and sufficient conditions can be used to decide if a matrix is positive definite.

Theorem 4.4 (Positive Definite Characterization) *The following statements are equivalent for a matrix* $\boldsymbol{A} \in \mathbb{C}^{n\times n}$.

1. $\boldsymbol{A}$ *is positive definite.*
2. $\boldsymbol{A}$ *is Hermitian with only positive eigenvalues.*
3. $\boldsymbol{A}$ *is Hermitian and all leading principal submatrices have a positive determinant.*
4. $\boldsymbol{A} = \boldsymbol{B}\boldsymbol{B}^*$ *for a nonsingular* $\boldsymbol{B} \in \mathbb{C}^{n\times n}$.

Proof

1 $\iff$ 2: This follows from Lemma 4.5.
1 $\implies$ 3: A positive definite matrix has positive eigenvalues, and since the determinant of a matrix equals the product of its eigenvalues (cf. Theorem 1.10) the determinant is positive. Every leading principal submatrix of a positive definite matrix is positive definite (cf. Lemma 4.4) and therefore has a positive determinant.
3 $\implies$ 4: Since a leading principal submatrix has a positive determinant it is nonsingular and Theorem 4.1 implies that $\boldsymbol{A}$ has a unique LDL* factorization and by Theorem 4.2 a unique Cholesky factorization $\boldsymbol{A} = \boldsymbol{B}\boldsymbol{B}^*$ with $\boldsymbol{B} = \boldsymbol{L}$.
4 $\implies$ 1: This follows from Lemma 4.2. □

Example 4.6 (Positive Definite Characterization) Consider the symmetric matrix $\boldsymbol{A} := \begin{bmatrix} 3 & 1 \\ 1 & 3 \end{bmatrix}$.

1. We have $\boldsymbol{x}^T\boldsymbol{A}\boldsymbol{x} = 2x_1^2 + 2x_2^2 + (x_1 + x_2)^2 > 0$ for all nonzero $\boldsymbol{x}$ showing that $\boldsymbol{A}$ is positive definite.
2. The eigenvalues of $\boldsymbol{A}$ are $\lambda_1 = 2$ and $\lambda_2 = 4$. They are positive showing that $\boldsymbol{A}$ is positive definite since it is symmetric.
3. We find $\det(\boldsymbol{A}_{[1]}) = 3$ and $\det(\boldsymbol{A}_{[2]}) = 8$ showing again that $\boldsymbol{A}$ is positive definite since it is also symmetric.
4. Finally $\boldsymbol{A}$ is positive definite since by Example 4.2 we have

$$\boldsymbol{A} = \boldsymbol{B}\boldsymbol{B}^*, \quad \boldsymbol{B} = \begin{bmatrix} 1 & 0 \\ 1/3 & 1 \end{bmatrix} \begin{bmatrix} \sqrt{3} & 0 \\ 0 & \sqrt{8/3} \end{bmatrix}.$$

4.3 Semi-Cholesky Factorization of a Banded Matrix

A positive semidefinite matrix has a factorization that is similar to the Cholesky factorization.

Definition 4.2 (Semi-Cholesky Factorization) A factorization $\boldsymbol{A} = \boldsymbol{L}\boldsymbol{L}^*$ of $\boldsymbol{A} \in \mathbb{C}^{n\times n}$, where $\boldsymbol{L}$ is lower triangular with nonnegative diagonal elements is called a **semi-Cholesky factorization**.

Note that a semi-Cholesky factorization of a positive definite matrix is necessarily a Cholesky factorization. For if $\boldsymbol{A}$ is positive definite then it is nonsingular and then $\boldsymbol{L}$ must be nonsingular. Thus the diagonal elements of $\boldsymbol{L}$ cannot be zero.

Theorem 4.5 (Characterization, Semi-Cholesky Factorization) *A matrix $\boldsymbol{A} \in \mathbb{C}^{n\times n}$ has a semi-Cholesky factorization $\boldsymbol{A} = \boldsymbol{L}\boldsymbol{L}^*$ if and only if it is positive semidefinite.*

Proof If $\boldsymbol{A} = \boldsymbol{L}\boldsymbol{L}^*$ is a semi-Cholesky factorization then $\boldsymbol{A}$ is Hermitian. Moreover, $\boldsymbol{x}^*\boldsymbol{A}\boldsymbol{x} = \|\boldsymbol{L}^*\boldsymbol{x}\|_2^2 \geq 0$ and $\boldsymbol{A}$ is positive semidefinite. For the converse we use induction on n. A positive semidefinite matrix of order one has a semi-Cholesky factorization since the only element in $\boldsymbol{A}$ is nonnegative. Suppose any positive semidefinite matrix of order $n-1$ has a semi-Cholesky factorization and suppose $\boldsymbol{A} \in \mathbb{C}^{n\times n}$ is positive semidefinite. We partition $\boldsymbol{A}$ as follows

$$\boldsymbol{A} = \begin{bmatrix} \alpha & \boldsymbol{v}^* \\ \boldsymbol{v} & \boldsymbol{B} \end{bmatrix}, \quad \alpha \in \mathbb{C},\ \boldsymbol{v} \in \mathbb{C}^{n-1},\ \boldsymbol{B} \in \mathbb{C}^{(n-1)\times(n-1)}. \tag{4.8}$$

There are two cases. Suppose first $\alpha = \boldsymbol{e}_1^*\boldsymbol{A}\boldsymbol{e}_1 > 0$. We claim that $\boldsymbol{C} := \boldsymbol{B} - \boldsymbol{v}\boldsymbol{v}^*/\alpha$ is positive semidefinite. $\boldsymbol{C}$ is Hermitian since $\boldsymbol{B}$ is. To show that $\boldsymbol{C}$ is positive semidefinite we consider any $\boldsymbol{y} \in \mathbb{C}^{n-1}$ and define $\boldsymbol{x}^* := [-\boldsymbol{y}^*\boldsymbol{v}/\alpha, \boldsymbol{y}^*] \in \mathbb{C}^n$. Then

$$\begin{aligned} 0 \leq \boldsymbol{x}^*\boldsymbol{A}\boldsymbol{x} &= [-\boldsymbol{y}^*\boldsymbol{v}/\alpha, \boldsymbol{y}^*] \begin{bmatrix} \alpha & \boldsymbol{v}^* \\ \boldsymbol{v} & \boldsymbol{B} \end{bmatrix} \begin{bmatrix} -\boldsymbol{v}^*\boldsymbol{y}/\alpha \\ \boldsymbol{y} \end{bmatrix} \\ &= [0, -(\boldsymbol{y}^*\boldsymbol{v})\boldsymbol{v}^*/\alpha + \boldsymbol{y}^*\boldsymbol{B}] \begin{bmatrix} -\boldsymbol{v}^*\boldsymbol{y}/\alpha \\ \boldsymbol{y} \end{bmatrix} \\ &= -\boldsymbol{y}^*\boldsymbol{v}\boldsymbol{v}^*\boldsymbol{y}/\alpha + \boldsymbol{y}^*\boldsymbol{B}\boldsymbol{y} = \boldsymbol{y}^*\boldsymbol{C}\boldsymbol{y}. \end{aligned} \tag{4.9}$$

So $\boldsymbol{C} \in \mathbb{C}^{(n-1)\times(n-1)}$ is positive semidefinite and by the induction hypothesis it has a semi-Cholesky factorization $\boldsymbol{C} = \boldsymbol{L}_1\boldsymbol{L}_1^*$. The matrix

$$\boldsymbol{L}^* := \begin{bmatrix} \beta & \boldsymbol{v}^*/\beta \\ \boldsymbol{0} & \boldsymbol{L}_1^* \end{bmatrix}, \quad \beta := \sqrt{\alpha}, \tag{4.10}$$

is upper triangular with nonnegative diagonal elements and

$$\boldsymbol{L}\boldsymbol{L}^* = \begin{bmatrix} \beta & \boldsymbol{0} \\ \boldsymbol{v}/\beta & \boldsymbol{L}_1 \end{bmatrix} \begin{bmatrix} \beta & \boldsymbol{v}^*/\beta \\ \boldsymbol{0} & \boldsymbol{L}_1^* \end{bmatrix} = \begin{bmatrix} \alpha & \boldsymbol{v}^* \\ \boldsymbol{v} & \boldsymbol{B} \end{bmatrix} = \boldsymbol{A}$$

is a semi-Cholesky factorization of $\boldsymbol{A}$.

If $\alpha = 0$ then part 4 of Theorem 4.3 implies that $\boldsymbol{v} = \boldsymbol{0}$. Moreover, $\boldsymbol{B} \in \mathbb{C}^{(n-1)\times(n-1)}$ in (4.8) is positive semidefinite and therefore has a semi-Cholesky factorization $\boldsymbol{B} = \boldsymbol{L}_1\boldsymbol{L}_1^*$. But then $\boldsymbol{L}\boldsymbol{L}^*$, where $\boldsymbol{L} = \begin{bmatrix} 0 & \boldsymbol{0}^* \\ \boldsymbol{0} & \boldsymbol{L}_1 \end{bmatrix}$ is a semi-Cholesky factorization of $\boldsymbol{A}$. Indeed, $\boldsymbol{L}$ is lower triangular and

$$\boldsymbol{L}\boldsymbol{L}^* = \begin{bmatrix} 0 & \boldsymbol{0}^* \\ \boldsymbol{0} & \boldsymbol{L}_1 \end{bmatrix}\begin{bmatrix} 0 & \boldsymbol{0}^* \\ \boldsymbol{0} & \boldsymbol{L}_1^* \end{bmatrix} = \begin{bmatrix} 0 & \boldsymbol{0}^* \\ \boldsymbol{0} & \boldsymbol{B} \end{bmatrix} = \boldsymbol{A}.$$

□

Recall that a matrix $\boldsymbol{A}$ is d-banded if $a_{ij} = 0$ for $|i - j| > d$. A (semi-) Cholesky factorization preserves bandwidth.

Theorem 4.6 (Bandwidth Semi-Cholesky Factor) *The semi-Cholesky factor* $\boldsymbol{L}$ *given by* (4.10) *has the same bandwidth as* $\boldsymbol{A}$.

Proof Suppose $\boldsymbol{A} \in \mathbb{C}^{n\times n}$ is d-banded. Then $\boldsymbol{v}^* = [\boldsymbol{u}^*, \boldsymbol{0}^*]$ in (4.8), where $\boldsymbol{u} \in \mathbb{C}^d$, and therefore $\boldsymbol{C} := \boldsymbol{B} - \boldsymbol{v}\boldsymbol{v}^*/\alpha$ differs from $\boldsymbol{B}$ only in the upper left $d \times d$ corner. It follows that $\boldsymbol{C}$ has the same bandwidth as $\boldsymbol{B}$ and $\boldsymbol{A}$. By induction on n, $\boldsymbol{C} = \boldsymbol{L}_1\boldsymbol{L}_1^*$, where $\boldsymbol{L}_1^*$ has the same bandwidth as $\boldsymbol{C}$. But then $\boldsymbol{L}$ in (4.10) has the same bandwidth as $\boldsymbol{A}$. □

Consider now implementing an algorithm based on the previous discussion. Since $\boldsymbol{A}$ is Hermitian we only need to use the lower part of $\boldsymbol{A}$. The first column of $\boldsymbol{L}$ is $[\beta, \boldsymbol{v}^*/\beta]^*$ if $\alpha > 0$. If $\alpha = 0$ then by 4 in Theorem 4.3 the first column of $\boldsymbol{A}$ is zero and this is also the first column of $\boldsymbol{L}$. We obtain

$$\boxed{\begin{array}{l} \text{if } A(1,1) > 0 \\ \quad A(1,1) = \sqrt{A(1,1)} \\ \quad A(2:n,1) = A(2:n,1)/A(1,1) \\ \quad \text{for } j = 2:n \\ \quad\quad A(j:n,j) = A(j:n,j) - A(j,1) * A(j:n,1) \end{array}} \qquad (4.11)$$

Here we store the first column of $\boldsymbol{L}$ in the first column of $\boldsymbol{A}$ and the lower part of $\boldsymbol{C} = \boldsymbol{B} - \boldsymbol{v}\boldsymbol{v}^*/\alpha$ in the lower part of $A(2:n, 2:n)$.

The code can be made more efficient when $\boldsymbol{A}$ is a d-banded matrix. We simply replace all occurrences of n by $\min(i+d, n)$. Continuing the reduction we arrive at the following algorithm, which take a d-banded positive semidefinite $\boldsymbol{A}$ and $d \geq 1$ as input, and returns a lower triangular matrix $\boldsymbol{L}$ so that $\boldsymbol{A} = \boldsymbol{L}\boldsymbol{L}^*$. This is the Cholesky factorization of $\boldsymbol{A}$ if $\boldsymbol{A}$ is positive definite and a semi-Cholesky factorization of $\boldsymbol{A}$ otherwise. The algorithm uses the MATLAB command `tril`:

```
function L=bandsemicholeskyL(A,d)
%L=bandsemicholeskyL(A,d)
n=length(A);
for k=1:n
   kp=min(n,k+d);
   if A(k,k)>0
      A(k,k)=sqrt(A(k,k));
      A((k+1):kp,k)=A((k+1):kp,k)/A(k,k);
       for j=k+1:kp
          A(j:kp,j)=A(j:kp,j)-A(j,k)*A(j:kp,k);
       end
   else
       A(k:kp,k)=zeros(kp-k+1,1);
   end
end
L=tril(A);
end
```

Listing 4.3 bandsemicholeskyL

In the algorithm we overwrite the lower triangle of $\boldsymbol{A}$ with the elements of $\boldsymbol{L}$. Column k of $\boldsymbol{L}$ is zero for those k where $\ell_{kk} = 0$. We reduce round-off noise by forcing those rows to be zero. In the semidefinite case no update is necessary and we "do nothing".

Deciding when a diagonal element is zero can be a problem in floating point arithmetic.

We end the section with some necessary and sufficient conditions for a matrix to be positive semidefinite.

Theorem 4.7 (Positive Semidefinite Characterization) *The following is equivalent for a matrix* $\boldsymbol{A} \in \mathbb{C}^{n\times n}$.

1. $\boldsymbol{A}$ *is positive semidefinite.*
2. $\boldsymbol{A}$ *is Hermitian with only nonnegative eigenvalues.*
3. $\boldsymbol{A}$ *is Hermitian and all principal submatrices have a nonnegative determinant.*
4. $\boldsymbol{A} = \boldsymbol{B}\boldsymbol{B}^*$ *for some* $\boldsymbol{B} \in \mathbb{C}^{n\times n}$.

Proof

$1 \iff 2$: This follows from Lemma 4.5.
$1 \iff 4$: This follows from Theorem 4.5
$1 \iff 3$: We refer to page 567 of [15], where it is shown that $4 \implies 3$ (and therefore $1 \implies 3$ since $1 \iff 4$) and $3 \implies 1$. □

Example 4.7 (Positive Semidefinite Characterization) Consider the symmetric matrix $\boldsymbol{A} := \begin{bmatrix} 1 & 1 \\ 1 & 1 \end{bmatrix}$.

1. We have $\boldsymbol{x}^*\boldsymbol{A}\boldsymbol{x} = x_1^2 + x_2^2 + x_1x_2 + x_2x_1 = (x_1 + x_2)^2 \geq 0$ for all $\boldsymbol{x} \in \mathbb{R}^2$ showing that $\boldsymbol{A}$ is positive semidefinite.
2. The eigenvalues of $\boldsymbol{A}$ are $\lambda_1 = 2$ and $\lambda_2 = 0$ and they are nonnegative showing that $\boldsymbol{A}$ is positive semidefinite since it is symmetric.
3. There are three principal sub matrices, and they have determinants $\det([a_{11}]) = 1$, $\det([a_{22}]) = 1$ and $\det(\boldsymbol{A}) = 0$ and showing again that $\boldsymbol{A}$ is positive semidefinite.
4. Finally $\boldsymbol{A}$ is positive semidefinite since $\boldsymbol{A} = \boldsymbol{B}\boldsymbol{B}^*$, where $\boldsymbol{B} = \begin{bmatrix} 1 & 0 \\ 1 & 0 \end{bmatrix}$.

In part 4 of Theorem 4.7 we require nonnegativity of all principal minors, while only positivity of leading principal minors was required for positive definite matrices (cf. Theorem 4.4). To see that nonnegativity of the leading principal minors is not enough consider the matrix $\boldsymbol{A} := \begin{bmatrix} 0 & 0 \\ 0 & -1 \end{bmatrix}$. The leading principal minors are nonnegative, but $\boldsymbol{A}$ is not positive semidefinite.

4.4 The Non-symmetric Real Case

In this section we say that a matrix $\boldsymbol{A} \in \mathbb{R}^{n \times n}$ is **positive semidefinite** if $\boldsymbol{x}^*\boldsymbol{A}\boldsymbol{x} \geq 0$ for all $\boldsymbol{x} \in \mathbb{R}^n$ and **positive definite** if $\boldsymbol{x}^*\boldsymbol{A}\boldsymbol{x} > 0$ for all nonzero $\boldsymbol{x} \in \mathbb{R}^n$. Thus we do not require $\boldsymbol{A}$ to be symmetric. This means that some of the eigenvalues can be complex (cf. Example 4.8). Note that a non-symmetric positive definite matrix is nonsingular, but in Exercise 4.3 you can show that a converse is not true.

We have the following theorem.

Theorem 4.8 (The Non-symmetric Case) *Suppose $\boldsymbol{A} \in \mathbb{R}^{n \times n}$ is positive definite. Then the following holds true.*

1. *Every principal submatrix of $\boldsymbol{A}$ is positive definite,*
2. *$\boldsymbol{A}$ has a unique LU factorization,*
3. *the real eigenvalues of $\boldsymbol{A}$ are positive,*
4. $\det(\boldsymbol{A}) > 0$,
5. $a_{ii}a_{jj} > a_{ij}a_{ji}$, *for $i \neq j$.*

Proof

1. The proof is the same as for Lemma 4.4.
2. Since all leading submatrices are positive definite they are nonsingular and the result follows from the LU Theorem 3.4.

3. Suppose $(\lambda, \boldsymbol{x})$ is an eigenpair of $\boldsymbol{A}$ and that λ is real. Since $\boldsymbol{A}$ is real we can choose $\boldsymbol{x}$ to be real. Multiplying $\boldsymbol{A}\boldsymbol{x} = \lambda\boldsymbol{x}$ by $\boldsymbol{x}^T$ and solving for λ we find $\lambda = \frac{\boldsymbol{x}^T\boldsymbol{A}\boldsymbol{x}}{\boldsymbol{x}^T\boldsymbol{x}} > 0$.
4. The determinant of $\boldsymbol{A}$ equals the product of its eigenvalues. The eigenvalues are either real and positive or occur in complex conjugate pairs. The product of two nonzero complex conjugate numbers is positive.
5. The principal submatrix $\begin{bmatrix} a_{ii} & a_{ij} \\ a_{ji} & a_{jj} \end{bmatrix}$ has a positive determinant. □

Example 4.8 (2 × 2 Positive Definite) A non-symmetric positive definite matrix can have complex eigenvalues. The family of matrices

$$\boldsymbol{A}[a] := \begin{bmatrix} 2 & 2-a \\ a & 1 \end{bmatrix}, \quad a \in \mathbb{R}$$

is positive definite for any $a \in \mathbb{R}$. Indeed, for any nonzero $\boldsymbol{x} \in \mathbb{R}^2$

$$\boldsymbol{x}^T\boldsymbol{A}\boldsymbol{x} = 2x_1^2 + (2-a)x_1x_2 + ax_2x_1 + x_2^2 = x_1^2 + (x_1 + x_2)^2 > 0.$$

The eigenvalues of $\boldsymbol{A}[a]$ are positive for $a \in [1 - \frac{\sqrt{5}}{2}, 1 + \frac{\sqrt{5}}{2}]$ and complex for other values of a.

4.5 Exercises Chap. 4

4.5.1 Exercises Sect. 4.2

Exercise 4.1 (Positive Definite Characterizations) Show directly that all 4 characterizations in Theorem 4.4 hold for the matrix $\begin{bmatrix} 2 & 1 \\ 1 & 2 \end{bmatrix}$.

Exercise 4.2 (L1U factorization (Exam 1982-1))]Find the L1U factorization of the following matrix $\boldsymbol{A} \in \mathbb{R}^{n \times n}$

$$\boldsymbol{A} = \begin{pmatrix} 1 & -1 & 0 & \cdots & 0 \\ -1 & 2 & -1 & \ddots & \vdots \\ 0 & \ddots & \ddots & \ddots & 0 \\ \vdots & \ddots & -1 & 2 & -1 \\ 0 & \cdots & 0 & -1 & 2 \end{pmatrix}.$$

Is $\boldsymbol{A}$ positive definite?

Exercise 4.3 (A Counterexample) In the non-symmetric case a nonsingular positive semidefinite matrix is not necessarily positive definite. Show this by considering the matrix $A := \begin{bmatrix} 1 & 0 \\ -2 & 1 \end{bmatrix}$.

Exercise 4.4 (Cholesky Update (Exam Exercise 2015-2))

a) Let $E \in \mathbb{R}^{n\times n}$ be of the form $E = I + uu^T$, where $u \in \mathbb{R}^n$. Show that E is symmetric and positive definite, and find an expression for E^{-1}.[1]

b) Let $A \in \mathbb{R}^{n\times n}$ be of the form $A = B + uu^T$, where $B \in \mathbb{R}^{n\times n}$ is symmetric and positive definite, and $u \in \mathbb{R}^n$. Show that A can be decomposed as

$$A = L(I + vv^T)L^T,$$

where L is nonsingular and lower triangular, and $v \in \mathbb{R}^n$.

c) Assume that the Cholesky decomposition of B is already computed. Outline a procedure to solve the system $Ax = b$, where A is of the form above.

Exercise 4.5 (Cholesky Update (Exam Exercise 2016-2)) Let $A \in \mathbb{R}^{n\times n}$ be a symmetric positive definite matrix with a known Cholesky factorization $A = LL^T$. Furthermore, let A_+ be a corresponding $(n+1)\times(n+1)$ matrix of the form

$$A_+ = \begin{bmatrix} A & a \\ a^T & \alpha \end{bmatrix},$$

where a is a vector in $\mathbb{R}^n$, and α is a real number. We assume that the matrix A_+ is symmetric positive definite.

a) Show that if $A_+ = L_+L_+^T$ is the Cholesky factorization of A_+, then L_+ is of the form

$$L_+ = \begin{bmatrix} L & 0 \\ y^T & \lambda \end{bmatrix},$$

i.e., that the leading principal $n \times n$ submatrix of L_+ is L.

b) Explain why $\alpha > \|L^{-1}a\|_2^2$.

c) Explain how you can compute L_+ when L is known.

4.6 Review Questions

4.6.1 What is the content of the LDL* theorem?

4.6.2 Is A^*A always positive definite?

[1]Hint: The matrix E^{-1} is of the form $E^{-1} = I + auu^T$ for some $a \in \mathbb{R}$.

4.6.3 Is the matrix $\begin{bmatrix} 10 & 4 & 3 \\ 4 & 0 & 2 \\ 3 & 2 & 5 \end{bmatrix}$ positive definite?

4.6.4 What class of matrices has a Cholesky factorization?

4.6.5 What is the bandwidth of the Cholesky factor of a band matrix?

4.6.6 For a symmetric matrix give 3 conditions that are equivalent to positive definiteness.

4.6.7 What class of matrices has a semi-Cholesky factorization?

Chapter 5
Orthonormal and Unitary Transformations

In Gaussian elimination and LU factorization we solve a linear system by transforming it to triangular form. These are not the only kind of transformations that can be used for such a task. Matrices with orthonormal columns, called unitary matrices can be used to reduce a square matrix to upper triangular form and more generally a rectangular matrix to upper triangular (also called upper trapezoidal) form. This lead to a decomposition of a rectangular matrix known as a **QR decomposition** and a reduced form which we refer to as a **QR factorization**. The QR decomposition and factorization will be used in later chapters to solve least squares- and eigenvalue problems.

Unitary transformations have the advantage that they preserve the Euclidian norm of a vector. This means that when a unitary transformation is applied to an inaccurate vector then the error will not grow. Thus a unitary transformation is said to be numerically stable. We consider two classes of unitary transformations known as Householder- and Givens transformations, respectively.

5.1 Inner Products, Orthogonality and Unitary Matrices

An **inner product** or **scalar product** in a vector space is a function mapping pairs of vectors into a scalar.

T. Lyche, *Numerical Linear Algebra and Matrix Factorizations*,
Texts in Computational Science and Engineering 22,
https://doi.org/10.1007/978-3-030-36468-7_5

5.1.1 *Real and Complex Inner Products*

Definition 5.1 (Inner Product) An **inner product** in a complex vector space $\mathcal{V}$ is a function $\mathcal{V} \times \mathcal{V} \to \mathbb{C}$ satisfying for all $\boldsymbol{x}, \boldsymbol{y}, \boldsymbol{z} \in \mathcal{V}$ and all $a, b \in \mathbb{C}$ the following conditions:

1. $\langle \boldsymbol{x}, \boldsymbol{x} \rangle \geq 0$ with equality if and only if $\boldsymbol{x} = \boldsymbol{0}$. (positivity)
2. $\langle \boldsymbol{x}, \boldsymbol{y} \rangle = \overline{\langle \boldsymbol{y}, \boldsymbol{x} \rangle}$ (skew symmetry)
3. $\langle a\boldsymbol{x} + b\boldsymbol{y}, \boldsymbol{z} \rangle = a\langle \boldsymbol{x}, \boldsymbol{z} \rangle + b\langle \boldsymbol{y}, \boldsymbol{z} \rangle$. (linearity)

The pair $(\mathcal{V}, \langle \cdot, \cdot \rangle)$ is called an **inner product space**.

Note the complex conjugate in 2. Since

$$\langle \boldsymbol{x}, a\boldsymbol{y} + b\boldsymbol{z} \rangle = \overline{\langle a\boldsymbol{y} + b\boldsymbol{z}, \boldsymbol{x} \rangle} = \overline{a\langle \boldsymbol{y}, \boldsymbol{x} \rangle + b\langle \boldsymbol{z}, \boldsymbol{x} \rangle} = \overline{a}\overline{\langle \boldsymbol{y}, \boldsymbol{x} \rangle} + \overline{b}\overline{\langle \boldsymbol{z}, \boldsymbol{x} \rangle}$$

we find

$$\langle \boldsymbol{x}, a\boldsymbol{y} + b\boldsymbol{z} \rangle = \overline{a}\langle \boldsymbol{x}, \boldsymbol{y} \rangle + \overline{b}\langle \boldsymbol{x}, \boldsymbol{z} \rangle, \quad \langle a\boldsymbol{x}, a\boldsymbol{y} \rangle = |a|^2\langle \boldsymbol{x}, \boldsymbol{y} \rangle. \tag{5.1}$$

An **inner product** in a real vector space $\mathcal{V}$ is real valued function satisfying Properties 1,2,3 in Definition 5.1, where we can replace skew symmetry by symmetry

$$\langle \boldsymbol{x}, \boldsymbol{y} \rangle = \langle \boldsymbol{y}, \boldsymbol{x} \rangle \quad \text{(symmetry)}.$$

In the real case we have linearity in both variables since we can remove the complex conjugates in (5.1).

Recall that (cf. (1.10)) the **standard inner product in $\mathbb{C}^n$** is given by

$$\langle \boldsymbol{x}, \boldsymbol{y} \rangle := \boldsymbol{y}^*\boldsymbol{x} = \boldsymbol{x}^T\overline{\boldsymbol{y}} = \sum_{j=1}^{n} x_j \overline{y_j}.$$

Note the complex conjugate on $\boldsymbol{y}$. It is clearly an inner product in $\mathbb{C}^n$.

The function

$$\|\cdot\| : \mathcal{V} \to \mathbb{R}, \qquad \boldsymbol{x} \longmapsto \|\boldsymbol{x}\| := \sqrt{\langle \boldsymbol{x}, \boldsymbol{x} \rangle} \tag{5.2}$$

is called the **inner product norm**.

The inner product norm for the standard inner product is the Euclidian norm $\|\boldsymbol{x}\| = \|\boldsymbol{x}\|_2 = \sqrt{\boldsymbol{x}^*\boldsymbol{x}}$.

The following inequality holds for any inner product.

Theorem 5.1 (Cauchy-Schwarz Inequality) *For any $\boldsymbol{x}, \boldsymbol{y}$ in a real or complex inner product space*

$$|\langle \boldsymbol{x}, \boldsymbol{y}\rangle| \le \|\boldsymbol{x}\|\|\boldsymbol{y}\|, \tag{5.3}$$

with equality if and only if $\boldsymbol{x}$ and $\boldsymbol{y}$ are linearly dependent.

Proof If $\boldsymbol{y} = \boldsymbol{0}$ then $0\boldsymbol{x} + \boldsymbol{y} = \boldsymbol{0}$ and $\boldsymbol{x}$ and $\boldsymbol{y}$ are linearly dependent. Moreover the inequality holds with equality since $\langle \boldsymbol{x}, \boldsymbol{y}\rangle = \langle \boldsymbol{x}, 0\boldsymbol{y}\rangle = 0\langle \boldsymbol{x}, \boldsymbol{y}\rangle = 0$ and $\|\boldsymbol{y}\| = 0$. So assume $\boldsymbol{y} \ne \boldsymbol{0}$. Define

$$z := \boldsymbol{x} - a\boldsymbol{y}, \qquad a := \frac{\langle \boldsymbol{x}, \boldsymbol{y}\rangle}{\langle \boldsymbol{y}, \boldsymbol{y}\rangle}.$$

By linearity $\langle z, \boldsymbol{y}\rangle = \langle \boldsymbol{x}, \boldsymbol{y}\rangle - a\langle \boldsymbol{y}, \boldsymbol{y}\rangle = 0$ so that by 2. and (5.1)

$$\langle a\boldsymbol{y}, z\rangle + \langle z, a\boldsymbol{y}\rangle = a\overline{\langle z, \boldsymbol{y}\rangle} + \overline{a}\langle z, \boldsymbol{y}\rangle = 0. \tag{5.4}$$

But then

$$\begin{aligned}
\|\boldsymbol{x}\|^2 &= \langle \boldsymbol{x}, \boldsymbol{x}\rangle = \langle z + a\boldsymbol{y}, z + a\boldsymbol{y}\rangle \\
&\overset{(5.4)}{=} \langle z, z\rangle + \langle a\boldsymbol{y}, a\boldsymbol{y}\rangle \overset{(5.1)}{=} \|z\|^2 + |a|^2\|\boldsymbol{y}\|^2 \\
&\ge |a|^2\|\boldsymbol{y}\|^2 = \frac{|\langle \boldsymbol{x}, \boldsymbol{y}\rangle|^2}{\|\boldsymbol{y}\|^2}.
\end{aligned}$$

Multiplying by $\|\boldsymbol{y}\|^2$ gives (5.3). We have equality if and only if $z = \boldsymbol{0}$, which means that $\boldsymbol{x}$ and $\boldsymbol{y}$ are linearly dependent. □

Theorem 5.2 (Inner Product Norm) *For all $\boldsymbol{x}, \boldsymbol{y}$ in an inner product space and all a in $\mathbb{C}$ we have*

1. $\|\boldsymbol{x}\| \ge 0$ *with equality if and only if* $\boldsymbol{x} = \boldsymbol{0}$. *(positivity)*
2. $\|a\boldsymbol{x}\| = |a|\,\|\boldsymbol{x}\|$. *(homogeneity)*
3. $\|\boldsymbol{x} + \boldsymbol{y}\| \le \|\boldsymbol{x}\| + \|\boldsymbol{y}\|$, *(subadditivity)*

where $\|\boldsymbol{x}\| := \sqrt{\langle \boldsymbol{x}, \boldsymbol{x}\rangle}$.

In general a function $\|\ \| : \mathbb{C}^n \to \mathbb{R}$ that satisfies these three properties is called a **vector norm**. A class of vector norms called p-norms will be studied in Chap. 8.

Proof The first statement is an immediate consequence of positivity, while the second one follows from (5.1). Expanding $\|\boldsymbol{x} + a\boldsymbol{y}\|^2 = \langle \boldsymbol{x} + a\boldsymbol{y}, \boldsymbol{x} + a\boldsymbol{y}\rangle$ using (5.1) we obtain

$$\|\boldsymbol{x}+a\boldsymbol{y}\|^2 = \|\boldsymbol{x}\|^2 + a\langle \boldsymbol{y}, \boldsymbol{x}\rangle + \overline{a}\langle \boldsymbol{x}, \boldsymbol{y}\rangle + |a|^2\|\boldsymbol{y}\|^2, \quad a \in \mathbb{C}, \quad \boldsymbol{x}, \boldsymbol{y} \in \mathcal{V}. \tag{5.5}$$

Now (5.5) with $a = 1$ and the Cauchy-Schwarz inequality implies

$$\|\boldsymbol{x} + \boldsymbol{y}\|^2 \le \|\boldsymbol{x}\|^2 + 2\|\boldsymbol{x}\|\|\boldsymbol{y}\| + \|\boldsymbol{y}\|^2 = (\|\boldsymbol{x}\| + \|\boldsymbol{y}\|)^2.$$

Taking square roots completes the proof. □

In the real case the Cauchy-Schwarz inequality implies that $-1 \le \frac{\langle \boldsymbol{x}, \boldsymbol{y}\rangle}{\|\boldsymbol{x}\|\|\boldsymbol{y}\|} \le 1$ for nonzero $\boldsymbol{x}$ and $\boldsymbol{y}$, so there is a unique angle θ in $[0, \pi]$ such that

$$\cos\theta = \frac{\langle \boldsymbol{x}, \boldsymbol{y}\rangle}{\|\boldsymbol{x}\|\|\boldsymbol{y}\|}. \tag{5.6}$$

This defines the **angle** between vectors in a real inner product space.

5.1.2 *Orthogonality*

Definition 5.2 (Orthogonality) Two vectors $\boldsymbol{x}, \boldsymbol{y}$ in a real or complex inner product space are **orthogonal** or **perpendicular**, denoted as $\boldsymbol{x} \perp \boldsymbol{y}$, if $\langle \boldsymbol{x}, \boldsymbol{y}\rangle = 0$. The vectors are **orthonormal** if in addition $\|\boldsymbol{x}\| = \|\boldsymbol{y}\| = 1$.

From the definitions (5.6), (5.20) of angle θ between two nonzero vectors in $\mathbb{R}^n$ or $\mathbb{C}^n$ it follows that $\boldsymbol{x} \perp \boldsymbol{y}$ if and only if $\theta = \pi/2$.

Theorem 5.3 (Pythagoras) *For a real or complex inner product space*

$$\|\boldsymbol{x} + \boldsymbol{y}\|^2 = \|\boldsymbol{x}\|^2 + \|\boldsymbol{y}\|^2, \quad \textit{if} \quad \boldsymbol{x} \perp \boldsymbol{y}. \tag{5.7}$$

Proof We set $a = 1$ in (5.5) and use the orthogonality. □

Definition 5.3 (Orthogonal- and Orthonormal Bases) A set of nonzero vectors $\{\boldsymbol{v}_1, \ldots, \boldsymbol{v}_k\}$ in a subspace $\mathcal{S}$ of a real or complex inner product space is an **orthogonal basis** for $\mathcal{S}$ if it is a basis for $\mathcal{S}$ and $\langle \boldsymbol{v}_i, \boldsymbol{v}_j\rangle = 0$ for $i \ne j$. It is an **orthonormal basis** for $\mathcal{S}$ if it is a basis for $\mathcal{S}$ and $\langle \boldsymbol{v}_i, \boldsymbol{v}_j\rangle = \delta_{ij}$ for all i, j.

A basis for a subspace of an inner product space can be turned into an orthogonal- or orthonormal basis for the subspace by the following construction (Fig. 5.1).

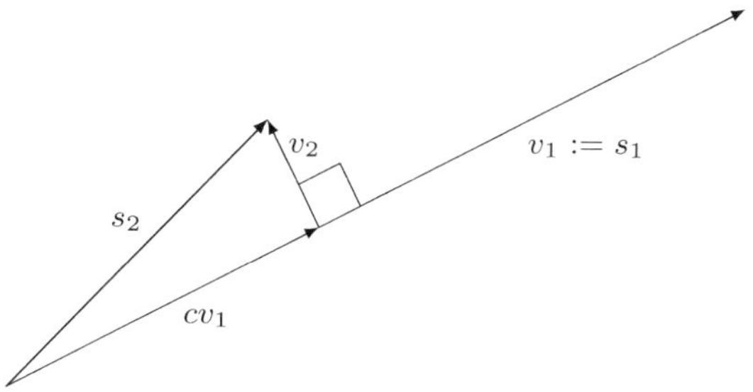

Fig. 5.1 The construction of $\boldsymbol{v}_1$ and $\boldsymbol{v}_2$ in Gram-Schmidt. The constant c is given by $c := \langle \boldsymbol{s}_2, \boldsymbol{v}_1\rangle / \langle \boldsymbol{v}_1, \boldsymbol{v}_1\rangle$

Theorem 5.4 (Gram-Schmidt) *Let $\{\boldsymbol{s}_1, \ldots, \boldsymbol{s}_k\}$ be a basis for a real or complex inner product space $(\mathcal{S}, \langle \cdot, \cdot \rangle)$. Define*

$$\boldsymbol{v}_1 := \boldsymbol{s}_1, \quad \boldsymbol{v}_j := \boldsymbol{s}_j - \sum_{i=1}^{j-1} \frac{\langle \boldsymbol{s}_j, \boldsymbol{v}_i\rangle}{\langle \boldsymbol{v}_i, \boldsymbol{v}_i\rangle} \boldsymbol{v}_i, \quad j = 2, \ldots, k. \tag{5.8}$$

Then $\{\boldsymbol{v}_1, \ldots, \boldsymbol{v}_k\}$ is an orthogonal basis for $\mathcal{S}$ and the normalized vectors

$$\{\boldsymbol{u}_1, \ldots, \boldsymbol{u}_k\} := \left\{ \frac{\boldsymbol{v}_1}{\|\boldsymbol{v}_1\|}, \ldots, \frac{\boldsymbol{v}_k}{\|\boldsymbol{v}_k\|} \right\}$$

form an orthonormal basis for $\mathcal{S}$.

Proof To show that $\{\boldsymbol{v}_1, \ldots, \boldsymbol{v}_k\}$ is an orthogonal basis for $\mathcal{S}$ we use induction on k. Define subspaces $S_j := \operatorname{span}\{\boldsymbol{s}_1, \ldots, \boldsymbol{s}_j\}$ for $j = 1, \ldots, k$. Clearly $\boldsymbol{v}_1 = \boldsymbol{s}_1$ is an orthogonal basis for S_1. Suppose for some $j \geq 2$ that $\boldsymbol{v}_1, \ldots, \boldsymbol{v}_{j-1}$ is an orthogonal basis for S_{j-1} and let $\boldsymbol{v}_j$ be given by (5.8) as a linear combination of $\boldsymbol{s}_j$ and $\boldsymbol{v}_1, \ldots, \boldsymbol{v}_{j-1}$. Now each of these $\boldsymbol{v}_i$ is a linear combination of $\boldsymbol{s}_1, \ldots, \boldsymbol{s}_i$, and we obtain $\boldsymbol{v}_j = \sum_{i=1}^{j} a_i \boldsymbol{s}_i$ for some $a_0, \ldots, a_j$ with $a_j = 1$. Since $\boldsymbol{s}_1, \ldots, \boldsymbol{s}_j$ are linearly independent and $a_j \neq 0$ we deduce that $\boldsymbol{v}_j \neq 0$. By the induction hypothesis

$$\langle \boldsymbol{v}_j, \boldsymbol{v}_l\rangle = \langle \boldsymbol{s}_j, \boldsymbol{v}_l\rangle - \sum_{i=1}^{j-1} \frac{\langle \boldsymbol{s}_j, \boldsymbol{v}_i\rangle}{\langle \boldsymbol{v}_i, \boldsymbol{v}_i\rangle} \langle \boldsymbol{v}_i, \boldsymbol{v}_l\rangle = \langle \boldsymbol{s}_j, \boldsymbol{v}_l\rangle - \frac{\langle \boldsymbol{s}_j, \boldsymbol{v}_l\rangle}{\langle \boldsymbol{v}_l, \boldsymbol{v}_l\rangle} \langle \boldsymbol{v}_l, \boldsymbol{v}_l\rangle = 0$$

for $l = 1, \ldots, j-1$. Thus $\boldsymbol{v}_1, \ldots, \boldsymbol{v}_j$ is an orthogonal basis for S_j.

If $\{\boldsymbol{v}_1, \ldots, \boldsymbol{v}_k\}$ is an orthogonal basis for $\mathcal{S}$ then clearly $\{\boldsymbol{u}_1, \ldots, \boldsymbol{u}_k\}$ is an orthonormal basis for $\mathcal{S}$. □

Sometimes we want to extend an orthogonal basis for a subspace to an orthogonal basis for a larger space.

Theorem 5.5 (Orthogonal Extension of Basis) *Suppose $\mathcal{S} \subset \mathcal{T}$ are finite dimensional subspaces of a vector space $\mathcal{V}$. An orthogonal basis for $\mathcal{S}$ can always be extended to an orthogonal basis for $\mathcal{T}$.*

Proof Suppose $\dim \mathcal{S} := k < \dim \mathcal{T} = n$. Using Theorem 1.3 we first extend an orthogonal basis $\boldsymbol{s}_1, \ldots, \boldsymbol{s}_k$ for $\mathcal{S}$ to a basis $\boldsymbol{s}_1, \ldots, \boldsymbol{s}_k, \boldsymbol{s}_{k+1}, \ldots, \boldsymbol{s}_n$ for $\mathcal{T}$, and then apply the Gram-Schmidt process to this basis obtaining an orthogonal basis $\boldsymbol{v}_1, \ldots, \boldsymbol{v}_n$ for $\mathcal{T}$. This is an extension of the basis for $\mathcal{S}$ since $\boldsymbol{v}_i = \boldsymbol{s}_i$ for $i = 1, \ldots, k$. We show this by induction. Clearly $\boldsymbol{v}_1 = \boldsymbol{s}_1$. Suppose for some $2 \le r < k$ that $\boldsymbol{v}_j = \boldsymbol{s}_j$ for $j = 1, \ldots, r-1$. Consider (5.8) for $j = r$. Since $\langle \boldsymbol{s}_r, \boldsymbol{v}_i \rangle = \langle \boldsymbol{s}_r, \boldsymbol{s}_i \rangle = 0$ for $i < r$ we obtain $\boldsymbol{v}_r = \boldsymbol{s}_r$. □

Letting $\mathcal{S} = \operatorname{span}(\boldsymbol{s}_1, \ldots, \boldsymbol{s}_k)$ and $\mathcal{T}$ be $\mathbb{R}^n$ or $\mathbb{C}^n$ we obtain

Corollary 5.1 (Extending Orthogonal Vectors to a Basis) *For $1 \le k < n$ a set $\{\boldsymbol{s}_1, \ldots, \boldsymbol{s}_k\}$ of nonzero orthogonal vectors in $\mathbb{R}^n$ or $\mathbb{C}^n$ can be extended to an orthogonal basis for the whole space.*

5.1.3 Sum of Subspaces and Orthogonal Projections

Suppose $\mathcal{S}$ and $\mathcal{T}$ are subspaces of a real or complex vector space $\mathcal{V}$ endowed with an inner product $\langle \boldsymbol{x}, \boldsymbol{y} \rangle$. We define

- **Sum**: $\mathcal{S} + \mathcal{T} := \{\boldsymbol{s} + \boldsymbol{t} : \boldsymbol{s} \in \mathcal{S} \text{ and } \boldsymbol{t} \in \mathcal{T}\}$,
- **direct sum** $\mathcal{S} \oplus \mathcal{T}$: a sum where $\mathcal{S} \cap \mathcal{T} = \{\boldsymbol{0}\}$,
- **orthogonal sum** $\mathcal{S} \overset{\perp}{\oplus} \mathcal{T}$: a sum where $\langle \boldsymbol{s}, \boldsymbol{t} \rangle = 0$ for all $\boldsymbol{s} \in \mathcal{S}$ and $\boldsymbol{t} \in \mathcal{T}$.

We note that

- $\mathcal{S} + \mathcal{T}$ is a vector space, a subspace of $\mathcal{V}$ which in this book will be $\mathbb{R}^n$ or $\mathbb{C}^n$ (cf. Example 1.2).
- Every $\boldsymbol{v} \in \mathcal{S} \oplus \mathcal{T}$ can be decomposed uniquely in the form $\boldsymbol{v} = \boldsymbol{s} + \boldsymbol{t}$, where $\boldsymbol{s} \in \mathcal{S}$ and $\boldsymbol{t} \in \mathcal{T}$. For if $\boldsymbol{v} = \boldsymbol{s}_1 + \boldsymbol{t}_1 = \boldsymbol{s}_2 + \boldsymbol{t}_2$ for $\boldsymbol{s}_1, \boldsymbol{s}_2 \in \mathcal{S}$ and $\boldsymbol{t}_1, \boldsymbol{t}_2 \in \mathcal{T}$, then $\boldsymbol{0} = \boldsymbol{s}_1 - \boldsymbol{s}_2 + \boldsymbol{t}_1 - \boldsymbol{t}_2$ or $\boldsymbol{s}_1 - \boldsymbol{s}_2 = \boldsymbol{t}_2 - \boldsymbol{t}_1$. It follows that $\boldsymbol{s}_1 - \boldsymbol{s}_2$ and $\boldsymbol{t}_2 - \boldsymbol{t}_1$ belong to both $\mathcal{S}$ and $\mathcal{T}$ and hence to $\mathcal{S} \cap \mathcal{T}$. But then $\boldsymbol{s}_1 - \boldsymbol{s}_2 = \boldsymbol{t}_2 - \boldsymbol{t}_1 = \boldsymbol{0}$ so $\boldsymbol{s}_1 = \boldsymbol{s}_2$ and $\boldsymbol{t}_2 = \boldsymbol{t}_1$.

  By (1.8) in the introduction chapter we have

$$\dim(\mathcal{S} \oplus \mathcal{T}) = \dim(\mathcal{S}) + \dim(\mathcal{T}).$$

  The subspaces $\mathcal{S}$ and $\mathcal{T}$ in a direct sum are called **complementary subspaces**.
- An orthogonal sum is a direct sum. For if $\boldsymbol{v} \in \mathcal{S} \cap \mathcal{T}$ then $\boldsymbol{v}$ is orthogonal to itself, $\langle \boldsymbol{v}, \boldsymbol{v} \rangle = 0$, which implies that $\boldsymbol{v} = 0$. We often write $\mathcal{T} := \mathcal{S}^{\perp}$.
- Suppose $\boldsymbol{v} = \boldsymbol{s}_0 + \boldsymbol{t}_0 \in \mathcal{S} \oplus \mathcal{T}$, where $\boldsymbol{s}_0 \in \mathcal{S}$ and $\boldsymbol{t}_0 \in \mathcal{T}$. The vector $\boldsymbol{s}_0$ is called the **oblique projection of $\boldsymbol{v}$ into $\mathcal{S}$ along $\mathcal{T}$**. Similarly, The vector $\boldsymbol{t}_0$ is

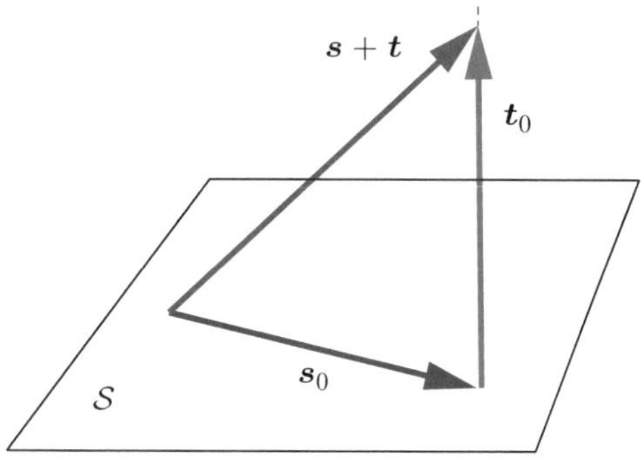

Fig. 5.2 The orthogonal projections of $\boldsymbol{s}+\boldsymbol{t}$ into $\mathcal{S}$ and $\mathcal{T}$

called the **oblique projection of** $\boldsymbol{v}$ **into** $\mathcal{T}$ **along** $\mathcal{S}$. If $\mathcal{S}\overset{\perp}{\oplus}\mathcal{T}$ is an orthogonal sum then $\boldsymbol{s}_0$ is called the **orthogonal projection** of $\boldsymbol{v}$ into $\mathcal{S}$. Similarly, $\boldsymbol{t}_0$ is called the **orthogonal projection** of $\boldsymbol{v}$ in $\mathcal{T}=\mathcal{S}^{\perp}$. The orthogonal projections are illustrated in Fig. 5.2.

Theorem 5.6 (Orthogonal Projection) *Let $\mathcal{S}$ and $\mathcal{T}$ be subspaces of a finite dimensional real or complex vector space $\mathcal{V}$ with an inner product $\langle\cdot,\cdot\rangle$. The orthogonal projections $\boldsymbol{s}_0$ of $\boldsymbol{v}\in\mathcal{S}\overset{\perp}{\oplus}\mathcal{T}$ into $\mathcal{S}$ and $\boldsymbol{t}_0$ of $\boldsymbol{v}\in\mathcal{S}\overset{\perp}{\oplus}\mathcal{T}$ into $\mathcal{T}$ satisfy $\boldsymbol{v}=\boldsymbol{s}_0+\boldsymbol{t}_0$, and*

$$\langle\boldsymbol{s}_0,\boldsymbol{s}\rangle=\langle\boldsymbol{v},\boldsymbol{s}\rangle,\quad \textit{for all } \boldsymbol{s}\in\mathcal{S},\quad \langle\boldsymbol{t}_0,\boldsymbol{t}\rangle=\langle\boldsymbol{v},\boldsymbol{t}\rangle,\quad \textit{for all } \boldsymbol{t}\in\mathcal{T}. \tag{5.9}$$

Moreover, if $\{\boldsymbol{v}_1,\ldots,\boldsymbol{v}_k\}$ is an orthogonal basis for $\mathbf{S}$ then

$$\boldsymbol{s}_0=\sum_{i=1}^{k}\frac{\langle\boldsymbol{v},\boldsymbol{v}_i\rangle}{\langle\boldsymbol{v}_i,\boldsymbol{v}_i\rangle}\boldsymbol{v}_i. \tag{5.10}$$

Proof We have $\langle\boldsymbol{s}_0,\boldsymbol{s}\rangle=\langle\boldsymbol{v}-\boldsymbol{t}_0,\boldsymbol{s}\rangle=\langle\boldsymbol{v},\boldsymbol{s}\rangle$, since $\langle\boldsymbol{t}_0,\boldsymbol{s}\rangle=0$ for all $\boldsymbol{s}\in\mathcal{S}$ and (5.9) follows. If $\boldsymbol{s}_0$ is given by (5.10) then for $j=1,\ldots,k$

$$\langle\boldsymbol{s}_0,\boldsymbol{v}_j\rangle=\langle\sum_{i=1}^{k}\frac{\langle\boldsymbol{v},\boldsymbol{v}_i\rangle}{\langle\boldsymbol{v}_i,\boldsymbol{v}_i\rangle}\boldsymbol{v}_i,\boldsymbol{v}_j\rangle=\sum_{i=1}^{k}\frac{\langle\boldsymbol{v},\boldsymbol{v}_i\rangle}{\langle\boldsymbol{v}_i,\boldsymbol{v}_i\rangle}\langle\boldsymbol{v}_i,\boldsymbol{v}_j\rangle=\langle\boldsymbol{v},\boldsymbol{v}_j\rangle.$$

By linearity (5.9) holds for all $\boldsymbol{s}\in\mathcal{S}$. By uniqueness it must be the orthogonal projections of $\boldsymbol{v}\in\mathcal{S}\overset{\perp}{\oplus}\mathcal{T}$ into $\mathcal{S}$. The proof for $\boldsymbol{t}_0$ is similar. □

Corollary 5.2 (Best Approximation) *Let $\mathcal{S}$ be a subspaces of a finite dimensional real or complex vector space $\mathcal{V}$ with an inner product $\langle \cdot, \cdot \rangle$ and corresponding norm $\|\boldsymbol{v}\| := \sqrt{\langle \boldsymbol{v}, \boldsymbol{v} \rangle}$. If $s_0 \in \mathcal{S}$ is the orthogonal projection of $\boldsymbol{v} \in \mathcal{V}$ then*

$$\|\boldsymbol{v} - \boldsymbol{s}_0\| < \|\boldsymbol{v} - \boldsymbol{s}\|, \text{ for all } \boldsymbol{s} \in \mathcal{S}, \ \boldsymbol{s} \neq \boldsymbol{s}_0. \tag{5.11}$$

Proof Let $\boldsymbol{s}_0 \neq \boldsymbol{s} \in \mathcal{S}$ and $0 \neq \boldsymbol{u} := \boldsymbol{s}_0 - \boldsymbol{s} \in \mathcal{S}$. It follows from (5.9) that $\langle \boldsymbol{v} - \boldsymbol{s}_0, \boldsymbol{u} \rangle = 0$. By (5.7) (Pythagoras) we obtain

$$\|\boldsymbol{v} - \boldsymbol{s}\|^2 = \|\boldsymbol{v} - \boldsymbol{s}_0 + \boldsymbol{u}\|^2 = \|\boldsymbol{v} - \boldsymbol{s}_0\|^2 + \|\boldsymbol{u}\|^2 > \|\boldsymbol{v} - \boldsymbol{s}_0\|^2.$$

□

5.1.4 Unitary and Orthogonal Matrices

In the rest of this chapter orthogonality is in terms of the **standard inner product in** $\mathbb{C}^n$ given by $\langle \boldsymbol{x}, \boldsymbol{y} \rangle := \boldsymbol{y}^* \boldsymbol{x} = \sum_{j=1}^n x_j \overline{y_j}$. For symmetric and Hermitian matrices we have the following characterization.

Lemma 5.1 *Let $A \in \mathbb{C}^{n \times n}$ and $\langle \boldsymbol{x}, \boldsymbol{y} \rangle$ be the standard inner product in $\mathbb{C}^n$. Then*

1. $\boldsymbol{A}^T = \boldsymbol{A} \iff \langle \boldsymbol{A}\boldsymbol{x}, \boldsymbol{y} \rangle = \langle \boldsymbol{x}, \overline{\boldsymbol{A}}\boldsymbol{y} \rangle$ *for all* $\boldsymbol{x}, \boldsymbol{y} \in \mathbb{C}^n$.
2. $\boldsymbol{A}^* = \boldsymbol{A} \iff \langle \boldsymbol{A}\boldsymbol{x}, \boldsymbol{y} \rangle = \langle \boldsymbol{x}, \boldsymbol{A}\boldsymbol{y} \rangle$ *for all* $\boldsymbol{x}, \boldsymbol{y} \in \mathbb{C}^n$.

Proof Suppose $\boldsymbol{A}^T = \boldsymbol{A}$ and $\boldsymbol{x}, \boldsymbol{y} \in \mathbb{C}^n$. Then

$$\langle \boldsymbol{x}, \overline{\boldsymbol{A}}\boldsymbol{y} \rangle = (\overline{\boldsymbol{A}}\boldsymbol{y})^* \boldsymbol{x} = \boldsymbol{y}^* \overline{\boldsymbol{A}}^* \boldsymbol{x} = \boldsymbol{y}^* \boldsymbol{A}^T \boldsymbol{x} = \boldsymbol{y}^* \boldsymbol{A}\boldsymbol{x} = \langle \boldsymbol{A}\boldsymbol{x}, \boldsymbol{y} \rangle.$$

For the converse we take $\boldsymbol{x} = \boldsymbol{e}_j$ and $\boldsymbol{y} = \boldsymbol{e}_i$ for some i, j and obtain

$$\boldsymbol{e}_i^T \boldsymbol{A}\boldsymbol{e}_j = \langle \boldsymbol{A}\boldsymbol{e}_j, \boldsymbol{e}_i \rangle = \langle \boldsymbol{e}_j, \overline{\boldsymbol{A}}\boldsymbol{e}_i \rangle = \boldsymbol{e}_i^T \boldsymbol{A}^T \boldsymbol{e}_j.$$

Thus, $\boldsymbol{A} = \boldsymbol{A}^T$ since they have the same i, j element for all i, j. The proof of 2. is similar. □

A square matrix $\boldsymbol{U} \in \mathbb{C}^{n \times n}$ is **unitary** if $\boldsymbol{U}^*\boldsymbol{U} = \boldsymbol{I}$. If $\boldsymbol{U}$ is real then $\boldsymbol{U}^T\boldsymbol{U} = \boldsymbol{I}$ and $\boldsymbol{U}$ is called an **orthogonal matrix**. Unitary and orthogonal matrices have orthonormal columns.

If $\boldsymbol{U}^*\boldsymbol{U} = \boldsymbol{I}$ the matrix $\boldsymbol{U}$ is nonsingular, $\boldsymbol{U}^{-1} = \boldsymbol{U}^*$ and therefore $\boldsymbol{U}\boldsymbol{U}^* = \boldsymbol{U}\boldsymbol{U}^{-1} = \boldsymbol{I}$ as well. Moreover, both the columns and rows of a unitary matrix of order n form orthonormal bases for $\mathbb{C}^n$. We also note that the product of two unitary matrices is unitary. Indeed, if $\boldsymbol{U}_1^*\boldsymbol{U}_1 = \boldsymbol{I}$ and $\boldsymbol{U}_2^*\boldsymbol{U}_2 = \boldsymbol{I}$ then $(\boldsymbol{U}_1\boldsymbol{U}_2)^*(\boldsymbol{U}_1\boldsymbol{U}_2) = \boldsymbol{U}_2^*\boldsymbol{U}_1^*\boldsymbol{U}_1\boldsymbol{U}_2 = \boldsymbol{I}$.

Theorem 5.7 (Unitary Matrix) *The matrix $\boldsymbol{U} \in \mathbb{C}^{n\times n}$ is unitary if and only if $\langle \boldsymbol{U}\boldsymbol{x}, \boldsymbol{U}\boldsymbol{y}\rangle = \langle \boldsymbol{x}, \boldsymbol{y}\rangle$ for all $\boldsymbol{x}, \boldsymbol{y} \in \mathbb{C}^n$. In particular, if $\boldsymbol{U}$ is unitary then $\|\boldsymbol{U}\boldsymbol{x}\|_2 = \|\boldsymbol{x}\|_2$ for all $\boldsymbol{x} \in \mathbb{C}^n$.*

Proof If $\boldsymbol{U}^*\boldsymbol{U} = \boldsymbol{I}$ and $\boldsymbol{x}, \boldsymbol{y} \in \mathbb{C}^n$ then

$$\langle \boldsymbol{U}\boldsymbol{x}, \boldsymbol{U}\boldsymbol{y}\rangle = (\boldsymbol{U}\boldsymbol{y})^*(\boldsymbol{U}\boldsymbol{x}) = \boldsymbol{y}^*\boldsymbol{U}^*\boldsymbol{U}\boldsymbol{x} = \boldsymbol{y}^*\boldsymbol{x} = \langle \boldsymbol{x}, \boldsymbol{y}\rangle.$$

Conversely, if $\langle \boldsymbol{U}\boldsymbol{x}, \boldsymbol{U}\boldsymbol{y}\rangle = \langle \boldsymbol{x}, \boldsymbol{y}\rangle$ for all $\boldsymbol{x}, \boldsymbol{y} \in \mathbb{C}^n$ then $\boldsymbol{U}^*\boldsymbol{U} = \boldsymbol{I}$ since for $i, j = 1, \ldots, n$

$$(\boldsymbol{U}^*\boldsymbol{U})_{i,j} = \boldsymbol{e}_i^*\boldsymbol{U}^*\boldsymbol{U}\boldsymbol{e}_j = (\boldsymbol{U}\boldsymbol{e}_i)^*(\boldsymbol{U}\boldsymbol{e}_j) = \langle \boldsymbol{U}\boldsymbol{e}_j, \boldsymbol{U}\boldsymbol{e}_i\rangle = \langle \boldsymbol{e}_j, \boldsymbol{e}_i\rangle = \boldsymbol{e}_i^*\boldsymbol{e}_j,$$

so that $(\boldsymbol{U}^*\boldsymbol{U})_{i,j} = \delta_{i,j}$ for all i, j. The last part of the theorem follows immediately by taking $\boldsymbol{y} = \boldsymbol{x}$: □

5.2 The Householder Transformation

We consider a unitary matrix with many useful properties.

Definition 5.4 (Householder Transformation) A matrix $\boldsymbol{H} \in \mathbb{C}^{n\times n}$ of the form

$$\boldsymbol{H} := \boldsymbol{I} - \boldsymbol{u}\boldsymbol{u}^*, \text{ where } \boldsymbol{u} \in \mathbb{C}^n \text{ and } \boldsymbol{u}^*\boldsymbol{u} = 2$$

is called a **Householder transformation**. The name **elementary reflector** is also used.

In the real case and for $n = 2$ we find

$$\boldsymbol{H} = \begin{bmatrix} 1 & 0 \\ 0 & 1 \end{bmatrix} - \begin{bmatrix} u_1 \\ u_2 \end{bmatrix} \begin{bmatrix} u_1 & u_2 \end{bmatrix} = \begin{bmatrix} 1 - u_1^2 & -u_1u_2 \\ -u_2u_1 & 1 - u_2^2 \end{bmatrix}.$$

A Householder transformation is Hermitian and unitary. Indeed, $\boldsymbol{H}^* = (\boldsymbol{I} - \boldsymbol{u}\boldsymbol{u}^*)^* = \boldsymbol{H}$ and

$$\boldsymbol{H}^*\boldsymbol{H} = \boldsymbol{H}^2 = (\boldsymbol{I} - \boldsymbol{u}\boldsymbol{u}^*)(\boldsymbol{I} - \boldsymbol{u}\boldsymbol{u}^*) = \boldsymbol{I} - 2\boldsymbol{u}\boldsymbol{u}^* + \boldsymbol{u}(\boldsymbol{u}^*\boldsymbol{u})\boldsymbol{u}^* = \boldsymbol{I}.$$

In the real case $\boldsymbol{H}$ is symmetric and orthogonal.

There are several ways to represent a Householder transformation. Householder used $\boldsymbol{I} - 2\boldsymbol{u}\boldsymbol{u}^*$, where $\boldsymbol{u}^*\boldsymbol{u} = 1$. For any nonzero $\boldsymbol{v} \in \mathbb{C}^n$ the matrix

$$\boldsymbol{H} := \boldsymbol{I} - 2\frac{\boldsymbol{v}\boldsymbol{v}^*}{\boldsymbol{v}^*\boldsymbol{v}} \tag{5.12}$$

is a Householder transformation. Indeed, $\boldsymbol{H} = \boldsymbol{I} - \boldsymbol{u}\boldsymbol{u}^*$, where $\boldsymbol{u} := \sqrt{2}\frac{\boldsymbol{v}}{\|\boldsymbol{v}\|_2}$ has length $\sqrt{2}$.

Two vectors can, under certain conditions, be mapped into each other by a Householder transformation.

Lemma 5.2 *Suppose $\boldsymbol{x}, \boldsymbol{y} \in \mathbb{C}^n$ are two vectors such that $\|\boldsymbol{x}\|_2 = \|\boldsymbol{y}\|_2$, $\boldsymbol{y}^*\boldsymbol{x}$ is real and $\boldsymbol{v} := \boldsymbol{x} - \boldsymbol{y} \neq \boldsymbol{0}$. Then $\big(\boldsymbol{I} - 2\frac{\boldsymbol{v}\boldsymbol{v}^*}{\boldsymbol{v}^*\boldsymbol{v}}\big)\boldsymbol{x} = \boldsymbol{y}$.*

Proof Since $\boldsymbol{x}^*\boldsymbol{x} = \boldsymbol{y}^*\boldsymbol{y}$ and $\mathrm{Re}(\boldsymbol{y}^*\boldsymbol{x}) = \boldsymbol{y}^*\boldsymbol{x}$ we have

$$\boldsymbol{v}^*\boldsymbol{v} = (\boldsymbol{x} - \boldsymbol{y})^*(\boldsymbol{x} - \boldsymbol{y}) = 2\boldsymbol{x}^*\boldsymbol{x} - 2\mathrm{Re}(\boldsymbol{y}^*\boldsymbol{x}) = 2\boldsymbol{v}^*\boldsymbol{x}. \tag{5.13}$$

But then $\big(\boldsymbol{I} - 2\frac{\boldsymbol{v}\boldsymbol{v}^*}{\boldsymbol{v}^*\boldsymbol{v}}\big)\boldsymbol{x} = \boldsymbol{x} - \frac{2\boldsymbol{v}^*\boldsymbol{x}}{\boldsymbol{v}^*\boldsymbol{v}}\boldsymbol{v} = \boldsymbol{x} - \boldsymbol{v} = \boldsymbol{y}$. □

There is a nice geometric interpretation of this Lemma. We have

$$\boldsymbol{H} = \boldsymbol{I} - \frac{2\boldsymbol{v}\boldsymbol{v}^*}{\boldsymbol{v}^*\boldsymbol{v}} = \boldsymbol{P} - \frac{\boldsymbol{v}\boldsymbol{v}^*}{\boldsymbol{v}^*\boldsymbol{v}}, \text{ where } \boldsymbol{P} := \boldsymbol{I} - \frac{\boldsymbol{v}\boldsymbol{v}^*}{\boldsymbol{v}^*\boldsymbol{v}},$$

and

$$\boldsymbol{P}\boldsymbol{x} = \boldsymbol{x} - \frac{\boldsymbol{v}^*\boldsymbol{x}}{\boldsymbol{v}^*\boldsymbol{v}}\boldsymbol{v} \overset{(5.13)}{=} \boldsymbol{x} - \frac{1}{2}\boldsymbol{v} = \frac{1}{2}(\boldsymbol{x} + \boldsymbol{y}).$$

If $\boldsymbol{x}, \boldsymbol{y} \in \mathbb{R}^n$ it follows that $\boldsymbol{H}\boldsymbol{x}$ is the reflected image of $\boldsymbol{x}$. The "mirror" $\mathcal{M} := \{\boldsymbol{w} \in \mathbb{R}^n : \boldsymbol{w}^*\boldsymbol{v} = 0\}$ contains the vector $(\boldsymbol{x} + \boldsymbol{y})/2$ and has normal $\boldsymbol{x} - \boldsymbol{y}$. This is illustrated for the real case in Fig. 5.3.

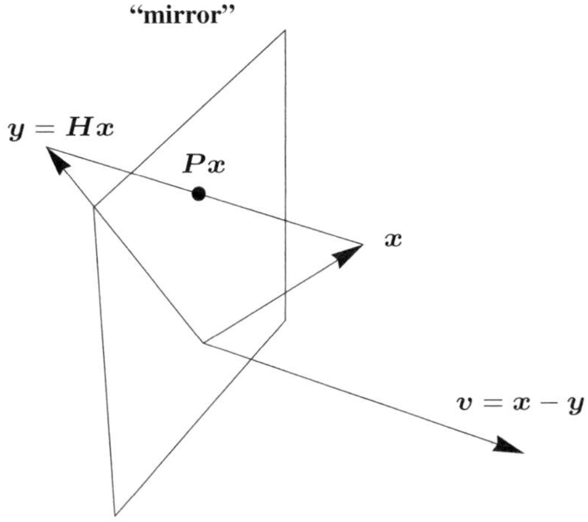

Fig. 5.3 The Householder transformation in Example 5.1

Example 5.1 (Reflector) Suppose $\boldsymbol{x} := [1, 0, 1]^T$ and $\boldsymbol{y} := [-1, 0, 1]^T$. Then $\boldsymbol{v} = \boldsymbol{x} - \boldsymbol{y} = [2, 0, 0]^T$ and

$$\boldsymbol{H} := \boldsymbol{I} - \frac{2\boldsymbol{v}\boldsymbol{v}^T}{\boldsymbol{v}^T\boldsymbol{v}} = \begin{bmatrix} 1\,0\,0 \\ 0\,1\,0 \\ 0\,0\,1 \end{bmatrix} - \frac{2}{4}\begin{bmatrix} 2 \\ 0 \\ 0 \end{bmatrix}\begin{bmatrix} 2\,0\,0 \end{bmatrix} = \begin{bmatrix} -1\,0\,0 \\ 0\;\,1\,0 \\ 0\;\,0\,1 \end{bmatrix},$$

$$\boldsymbol{P} := \boldsymbol{I} - \frac{\boldsymbol{v}\boldsymbol{v}^T}{\boldsymbol{v}^T\boldsymbol{v}} = \begin{bmatrix} 1\,0\,0 \\ 0\,1\,0 \\ 0\,0\,1 \end{bmatrix} - \frac{1}{4}\begin{bmatrix} 2 \\ 0 \\ 0 \end{bmatrix}\begin{bmatrix} 2\,0\,0 \end{bmatrix} = \begin{bmatrix} 0\,0\,0 \\ 0\,1\,0 \\ 0\,0\,1 \end{bmatrix}.$$

The set

$$\mathcal{M} := \{\boldsymbol{w} \in \mathbb{R}^3 : \boldsymbol{w}^T\boldsymbol{v} = 0\} = \{\begin{bmatrix} w_1 \\ w_2 \\ w_3 \end{bmatrix} : 2w_1 = 0\}$$

is the yz plane (cf. Fig. 5.3), $\boldsymbol{H}\boldsymbol{x} = [-1, 0, 1]^T = \boldsymbol{y}$, and $\boldsymbol{P}\boldsymbol{x} = [0, 0, 1]^T = (\boldsymbol{x} + \boldsymbol{y})/2 \in \mathcal{M}$.

Householder transformations can be used to produce zeros in vectors. In the following Theorem we map any vector in $\mathbb{C}^n$ into a multiple of the first unit vector.

Theorem 5.8 (Zeros in Vectors) *For any $\boldsymbol{x} \in \mathbb{C}^n$ there is a Householder transformation $\boldsymbol{H} \in \mathbb{C}^{n\times n}$ such that*

$$\boldsymbol{H}\boldsymbol{x} = a\boldsymbol{e}_1, \quad a = -\rho\|\boldsymbol{x}\|_2, \quad \rho := \begin{cases} x_1/|x_1|, & \text{if } x_1 \neq 0, \\ 1, & \text{otherwise.} \end{cases}$$

Proof If $\boldsymbol{x} = 0$ then $\boldsymbol{H}\boldsymbol{x} = \boldsymbol{0}$ and $a = 0$. Any $\boldsymbol{u}$ with $\|\boldsymbol{u}\|_2 = \sqrt{2}$ will work, and we choose $\boldsymbol{u} := \sqrt{2}\boldsymbol{e}_1$ in this case. For $\boldsymbol{x} \neq \boldsymbol{0}$ we define

$$\boldsymbol{u} := \frac{\boldsymbol{z} + \boldsymbol{e}_1}{\sqrt{1 + z_1}}, \text{ where } \boldsymbol{z} := \overline{\rho}\boldsymbol{x}/\|\boldsymbol{x}\|_2. \tag{5.14}$$

Since $|\rho| = 1$ we have $\rho\|\boldsymbol{x}\|_2\boldsymbol{z} = |\rho|^2\boldsymbol{x} = \boldsymbol{x}$. Moreover, $\|\boldsymbol{z}\|_2 = 1$ and $z_1 = |x_1|/\|\boldsymbol{x}\|_2$ is real so that $\boldsymbol{u}^*\boldsymbol{u} = \frac{(\boldsymbol{z}+\boldsymbol{e}_1)^*(\boldsymbol{z}+\boldsymbol{e}_1)}{1+z_1} = \frac{2+2z_1}{1+z_1} = 2$. Finally,

$$\boldsymbol{H}\boldsymbol{x} = \boldsymbol{x} - (\boldsymbol{u}^*\boldsymbol{x})\boldsymbol{u} = \rho\|\boldsymbol{x}\|_2(\boldsymbol{z} - (\boldsymbol{u}^*\boldsymbol{z})\boldsymbol{u}) = \rho\|\boldsymbol{x}\|_2(\boldsymbol{z} - \frac{(\boldsymbol{z}^* + \boldsymbol{e}_1^*)\boldsymbol{z}}{1 + z_1}(\boldsymbol{z} + \boldsymbol{e}_1))$$

$$= \rho\|\boldsymbol{x}\|_2(\boldsymbol{z} - (\boldsymbol{z} + \boldsymbol{e}_1)) = -\rho\|\boldsymbol{x}\|_2\boldsymbol{e}_1 = a\boldsymbol{e}_1.$$

□

The formulas in Theorem 5.8 are implemented in the following algorithm adapted from [17]. To any given $\boldsymbol{x} \in \mathbb{C}^n$, a number a and a vector $\boldsymbol{u}$ with $\boldsymbol{u}^*\boldsymbol{u} = 2$ is computed so that $(\boldsymbol{I} - \boldsymbol{u}\boldsymbol{u}^*)\boldsymbol{x} = a\boldsymbol{e}_1$:

```
function [u,a]=housegen(x)
% [u,a]=housegen(x)
a=norm(x);
if a==0
   u=x; u(1)=sqrt(2); return;
end
if x(1)== 0
   r=1;
else
   r=x(1)/abs(x(1));
end
u=conj(r)*x/a;
u(1)=u(1)+1;
u=u/sqrt(u(1));
a=-r*a;
end
```

Listing 5.1 housegen

Note that

- In Theorem 5.8 the first component of $\boldsymbol{z}$ is $z_1 = |x_1|/\|\boldsymbol{x}\|_2 \geq 0$. Since $\|\boldsymbol{z}\|_2 = 1$ we have $1 \leq 1 + z_1 \leq 2$. It follows that we avoid cancelation error when computing $1 + z_1$ and $\boldsymbol{u}$ and a are computed in a numerically stable way.
- In order to compute $\boldsymbol{H}\boldsymbol{x}$ for a vector $\boldsymbol{x}$ we do not need to form the matrix $\boldsymbol{H}$. Indeed, $\boldsymbol{H}\boldsymbol{x} = (\boldsymbol{I} - \boldsymbol{u}\boldsymbol{u}^*)\boldsymbol{x} = \boldsymbol{x} - (\boldsymbol{u}^*\boldsymbol{x})\boldsymbol{u}$. If $\boldsymbol{u}, \boldsymbol{x} \in \mathbb{R}^m$ this requires $2m$ operations to find $\boldsymbol{u}^T\boldsymbol{x}$, m operations for $(\boldsymbol{u}^T\boldsymbol{x})\boldsymbol{u}$ and m operations for the final subtraction of the two vectors, a total of $4m$ arithmetic operations. If $\boldsymbol{A} \in \mathbb{R}^{m\times n}$ then $4mn$ operations are required for $\boldsymbol{H}\boldsymbol{A} = \boldsymbol{A} - (\boldsymbol{u}^T\boldsymbol{A})\boldsymbol{u}$, i.e., $4m$ operations for each of the n columns of $\boldsymbol{A}$.
- Householder transformations can also be used to zero out only the lower part of a vector. Suppose $\boldsymbol{x}^T := [\boldsymbol{y}, \boldsymbol{z}]^T$, where $\boldsymbol{y} \in \mathbb{C}^k$, $\boldsymbol{z} \in \mathbb{C}^{n-k}$ for some $1 \leq k < n$. The command $[\hat{\boldsymbol{u}}, a] :=$ housegen($\boldsymbol{z}$) defines a Householder transformation $\hat{\boldsymbol{H}} = \boldsymbol{I} - \hat{\boldsymbol{u}}\hat{\boldsymbol{u}}^*$ so that $\hat{\boldsymbol{H}}\boldsymbol{z} = a\boldsymbol{e}_1 \in \mathbb{C}^{n-k}$. With $\boldsymbol{u} := \left[\begin{smallmatrix}\boldsymbol{0}\\ \hat{\boldsymbol{u}}\end{smallmatrix}\right] \in \mathbb{C}^n$ we see that $\boldsymbol{u}^*\boldsymbol{u} = \hat{\boldsymbol{u}}^*\hat{\boldsymbol{u}} = 2$, and

$$\boldsymbol{H}\boldsymbol{x} = \begin{bmatrix}\boldsymbol{y}\\ a\boldsymbol{e}_1\end{bmatrix}, \text{ where } \boldsymbol{H} := \boldsymbol{I} - \boldsymbol{u}\boldsymbol{u}^* = \begin{bmatrix}\boldsymbol{I} & \boldsymbol{0}\\ \boldsymbol{0} & \boldsymbol{I}\end{bmatrix} - \begin{bmatrix}\boldsymbol{0}\\ \hat{\boldsymbol{u}}\end{bmatrix}\begin{bmatrix}\boldsymbol{0} & \hat{\boldsymbol{u}}^*\end{bmatrix} = \begin{bmatrix}\boldsymbol{I} & \boldsymbol{0}\\ \boldsymbol{0} & \hat{\boldsymbol{H}}\end{bmatrix},$$

defines a Householder transformation that produces zeros in the lower part of $\boldsymbol{x}$.

5.3 Householder Triangulation

We say that a matrix $\boldsymbol{R} \in \mathbb{C}^{m\times n}$ is **upper trapezoidal**, if $r_{i,j} = 0$ for $j < i$ and $i = 2, 3 \ldots, m$. Upper trapezoidal matrices corresponding to $m < n$, $m = n$, and $m > n$ look as follows:

$$\begin{bmatrix} x & x & x & x \\ 0 & x & x & x \\ 0 & 0 & x & x \end{bmatrix}, \quad \begin{bmatrix} x & x & x & x \\ 0 & x & x & x \\ 0 & 0 & x & x \\ 0 & 0 & 0 & x \end{bmatrix}, \quad \begin{bmatrix} x & x & x \\ 0 & x & x \\ 0 & 0 & x \\ 0 & 0 & 0 \end{bmatrix}.$$

In this section we consider a method for bringing a matrix to upper trapezoidal form using Householder transformations.

5.3.1 The Algorithm

We treat the cases $m > n$ and $m \le n$ separately and consider first $m > n$. We describe how to find a sequence $\boldsymbol{H}_1, \ldots, \boldsymbol{H}_n$ of Householder transformations such that

$$\boldsymbol{A}_{n+1} := \boldsymbol{H}_n \boldsymbol{H}_{n-1} \cdots \boldsymbol{H}_1 \boldsymbol{A} = \begin{bmatrix} \boldsymbol{R}_1 \\ \boldsymbol{0} \end{bmatrix} = \boldsymbol{R},$$

and where $\boldsymbol{R}_1$ is square and upper triangular. We define

$$\boldsymbol{A}_1 := \boldsymbol{A}, \quad \boldsymbol{A}_{k+1} = \boldsymbol{H}_k \boldsymbol{A}_k, \quad k = 1, 2, \ldots, n.$$

Suppose $\boldsymbol{A}_k$ has the following form

$$\boldsymbol{A}_k = \left[\begin{array}{ccc|ccccc} a_{1,1}^{(1)} & \cdots & a_{1,k-1}^{(1)} & a_{1,k}^{(1)} & \cdots & a_{1,j}^{(1)} & \cdots & a_{1,n}^{(1)} \\ & \ddots & \vdots & \vdots & & \vdots & & \vdots \\ & & a_{k-1,k-1}^{(k-1)} & a_{k-1,k}^{(k-1)} & \cdots & a_{k-1,j}^{(k-1)} & \cdots & a_{k-1,n}^{(k-1)} \\ \hline & & & a_{k,k}^{(k)} & \cdots & a_{k,j}^{(k)} & \cdots & a_{k,n}^{(k)} \\ & & & \vdots & & \vdots & & \vdots \\ & & & a_{i,k}^{(k)} & \cdots & a_{i,j}^{(k)} & \cdots & a_{i,n}^{(k)} \\ & & & \vdots & & \vdots & & \vdots \\ & & & a_{m,k}^{(k)} & \cdots & a_{m,j}^{(k)} & \cdots & a_{m,n}^{(k)} \end{array}\right] \tag{5.15}$$

$$= \begin{bmatrix} \boldsymbol{B}_k & \boldsymbol{C}_k \\ \boldsymbol{0} & \boldsymbol{D}_k \end{bmatrix}.$$

Thus $\boldsymbol{A}_k$ is upper trapezoidal in its first $k-1$ columns (which is true for $k = 1$).

Let $\hat{\boldsymbol{H}}_k := \boldsymbol{I} - \hat{\boldsymbol{u}}_k \hat{\boldsymbol{u}}_k^*$ be a Householder transformation that maps the first column $[a_{k,k}^{(k)}, \ldots, a_{m,k}^{(k)}]^T$ of $\boldsymbol{D}_k$ to a multiple of $\boldsymbol{e}_1$, $\hat{\boldsymbol{H}}_k(\boldsymbol{D}_k \boldsymbol{e}_1) = \alpha_k \boldsymbol{e}_1$. Using Algorithm 5.1 we have $[\hat{\boldsymbol{u}}_k, \alpha_k] = \texttt{housegen}(\boldsymbol{D}_k \boldsymbol{e}_1)$. Then $\boldsymbol{H}_k := \begin{bmatrix} \boldsymbol{I}_{k-1} & \boldsymbol{0} \\ \boldsymbol{0} & \hat{\boldsymbol{H}}_k \end{bmatrix}$ is a Householder transformation and

$$\boldsymbol{A}_{k+1} := \boldsymbol{H}_k \boldsymbol{A}_k = \begin{bmatrix} \boldsymbol{B}_k & \boldsymbol{C}_k \\ \boldsymbol{0} & \hat{\boldsymbol{H}}_k \boldsymbol{D}_k \end{bmatrix} = \begin{bmatrix} \boldsymbol{B}_{k+1} & \boldsymbol{C}_{k+1} \\ \boldsymbol{0} & \boldsymbol{D}_{k+1} \end{bmatrix},$$

where $\boldsymbol{B}_{k+1} \in \mathbb{C}^{k \times k}$ is upper triangular and $\boldsymbol{D}_{k+1} \in \mathbb{C}^{(m-k)\times(n-k)}$. Thus $\boldsymbol{A}_{k+1}$ is upper trapezoidal in its first k columns and the reduction has been carried one step further. At the end $\boldsymbol{R} := \boldsymbol{A}_{n+1} = \begin{bmatrix} \boldsymbol{R}_1 \\ \boldsymbol{0} \end{bmatrix}$, where $\boldsymbol{R}_1$ is upper triangular.

The process can also be applied to $\boldsymbol{A} \in \mathbb{C}^{m \times n}$ if $m \le n$. If $m = 1$ then $\boldsymbol{A}$ is already in upper trapezoidal form. Suppose $m > 1$. In this case $m - 1$ Householder transformations will suffice and $\boldsymbol{H}_{m-1} \cdots \boldsymbol{H}_1 \boldsymbol{A}$ is upper trapezoidal.

In an algorithm we can store most of the vector $\hat{\boldsymbol{u}}_k = [u_{kk}, \ldots, u_{mk}]^T$ and the matrix and $\boldsymbol{A}_k$ in $\boldsymbol{A}$. However, the elements $u_{k,k}$ and $\alpha_k = r_{k,k}$ have to compete for the diagonal in $\boldsymbol{A}$. For $m = 4$ and $n = 3$ the two possibilities look as follows:

$$\boldsymbol{A} = \begin{bmatrix} u_{11} & r_{12} & r_{13} \\ u_{21} & u_{22} & r_{23} \\ u_{31} & u_{32} & u_{33} \\ u_{41} & u_{42} & u_{43} \end{bmatrix} \text{ or } \boldsymbol{A} = \begin{bmatrix} r_{11} & r_{12} & r_{13} \\ u_{21} & r_{22} & r_{23} \\ u_{31} & u_{32} & r_{33} \\ u_{41} & u_{42} & u_{43} \end{bmatrix}.$$

The following algorithm for Householder triangulation takes $\boldsymbol{A} \in \mathbb{C}^{m \times n}$ and $\boldsymbol{B} \in \mathbb{C}^{m \times r}$ as input, and uses housegen to compute Householder transformations $\boldsymbol{H}_1, \ldots, \boldsymbol{H}_s$ so that $\boldsymbol{R} = \boldsymbol{H}_s \cdots \boldsymbol{H}_1 \boldsymbol{A}$ is upper trapezoidal, and $\boldsymbol{C} = \boldsymbol{H}_s \cdots \boldsymbol{H}_1 \boldsymbol{B}$. The matrices $\boldsymbol{R}$ and $\boldsymbol{C}$ are returned. If $\boldsymbol{B}$ is the empty matrix then $\boldsymbol{C}$ is the empty matrix with m rows and 0 columns. $r_{k,k}$ is stored in $\boldsymbol{A}$, and $u_{k,k}$ is stored in a separate vector. We will see that the algorithm can be used to solve linear systems and least squares problems with right hand side(s) $\boldsymbol{B}$, and to compute the product of the Householder transformations by choosing $\boldsymbol{B} = \boldsymbol{I}$.

```
function [R,C] = housetriang(A,B)
% [R,C] = housetriang(A,B)
[m,n]=size(A); r=size(B,2); A=[A,B];
for k=1:min(n,m-1)
    [v,A(k,k)]=housegen(A(k:m,k));
    C=A(k:m,k+1:n+r); A(k:m,k+1:n+r)=C-v*(v'*C);
end
R=triu(A(:,1:n)); C=A(:,n+1:n+r);
end
```

Listing 5.2 housetriang

Here $v = \hat{\boldsymbol{u}}_k$ and the update is computed as $\hat{\boldsymbol{H}}_k \boldsymbol{C} = (\boldsymbol{I} - \boldsymbol{v}\boldsymbol{v}^*)\boldsymbol{C} = \boldsymbol{C} - \boldsymbol{v}(\boldsymbol{v}^*\boldsymbol{C})$. The MATLAB command `triu` extracts the upper triangular part of $\boldsymbol{A}$ introducing zeros in rows $n+1, \ldots, m$.

5.3.2 *The Number of Arithmetic Operations*

The bulk of the work in Algorithm 5.2 is the computation of $\boldsymbol{C} - \boldsymbol{v} * (\boldsymbol{v}^* * \boldsymbol{C})$ for each k. Since in Algorithm 5.2, $\boldsymbol{C} \in \mathbb{C}^{(m-k+1)\times(n+r-k)}$ and $m \geq n$ the cost of computing the update $\boldsymbol{C} - \boldsymbol{v} * (\boldsymbol{v}^T * \boldsymbol{C})$ in the real case is $4(m-k+1)(n+r-k)$ arithmetic operations. This implies that the work in Algorithm 5.2 can be estimated as

$$\int_0^n 4(m-k)(n+r-k)dk = 2m(n+r)^2 - \frac{2}{3}(n+r)^3. \tag{5.16}$$

For $m = n$ and $r = 0$ this gives $4n^3/3 = 2G_n$ for the number of arithmetic operations to bring a matrix $\boldsymbol{A} \in \mathbb{R}^{n\times n}$ to upper triangular form using Householder transformations.

5.3.3 *Solving Linear Systems Using Unitary Transformations*

Consider now the linear system $\boldsymbol{A}\boldsymbol{x} = \boldsymbol{b}$, where $\boldsymbol{A}$ is square. Using Algorithm 5.2 we obtain an upper triangular system $\boldsymbol{R}\boldsymbol{x} = \boldsymbol{c}$ that is upper triangular and nonsingular if $\boldsymbol{A}$ is nonsingular. Thus, it can be solved by back substitution and we have a method for solving linear systems that is an alternative to Gaussian elimination. The two methods are similar since they both reduce $\boldsymbol{A}$ to upper triangular form using certain transformations and they both work for nonsingular systems.

Which method is better? Here is a very brief discussion.

- Advantages with Householder:
  - Row interchanges are not necessary, but see [3].
  - Numerically stable.
- Advantages with Gauss
  - Half the number of arithmetic operations compared to Householder.
  - Row interchanges are often not necessary.
  - Usually stable (but no guarantee).

Linear systems can be constructed where Gaussian elimination will fail numerically even if row interchanges are used, see [21]. On the other hand the transformations used in Householder triangulation are unitary so the method is quite stable. So why is Gaussian elimination more popular than Householder triangulation? One reason is that the number of arithmetic operations in (5.16) when $m = n$ is $4n^3/3 = 2G_n$, which is twice the number for Gaussian elimination. Numerical stability can be a problem with Gaussian elimination, but years and years of experience shows that it works well for most practical problems and pivoting is often not necessary. Also Gaussian elimination often wins for banded and sparse problems.

5.4 The QR Decomposition and QR Factorization

Gaussian elimination without row interchanges results in an LU factorization $\boldsymbol{A} = \boldsymbol{LU}$ of $\boldsymbol{A} \in \mathbb{C}^{n\times n}$. Consider Householder triangulation of $\boldsymbol{A}$. Applying Algorithm 5.2 gives $\boldsymbol{R} = \boldsymbol{H}_{n-1}\cdots\boldsymbol{H}_1\boldsymbol{A}$ implying the factorization $\boldsymbol{A} = \boldsymbol{QR}$, where $\boldsymbol{Q} = \boldsymbol{H}_1\cdots\boldsymbol{H}_{n-1}$ is unitary and $\boldsymbol{R}$ is upper triangular. This is known as a QR factorization of $\boldsymbol{A}$.

5.4.1 *Existence*

For a rectangular matrix we define the following.

Definition 5.5 (QR Decomposition) Let $\boldsymbol{A} \in \mathbb{C}^{m\times n}$ with $m, n \in \mathbb{N}$. We say that $\boldsymbol{A} = \boldsymbol{QR}$ is a **QR decomposition** of $\boldsymbol{A}$ if $\boldsymbol{Q} \in \mathbb{C}^{m,m}$ is square and unitary and $\boldsymbol{R} \in \mathbb{C}^{m\times n}$ is upper trapezoidal. If $m \geq n$ then $\boldsymbol{R}$ takes the form

$$\boldsymbol{R} = \begin{bmatrix} \boldsymbol{R}_1 \\ \boldsymbol{0}_{m-n,n} \end{bmatrix}$$

where $\boldsymbol{R}_1 \in \mathbb{C}^{n\times n}$ is upper triangular and $\boldsymbol{0}_{m-n,n}$ is the zero matrix with $m-n$ rows and n columns. For $m \geq n$ we call $\boldsymbol{A} = \boldsymbol{Q}_1\boldsymbol{R}_1$ a **QR factorization** of $\boldsymbol{A}$ if $\boldsymbol{Q}_1 \in \mathbb{C}^{m\times n}$ has orthonormal columns and $\boldsymbol{R}_1 \in \mathbb{C}^{n\times n}$ is upper triangular.

Suppose $m \geq n$. A QR factorization is obtained from a QR decomposition $\boldsymbol{A} = \boldsymbol{QR}$ by simply using the first n columns of $\boldsymbol{Q}$ and the first n rows of $\boldsymbol{R}$. Indeed, if we partition $\boldsymbol{Q}$ as $[\boldsymbol{Q}_1, \boldsymbol{Q}_2]$ and $\boldsymbol{R} = \left[\begin{smallmatrix} \boldsymbol{R}_1 \\ \boldsymbol{0} \end{smallmatrix}\right]$, where $\boldsymbol{Q}_1 \in \mathbb{R}^{m\times n}$ and $\boldsymbol{R}_1 \in \mathbb{R}^{n\times n}$ then $\boldsymbol{A} = \boldsymbol{Q}_1\boldsymbol{R}_1$ is a QR factorization of $\boldsymbol{A}$. On the other hand a QR factorization $\boldsymbol{A} = \boldsymbol{Q}_1\boldsymbol{R}_1$ of $\boldsymbol{A}$ can be turned into a QR decomposition by extending the set of columns $\{\boldsymbol{q}_1, \ldots, \boldsymbol{q}_n\}$ of $\boldsymbol{Q}_1$ into an orthonormal basis $\{\boldsymbol{q}_1, \ldots, \boldsymbol{q}_n, \boldsymbol{q}_{n+1}, \ldots, \boldsymbol{q}_m\}$ for $\mathbb{R}^m$ and adding $m-n$ rows of zeros to $\boldsymbol{R}_1$. We then obtain a QR decomposition $\boldsymbol{A} = \boldsymbol{QR}$, where $\boldsymbol{Q} = [\boldsymbol{q}_1, \ldots, \boldsymbol{q}_m]$ and $\boldsymbol{R} = \left[\begin{smallmatrix} \boldsymbol{R}_1 \\ \boldsymbol{0} \end{smallmatrix}\right]$.

Example 5.2 (QR Decomposition and Factorization) Consider the factorization

$$A = \begin{bmatrix} 1 & 3 & 1 \\ 1 & 3 & 7 \\ 1 & -1 & -4 \\ 1 & -1 & 2 \end{bmatrix} = \frac{1}{2} \begin{bmatrix} 1 & 1 & -1 & -1 \\ 1 & 1 & 1 & 1 \\ 1 & -1 & -1 & 1 \\ 1 & -1 & 1 & -1 \end{bmatrix} \times \begin{bmatrix} 2 & 2 & 3 \\ 0 & 4 & 5 \\ 0 & 0 & 6 \\ 0 & 0 & 0 \end{bmatrix} = QR.$$

Since $Q^T Q = I$ and R is upper trapezoidal, this is a QR decomposition of A. A QR factorization $A = Q_1 R_1$ is obtained by dropping the last column of Q and the last row of R, so that

$$A = \frac{1}{2} \begin{bmatrix} 1 & 1 & -1 \\ 1 & 1 & 1 \\ 1 & -1 & -1 \\ 1 & -1 & 1 \end{bmatrix} \times \begin{bmatrix} 2 & 2 & 3 \\ 0 & 4 & 5 \\ 0 & 0 & 6 \end{bmatrix} = Q_1 R_1.$$

Consider existence and uniqueness.

Theorem 5.9 (Existence of QR Decomposition) *Any matrix $A \in \mathbb{C}^{m \times n}$ with $m, n \in \mathbb{N}$ has a QR decomposition.*

Proof If $m = 1$ then A is already in upper trapezoidal form and $A = [1]A$ is a QR decomposition of A. Suppose $m > 1$ and set $s := \min(m-1, n)$. Note that the function `housegen(x)` returns the vector u in a Householder transformation for any vector x. With $B = I$ in Algorithm 5.2 we obtain $R = CA$ and $C = H_s \cdots H_2 H_1$. Thus $A = QR$ is a QR decomposition of A since $Q := C^* = H_1 \cdots H_s$ is a product of unitary matrices and therefore unitary. □

Theorem 5.10 (Uniqueness of QR Factorization) *If $m \geq n$ then the QR factorization is unique if A has linearly independent columns and R has positive diagonal elements.*

Proof Let $A = Q_1 R_1$ be a QR factorization of $A \in \mathbb{C}^{m \times n}$. Now $A^* A = R_1^* Q_1^* Q_1 R_1 = R_1^* R_1$. By Lemma 4.2 the matrix $A^* A$ is positive definite, the matrix R_1 is nonsingular, and if its diagonal elements are positive then $R_1^* R_1$ is the Cholesky factorization of $A^* A$. Since the Cholesky factorization is unique it follows that R_1 is unique and since necessarily $Q_1 = A R_1^{-1}$, it must also be unique. □

Example 5.3 (QR Decomposition and Factorization) Consider finding the QR decomposition and factorization of the matrix $A = \left[\begin{smallmatrix} 2 & -1 \\ -1 & 2 \end{smallmatrix}\right]$ using the method of the uniqueness proof of Theorem 5.10. We find $B := A^T A = \left[\begin{smallmatrix} 5 & -4 \\ -4 & 5 \end{smallmatrix}\right]$. The Cholesky factorization of $B = R^T R$ is given by $R = \frac{1}{\sqrt{5}} \left[\begin{smallmatrix} 5 & -4 \\ 0 & 3 \end{smallmatrix}\right]$. Now $R^{-1} = \frac{1}{3\sqrt{5}} \left[\begin{smallmatrix} 3 & 4 \\ 0 & 5 \end{smallmatrix}\right]$ so $Q = AR^{-1} = \frac{1}{\sqrt{5}} \left[\begin{smallmatrix} 2 & 1 \\ -1 & 2 \end{smallmatrix}\right]$. Since A is square $A = QR$ is both the QR decomposition and QR factorization of A.

5.5 QR and Gram-Schmidt

The Gram-Schmidt orthogonalization of the columns of $\boldsymbol{A}$ can be used to find the QR factorization of $\boldsymbol{A}$.

Theorem 5.11 (QR and Gram-Schmidt) *Suppose $\boldsymbol{A} \in \mathbb{R}^{m\times n}$ has rank n and let $\boldsymbol{v}_1, \ldots, \boldsymbol{v}_n$ be the result of applying Gram Schmidt to the columns $\boldsymbol{a}_1, \ldots, \boldsymbol{a}_n$ of $\boldsymbol{A}$, i.e.,*

$$\boldsymbol{v}_1 = \boldsymbol{a}_1, \quad \boldsymbol{v}_j = \boldsymbol{a}_j - \sum_{i=1}^{j-1} \frac{\boldsymbol{a}_j^T \boldsymbol{v}_i}{\boldsymbol{v}_i^T \boldsymbol{v}_i} \boldsymbol{v}_i, \quad \textit{for } j = 2, \ldots, n. \tag{5.17}$$

Let

$$\boldsymbol{Q}_1 := [\boldsymbol{q}_1, \ldots, \boldsymbol{q}_n], \quad \boldsymbol{q}_j := \frac{\boldsymbol{v}_j}{\|\boldsymbol{v}_j\|_2}, \quad j = 1, \ldots, n \textit{ and}$$

$$\boldsymbol{R}_1 := \begin{bmatrix} \|\boldsymbol{v}_1\|_2 & \boldsymbol{a}_2^T\boldsymbol{q}_1 & \boldsymbol{a}_3^T\boldsymbol{q}_1 & \cdots & \boldsymbol{a}_{n-1}^T\boldsymbol{q}_1 & \boldsymbol{a}_n^T\boldsymbol{q}_1 \\ 0 & \|\boldsymbol{v}_2\|_2 & \boldsymbol{a}_3^T\boldsymbol{q}_2 & \cdots & \boldsymbol{a}_{n-1}^T\boldsymbol{q}_2 & \boldsymbol{a}_n^T\boldsymbol{q}_2 \\ & 0 & \|\boldsymbol{v}_3\|_2 & \cdots & \boldsymbol{a}_{n-1}^T\boldsymbol{q}_3 & \boldsymbol{a}_n^T\boldsymbol{q}_3 \\ & & \ddots & \ddots & \vdots & \vdots \\ & & & \ddots & \|\boldsymbol{v}_{n-1}\|_2 & \boldsymbol{a}_n^T\boldsymbol{q}_{n-1} \\ & & & & 0 & \|\boldsymbol{v}_n\|_2 \end{bmatrix}. \tag{5.18}$$

Then $\boldsymbol{A} = \boldsymbol{Q}_1\boldsymbol{R}_1$ is the unique QR factorization of $\boldsymbol{A}$.

Proof Let $\boldsymbol{Q}_1$ and $\boldsymbol{R}_1$ be given by (5.18). The matrix $\boldsymbol{Q}_1$ is well defined and has orthonormal columns, since $\{\boldsymbol{q}_1, \ldots, \boldsymbol{q}_n\}$ is an orthonormal basis for $\mathcal{R}(\boldsymbol{A})$ by Theorem 5.4. By (5.17)

$$\boldsymbol{a}_j = \boldsymbol{v}_j + \sum_{i=1}^{j-1} \frac{\boldsymbol{a}_j^T \boldsymbol{v}_i}{\boldsymbol{v}_i^T \boldsymbol{v}_i} \boldsymbol{v}_i = r_{jj}\boldsymbol{q}_j + \sum_{i=1}^{j-1} \boldsymbol{q}_i r_{ij} = \boldsymbol{Q}_1\boldsymbol{R}_1\boldsymbol{e}_j, \; j = 1, \ldots, n.$$

Clearly $\boldsymbol{R}_1$ has positive diagonal elements and the factorization is unique. □

Example 5.4 (QR Using Gram-Schmidt) Consider finding the QR decomposition and factorization of the matrix $\boldsymbol{A} = \left[\begin{smallmatrix} 2 & -1 \\ -1 & 2 \end{smallmatrix}\right] = [\boldsymbol{a}_1, \boldsymbol{a}_2]$ using Gram-Schmidt. Using (5.17) we find $\boldsymbol{v}_1 = \boldsymbol{a}_1$ and $\boldsymbol{v}_2 = \boldsymbol{a}_2 - \frac{\boldsymbol{a}_2^T\boldsymbol{v}_1}{\boldsymbol{v}_1^T\boldsymbol{v}_1}\boldsymbol{v}_1 = \frac{3}{5}\left[\begin{smallmatrix} 1 \\ 2 \end{smallmatrix}\right]$. Thus $\boldsymbol{Q} = [\boldsymbol{q}_1, \boldsymbol{q}_2]$, where $\boldsymbol{q}_1 = \frac{1}{\sqrt{5}}\left[\begin{smallmatrix} 2 \\ -1 \end{smallmatrix}\right]$ and $\boldsymbol{q}_2 = \frac{1}{\sqrt{5}}\left[\begin{smallmatrix} 1 \\ 2 \end{smallmatrix}\right]$. By (5.18) we find

$$\boldsymbol{R}_1 = \boldsymbol{R} = \begin{bmatrix} \|\boldsymbol{v}_1\|_2 & \boldsymbol{a}_2^T\boldsymbol{q}_1 \\ 0 & \|\boldsymbol{v}_2\|_2 \end{bmatrix} = \frac{1}{\sqrt{5}} \begin{bmatrix} 5 & -4 \\ 0 & 3 \end{bmatrix}$$

and this agrees with what we found in Example 5.3.

Warning The Gram-Schmidt orthogonalization process should not be used to compute the QR factorization numerically. The columns of $\boldsymbol{Q}_1$ computed in floating point arithmetic using Gram-Schmidt orthogonalization will often be far from orthogonal. There is a modified version of Gram-Schmidt which behaves better numerically, see [2]. Here we only considered Householder transformations (cf. Algorithm 5.2).

5.6 Givens Rotations

In some applications, the matrix we want to triangulate has a special structure. Suppose for example that $\boldsymbol{A} \in \mathbb{R}^{n \times n}$ is square and upper Hessenberg as illustrated by a **Wilkinson diagram** for $n = 4$

$$\boldsymbol{A} = \begin{bmatrix} x & x & x & x \\ x & x & x & x \\ 0 & x & x & x \\ 0 & 0 & x & x \end{bmatrix}.$$

Only one element in each column needs to be annihilated and a full Householder transformation will be inefficient. In this case we can use a simpler transformation.

Definition 5.6 (Givens Rotation, Plane Rotation) A **plane rotation** (also called a **Given's rotation**) is a matrix $\boldsymbol{P} \in \mathbb{R}^{2,2}$ of the form

$$\boldsymbol{P} := \begin{bmatrix} c & s \\ -s & c \end{bmatrix}, \text{ where } c^2 + s^2 = 1.$$

A plane rotation is an orthogonal matrix and there is a unique angle $\theta \in [0, 2\pi)$ such that $c = \cos\theta$ and $s = \sin\theta$. Moreover, the identity matrix is a plane rotation corresponding to $\theta = 0$. A vector $\boldsymbol{x}$ in the plane is rotated an angle θ clockwise by $\boldsymbol{P} = \boldsymbol{R}$. See Exercise 5.16 and Fig. 5.4.

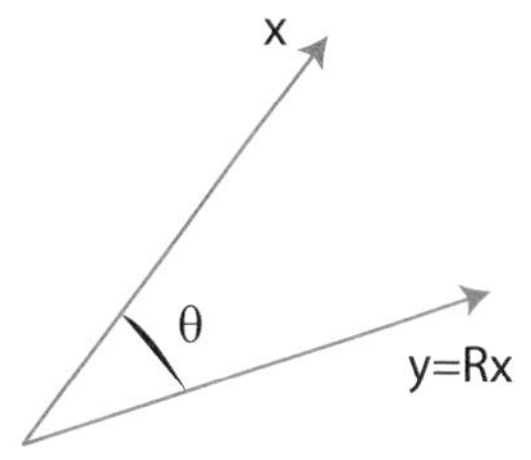

Fig. 5.4 A plane rotation

A Givens rotation can be used to introduce one zero in a vector. Consider first the case of a 2-vector. Suppose

$$\boldsymbol{x} = \begin{bmatrix} x_1 \\ x_2 \end{bmatrix} \neq \boldsymbol{0}, \quad c := \frac{x_1}{r}, \quad s := \frac{x_2}{r}, \quad r := \|\boldsymbol{x}\|_2.$$

If $\boldsymbol{x} \in \mathbb{R}^2$ then

$$\boldsymbol{P}\boldsymbol{x} = \frac{1}{r}\begin{bmatrix} x_1 & x_2 \\ -x_2 & x_1 \end{bmatrix}\begin{bmatrix} x_1 \\ x_2 \end{bmatrix} = \frac{1}{r}\begin{bmatrix} x_1^2 + x_2^2 \\ 0 \end{bmatrix} = \begin{bmatrix} r \\ 0 \end{bmatrix},$$

and we have introduced a zero in $\boldsymbol{x}$. We can take $\boldsymbol{P} = \boldsymbol{I}$ when $\boldsymbol{x} = \boldsymbol{0}$.

For an n-vector $\boldsymbol{x} \in \mathbb{R}^n$ and $1 \le i < j \le n$ we define a **rotation in the i, j-plane** as a matrix $\boldsymbol{P}_{ij} = (p_{kl}) \in \mathbb{R}^{n \times n}$ by $p_{kl} = \delta_{kl}$ except for positions ii, jj, ij, ji, which are given by

$$\begin{bmatrix} p_{ii} & p_{ij} \\ p_{ji} & p_{jj} \end{bmatrix} = \begin{bmatrix} c & s \\ -s & c \end{bmatrix}, \quad \text{where } c^2 + s^2 = 1.$$

Thus, for $n = 4$,

$$\boldsymbol{P}_{12} = \begin{bmatrix} c & s & 0 & 0 \\ -s & c & 0 & 0 \\ 0 & 0 & 1 & 0 \\ 0 & 0 & 0 & 1 \end{bmatrix}, \quad \boldsymbol{P}_{13} = \begin{bmatrix} c & 0 & s & 0 \\ 0 & 1 & 0 & 0 \\ -s & 0 & c & 0 \\ 0 & 0 & 0 & 1 \end{bmatrix}, \quad \boldsymbol{P}_{23} = \begin{bmatrix} 1 & 0 & 0 & 0 \\ 0 & s & c & 0 \\ 0 & -s & c & 0 \\ 0 & 0 & 0 & 1 \end{bmatrix}.$$

Premultiplying a matrix by a rotation in the i, j-plane changes only rows i and j of the matrix, while post multiplying the matrix by such a rotation only changes column i and j. In particular, if $\boldsymbol{B} = \boldsymbol{P}_{ij}\boldsymbol{A}$ and $\boldsymbol{C} = \boldsymbol{A}\boldsymbol{P}_{ij}$ then $\boldsymbol{B}(k, :) = \boldsymbol{A}(k, :)$, $\boldsymbol{C}(:, k) = \boldsymbol{A}(:, k)$ for all $k \neq i, j$ and

$$\begin{bmatrix} \boldsymbol{B}(i, :) \\ \boldsymbol{B}(j, :) \end{bmatrix} = \begin{bmatrix} c & s \\ -s & c \end{bmatrix}\begin{bmatrix} \boldsymbol{A}(i, :) \\ \boldsymbol{A}(j, :) \end{bmatrix}, \quad \begin{bmatrix} \boldsymbol{C}(:, i) & \boldsymbol{C}(:, j) \end{bmatrix} = \begin{bmatrix} \boldsymbol{A}(:, i) & \boldsymbol{A}(:, j) \end{bmatrix}\begin{bmatrix} c & s \\ -s & c \end{bmatrix}. \tag{5.19}$$

Givens rotations can be used as an alternative to Householder transformations for solving linear systems. It can be shown that for a dense system of order n the number of arithmetic operations is asymptotically $2n^3$, corresponding to the work of 3 Gaussian eliminations, while, the work using Householder transformations corresponds to 2 Gaussian eliminations. However, for matrices with a special structure Givens rotations can be used to advantage. As an example consider an upper Hessenberg matrix $\boldsymbol{A} \in \mathbb{R}^{n \times n}$. It can be transformed to upper triangular form using rotations $\boldsymbol{P}_{i,i+1}$ for $i = 1, \ldots, n-1$. For $n = 4$ the process can be illustrated as follows.

$$A = \begin{bmatrix} x & x & x & x \\ x & x & x & x \\ 0 & x & x & x \\ 0 & 0 & x & x \end{bmatrix} \xrightarrow{P_{12}} \begin{bmatrix} r_{11} & r_{12} & r_{13} & r_{14} \\ \mathbf{0} & x & x & x \\ 0 & x & x & x \\ 0 & 0 & x & x \end{bmatrix} \xrightarrow{P_{23}} \begin{bmatrix} r_{11} & r_{12} & r_{13} & r_{14} \\ 0 & r_{22} & r_{23} & r_{24} \\ 0 & \mathbf{0} & x & x \\ 0 & 0 & x & x \end{bmatrix} \xrightarrow{P_{34}} \begin{bmatrix} r_{11} & r_{12} & r_{13} & r_{14} \\ 0 & r_{22} & r_{23} & r_{24} \\ 0 & 0 & r_{33} & r_{34} \\ 0 & 0 & \mathbf{0} & r_{44} \end{bmatrix}.$$

For an algorithm see Exercise 5.18. This reduction is used in the QR method discussed in Chap. 15.

5.7 Exercises Chap. 5

5.7.1 Exercises Sect. 5.1

Exercise 5.1 (The A^*A Inner Product) Suppose $\boldsymbol{A} \in \mathbb{C}^{m\times n}$ has linearly independent columns. Show that $\langle \boldsymbol{x}, \boldsymbol{y}\rangle := \boldsymbol{y}^*\boldsymbol{A}^*\boldsymbol{A}\boldsymbol{x}$ defines an inner product on $\mathbb{C}^n$.

Exercise 5.2 (Angle Between Vectors in Complex Case) Show that in the complex case there is a unique angle θ in $[0, \pi/2]$ such that

$$\cos\theta = \frac{|\langle \boldsymbol{x}, \boldsymbol{y}\rangle|}{\|\boldsymbol{x}\|\|\boldsymbol{y}\|}. \tag{5.20}$$

Exercise 5.3 (x^TAy Inequality (Exam Exercise 1979-3)) Suppose $\boldsymbol{A} \in \mathbb{R}^{n\times n}$ is symmetric positive definite. Show that

$$|\boldsymbol{x}^T\boldsymbol{A}\boldsymbol{y}|^2 \le \boldsymbol{x}^T\boldsymbol{A}\boldsymbol{x}\, \boldsymbol{y}^T\boldsymbol{A}\boldsymbol{y}$$

for all $\boldsymbol{x}, \boldsymbol{y} \in \mathbb{R}^n$, with equality if and only if $\boldsymbol{x}$ and $\boldsymbol{y}$ are linearly dependent.

5.7.2 Exercises Sect. 5.2

Exercise 5.4 (What Does Algorithm Housegen Do When $x = e_1$?) Determine $\boldsymbol{H}$ in Algorithm 5.1 when $\boldsymbol{x} = \boldsymbol{e}_1$.

Exercise 5.5 (Examples of Householder Transformations) If $\boldsymbol{x}, \boldsymbol{y} \in \mathbb{R}^n$ with $\|\boldsymbol{x}\|_2 = \|\boldsymbol{y}\|_2$ and $\boldsymbol{v} := \boldsymbol{x} - \boldsymbol{y} \ne \boldsymbol{0}$ then it follows from Example 5.1 that $\big(\boldsymbol{I} - 2\frac{\boldsymbol{v}\boldsymbol{v}^T}{\boldsymbol{v}^T\boldsymbol{v}}\big)\boldsymbol{x} = \boldsymbol{y}$. Use this to construct a Householder transformation $\boldsymbol{H}$ such that $\boldsymbol{H}\boldsymbol{x} = \boldsymbol{y}$ in the following cases.

a) $\boldsymbol{x} = \begin{bmatrix} 3 \\ 4 \end{bmatrix}, \quad \boldsymbol{y} = \begin{bmatrix} 5 \\ 0 \end{bmatrix}.$

b) $\boldsymbol{x} = \begin{bmatrix} 2 \\ 2 \\ 1 \end{bmatrix}, \quad \boldsymbol{y} = \begin{bmatrix} 0 \\ 3 \\ 0 \end{bmatrix}.$

Exercise 5.6 (2 × 2 **Householder Transformation**) Show that a real 2 × 2 Householder transformation can be written in the form

$$\boldsymbol{H} = \begin{bmatrix} -\cos\phi & \sin\phi \\ \sin\phi & \cos\phi \end{bmatrix}.$$

Find $\boldsymbol{Hx}$ if $\boldsymbol{x} = [\cos\phi, \sin\phi]^T$.

Exercise 5.7 (Householder Transformation (Exam Exercise 2010-1))

a) Suppose $\boldsymbol{x}, \boldsymbol{y} \in \mathbb{R}^n$ with $\|\boldsymbol{x}\|_2 = \|\boldsymbol{y}\|_2$ and $\boldsymbol{v} := \boldsymbol{x} - \boldsymbol{y} \neq 0$. Show that

$$\boldsymbol{Hx} = \boldsymbol{y}, \quad \text{where} \quad \boldsymbol{H} := \boldsymbol{I} - 2\frac{\boldsymbol{v}\boldsymbol{v}^T}{\boldsymbol{v}^T\boldsymbol{v}}.$$

b) Let $\boldsymbol{B} \in \mathbb{R}^{4,4}$ be given by

$$\boldsymbol{B} := \begin{bmatrix} 0 & 1 & 0 & 0 \\ 0 & 0 & 1 & 0 \\ 0 & 0 & 0 & 1 \\ \epsilon & 0 & 0 & 0 \end{bmatrix},$$

where $0 < \epsilon < 1$. Compute a Householder transformation $\boldsymbol{H}$ and a matrix $\boldsymbol{B}_1$ such that the first column of $\boldsymbol{B}_1 := \boldsymbol{HBH}$ has a zero in the last two positions.

5.7.3 *Exercises Sect. 5.4*

Exercise 5.8 (QR Decomposition)

$$\boldsymbol{A} = \begin{bmatrix} 1 & 2 \\ 1 & 2 \\ 1 & 0 \\ 1 & 0 \end{bmatrix}, \quad \boldsymbol{Q} = \frac{1}{2}\begin{bmatrix} 1 & 1 & 1 & 1 \\ 1 & 1 & -1 & -1 \\ 1 & -1 & -1 & 1 \\ 1 & -1 & 1 & -1 \end{bmatrix}, \quad \boldsymbol{R} = \begin{bmatrix} 2 & 2 \\ 0 & 2 \\ 0 & 0 \\ 0 & 0 \end{bmatrix}.$$

Show that $\boldsymbol{Q}$ is orthonormal and that $\boldsymbol{QR}$ is a QR decomposition of $\boldsymbol{A}$. Find a QR factorization of $\boldsymbol{A}$.

Exercise 5.9 (Householder Triangulation)

a) Let

$$\boldsymbol{A} := \begin{bmatrix} 1 & 0 & 1 \\ -2 & -1 & 0 \\ 2 & 2 & 1 \end{bmatrix}.$$

Find Householder transformations $\boldsymbol{H}_1, \boldsymbol{H}_2 \in \mathbb{R}^{3\times 3}$ such that $\boldsymbol{H}_2\boldsymbol{H}_1\boldsymbol{A}$ is upper triangular.

b) Find the QR factorization of $\boldsymbol{A}$, when $\boldsymbol{R}$ has positive diagonal elements.

Exercise 5.10 (Hadamard's Inequality) In this exercise we use the QR factorization to prove a classical determinant inequality. For any $\boldsymbol{A} = [\boldsymbol{a}_1, \dots, \boldsymbol{a}_n] \in \mathbb{C}^{n\times n}$ we have

$$|\det(\boldsymbol{A})| \le \prod_{j=1}^{n} \|\boldsymbol{a}_j\|_2. \tag{5.21}$$

Equality holds if and only if $\boldsymbol{A}$ has a zero column or the columns of $\boldsymbol{A}$ are orthogonal.

a) Show that if $\boldsymbol{Q}$ is unitary then $|\det(\boldsymbol{Q})| = 1$.
b) Let $\boldsymbol{A} = \boldsymbol{Q}\boldsymbol{R}$ be a QR factorization of $\boldsymbol{A}$ and let $\boldsymbol{R} = [\boldsymbol{r}_1, \dots, \boldsymbol{r}_n]$. Show that $(\boldsymbol{A}^*\boldsymbol{A})_{jj} = \|\boldsymbol{a}_j\|_2^2 = (\boldsymbol{R}^*\boldsymbol{R})_{jj} = \|\boldsymbol{r}_j\|_2^2$.
c) Show that $|\det(\boldsymbol{A})| = \prod_{j=1}^{n} |r_{jj}| \le \prod_{j=1}^{n} \|\boldsymbol{a}_j\|_2$.
d) Show that we have equality if $\boldsymbol{A}$ has a zero column,
e) Suppose the columns of $\boldsymbol{A}$ are nonzero. Show that we have equality if and only if the columns of $\boldsymbol{A}$ are orthogonal.[1]

Exercise 5.11 ($\boldsymbol{QL}$ Factorization (Exam Exercise 1982-2)) Suppose $\boldsymbol{B} \in \mathbb{R}^{n\times n}$ is symmetric and positive definite. It can be shown that $\boldsymbol{B}$ has a factorization of the form $\boldsymbol{B} = \boldsymbol{L}^T\boldsymbol{L}$, where $\boldsymbol{L}$ is lower triangular with positive diagonal elements (you should not show this). Note that this is different from the Cholesky factorization $\boldsymbol{B} = \boldsymbol{L}\boldsymbol{L}^T$.

a) Suppose $\boldsymbol{B} = \boldsymbol{L}^T\boldsymbol{L}$. Write down the equations to determine the elements $l_{i,j}$ of $\boldsymbol{L}$, in the order $i = n, n-1, \dots, 1$ and $j = i, 1, 2 \dots, i-1$.
b) Explain (without making a detailed algorithm) how the $\boldsymbol{L}^T\boldsymbol{L}$ factorization can be used to solve the linear system $\boldsymbol{B}\boldsymbol{x} = \boldsymbol{c}$. Compute $\|\boldsymbol{L}\|_F$. Is the algorithm stable?
c) Show that every nonsingular matrix $\boldsymbol{A} \in \mathbb{R}^{n\times n}$ can be factored in the form $\boldsymbol{A} = \boldsymbol{Q}\boldsymbol{L}$, where $\boldsymbol{Q} \in \mathbb{R}^{n\times n}$ is orthogonal and $\boldsymbol{L} \in \mathbb{R}^{n\times n}$ is lower triangular with positive diagonal elements.
d) Show that the $\boldsymbol{QL}$ factorization in c) is unique.

Exercise 5.12 (QL-Factorization (Exam Exercise 1982-3)) In this exercise we will develop an algorithm to find a QL-factorization of $\boldsymbol{A} \in \mathbb{R}^{n\times n}$ (cf. Exam exercise 1982-2) using Householder transformations.

a) Given vectors $\boldsymbol{a} := [a_1, \dots, a_n]^T \in \mathbb{R}^n$ and $\boldsymbol{e}_n := [0, \dots, 0, 1]^T$. Find $\boldsymbol{v} \in \mathbb{R}^n$ such that the Householder transformation $\boldsymbol{H} := \boldsymbol{I} - 2\frac{\boldsymbol{v}\boldsymbol{v}^*}{\boldsymbol{v}^*\boldsymbol{v}}$ satisfies $\boldsymbol{H}\boldsymbol{a} = -s\boldsymbol{e}_n$, where $|s| = \|\boldsymbol{a}\|_2$. How should we choose the sign of s?

---

[1] Show that we have equality $\iff$ $\boldsymbol{R}$ is diagonal $\iff$ $\boldsymbol{A}^*\boldsymbol{A}$ is diagonal.

b) Let $1 \le r \le n$, $\boldsymbol{v}_r \in \mathbb{R}^r$, $\boldsymbol{v}_r \ne \boldsymbol{0}$, and

$$\boldsymbol{V}_r := \boldsymbol{I}_r - 2\frac{\boldsymbol{v}_r\boldsymbol{v}_r^*}{\boldsymbol{v}_r^*\boldsymbol{v}_r} = \boldsymbol{I}_r - \boldsymbol{u}_r\boldsymbol{u}_r^*, \text{ with } \boldsymbol{u}_r := \sqrt{2}\frac{\boldsymbol{v}_r}{\|\boldsymbol{v}_r\|_2}.$$

Show that $\boldsymbol{H} := \begin{bmatrix} \boldsymbol{V}_r & \boldsymbol{0} \\ \boldsymbol{0} & \boldsymbol{I}_{n-r} \end{bmatrix}$ is a Householder transformation. Show also that if $a_{i,j} = 0$ for $i = 1, \ldots, r$ and $j = r+1, \ldots, n$ then the last r columns of $\boldsymbol{A}$ and $\boldsymbol{HA}$ are the same.

c) Explain, without making a detailed algorithm, how we to a given matrix $\boldsymbol{A} \in \mathbb{R}^{n\times n}$ can find Householder transformations $\boldsymbol{H}_1, \ldots, \boldsymbol{H}_{n-1}$ such that $\boldsymbol{H}_{n-1}, \ldots, \boldsymbol{H}_1\boldsymbol{A}$ is lower triangular. Give a $\boldsymbol{QL}$ factorization of $\boldsymbol{A}$.

Exercise 5.13 (QR Fact. of Band Matrices (Exam Exercise 2006-2)) Let $\boldsymbol{A} \in \mathbb{R}^{n\times n}$ be a nonsingular symmetric band matrix with bandwidth $d \le n-1$, so that $a_{ij} = 0$ for all i, j with $|i-j| > d$. We define $\boldsymbol{B} := \boldsymbol{A}^T\boldsymbol{A}$ and let $\boldsymbol{A} = \boldsymbol{QR}$ be the QR factorization of $\boldsymbol{A}$ where $\boldsymbol{R}$ has positive diagonal entries.

a) Show that $\boldsymbol{B}$ is symmetric.
b) Show that $\boldsymbol{B}$ has bandwidth $\le 2d$.
c) Write a MATLAB `function B=ata(A,d)` which computes $\boldsymbol{B}$. You shall exploit the symmetry and the function should only use $\mathcal{O}(cn^2)$ flops, where c only depends on d.
d) Estimate the number of arithmetic operations in your algorithm.
e) Show that $\boldsymbol{A}^T\boldsymbol{A} = \boldsymbol{R}^T\boldsymbol{R}$.
f) Explain why $\boldsymbol{R}$ has upper bandwidth $2d$.
g) We consider 3 methods for finding the QR factorization of the band matrix $\boldsymbol{A}$, where we assume that n is much bigger than d. The methods are based on

1. Gram-Schmidt orthogonalization,
2. Householder transformations,
3. Givens rotations.

Which method would you recommend for a computer program using floating point arithmetic? Give reasons for your answer.

Exercise 5.14 (Find QR Factorization (Exam Exercise 2008-2)) Let

$$\boldsymbol{A} := \begin{bmatrix} 2 & 1 \\ 2 & -3 \\ -2 & -1 \\ -2 & 3 \end{bmatrix}$$

a) Find the Cholesky factorization of $\boldsymbol{A}^T\boldsymbol{A}$.
b) Find the QR factorization of $\boldsymbol{A}$.

5.7.4 Exercises Sect. 5.5

Exercise 5.15 (QR Using Gram-Schmidt, II) Construct $\boldsymbol{Q}_1$ and $\boldsymbol{R}_1$ in Example 5.2 using Gram-Schmidt orthogonalization.

5.7.5 Exercises Sect. 5.6

Exercise 5.16 (Plane Rotation) Show that if $\boldsymbol{x} = \begin{bmatrix} r\cos\alpha \\ r\sin\alpha \end{bmatrix}$ then $\boldsymbol{P}\boldsymbol{x} = \begin{bmatrix} r\cos(\alpha-\theta) \\ r\sin(\alpha-\theta) \end{bmatrix}$.

Exercise 5.17 (Updating the QR Decomposition) Let $\boldsymbol{H} \in \mathbb{R}^{4,4}$ be upper Hessenberg. Find Givens rotation matrices $\boldsymbol{G}_1, \boldsymbol{G}_2, \boldsymbol{G}_3$ such that

$$\boldsymbol{G}_3\boldsymbol{G}_2\boldsymbol{G}_1\boldsymbol{H} = \boldsymbol{R}$$

is upper triangular. (Here each $\boldsymbol{G}_k = \boldsymbol{P}_{i,j}$ for suitable i, j, c og s, and for each k you are meant to find suitable i and j.)

Exercise 5.18 (Solving Upper Hessenberg System Using Rotations) Let $\boldsymbol{A} \in \mathbb{R}^{n\times n}$ be upper Hessenberg and nonsingular, and let $\boldsymbol{b} \in \mathbb{R}^n$. The following algorithm solves the linear system $\boldsymbol{A}\boldsymbol{x} = \boldsymbol{b}$ using rotations $\boldsymbol{P}_{k,k+1}$ for $k = 1, \ldots, n-1$. It uses the back solve Algorithm 3.2. Determine the number of arithmetic operations of this algorithm.

```
function x=rothesstri(A,b)
% x=rothesstri(A,b)
n=length(A); A=[A b];
for k=1:n-1
   r=norm([A(k,k),A(k+1,k)]);
   if r>0
      c=A(k,k)/r; s=A(k+1,k)/r;
      A([k k+1],k+1:n+1) ...
            =[c s;-s c]*A([k k+1],k+1:n+1);
   end
   A(k,k)=r; A(k+1,k)=0;
end
x=rbacksolve(A(:,1:n),A(:,n+1),n);
end
```

Listing 5.3 rothesstri

Exercise 5.19 (A Givens Transformation (Exam Exercise 2013-2)) A Givens rotation of order 2 has the form $\boldsymbol{G} := \begin{bmatrix} c & s \\ -s & c \end{bmatrix} \in \mathbb{R}^{2\times 2}$, where $s^2 + c^2 = 1$.

a) Is $\boldsymbol{G}$ symmetric and unitary?

b) Given $x_1, x_2 \in \mathbb{R}$ and set $r := \sqrt{x_1^2 + x_2^2}$. Find $\boldsymbol{G}$ and y_1, y_2 so that $y_1 = y_2$, where $\begin{bmatrix} y_1 \\ y_2 \end{bmatrix} = \boldsymbol{G} \begin{bmatrix} x_1 \\ x_2 \end{bmatrix}$.

Exercise 5.20 (Givens Transformations (Exam Exercise 2016-3)) Recall that a rotation in the ij-plane is an $m \times m$-matrix, denoted $\boldsymbol{P}_{i,j}$, which differs from the identity matrix only in the entries ii, ij, ji, jj, which equal

$$\begin{bmatrix} p_{ii} & p_{ij} \\ p_{ji} & p_{jj} \end{bmatrix} = \begin{bmatrix} \cos\theta & \sin\theta \\ -\sin\theta & \cos\theta \end{bmatrix},$$

i.e., these four entries are those of a Givens rotation.

a) For $\theta \in \mathbb{R}$, let $\boldsymbol{P}$ be a Givens rotation of the form

$$\boldsymbol{P} = \begin{bmatrix} \cos\theta & \sin\theta \\ -\sin\theta & \cos\theta \end{bmatrix}$$

and let $\boldsymbol{x}$ be a fixed vector in $\mathbb{R}^2$. Show that there exists a unique $\theta \in (-\pi/2, \pi/2]$ so that $\boldsymbol{P}\boldsymbol{x} = \pm\|\boldsymbol{x}\|_2 \boldsymbol{e}_1$, where $\boldsymbol{e}_1 = (1, 0)^T$.

b) Show that, for any vector $\boldsymbol{w} \in \mathbb{R}^m$, one can find rotations in the 12-plane, 23-plane, ..., $(m-1)m$-plane, so that

$$\boldsymbol{P}_{1,2}\boldsymbol{P}_{2,3}\cdots\boldsymbol{P}_{m-2,m-1}\boldsymbol{P}_{m-1,m}\boldsymbol{w} = \begin{bmatrix} \alpha \\ 0 \\ \vdots \\ 0 \end{bmatrix},$$

where $\alpha = \pm\|\boldsymbol{w}\|_2$.

c) Assume that $m \geq n$. Recall that an $m \times n$-matrix $\boldsymbol{A}$ with entries $a_{i,j}$ is called upper trapezoidal if there are no nonzero entries below the main diagonal

$$(a_{1,1}, a_{2,2}, \ldots, a_{n,n})$$

(for $m = n$, upper trapezoidal is the same as upper triangular). Recall also that an $m \times n$-matrix is said to be in upper Hessenberg form if there are no nonzero entries below the subdiagonal

$$(a_{2,1}, a_{3,2}, \ldots, a_{n,n-1}).$$

Explain that, if an $m \times n$-matrix $\boldsymbol{H}$ is in upper Hessenberg form, one can find plane rotations so that

$$\boldsymbol{P}_{m-1,m}\boldsymbol{P}_{m-2,m-1}\cdots\boldsymbol{P}_{2,3}\boldsymbol{P}_{1,2}\boldsymbol{H}$$

is upper trapezoidal.

d) Let again $\boldsymbol{A}$ be an $m \times n$-matrix with $m \geq n$, and let $\boldsymbol{A}_-$ be the matrix obtained by removing column k in $\boldsymbol{A}$. Explain how you can find a QR Decomposition of $\boldsymbol{A}_-$, when we already have a QR decomposition $\boldsymbol{A} = \boldsymbol{Q}\boldsymbol{R}$ of $\boldsymbol{A}$.[2]

Exercise 5.21 (Cholesky and Givens (Exam Exercise 2018-2)) Assume that $\boldsymbol{A}$ is $n \times n$ symmetric positive definite, and with Cholesky factorization $\boldsymbol{A} = \boldsymbol{L}\boldsymbol{L}^*$. Assume also that z is a given column vector of length n.

a) Explain why $\boldsymbol{A} + zz^*$ has a unique Cholesky factorization.

b) Assume that we are given a QR decomposition

$$\begin{bmatrix} \boldsymbol{L}^* \\ z^* \end{bmatrix} = \boldsymbol{Q}\begin{bmatrix} \boldsymbol{R} \\ 0 \end{bmatrix},$$

with $\boldsymbol{R}$ square and upper triangular. Explain why $\boldsymbol{R}$ is nonsingular. Explain also that, if $\boldsymbol{R}$ also has nonnegative diagonal entries, then $\boldsymbol{A} + zz^*$ has the Cholesky factorization $\boldsymbol{R}^*\boldsymbol{R}$.

c) Explain how one can find plane rotations $P_{i_1,n+1}, P_{i_2,n+1},\ldots,P_{i_n,n+1}$ so that

$$P_{i_1,n+1}P_{i_2,n+1}\cdots P_{i_n,n+1}\begin{bmatrix} \boldsymbol{L}^* \\ z^* \end{bmatrix} = \begin{bmatrix} \boldsymbol{R}' \\ 0 \end{bmatrix}, \tag{5.22}$$

with $\boldsymbol{R}'$ upper triangular, and explain how to obtain a QR decomposition of $\begin{bmatrix} \boldsymbol{L}^* \\ z^* \end{bmatrix}$ from this. In particular you should write down the numbers $i_1, \ldots, i_n$. Is it possible to choose the plane rotations so that $\boldsymbol{R}'$ in (5.22) also has positive diagonal entries?

5.8 Review Questions

5.8.1 What is a Householder transformation?

5.8.2 Why are they good for numerical work?

5.8.3 What are the main differences between solving a linear system by Gaussian elimination and Householder transformations?

[2] Consider the matrix $\boldsymbol{Q}^T\boldsymbol{A}_-$.

5.8.4 What are the differences between a QR decomposition and a QR factorization?
5.8.5 Does any matrix have a QR decomposition?
5.8.6 What is a Givens transformation?
5.8.7 Is a unitary matrix always well conditioned?

Part II
Eigenpairs and Singular Values

We turn now to eigenpairs of matrices, i.e., eigenvalues and corresponding eigenvectors. The eigenpairs of a matrix are easily determined if it is diagonal. Indeed, the eigenvalues are the diagonal elements and the eigenvectors are unit vectors. We will see that not all matrices can be reduced to diagonal form using eigenvalue preserving transformations known as similarity transformations. This raises the question: how close to a diagonal matrix can we reduce a general matrix using similarity transformations? We give one answer to this question, the Jordan factorization or the Jordan canonical form. We also characterize matrices which can be diagonalized using unitary similarity transformations, and study the subclass of Hermitian matrices. Numerical methods for determining eigenvalues and eigenvectors will be considered in Chaps. 14 and 15.

In the second chapter in this part we consider the important singular value decomposition of a rectangular matrix. This decomposition will play a central role in several of the remaining chapters in this book.

Chapter 6
Eigenpairs and Similarity Transformations

We have seen that a Hermitian matrix is positive definite if and only if it has positive eigenvalues. Eigenvalues and some related quantities called singular values occur in many branches of applied mathematics and are also needed for a deeper study of linear systems and least squares problems. In this and the next chapter we study eigenvalues and singular values. Recall that if $\boldsymbol{A} \in \mathbb{C}^{n\times n}$ is a square matrix, $\lambda \in \mathbb{C}$ and $\boldsymbol{x} \in \mathbb{C}^n$ then $(\lambda, \boldsymbol{x})$ is an **eigenpair** for $\boldsymbol{A}$ if $\boldsymbol{A}\boldsymbol{x} = \lambda\boldsymbol{x}$ and $\boldsymbol{x}$ is nonzero. The scalar λ is called an **eigenvalue** and $\boldsymbol{x}$ is said to be an **eigenvector**. The set of eigenvalues is called the **spectrum** of $\boldsymbol{A}$ and is denoted by $\sigma(\boldsymbol{A})$. For example, $\sigma(\boldsymbol{I}) = \{1, \ldots, 1\} = \{1\}$. The eigenvalues are the roots of the **characteristic polynomial** of $\boldsymbol{A}$ given for $\lambda \in \mathbb{C}$ by

$$\pi_{\boldsymbol{A}}(\lambda) = \det(\boldsymbol{A} - \lambda\boldsymbol{I}).$$

The equation $\det(\boldsymbol{A} - \lambda\boldsymbol{I}) = 0$ is called the **characteristic equation** of $\boldsymbol{A}$. Equivalently the characteristic equation can be written $\det(\lambda\boldsymbol{I} - \boldsymbol{A}) = 0$.

6.1 Defective and Nondefective Matrices

For the eigenvectors we will see that it is important to know if the eigenvectors of a matrix of order n form a basis for $\mathbb{C}^n$. We say that $\boldsymbol{A}$ is **defective** if the eigenvectors do not form a basis for $\mathbb{C}^n$ and **nondefective** otherwise.

We have the following sufficient condition for a matrix to be nondefective.

Theorem 6.1 (Distinct Eigenvalues) *A matrix with distinct eigenvalues is nondefective, i.e., its eigenvectors are linearly independent.*

T. Lyche, *Numerical Linear Algebra and Matrix Factorizations*, Texts in Computational Science and Engineering 22, https://doi.org/10.1007/978-3-030-36468-7_6

Proof The proof is by contradiction. Suppose $\boldsymbol{A}$ has eigenpairs $(\lambda_k, \boldsymbol{x}_k)$, $k = 1, \ldots, n$, with linearly dependent eigenvectors. Let m be the smallest integer such that $\{\boldsymbol{x}_1, \ldots, \boldsymbol{x}_m\}$ is linearly dependent. Thus $\sum_{j=1}^m c_j \boldsymbol{x}_j = \boldsymbol{0}$, where at least one c_j is nonzero. We must have $m \geq 2$ since eigenvectors are nonzero. We find

$$\sum_{j=1}^m c_j \boldsymbol{x}_j = \boldsymbol{0} \Rightarrow \sum_{j=1}^m c_j \boldsymbol{A}\boldsymbol{x}_j = \sum_{j=1}^m c_j \lambda_j \boldsymbol{x}_j = \boldsymbol{0}.$$

From the last relation we subtract $\sum_{j=1}^m c_j \lambda_m \boldsymbol{x}_j = \boldsymbol{0}$ and find $\sum_{j=1}^{m-1} c_j (\lambda_j - \lambda_m)\boldsymbol{x}_j = \boldsymbol{0}$. But since $\lambda_j - \lambda_m \neq 0$ for $j = 1, \ldots, m-1$ and at least one $c_j \neq 0$ for $j < m$ we see that $\{\boldsymbol{x}_1, \ldots, \boldsymbol{x}_{m-1}\}$ is linearly dependent, contradicting the minimality of m. □

If some of the eigenvalues occur with multiplicity higher than one then the matrix can be either defective or nondefective.

Example 6.1 (Defective and Nondefective Matrices) Consider the matrices

$$\boldsymbol{I} := \begin{bmatrix} 1 & 0 \\ 0 & 1 \end{bmatrix}, \quad \boldsymbol{J} := \begin{bmatrix} 1 & 1 \\ 0 & 1 \end{bmatrix}.$$

Since $\boldsymbol{I}\boldsymbol{x} = \boldsymbol{x}$ and $\lambda_1 = \lambda_2 = 1$ any vector $\boldsymbol{x} \in \mathbb{C}^2$ is an eigenvector for $\boldsymbol{I}$. In particular the two unit vectors $\boldsymbol{e}_1$ and $\boldsymbol{e}_2$ are eigenvectors and form an orthonormal basis for $\mathbb{C}^2$. We conclude that the identity matrix is nondefective. The matrix $\boldsymbol{J}$ also has the eigenvalue one with multiplicity two, but since $\boldsymbol{J}\boldsymbol{x} = \boldsymbol{x}$ if and only if $x_2 = 0$, any eigenvector must be a multiple of $\boldsymbol{e}_1$. Thus $\boldsymbol{J}$ is defective.

If the eigenvectors $\boldsymbol{x}_1, \ldots, \boldsymbol{x}_n$ form a basis for $\mathbb{C}^n$ then any $\boldsymbol{x} \in \mathbb{C}^n$ can be written

$$\boldsymbol{x} = \sum_{j=1}^n c_j \boldsymbol{x}_j \text{ for some scalars } c_1, \ldots, c_n.$$

We call this an **eigenvector expansion** of $\boldsymbol{x}$. Thus to any nondefective matrix there corresponds an eigenvector expansion.

Example 6.2 (Eigenvector Expansion Example) Eigenpairs of $\begin{bmatrix} 2 & -1 \\ -1 & 2 \end{bmatrix}$ are $(1, [1, 1]^T)$ and $(3, [1, -1]^T)$. Any $\boldsymbol{x} = [x_1, x_2]^T \in \mathbb{C}^2$ has the eigenvector expansion

$$\boldsymbol{x} = \frac{x_1 + x_2}{2} \begin{bmatrix} 1 \\ 1 \end{bmatrix} + \frac{x_1 - x_2}{2} \begin{bmatrix} 1 \\ -1 \end{bmatrix}.$$

6.1.1 Similarity Transformations

We need a transformation that can be used to simplify a matrix without changing the eigenvalues.

Definition 6.1 (Similar Matrices) Two matrices $\boldsymbol{A}, \boldsymbol{B} \in \mathbb{C}^{n\times n}$ are said to be **similar** if there is a nonsingular matrix $\boldsymbol{S} \in \mathbb{C}^{n\times n}$ such that $\boldsymbol{B} = \boldsymbol{S}^{-1}\boldsymbol{A}\boldsymbol{S}$. The transformation $\boldsymbol{A} \to \boldsymbol{B}$ is called a **similarity transformation**. The columns of $\boldsymbol{S}$ are denoted by $\boldsymbol{s}_1, \boldsymbol{s}_2, \ldots, \boldsymbol{s}_n$.

We note that

1. Similar matrices have the same eigenvalues, they even have the same characteristic polynomial. Indeed, by the product rule for determinants $\det(\boldsymbol{A}\boldsymbol{C}) = \det(\boldsymbol{A})\det(\boldsymbol{C})$ so that

$$\begin{aligned}\pi_{\boldsymbol{B}}(\lambda) &= \det(\boldsymbol{S}^{-1}\boldsymbol{A}\boldsymbol{S} - \lambda\boldsymbol{I}) = \det\big(\boldsymbol{S}^{-1}(\boldsymbol{A} - \lambda\boldsymbol{I})\boldsymbol{S}\big) \\ &= \det(\boldsymbol{S}^{-1})\det(\boldsymbol{A} - \lambda\boldsymbol{I})\det(\boldsymbol{S}) = \det(\boldsymbol{S}^{-1}\boldsymbol{S})\det(\boldsymbol{A} - \lambda\boldsymbol{I}) = \pi_{\boldsymbol{A}}(\lambda),\end{aligned}$$

since $\det(\boldsymbol{I}) = 1$.
2. $(\lambda, \boldsymbol{x})$ is an eigenpair for $\boldsymbol{S}^{-1}\boldsymbol{A}\boldsymbol{S}$ if and only if $(\lambda, \boldsymbol{S}\boldsymbol{x})$ is an eigenpair for $\boldsymbol{A}$. In fact $(\boldsymbol{S}^{-1}\boldsymbol{A}\boldsymbol{S})\boldsymbol{x} = \lambda\boldsymbol{x}$ if and only if $\boldsymbol{A}(\boldsymbol{S}\boldsymbol{x}) = \lambda(\boldsymbol{S}\boldsymbol{x})$.
3. If $\boldsymbol{S}^{-1}\boldsymbol{A}\boldsymbol{S} = \boldsymbol{D} = \operatorname{diag}(\lambda_1, \ldots, \lambda_n)$ we can partition $\boldsymbol{A}\boldsymbol{S} = \boldsymbol{S}\boldsymbol{D}$ by columns to obtain $[\boldsymbol{A}\boldsymbol{s}_1, \ldots, \boldsymbol{A}\boldsymbol{s}_n] = [\lambda_1\boldsymbol{s}_1, \ldots, \lambda_n\boldsymbol{s}_n]$. Thus the columns of $\boldsymbol{S}$ are eigenvectors of $\boldsymbol{A}$. Moreover, $\boldsymbol{A}$ is nondefective since $\boldsymbol{S}$ is nonsingular. Conversely, if $\boldsymbol{A}$ is nondefective then it can be diagonalized by a similarity transformation $\boldsymbol{S}^{-1}\boldsymbol{A}\boldsymbol{S}$, where the columns of $\boldsymbol{S}$ are eigenvectors of $\boldsymbol{A}$.
4. For any square matrices $\boldsymbol{A}, \boldsymbol{C} \in \mathbb{C}^{n\times n}$ the two products $\boldsymbol{A}\boldsymbol{C}$ and $\boldsymbol{C}\boldsymbol{A}$ have the same characteristic polynomial. More generally, for rectangular matrices $\boldsymbol{A} \in \mathbb{C}^{m\times n}$ and $\boldsymbol{C} \in \mathbb{C}^{n\times m}$, with say $m > n$, the bigger matrix has $m - n$ extra zero eigenvalues

$$\pi_{\boldsymbol{A}\boldsymbol{C}}(\lambda) = \lambda^{m-n}\pi_{\boldsymbol{C}\boldsymbol{A}}(\lambda), \quad \lambda \in \mathbb{C}. \tag{6.1}$$

To show this define for any $m, n \in \mathbb{N}$ block triangular matrices of order $n + m$ by

$$\boldsymbol{E} := \begin{bmatrix} \boldsymbol{A}\boldsymbol{C} & \boldsymbol{0} \\ \boldsymbol{C} & \boldsymbol{0} \end{bmatrix}, \quad \boldsymbol{F} := \begin{bmatrix} \boldsymbol{0} & \boldsymbol{0} \\ \boldsymbol{C} & \boldsymbol{C}\boldsymbol{A} \end{bmatrix}, \quad \boldsymbol{S} = \begin{bmatrix} \boldsymbol{I} & \boldsymbol{A} \\ \boldsymbol{0} & \boldsymbol{I} \end{bmatrix}.$$

The matrix $\boldsymbol{S}$ is nonsingular with $\boldsymbol{S}^{-1} = \begin{bmatrix} \boldsymbol{I} & -\boldsymbol{A} \\ \boldsymbol{0} & \boldsymbol{I} \end{bmatrix}$. Moreover, $\boldsymbol{E}\boldsymbol{S} = \boldsymbol{S}\boldsymbol{F}$ so $\boldsymbol{E}$ and $\boldsymbol{F}$ are similar and therefore have the same characteristic polynomials. Moreover, this polynomial is the product of the characteristic polynomial of the diagonal blocks. But then $\pi_{\boldsymbol{E}}(\lambda) = \lambda^n\pi_{\boldsymbol{A}\boldsymbol{C}}(\lambda) = \pi_{\boldsymbol{F}}(\lambda) = \lambda^m\pi_{\boldsymbol{C}\boldsymbol{A}}(\lambda)$. This implies the statements for $m \geq n$.

6.1.2 Algebraic and Geometric Multiplicity of Eigenvalues

Linear independence of eigenvectors depends on the multiplicity of the eigenvalues in a nontrivial way. For multiple eigenvalues we need to distinguish between two kinds of multiplicities.

Suppose $\boldsymbol{A} \in \mathbb{C}^{n \times n}$ has k distinct eigenvalues $\lambda_1, \ldots, \lambda_k$ with multiplicities $a_1, \ldots, a_k$ so that

$$\pi_{\boldsymbol{A}}(\lambda) := \det(\boldsymbol{A} - \lambda \boldsymbol{I}) = (\lambda_1 - \lambda)^{a_1} \cdots (\lambda_k - \lambda)^{a_k}, \quad \lambda_i \neq \lambda_j,\ i \neq j,\ \sum_{i=1}^{k} a_i = n. \tag{6.2}$$

The positive integer $a_i = a(\lambda_i) = a_{\boldsymbol{A}}(\lambda_i)$ is called the **multiplicity**, or more precisely the **algebraic multiplicity** of the eigenvalue λ_i. The multiplicity of an eigenvalue is simple (double, triple) if a_i is equal to one (two, three).

To define a second kind of multiplicity we consider for each $\lambda \in \sigma(\boldsymbol{A})$ the nullspace

$$\mathcal{N}(\boldsymbol{A} - \lambda \boldsymbol{I}) := \{\boldsymbol{x} \in \mathbb{C}^n : (\boldsymbol{A} - \lambda \boldsymbol{I})\boldsymbol{x} = \boldsymbol{0}\} \tag{6.3}$$

of $\boldsymbol{A} - \lambda \boldsymbol{I}$. The nullspace is a subspace of $\mathbb{C}^n$ consisting of all eigenvectors of $\boldsymbol{A}$ corresponding to the eigenvalue λ. The dimension of the subspace must be at least one since $\boldsymbol{A} - \lambda \boldsymbol{I}$ is singular.

Definition 6.2 (Geometric Multiplicity) The **geometric multiplicity** $g = g(\lambda) = g_{\boldsymbol{A}}(\lambda)$ of an eigenvalue λ of $\boldsymbol{A}$ is the dimension of the nullspace $\mathcal{N}(\boldsymbol{A} - \lambda \boldsymbol{I})$.

Example 6.3 (Geometric Multiplicity) The $n \times n$ identity matrix $\boldsymbol{I}$ has the eigenvalue $\lambda = 1$ with $\pi_{\boldsymbol{I}}(\lambda) = (1 - \lambda)^n$. Since $\boldsymbol{I} - \lambda \boldsymbol{I}$ is the zero matrix when $\lambda = 1$, the nullspace of $\boldsymbol{I} - \lambda \boldsymbol{I}$ is all of n-space and it follows that $a = g = n$. On the other hand we saw in Example 6.1 that the matrix $\boldsymbol{J} := \begin{bmatrix} 1 & 1 \\ 0 & 1 \end{bmatrix}$ has the eigenvalue $\lambda = 1$ with $a = 2$ and any eigenvector is a multiple of $\boldsymbol{e}_1$. Thus $g = 1$.

Theorem 6.2 (Geometric Multiplicity of Similar Matrices) *Similar matrices have the same eigenvalues with the same algebraic and geometric multiplicities.*

Proof Similar matrices have the same characteristic polynomials and only the invariance of geometric multiplicity needs to be shown. Suppose $\lambda \in \sigma(\boldsymbol{A})$, $\dim \mathcal{N}(\boldsymbol{S}^{-1}\boldsymbol{A}\boldsymbol{S} - \lambda \boldsymbol{I}) = k$, and $\dim \mathcal{N}(\boldsymbol{A} - \lambda \boldsymbol{I}) = \ell$. We need to show that $k = \ell$. Suppose $\boldsymbol{v}_1, \ldots, \boldsymbol{v}_k$ is a basis for $\mathcal{N}(\boldsymbol{S}^{-1}\boldsymbol{A}\boldsymbol{S} - \lambda \boldsymbol{I})$. Then $\boldsymbol{S}^{-1}\boldsymbol{A}\boldsymbol{S}\boldsymbol{v}_i = \lambda \boldsymbol{v}_i$ or $\boldsymbol{A}\boldsymbol{S}\boldsymbol{v}_i = \lambda \boldsymbol{S}\boldsymbol{v}_i$, $i = 1, \ldots, k$. But then $\{\boldsymbol{S}\boldsymbol{v}_1, \ldots, \boldsymbol{S}\boldsymbol{v}_k\} \subset \mathcal{N}(\boldsymbol{A} - \lambda \boldsymbol{I})$, which implies that $k \leq \ell$. Similarly, if $\boldsymbol{w}_1, \ldots, \boldsymbol{w}_\ell$ is a basis for $\mathcal{N}(\boldsymbol{A} - \lambda \boldsymbol{I})$ then $\{\boldsymbol{S}^{-1}\boldsymbol{w}_1, \ldots, \boldsymbol{S}^{-1}\boldsymbol{w}_\ell\} \subset \mathcal{N}(\boldsymbol{S}^{-1}\boldsymbol{A}\boldsymbol{S} - \lambda \boldsymbol{I})$. which implies that $k \geq \ell$. We conclude that $k = \ell$. □

For a proof of the following theorem see the next section.[1]

Theorem 6.3 (Geometric Multiplicity) *We have*

1. *The geometric multiplicity of an eigenvalue is always bounded above by the algebraic multiplicity of the eigenvalue.*
2. *The number of linearly independent eigenvectors of a matrix equals the sum of the geometric multiplicities of the eigenvalues.*
3. *A matrix $\boldsymbol{A} \in \mathbb{C}^{n\times n}$ has n linearly independent eigenvectors if and only if the algebraic and geometric multiplicity of all eigenvalues are the same.*

6.2 The Jordan Factorization

We have seen that a nondefective matrix can be diagonalized by its eigenvectors, while a defective matrix does not enjoy this property. The following question arises. How close to a diagonal matrix can we reduce a general matrix by a similarity transformation? We give one answer to this question, called the Jordan factorization, or the Jordan canonical form, in Theorem 6.4. For a proof, see for example [10]. The Jordan factorization is an important tool in matrix analysis and it has applications to systems of differential equations, see [8].

The Jordan factorization involves bidiagonal matrices called Jordan blocks.

Definition 6.3 (Jordan Block)
A **Jordan block** of order m, denoted $\boldsymbol{J}_m(\lambda)$ is an $m \times m$ matrix of the form

$$\boldsymbol{J}_m(\lambda) := \begin{bmatrix} \lambda & 1 & 0 & \cdots & 0 & 0 \\ 0 & \lambda & 1 & \cdots & 0 & 0 \\ 0 & 0 & \lambda & \cdots & 0 & 0 \\ \vdots & & & & & \vdots \\ 0 & 0 & 0 & \cdots & \lambda & 1 \\ 0 & 0 & 0 & \cdots & 0 & \lambda \end{bmatrix} = \lambda \boldsymbol{I}_m + \boldsymbol{E}_m, \quad \boldsymbol{E}_m := \begin{bmatrix} 0 & 1 & 0 & \cdots & 0 & 0 \\ 0 & 0 & 1 & \cdots & 0 & 0 \\ 0 & 0 & 0 & \cdots & 0 & 0 \\ \vdots & & & & & \vdots \\ 0 & 0 & 0 & \cdots & 0 & 1 \\ 0 & 0 & 0 & \cdots & 0 & 0 \end{bmatrix}. \tag{6.4}$$

A 3×3 Jordan block has the form $\boldsymbol{J}_3(\lambda) = \begin{bmatrix} \lambda & 1 & 0 \\ 0 & \lambda & 1 \\ 0 & 0 & \lambda \end{bmatrix}$. Since a Jordan block is upper triangular λ is an eigenvalue of $\boldsymbol{J}_m(\lambda)$ and any eigenvector must be a multiple of $\boldsymbol{e}_1$. Indeed, if $\boldsymbol{J}_m(\lambda)\boldsymbol{v} = \lambda\boldsymbol{v}$ for some $\boldsymbol{v} = [v_1, \ldots, v_m]$ then $\lambda v_{i-1} + v_i = \lambda v_{i-1}$, $i = 2, \ldots, m$ which shows that $v_2 = \cdots = v_m = 0$. Thus, the eigenvalue λ of $\boldsymbol{J}_m(\lambda)$ has algebraic multiplicity $a = m$ and geometric multiplicity $g = 1$.

The Jordan factorization is a factorization of a matrix into Jordan blocks.

Theorem 6.4 (The Jordan Factorization of a Matrix)
Suppose $\boldsymbol{A} \in \mathbb{C}^{n\times n}$ has k distinct eigenvalues $\lambda_1, \ldots, \lambda_k$ of algebraic multiplicities $a_1, \ldots, a_k$ and geometric multiplicities $g_1, \ldots, g_k$. There is a nonsingular matrix

[1]This can also be shown without using the Jordan factorization, see [9].

$S \in \mathbb{C}^{n \times n}$ *such that*

$$J := S^{-1}AS = \operatorname{diag}(U_1, \ldots, U_k), \text{ with } U_i \in \mathbb{C}^{a_i \times a_i}, \tag{6.5}$$

where each U_i *is block diagonal having* g_i *Jordan blocks along the diagonal*

$$U_i = \operatorname{diag}(J_{m_{i,1}}(\lambda_i), \ldots, J_{m_{i,g_i}}(\lambda_i)). \tag{6.6}$$

Here $m_{i,1}, \ldots, m_{i,g_i}$ *are positive integers and they are unique if they are ordered so that* $m_{i,1} \geq m_{i,2} \geq \cdots \geq m_{i,g_i}$. *Moreover,* $a_i = \sum_{j=1}^{g_i} m_{i,j}$ *for all* i.

We note that

1. The matrices S and J in (6.5) are called **Jordan factors**. We also call J the **Jordan factorization** of A.
2. The columns of S are called **principal vectors** or **generalized eigenvectors**. They satisfy the matrix equation $AS = SJ$.
3. Each U_i is upper triangular with the eigenvalue λ_i on the diagonal and consists of g_i Jordan blocks. These Jordan blocks can be taken in any order and it is customary to refer to any such block diagonal matrix as the Jordan factorization of A.

Example 6.4 (Jordan Factorization) As an example consider the Jordan factorization

$$J := \operatorname{diag}(U_1, U_2) = \begin{bmatrix} 2&1&0&&&&& \\ 0&2&1&&&&& \\ 0&0&2&&&&& \\ &&&2&1&&& \\ &&&0&2&&& \\ &&&&&2&& \\ &&&&&&3&1 \\ &&&&&&0&3 \end{bmatrix} \in \mathbb{R}^{8 \times 8}. \tag{6.7}$$

We encountered this matrix in Exercise 6.1. The eigenvalues together with their algebraic and geometric multiplicities can be read off directly from the Jordan factorization.

- $U_1 = \operatorname{diag}(J_3(2), J_2(2), J_1(2))$ and $U_2 = J_2(3)$.
- 2 is an eigenvalue of algebraic multiplicity 6 and geometric multiplicity 3, the number of Jordan blocks corresponding to $\lambda = 2$.
- 3 is an eigenvalue of algebraic multiplicity 2 and geometric multiplicity 1.

The columns of $S = [s_1, \ldots, s_8]$ are determined from the columns of J as follows

$$\begin{aligned} As_1 &= 2s_1, & As_2 &= s_1 + 2s_2, & As_3 &= s_2 + 2s_3, \\ As_4 &= 2s_4, & As_5 &= s_4 + 2s_5, && \\ As_6 &= 2s_6, &&&& \\ As_7 &= 3s_7, & As_8 &= s_7 + 3s_8. && \end{aligned}$$

We see that the generalized eigenvector corresponding to the first column in a Jordan block is an eigenvector of $\boldsymbol{A}$. The remaining generalized eigenvectors are not eigenvectors.

The matrix

$$\boldsymbol{J} := \begin{bmatrix} 3 & 1 & & & & & & \\ 0 & 3 & & & & & & \\ & & 2 & 1 & & & & \\ & & 0 & 2 & & & & \\ & & & & 2 & & & \\ & & & & & 2 & 1 & 0 \\ & & & & & 0 & 2 & 1 \\ & & & & & 0 & 0 & 2 \end{bmatrix}$$

is also a Jordan factorization of $\boldsymbol{A}$. In any Jordan factorization of this $\boldsymbol{A}$ the sizes of the 4 Jordan blocks $\boldsymbol{J}_2(3)$, $\boldsymbol{J}_2(2)$, $\boldsymbol{J}_1(2)$, $\boldsymbol{J}_3(2)$ are uniquely given.

Proof of Theorem 6.3

1. The algebraic multiplicity a_i of an eigenvalue λ_i is equal to the size of the corresponding $\boldsymbol{U}_i$. Moreover each $\boldsymbol{U}_i$ contains g_i Jordan blocks of size $m_{i,j} \geq 1$. Thus $g_i \leq a_i$.
2. Since $\boldsymbol{A}$ and $\boldsymbol{J}$ are similar the geometric multiplicities of the eigenvalues of these matrices are the same, and it is enough to prove statement 2 for the Jordan factor $\boldsymbol{J}$. We show this only for the matrix $\boldsymbol{J}$ given by (6.7). The general case should then be clear. There are only 4 eigenvectors of $\boldsymbol{J}$, namely $\boldsymbol{e}_1, \boldsymbol{e}_4, \boldsymbol{e}_6, \boldsymbol{e}_7$ corresponding to the 4 Jordan blocks. These 4 vectors are clearly linearly independent. Moreover there are $k = 2$ distinct eigenvalues and $g_1 + g_2 = 3 + 1 = 4$.
3. Since $g_i \leq a_i$ for all i and $\sum_i a_i = n$ we have $\sum_i g_i = n$ if and only if $a_i = g_i$ for $i = 1, \ldots, k$.

6.3 The Schur Factorization and Normal Matrices

6.3.1 *The Schur Factorization*

We turn now to **unitary similarity transformations** $\boldsymbol{S}^{-1}\boldsymbol{A}\boldsymbol{S}$, where $\boldsymbol{S} = \boldsymbol{U}$ is unitary. Thus $\boldsymbol{S}^{-1} = \boldsymbol{U}^*$ and a unitary similarity transformation takes the form $\boldsymbol{U}^*\boldsymbol{A}\boldsymbol{U}$.

6.3.2 *Unitary and Orthogonal Matrices*

Although not every matrix can be diagonalized it can be brought into **triangular form** by a **unitary** similarity transformation.

Theorem 6.5 (Schur Factorization) *For each $\boldsymbol{A} \in \mathbb{C}^{n\times n}$ there exists a unitary matrix $\boldsymbol{U} \in \mathbb{C}^{n\times n}$ such that $\boldsymbol{R} := \boldsymbol{U}^*\boldsymbol{A}\boldsymbol{U}$ is upper triangular.*

The matrices $\boldsymbol{U}$ and $\boldsymbol{R}$ in the Schur factorization are called **Schur factors**. We call $\boldsymbol{A} = \boldsymbol{U}\boldsymbol{R}\boldsymbol{U}^*$ the **Schur factorization** of $\boldsymbol{A}$.

Proof We use induction on n. For $n = 1$ the matrix $\boldsymbol{U}$ is the 1×1 identity matrix. Assume that the theorem is true for all $k \times k$ matrices, and suppose $\boldsymbol{A} \in \mathbb{C}^{n\times n}$, where $n := k + 1$. Let $(\lambda_1, \boldsymbol{v}_1)$ be an eigenpair for $\boldsymbol{A}$ with $\|\boldsymbol{v}_1\|_2 = 1$. By Theorem 5.5 we can extend $\boldsymbol{v}_1$ to an orthonormal basis $\{\boldsymbol{v}_1, \boldsymbol{v}_2, \dots, \boldsymbol{v}_n\}$ for $\mathbb{C}^n$. The matrix $\boldsymbol{V} := [\boldsymbol{v}_1, \dots, \boldsymbol{v}_n] \in \mathbb{C}^{n\times n}$ is unitary, and

$$\boldsymbol{V}^*\boldsymbol{A}\boldsymbol{V}\boldsymbol{e}_1 = \boldsymbol{V}^*\boldsymbol{A}\boldsymbol{v}_1 = \lambda_1\boldsymbol{V}^*\boldsymbol{v}_1 = \lambda_1\boldsymbol{e}_1.$$

It follows that

$$\boldsymbol{V}^*\boldsymbol{A}\boldsymbol{V} = \left[\begin{array}{c|c}\lambda_1 & \boldsymbol{x}^* \\ \hline \boldsymbol{0} & \boldsymbol{M}\end{array}\right], \text{ for some } \boldsymbol{M} \in \mathbb{C}^{k\times k} \text{ and } \boldsymbol{x} \in \mathbb{C}^k. \tag{6.8}$$

By the induction hypothesis there is a unitary matrix $\boldsymbol{W}_1 \in \mathbb{C}^{(n-1)\times(n-1)}$ such that $\boldsymbol{W}_1^*\boldsymbol{M}\boldsymbol{W}_1$ is upper triangular. Define

$$\boldsymbol{W} = \left[\begin{array}{c|c}1 & \boldsymbol{0}^* \\ \hline \boldsymbol{0} & \boldsymbol{W}_1\end{array}\right] \text{ and } \boldsymbol{U} = \boldsymbol{V}\boldsymbol{W}.$$

Then $\boldsymbol{W}$ and $\boldsymbol{U}$ are unitary and

$$\begin{aligned}\boldsymbol{U}^*\boldsymbol{A}\boldsymbol{U} = \boldsymbol{W}^*(\boldsymbol{V}^*\boldsymbol{A}\boldsymbol{V})\boldsymbol{W} &= \left[\begin{array}{c|c}1 & \boldsymbol{0}^* \\ \hline \boldsymbol{0} & \boldsymbol{W}_1^*\end{array}\right]\left[\begin{array}{c|c}\lambda_1 & \boldsymbol{x}^* \\ \hline \boldsymbol{0} & \boldsymbol{M}\end{array}\right]\left[\begin{array}{c|c}1 & \boldsymbol{0}^* \\ \hline \boldsymbol{0} & \boldsymbol{W}_1\end{array}\right] \\ &= \left[\begin{array}{c|c}\lambda_1 & \boldsymbol{x}^*\boldsymbol{W}_1 \\ \hline \boldsymbol{0} & \boldsymbol{W}_1^*\boldsymbol{M}\boldsymbol{W}_1\end{array}\right]\end{aligned}$$

is upper triangular. □

If $\boldsymbol{A}$ has complex eigenvalues then $\boldsymbol{U}$ will be complex even if $\boldsymbol{A}$ is real. The following is a real version of Theorem 6.5.

Theorem 6.6 (Schur Form, Real Eigenvalues) *For each $\boldsymbol{A} \in \mathbb{R}^{n\times n}$ with real eigenvalues there exists an orthogonal matrix $\boldsymbol{U} \in \mathbb{R}^{n\times n}$ such that $\boldsymbol{U}^T\boldsymbol{A}\boldsymbol{U}$ is upper triangular.*

Proof Consider the proof of Theorem 6.5. Since $\boldsymbol{A}$ and λ_1 are real the eigenvector $\boldsymbol{v}_1$ is real and the matrix $\boldsymbol{W}$ is real and $\boldsymbol{W}^T\boldsymbol{W} = \boldsymbol{I}$. By the induction hypothesis $\boldsymbol{V}$ is real and $\boldsymbol{V}^T\boldsymbol{V} = \boldsymbol{I}$. But then also $\boldsymbol{U} = \boldsymbol{V}\boldsymbol{W}$ is real and $\boldsymbol{U}^T\boldsymbol{U} = \boldsymbol{I}$. □

A real matrix with some complex eigenvalues can only be reduced to block triangular form by a real unitary similarity transformation. We consider this in Sect. 6.3.5.

Example 6.5 (Deflation Example) By using the unitary transformation $\boldsymbol{V}$ on the $n \times n$ matrix $\boldsymbol{A}$, we obtain a matrix $\boldsymbol{M}$ of order $n-1$. $\boldsymbol{M}$ has the same eigenvalues as $\boldsymbol{A}$ except λ. Thus we can find another eigenvalue of $\boldsymbol{A}$ by working with a smaller matrix $\boldsymbol{M}$. This is an example of a **deflation** technique which is very useful in numerical work. The second derivative matrix $\boldsymbol{T} := \begin{bmatrix} 2 & -1 & 0 \\ -1 & 2 & -1 \\ 0 & -1 & 2 \end{bmatrix}$ has an eigenpair $(2, \boldsymbol{x}_1)$, where $\boldsymbol{x}_1 = [-1, 0, 1]^T$. Find the remaining eigenvalues using deflation. For this we extend $\boldsymbol{x}_1$ to a basis $\{\boldsymbol{x}_1, \boldsymbol{x}_2, \boldsymbol{x}_3\}$ for $\mathbb{R}^3$ by defining $\boldsymbol{x}_2 = [0, 1, 0]^T$, $\boldsymbol{x}_3 = [1, 0, 1]^T$. This is already an orthogonal basis and normalizing we obtain the orthogonal matrix

$$\boldsymbol{V} = \begin{bmatrix} -\frac{1}{\sqrt{2}} & 0 & \frac{1}{\sqrt{2}} \\ 0 & 1 & 0 \\ \frac{1}{\sqrt{2}} & 0 & \frac{1}{\sqrt{2}} \end{bmatrix}.$$

We obtain (6.8) with $\lambda = 2$ and

$$\boldsymbol{M} = \begin{bmatrix} 2 & -\sqrt{2} \\ -\sqrt{2} & 2 \end{bmatrix}.$$

We can now find the remaining eigenvalues of $\boldsymbol{A}$ from the 2×2 matrix $\boldsymbol{M}$.

6.3.3 *Normal Matrices*

A matrix $\boldsymbol{A} \in \mathbb{C}^{n \times n}$ is **normal** if $\boldsymbol{A}^*\boldsymbol{A} = \boldsymbol{A}\boldsymbol{A}^*$. In this section we show that a matrix has orthogonal eigenvectors if and only if it is normal.

Examples of normal matrices are

1. $\boldsymbol{A}^* = \boldsymbol{A}$, (Hermitian)
2. $\boldsymbol{A}^* = -\boldsymbol{A}$, (Skew-Hermitian)
3. $\boldsymbol{A}^* = \boldsymbol{A}^{-1}$, (Unitary)
4. $\boldsymbol{A} = \mathrm{diag}(d_1, \ldots, d_n)$. (Diagonal)

Clearly the matrices in 1. 2. 3. are normal. If $\boldsymbol{A}$ is diagonal then

$$\boldsymbol{A}^*\boldsymbol{A} = \mathrm{diag}(\overline{d_1}d_1, \ldots, \overline{d_n}d_n) = \mathrm{diag}(|d_1|^2, \ldots, |d_n|^2) = \boldsymbol{A}\boldsymbol{A}^*,$$

and $\boldsymbol{A}$ is normal. The 2. derivative matrix $\boldsymbol{T}$ in (2.27) is symmetric and therefore normal. The eigenvalues of a normal matrix can be complex (cf. Exercise 6.21). However in the Hermitian case the eigenvalues are real (cf. Lemma 2.3).

The following theorem shows that $\boldsymbol{A}$ has a set of orthogonal eigenvectors if and only if it is normal.

Theorem 6.7 (Spectral Theorem for Normal Matrices) *A matrix* $\boldsymbol{A} \in \mathbb{C}^{n\times n}$ *is normal if and only if there exists a unitary matrix* $\boldsymbol{U} \in \mathbb{C}^{n\times n}$ *such that* $\boldsymbol{U}^*\boldsymbol{A}\boldsymbol{U} = \boldsymbol{D}$ *is diagonal. If* $\boldsymbol{D} = \operatorname{diag}(\lambda_1, \dots, \lambda_n)$ *and* $\boldsymbol{U} = [\boldsymbol{u}_1, \dots, \boldsymbol{u}_n]$ *then* $(\lambda_j, \boldsymbol{u}_j)$, $j = 1, \dots, n$ *are orthonormal eigenpairs for* $\boldsymbol{A}$.

Proof If $\boldsymbol{B} = \boldsymbol{U}^*\boldsymbol{A}\boldsymbol{U}$, with $\boldsymbol{B}$ diagonal, and $\boldsymbol{U}^*\boldsymbol{U} = \boldsymbol{I}$, then $\boldsymbol{A} = \boldsymbol{U}\boldsymbol{B}\boldsymbol{U}^*$ and

$$\begin{aligned}\boldsymbol{A}\boldsymbol{A}^* &= (\boldsymbol{U}\boldsymbol{B}\boldsymbol{U}^*)(\boldsymbol{U}\boldsymbol{B}^*\boldsymbol{U}^*) = \boldsymbol{U}\boldsymbol{B}\boldsymbol{B}^*\boldsymbol{U}^* \text{ and}\\ \boldsymbol{A}^*\boldsymbol{A} &= (\boldsymbol{U}\boldsymbol{B}^*\boldsymbol{U}^*)(\boldsymbol{U}\boldsymbol{B}\boldsymbol{U}^*) = \boldsymbol{U}\boldsymbol{B}^*\boldsymbol{B}\boldsymbol{U}^*.\end{aligned}$$

Now $\boldsymbol{B}\boldsymbol{B}^* = \boldsymbol{B}^*\boldsymbol{B}$ since $\boldsymbol{B}$ is diagonal, and $\boldsymbol{A}$ is normal.

Conversely, suppose $\boldsymbol{A}^*\boldsymbol{A} = \boldsymbol{A}\boldsymbol{A}^*$. By Theorem 6.5 we can find $\boldsymbol{U}$ with $\boldsymbol{U}^*\boldsymbol{U} = \boldsymbol{I}$ such that $\boldsymbol{B} := \boldsymbol{U}^*\boldsymbol{A}\boldsymbol{U}$ is upper triangular. Since $\boldsymbol{A}$ is normal $\boldsymbol{B}$ is normal. Indeed,

$$\boldsymbol{B}\boldsymbol{B}^* = \boldsymbol{U}^*\boldsymbol{A}\boldsymbol{U}\boldsymbol{U}^*\boldsymbol{A}^*\boldsymbol{U} = \boldsymbol{U}^*\boldsymbol{A}\boldsymbol{A}^*\boldsymbol{U} = \boldsymbol{U}^*\boldsymbol{A}^*\boldsymbol{A}\boldsymbol{U} = \boldsymbol{B}^*\boldsymbol{B}.$$

The proof is complete if we can show that an upper triangular normal matrix $\boldsymbol{B}$ must be diagonal. The diagonal elements e_{ii} in $\boldsymbol{E} := \boldsymbol{B}^*\boldsymbol{B}$ and f_{ii} in $\boldsymbol{F} := \boldsymbol{B}\boldsymbol{B}^*$ are given by

$$e_{ii} = \sum_{k=1}^{n} \overline{b}_{ki} b_{ki} = \sum_{k=1}^{i} |b_{ki}|^2 \text{ and } f_{ii} = \sum_{k=1}^{n} b_{ik}\overline{b}_{ik} = \sum_{k=i}^{n} |b_{ik}|^2.$$

The result now follows by equating e_{ii} and f_{ii} for $i = 1, 2, \dots, n$. In particular for $i = 1$ we have $|b_{11}|^2 = |b_{11}|^2 + |b_{12}|^2 + \cdots + |b_{1n}|^2$, so $b_{1k} = 0$ for $k = 2, 3, \dots, n$. Suppose $\boldsymbol{B}$ is diagonal in its first $i - 1$ rows so that $b_{jk} = 0$ for $j = 1, \dots, i-1$, $k = j+1, \dots, n$. Then

$$e_{ii} = \sum_{k=1}^{i} |b_{ki}|^2 = |b_{ii}|^2 = \sum_{k=i}^{n} |b_{ik}|^2 = f_{ii}$$

and it follows that $b_{ik} = 0$, $k = i+1, \dots, n$. By induction on the rows we see that $\boldsymbol{B}$ is diagonal. The last part of the theorem follows from Sect. 6.1.1. □

Example 6.6 The orthogonal diagonalization of $\boldsymbol{A} = \left[\begin{smallmatrix} 2 & -1 \\ -1 & 2 \end{smallmatrix}\right]$ is $\boldsymbol{U}^T\boldsymbol{A}\boldsymbol{U} = \operatorname{diag}(1, 3)$, where $\boldsymbol{U} = \frac{1}{\sqrt{2}}\left[\begin{smallmatrix} 1 & 1 \\ 1 & -1 \end{smallmatrix}\right]$.

6.3.4 *The Rayleigh Quotient*

The Rayleigh quotient is a useful tool when studying eigenvalues.

Definition 6.4 (Rayleigh Quotient) For $A \in \mathbb{C}^{n\times n}$ and a nonzero $\boldsymbol{x}$ the number

$$R(\boldsymbol{x}) = R_A(\boldsymbol{x}) := \frac{\boldsymbol{x}^*\boldsymbol{A}\boldsymbol{x}}{\boldsymbol{x}^*\boldsymbol{x}}$$

is called a **Rayleigh quotient**.

If $(\lambda, \boldsymbol{x})$ is an eigenpair for $\boldsymbol{A}$ then $R(\boldsymbol{x}) = \frac{\boldsymbol{x}^*\boldsymbol{A}\boldsymbol{x}}{\boldsymbol{x}^*\boldsymbol{x}} = \lambda$.

Equation (6.9) in the following theorem shows that the Rayleigh quotient of a normal matrix is a **convex combination** of its eigenvalues.

Theorem 6.8 (Convex Combination of the Eigenvalues) *Suppose* $\boldsymbol{A} \in \mathbb{C}^{n\times n}$ *is normal with orthonormal eigenpairs* $(\lambda_j, \boldsymbol{u}_j)$, *for* $j = 1, 2, \ldots, n$. *Then the Rayleigh quotient is a convex combination of the eigenvalues of* $\boldsymbol{A}$

$$R_A(\boldsymbol{x}) = \frac{\sum_{i=1}^n \lambda_i |c_i|^2}{\sum_{j=1}^n |c_j|^2}, \quad \boldsymbol{x} \neq \boldsymbol{0}, \quad \boldsymbol{x} = \sum_{j=1}^n c_j \boldsymbol{u}_j. \tag{6.9}$$

Proof By orthonormality of the eigenvectors $\boldsymbol{x}^*\boldsymbol{x} = \sum_{i=1}^n \sum_{j=1}^n \overline{c}_i \overline{\boldsymbol{u}}_i c_j \boldsymbol{u}_j = \sum_{j=1}^n |c_j|^2$. Similarly, $\boldsymbol{x}^*\boldsymbol{A}\boldsymbol{x} = \sum_{i=1}^n \sum_{j=1}^n \overline{c}_i \overline{\boldsymbol{u}}_i c_j \lambda_j \boldsymbol{u}_j = \sum_{i=1}^n \lambda_i |c_i|^2$. and (6.9) follows. This is clearly a combination of nonnegative quantities and a convex combination since $\sum_{i=1}^n |c_i|^2 / \sum_{j=1}^n |c_j|^2 = 1$. □

6.3.5 *The Quasi-Triangular Form*

How far can we reduce a real matrix $\boldsymbol{A}$ with some complex eigenvalues by a real unitary similarity transformation? To study this we note that the complex eigenvalues of a real matrix occur in conjugate pairs, $\lambda = \mu + i\nu$, $\overline{\lambda} = \mu - i\nu$, where μ, ν are real. The real 2×2 matrix

$$\boldsymbol{M} = \begin{bmatrix} \mu & \nu \\ -\nu & \mu \end{bmatrix} \tag{6.10}$$

has eigenvalues $\lambda = \mu + i\nu$ and $\overline{\lambda} = \mu - i\nu$.

Definition 6.5 (Quasi-Triangular Matrix) We say that a matrix is **quasi-triangular** if it is block triangular with only 1×1 and 2×2 blocks on the diagonal. Moreover, no 2×2 block should have real eigenvalues.

As an example consider the matrix

$$R := \begin{bmatrix} D_1 & R_{1,2} & R_{1,3} \\ 0 & D_2 & R_{2,3} \\ 0 & 0 & D_3 \end{bmatrix}, \quad D_1 := \begin{bmatrix} 2 & 1 \\ -1 & 2 \end{bmatrix}, \quad D_2 := \begin{bmatrix} 1 \end{bmatrix}, \quad D_3 := \begin{bmatrix} 3 & 2 \\ -1 & 1 \end{bmatrix}.$$

Since R is block triangular the characteristic polynomial of R is given by $\pi_R = \pi_{D_1}\pi_{D_2}\pi_{D_3}$. We find

$$\pi_{D_1}(\lambda) = \pi_{D_3}(\lambda) = \lambda^2 - 4\lambda + 5, \quad \pi_{D_2}(\lambda) = \lambda - 1,$$

and the eigenvalues D_1 and D_3 are $\lambda_1 = 2+i$, $\lambda_2 = 2-i$, while D_2 obviously has the eigenvalue $\lambda = 1$.

Any $A \in \mathbb{R}^{n\times n}$ can be reduced to quasi-triangular form by a real orthogonal similarity transformation. For a proof see [16]. We will encounter the quasi-triangular form in Chap. 15.

6.3.6 *Hermitian Matrices*

The special cases where A is Hermitian, or real and symmetric, deserve special attention.

Theorem 6.9 (Spectral Theorem, Complex Form) *Suppose $A \in \mathbb{C}^{n\times n}$ is Hermitian. Then A has real eigenvalues $\lambda_1, \dots, \lambda_n$. Moreover, there is a unitary matrix $U \in \mathbb{C}^{n\times n}$ such that*

$$U^*AU = \mathrm{diag}(\lambda_1, \dots, \lambda_n).$$

For any such U the columns $\{u_1, \dots, u_n\}$ of U are orthonormal eigenvectors of A and $Au_j = \lambda_j u_j$ for $j = 1, \dots, n$.

Proof That the eigenvalues are real was shown in Lemma 2.3. The rest follows from Theorem 6.7. □

There is also a real version.

Theorem 6.10 (Spectral Theorem (Real Form)) *Suppose $A \in \mathbb{R}^{n\times n}$ is symmetric. Then A has real eigenvalues $\lambda_1, \lambda_2, \dots, \lambda_n$. Moreover, there is an orthogonal matrix $U \in \mathbb{R}^{n\times n}$ such that*

$$U^TAU = \mathrm{diag}(\lambda_1, \lambda_2, \dots, \lambda_n).$$

For any such U the columns $\{u_1, \dots, u_n\}$ of U are orthonormal eigenvectors of A and $Au_j = \lambda_j u_j$ for $j = 1, \dots, n$.

Proof Since a real symmetric matrix has real eigenvalues and eigenvectors this follows from Theorem 6.9. □

6.4 Minmax Theorems

There are some useful characterizations of the eigenvalues of a Hermitian matrix in terms of the Rayleigh quotient $R(\boldsymbol{x}) = R_A(\boldsymbol{x}) := \frac{\boldsymbol{x}^* A \boldsymbol{x}}{\boldsymbol{x}^* \boldsymbol{x}}$. First we show

Theorem 6.11 (Minmax) *Suppose $A \in \mathbb{C}^{n \times n}$ is Hermitian with orthonormal eigenpairs $(\lambda_j, \boldsymbol{u}_j)$, $1 \le j \le n$, ordered so that $\lambda_1 \ge \cdots \ge \lambda_n$. Let $1 \le k \le n$. For any subspace $\mathcal{S}$ of $\mathbb{C}^n$ of dimension $n-k+1$*

$$\lambda_k \le \max_{\substack{\boldsymbol{x} \in \mathcal{S} \\ \boldsymbol{x} \ne \boldsymbol{0}}} R(\boldsymbol{x}), \tag{6.11}$$

with equality for $\mathcal{S} = \tilde{\mathcal{S}} := \operatorname{span}(\boldsymbol{u}_k, \ldots, \boldsymbol{u}_n)$ and $\boldsymbol{x} = \boldsymbol{u}_k$.

Proof Let $\mathcal{S}$ be any subspace of $\mathbb{C}^n$ of dimension $n-k+1$ and define $\mathcal{S}' := \operatorname{span}(\boldsymbol{u}_1, \ldots, \boldsymbol{u}_k)$. It is enough to find $\boldsymbol{y} \in \mathcal{S}$ so that $R(\boldsymbol{y}) \ge \lambda_k$. Now $\mathcal{S} + \mathcal{S}' := \{\boldsymbol{s} + \boldsymbol{s}' : \boldsymbol{s} \in \mathcal{S}, \boldsymbol{s}' \in \mathcal{S}'\}$ is a subspace of $\mathbb{C}^n$ and by (1.7)

$$\dim(\mathcal{S} \cap \mathcal{S}') = \dim(\mathcal{S}) + \dim(\mathcal{S}') - \dim(\mathcal{S} + \mathcal{S}') \ge (n-k+1) + k - n = 1.$$

It follows that $\mathcal{S} \cap \mathcal{S}'$ is nonempty. Let $\boldsymbol{y} \in \mathcal{S} \cap \mathcal{S}' = \sum_{j=1}^{k} c_j \boldsymbol{u}_j$ with $\sum_{j=1}^{k} |c_j|^2 = 1$. Defining $c_j = 0$ for $k+1 \le j \le n$, we obtain by Theorem 6.8

$$\max_{\substack{\boldsymbol{x} \in \mathcal{S} \\ \boldsymbol{x} \ne \boldsymbol{0}}} R(\boldsymbol{x}) \ge R(\boldsymbol{y}) = \sum_{j=1}^{n} \lambda_j |c_j|^2 = \sum_{j=1}^{k} \lambda_j |c_j|^2 \ge \sum_{j=1}^{k} \lambda_k |c_j|^2 = \lambda_k,$$

and (6.11) follows. To show equality suppose $\boldsymbol{z} \in \mathcal{S} = \tilde{\mathcal{S}}$. Now $\boldsymbol{z} = \sum_{j=k}^{n} d_j \boldsymbol{u}_j$ for some $d_k, \ldots, d_n$ with $\sum_{j=k}^{n} |d_j|^2 = 1$ and by Lemma 6.8 $R(\boldsymbol{z}) = \sum_{j=k}^{n} \lambda_j |d_j|^2 \le \lambda_k$. Since $\boldsymbol{z} \in \tilde{\mathcal{S}}$ is arbitrary we have $\max_{\substack{\boldsymbol{x} \in \tilde{\mathcal{S}} \\ \boldsymbol{x} \ne \boldsymbol{0}}} R(\boldsymbol{x}) \le \lambda_k$ and equality in (6.11) follows for $\mathcal{S} = \tilde{\mathcal{S}}$. Moreover, $R(\boldsymbol{u}_k) = \lambda_k$. □

There is also a maxmin version of this result.

Theorem 6.12 (Maxmin) *Suppose $A \in \mathbb{C}^{n \times n}$ is Hermitian with eigenvalues $\lambda_1, \ldots, \lambda_n$, ordered so that $\lambda_1 \ge \cdots \ge \lambda_n$. Let $1 \le k \le n$. For any subspace $\mathcal{S}$ of $\mathbb{C}^n$ of dimension k*

$$\lambda_k \ge \min_{\substack{\boldsymbol{x} \in \mathcal{S} \\ \boldsymbol{x} \ne \boldsymbol{0}}} R(\boldsymbol{x}), \tag{6.12}$$

with equality for $\mathcal{S} = \tilde{\mathcal{S}} := \operatorname{span}(\boldsymbol{u}_1, \ldots, \boldsymbol{u}_k)$ *and* $\boldsymbol{x} = \boldsymbol{u}_k$. *Here* $(\lambda_j, \boldsymbol{u}_j)$, $1 \le j \le n$ *are orthonormal eigenpairs for* $\boldsymbol{A}$.

Proof The proof is very similar to the proof of Theorem 6.11. We define $\mathcal{S}' := \operatorname{span}(\boldsymbol{u}_k, \ldots, \boldsymbol{u}_n)$ and show that $R(\boldsymbol{y}) \le \lambda_k$ for some $\boldsymbol{y} \in \mathcal{S} \cap \mathcal{S}'$. It is easy to see that $R(\boldsymbol{y}) \ge \lambda_k$ for any $\boldsymbol{y} \in \tilde{\mathcal{S}}$. □

These theorems immediately lead to classical minmax and maxmin characterizations.

Corollary 6.1 (The Courant-Fischer Theorem) *Suppose* $\boldsymbol{A} \in \mathbb{C}^{n\times n}$ *is Hermitian with eigenvalues* $\lambda_1, \ldots, \lambda_n$, *ordered so that* $\lambda_1 \ge \cdots \ge \lambda_n$. *Then*

$$\lambda_k = \min_{\dim(\mathcal{S})=n-k+1} \max_{\substack{\boldsymbol{x}\in\mathcal{S}\\ \boldsymbol{x}\neq\boldsymbol{0}}} R(\boldsymbol{x}) = \max_{\dim(\mathcal{S})=k} \min_{\substack{\boldsymbol{x}\in\mathcal{S}\\ \boldsymbol{x}\neq\boldsymbol{0}}} R(\boldsymbol{x}), \quad k = 1, \ldots, n. \tag{6.13}$$

Using Theorem 6.11 we can prove inequalities of eigenvalues without knowing the eigenvectors and we can get both upper and lower bounds.

Theorem 6.13 (Eigenvalue Perturbation for Hermitian Matrices) *Let* $\boldsymbol{A}, \boldsymbol{B} \in \mathbb{C}^{n\times n}$ *be Hermitian with eigenvalues* $\alpha_1 \ge \alpha_2 \ge \cdots \ge \alpha_n$ *and* $\beta_1 \ge \beta_2 \ge \cdots \ge \beta_n$. *Then*

$$\alpha_k + \varepsilon_n \le \beta_k \le \alpha_k + \varepsilon_1, \ \textit{for}\ k = 1, \ldots, n, \tag{6.14}$$

where $\varepsilon_1 \ge \varepsilon_2 \ge \cdots \ge \varepsilon_n$ *are the eigenvalues of* $\boldsymbol{E} := \boldsymbol{B} - \boldsymbol{A}$.

Proof Since $\boldsymbol{E}$ is a difference of Hermitian matrices it is Hermitian and the eigenvalues are real. Let $(\alpha_j, \boldsymbol{u}_j)$, $j = 1, \ldots, n$ be orthonormal eigenpairs for $\boldsymbol{A}$ and let $\mathcal{S} := \operatorname{span}\{\boldsymbol{u}_k, \ldots, \boldsymbol{u}_n\}$. By Theorem 6.11 we obtain

$$\begin{aligned}
\beta_k &\le \max_{\substack{\boldsymbol{x}\in\mathcal{S}\\ \boldsymbol{x}\neq\boldsymbol{0}}} R_{\boldsymbol{B}}(\boldsymbol{x}) \le \max_{\substack{\boldsymbol{x}\in\mathcal{S}\\ \boldsymbol{x}\neq\boldsymbol{0}}} R_{\boldsymbol{A}}(\boldsymbol{x}) + \max_{\substack{\boldsymbol{x}\in\mathcal{S}\\ \boldsymbol{x}\neq\boldsymbol{0}}} R_{\boldsymbol{E}}(\boldsymbol{x}) \\
&\le \max_{\substack{\boldsymbol{x}\in\mathcal{S}\\ \boldsymbol{x}\neq\boldsymbol{0}}} R_{\boldsymbol{A}}(\boldsymbol{x}) + \max_{\substack{\boldsymbol{x}\in\mathbb{C}^n\\ \boldsymbol{x}\neq\boldsymbol{0}}} R_{\boldsymbol{E}}(\boldsymbol{x}) = \alpha_k + \varepsilon_1,
\end{aligned}$$

and this proves the upper inequality. For the lower one we define $\boldsymbol{D} := -\boldsymbol{E}$ and observe that $-\varepsilon_n$ is the largest eigenvalue of $\boldsymbol{D}$. Since $\boldsymbol{A} = \boldsymbol{B}+\boldsymbol{D}$ it follows from the result just proved that $\alpha_k \le \beta_k - \varepsilon_n$, which is the same as the lower inequality. □

In many applications of this result the eigenvalues of the matrix $\boldsymbol{E}$ will be small and then the theorem states that the eigenvalues of $\boldsymbol{B}$ are close to those of $\boldsymbol{A}$. Moreover, it associates a unique eigenvalue of $\boldsymbol{A}$ with each eigenvalue of $\boldsymbol{B}$.

6.4.1 The Hoffman-Wielandt Theorem

We can also give a bound involving all eigenvalues. The following theorem shows that the eigenvalue problem for a normal matrix is well conditioned.

Theorem 6.14 (Hoffman-Wielandt Theorem) *Suppose $\boldsymbol{A}, \boldsymbol{B} \in \mathbb{C}^{n\times n}$ are both normal matrices with eigenvalues $\lambda_1, \ldots, \lambda_n$ and $\mu_1, \ldots, \mu_n$, respectively. Then there is a permutation $i_1, \ldots, i_n$ of $1, 2, \ldots, n$ such that*

$$\sum_{j=1}^{n} |\mu_{i_j} - \lambda_j|^2 \le \sum_{i=1}^{n}\sum_{j=1}^{n} |a_{ij} - b_{ij}|^2. \tag{6.15}$$

For a proof of this theorem see [19, p. 190]. For a Hermitian matrix we can use the identity permutation if we order both set of eigenvalues in nonincreasing or nondecreasing order.

6.5 Left Eigenvectors

Definition 6.6 (Left and Right Eigenpairs) Suppose $\boldsymbol{A} \in \mathbb{C}^{n\times n}$ is a square matrix, $\lambda \in \mathbb{C}$ and $\boldsymbol{y} \in \mathbb{C}^n$ is nonzero. We say that $(\lambda, \boldsymbol{y})$ is a **left eigenpair** for $\boldsymbol{A}$ if $\boldsymbol{y}^*\boldsymbol{A} = \lambda\boldsymbol{y}^*$ or equivalently $\boldsymbol{A}^*\boldsymbol{y} = \overline{\lambda}\boldsymbol{y}$. We say that $(\lambda, \boldsymbol{y})$ is a **right eigenpair** for $\boldsymbol{A}$ if $\boldsymbol{A}\boldsymbol{y} = \lambda\boldsymbol{y}$. If $(\lambda, \boldsymbol{y})$ is a left eigenpair then λ is called a **left eigenvalue** and $\boldsymbol{y}$ a **left eigenvector**. Similarly if $(\lambda, \boldsymbol{y})$ is a right eigenpair then λ is called a **right eigenvalue** and $\boldsymbol{y}$ a **right eigenvector**.

In this book an eigenpair will always mean a right eigenpair. A left eigenvector is an eigenvector of $\boldsymbol{A}^*$. If λ is a left eigenvalue of $\boldsymbol{A}$ then $\overline{\lambda}$ is an eigenvalue of $\boldsymbol{A}^*$ and then λ is an eigenvalue of $\boldsymbol{A}$ (cf. Exercise 6.3). Thus left and right eigenvalues are identical, but left and right eigenvectors are in general different. For a Hermitian matrix the right and left eigenpairs are the same.

Using right and left linearly independent eigenpairs we get some useful eigenvector expansions.

Theorem 6.15 (Biorthogonal Eigenvector Expansion) *If $\boldsymbol{A} \in \mathbb{C}^{n\times n}$ has linearly independent right eigenvectors $\{\boldsymbol{x}_1, \ldots, \boldsymbol{x}_n\}$ then there exists a set of left eigenvectors $\{\boldsymbol{y}_1, \ldots, \boldsymbol{y}_n\}$ with $\boldsymbol{y}_i^*\boldsymbol{x}_j = \delta_{i,j}$. Conversely, if $\boldsymbol{A} \in \mathbb{C}^{n\times n}$ has linearly independent left eigenvectors $\{\boldsymbol{y}_1, \ldots, \boldsymbol{y}_n\}$ then there exists a set of right eigenvectors $\{\boldsymbol{x}_1, \ldots, \boldsymbol{x}_n\}$ with $\boldsymbol{y}_i^*\boldsymbol{x}_j = \delta_{i,j}$. For any scaling of these sets we have the eigenvector expansions*

$$\boldsymbol{v} = \sum_{j=1}^{n} \frac{\boldsymbol{y}_j^*\boldsymbol{v}}{\boldsymbol{y}_j^*\boldsymbol{x}_j}\boldsymbol{x}_j = \sum_{k=1}^{n} \frac{\boldsymbol{x}_k^*\boldsymbol{v}}{\boldsymbol{y}_k^*\boldsymbol{x}_k}\boldsymbol{y}_k, \quad \boldsymbol{v} \in \mathbb{C}^n. \tag{6.16}$$

Proof For any right eigenpairs $(\lambda_1, \boldsymbol{x}_1), \ldots, (\lambda_n, \boldsymbol{x}_n)$ and left eigenpairs $(\lambda_1, \boldsymbol{y}_1), \ldots, (\lambda_n, \boldsymbol{y}_n)$ of $\boldsymbol{A}$ we have $\boldsymbol{AX} = \boldsymbol{XD}$, $\boldsymbol{Y}^*\boldsymbol{A} = \boldsymbol{DY}^*$, where

$$\boldsymbol{X} := [\boldsymbol{x}_1, \ldots, \boldsymbol{x}_n], \quad \boldsymbol{Y} := [\boldsymbol{y}_1, \ldots, \boldsymbol{y}_n], \quad \boldsymbol{D} := \operatorname{diag}(\lambda_1, \ldots, \lambda_n).$$

Suppose $\boldsymbol{X}$ is nonsingular. Then $\boldsymbol{AX} = \boldsymbol{XD} \implies \boldsymbol{A} = \boldsymbol{XDX}^{-1} \implies \boldsymbol{X}^{-1}\boldsymbol{A} = \boldsymbol{DX}^{-1}$ and it follows that $\boldsymbol{Y}^* := \boldsymbol{X}^{-1}$ contains a collection of left eigenvectors such that $\boldsymbol{Y}^*\boldsymbol{X} = \boldsymbol{I}$. Thus the columns of $\boldsymbol{Y}$ are linearly independent and $\boldsymbol{y}_i^*\boldsymbol{x}_j = \delta_{i,j}$. Similarly, if $\boldsymbol{Y}$ is nonsingular then $\boldsymbol{AY}^{-*} = \boldsymbol{Y}^{-*}\boldsymbol{D}$ and it follows that $\boldsymbol{X} := \boldsymbol{Y}^{-*}$ contains a collection of linearly independent right eigenvectors such that $\boldsymbol{Y}^*\boldsymbol{X} = \boldsymbol{I}$. If $\boldsymbol{v} = \sum_{j=1}^n c_j\boldsymbol{x}_j$ then $\boldsymbol{y}_i^*\boldsymbol{v} = \sum_{j=1}^n c_j\boldsymbol{y}_i^*\boldsymbol{x}_j = c_i\boldsymbol{y}_i^*\boldsymbol{x}_i$, so $c_i = \boldsymbol{y}_i^*\boldsymbol{v}/\boldsymbol{y}_i^*\boldsymbol{x}_i$ for $i = 1, \ldots, n$ and the first expansion in (6.16) follows. The second expansion follows similarly. □

For a Hermitian matrix the right eigenvectors $\{\boldsymbol{x}_1, \ldots, \boldsymbol{x}_n\}$ are also left eigenvectors and (6.16) takes the form

$$\boldsymbol{v} = \sum_{j=1}^n \frac{\boldsymbol{x}_j^*\boldsymbol{v}}{\boldsymbol{x}_j^*\boldsymbol{x}_j}\boldsymbol{x}_j. \tag{6.17}$$

6.5.1 Biorthogonality

Left- and right eigenvectors corresponding to distinct eigenvalues are orthogonal.

Theorem 6.16 (Biorthogonality) *Suppose $(\mu, \boldsymbol{y})$ and $(\lambda, \boldsymbol{x})$ are left and right eigenpairs of $\boldsymbol{A} \in \mathbb{C}^{n\times n}$. If $\lambda \neq \mu$ then $\boldsymbol{y}^*\boldsymbol{x} = 0$.*

Proof Using the eigenpair relation in two ways we obtain $\boldsymbol{y}^*\boldsymbol{Ax} = \lambda\boldsymbol{y}^*\boldsymbol{x} = \mu\boldsymbol{y}^*\boldsymbol{x}$ and we conclude that $\boldsymbol{y}^*\boldsymbol{x} = 0$. □

Right and left eigenvectors corresponding to the same eigenvalue are sometimes orthogonal, sometimes not.

Theorem 6.17 (Simple Eigenvalue) *Suppose $(\lambda, \boldsymbol{x})$ and $(\lambda, \boldsymbol{y})$ are right and left eigenpairs of $\boldsymbol{A} \in \mathbb{C}^{n\times n}$. If λ has algebraic multiplicity one then $\boldsymbol{y}^*\boldsymbol{x} \neq 0$.*

Proof Assume that $\|\boldsymbol{x}\|_2 = 1$. We have (cf. (6.8))

$$\boldsymbol{V}^*\boldsymbol{AV} = \left[\begin{array}{c|c} \lambda & \boldsymbol{z}^* \\ \hline \boldsymbol{0} & \boldsymbol{M} \end{array}\right],$$

where V is unitary and $Ve_1 = x$. We show that if $y^*x = 0$ then λ is also an eigenvalue of M contradicting the multiplicity assumption of λ. Let $u := V^*y$. Then

$$(V^*A^*V)u = V^*A^*y = \overline{\lambda}V^*y = \overline{\lambda}u,$$

so $(\overline{\lambda}, u)$ is an eigenpair of V^*A^*V. But then $y^*x = u^*V^*Ve_1 = u^*e_1$. Suppose that $u^*e_1 = 0$, i.e., $u = \left[\begin{smallmatrix} 0 \\ v \end{smallmatrix}\right]$ for some nonzero $v \in \mathbb{C}^{n-1}$. Then

$$V^*A^*Vu = \left[\begin{array}{c|c} \overline{\lambda} & 0^* \\ \hline z & M^* \end{array}\right]\begin{bmatrix} 0 \\ v \end{bmatrix} = \begin{bmatrix} 0 \\ M^*v \end{bmatrix} = \overline{\lambda}\begin{bmatrix} 0 \\ v \end{bmatrix}$$

and λ is an eigenvalue of M. □

The case with multiple eigenvalues is more complicated. For example, the matrix $A := \left[\begin{smallmatrix} 1 & 1 \\ 0 & 1 \end{smallmatrix}\right]$ has one eigenvalue $\lambda = 1$ of algebraic multiplicity two, one right eigenvector $x = e_1$ and one left eigenvector $y = e_2$. Thus x and y are orthogonal.

6.6 Exercises Chap. 6

6.6.1 Exercises Sect. 6.1

Exercise 6.1 (Eigenvalues of a Block Triangular Matrix) What are the eigenvalues of the matrix

$$\begin{bmatrix} 2&1&0&0&0&0&0&0 \\ 0&2&1&0&0&0&0&0 \\ 0&0&2&0&0&0&0&0 \\ 0&0&0&2&0&0&0&0 \\ 0&0&0&1&2&0&0&0 \\ 0&0&0&0&0&2&0&0 \\ 0&0&0&0&0&0&3&0 \\ 0&0&0&0&0&0&1&3 \end{bmatrix} \in \mathbb{R}^{8,8}?$$

Exercise 6.2 (Characteristic Polynomial of Transpose) We have $\det(B^T) = \det(B)$ and $\det(\overline{B}) = \overline{\det(B)}$for any square matrix B. Use this to show that

a) $\pi_{A^T} = \pi_A$,
b) $\pi_{A^*}(\overline{\lambda}) = \overline{\pi_A(\lambda)}$.

Exercise 6.3 (Characteristic Polynomial of Inverse) Suppose (λ, x) is an eigenpair for $A \in \mathbb{C}^{n \times n}$. Show that

a) If A is nonsingular then (λ^{-1}, x) is an eigenpair for A^{-1}.
b) (λ^k, x) is an eigenpair for A^k for $k \in \mathbb{Z}$.

Exercise 6.4 (The Power of the Eigenvector Expansion) Show that if $\boldsymbol{A} \in \mathbb{C}^{n\times n}$ is nondefective with eigenpairs $(\lambda_j, \boldsymbol{x}_j)$, $j = 1, \ldots, n$ then for any $\boldsymbol{x} \in \mathbb{C}^n$ and $k \in \mathbb{N}$

$$\boldsymbol{A}^k\boldsymbol{x} = \sum_{j=1}^{n} c_j\lambda_j^k\boldsymbol{x}_j \text{ for some scalars } c_1, \ldots, c_n. \tag{6.18}$$

Show that if $\boldsymbol{A}$ is nonsingular then (6.18) holds for all $k \in \mathbb{Z}$.

Exercise 6.5 (Eigenvalues of an Idempotent Matrix) Let $\lambda \in \sigma(\boldsymbol{A})$ where $\boldsymbol{A}^2 = \boldsymbol{A} \in \mathbb{C}^{n\times n}$. Show that $\lambda = 0$ or $\lambda = 1$. (A matrix is called **idempotent** if $\boldsymbol{A}^2 = \boldsymbol{A}$).

Exercise 6.6 (Eigenvalues of a Nilpotent Matrix) Let $\lambda \in \sigma(\boldsymbol{A})$ where $\boldsymbol{A}^k = 0$ for some $k \in \mathbb{N}$. Show that $\lambda = 0$. (A matrix $\boldsymbol{A} \in \mathbb{C}^{n\times n}$ such that $\boldsymbol{A}^k = 0$ for some $k \in \mathbb{N}$ is called **nilpotent**).

Exercise 6.7 (Eigenvalues of a Unitary Matrix) Let $\lambda \in \sigma(\boldsymbol{A})$, where $\boldsymbol{A}^*\boldsymbol{A} = \boldsymbol{I}$. Show that $|\lambda| = 1$.

Exercise 6.8 (Nonsingular Approximation of a Singular Matrix) Suppose $\boldsymbol{A} \in \mathbb{C}^{n\times n}$ is singular. Then we can find $\epsilon_0 > 0$ such that $\boldsymbol{A} + \epsilon\boldsymbol{I}$ is nonsingular for all $\epsilon \in \mathbb{C}$ with $|\epsilon| < \epsilon_0$. Hint: $\det(\boldsymbol{A}) = \lambda_1\lambda_2\cdots\lambda_n$, where λ_i are the eigenvalues of $\boldsymbol{A}$.

Exercise 6.9 (Companion Matrix) For $q_0, \ldots, q_{n-1} \in \mathbb{C}$ let $p(\lambda) = \lambda^n + q_{n-1}\lambda^{n-1} + \cdots + q_0$ be a polynomial of degree n in λ. We derive two matrices that have $(-1)^n p$ as its characteristic polynomial.

a) Show that $p = (-1)^n\pi_{\boldsymbol{A}}$ where

$$\boldsymbol{A} = \begin{bmatrix} -q_{n-1} & -q_{n-2} & \cdots & -q_1 & -q_0 \\ 1 & 0 & \cdots & 0 & 0 \\ 0 & 1 & \cdots & 0 & 0 \\ \vdots & \vdots & \ddots & \vdots & \vdots \\ 0 & 0 & \cdots & 1 & 0 \end{bmatrix}.$$

$\boldsymbol{A}$ is called a **companion matrix** of p.

b) Show that $p = (-1)^n\pi_{\boldsymbol{B}}$ where

$$\boldsymbol{B} = \begin{bmatrix} 0 & 0 & \cdots & 0 & -q_0 \\ 1 & 0 & \cdots & 0 & -q_1 \\ 0 & 1 & \cdots & 0 & -q_2 \\ \vdots & \vdots & \ddots & \vdots & \vdots \\ 0 & 0 & \cdots & 1 & -q_{n-1} \end{bmatrix}.$$

Thus $\boldsymbol{B}$ can also be regarded as a companion matrix for p.

Exercise 6.10 (Find Eigenpair Example) Find eigenvalues and eigenvectors of $A = \begin{bmatrix} 1 & 2 & 3 \\ 0 & 2 & 3 \\ 0 & 0 & 2 \end{bmatrix}$. Is A defective?

Exercise 6.11 (Right or Wrong? (Exam Exercise 2005-1)) Decide if the following statements are right or wrong. Give supporting arguments for your decisions.

a) The matrix

$$A = \frac{1}{6}\begin{bmatrix} 3 & 4 \\ 4 & -3 \end{bmatrix}$$

is orthogonal?

b) Let

$$A = \begin{bmatrix} a & 1 \\ 0 & a \end{bmatrix}$$

where $a \in \mathbb{R}$. There is a nonsingular matrix $Y \in \mathbb{R}^{2\times 2}$ and a diagonal matrix $D \in \mathbb{R}^{2\times 2}$ such that $A = YDY^{-1}$?

Exercise 6.12 (Eigenvalues of Tridiagonal Matrix (Exam Exercise 2009-3)) Let $A \in \mathbb{R}^{n,n}$ be tridiagonal (i.e. $a_{ij} = 0$ when $|i - j| > 1$) and suppose also that $a_{i+1,i}a_{i,i+1} > 0$ for $i = 1, \ldots, n-1$. Show that the eigenvalues of A are real.[2]

6.6.2 *Exercises Sect. 6.2*

Exercise 6.13 (Jordan Example)
For the Jordan factorization of the matrix $A = \begin{bmatrix} 3 & 0 & 1 \\ -4 & 1 & -2 \\ -4 & 0 & -1 \end{bmatrix}$ we have $J = \begin{bmatrix} 1 & 1 & 0 \\ 0 & 1 & 0 \\ 0 & 0 & 1 \end{bmatrix}$. Find S.

Exercise 6.14 (A Nilpotent Matrix) Show that $(J_m(\lambda) - \lambda I)^r = \begin{bmatrix} \mathbf{0} & I_{m-r} \\ \mathbf{0} & \mathbf{0} \end{bmatrix}$ for $1 \le r \le m-1$ and conclude that $(J_m(\lambda) - \lambda I)^m = 0$.

Exercise 6.15 (Properties of the Jordan Factorization)
Let J be the Jordan factorization of a matrix $A \in \mathbb{C}^{n\times n}$ as given in Theorem 6.4. Then for $r = 0, 1, 2, \ldots, m = 2, 3, \ldots,$ and any $\lambda \in \mathbb{C}$

a) $A^r = SJ^rS^{-1}$,
b) $J^r = \operatorname{diag}(U_1^r, \ldots, U_k^r)$,

[2]Hint: show that there is a diagonal matrix D such that $D^{-1}AD$ is symmetric.

c) $\boldsymbol{U}_i^r = \text{diag}(\boldsymbol{J}_{m_{i,1}}(\lambda_i)^r, \ldots, \boldsymbol{J}_{m_{i,g_i}}(\lambda_i)^r)$,
d) $\boldsymbol{J}_m(\lambda)^r = (\boldsymbol{E}_m + \lambda \boldsymbol{I}_m)^r = \sum_{k=0}^{\min\{r,m-1\}} \binom{r}{k} \lambda^{r-k} \boldsymbol{E}_m^k$.

Exercise 6.16 (Powers of a Jordan Block) Find $\boldsymbol{J}^{100}$ and $\boldsymbol{A}^{100}$ for the matrix in Exercise 6.13.

Exercise 6.17 (The Minimal Polynomial) Let $\boldsymbol{J}$ be the Jordan factorization of a matrix $\boldsymbol{A} \in \mathbb{C}^{n \times n}$ as given in Theorem 6.4. The polynomial

$$\mu_{\boldsymbol{A}}(\lambda) := \prod_{i=1}^{k} (\lambda_i - \lambda)^{m_i} \text{ where } m_i := \max_{1 \le j \le g_i} m_{i,j}, \tag{6.19}$$

is called the **minimal polynomial** of $\boldsymbol{A}$. We define the matrix polynomial $\mu_{\boldsymbol{A}}(\boldsymbol{A})$ by replacing the factors $\lambda_i - \lambda$ by $\lambda_i \boldsymbol{I} - \boldsymbol{A}$.

a) We have $\pi_{\boldsymbol{A}}(\lambda) = \prod_{i=1}^{k} \prod_{j=1}^{g_i} (\lambda_i - \lambda)^{m_{i,j}}$. Use this to show that the minimal polynomial divides the characteristic polynomial, i.e., $\pi_{\boldsymbol{A}} = \mu_{\boldsymbol{A}} \nu_{\boldsymbol{A}}$ for some polynomial $\nu_{\boldsymbol{A}}$.
b) Show that $\mu_{\boldsymbol{A}}(\boldsymbol{A}) = \boldsymbol{0} \iff \mu_{\boldsymbol{A}}(\boldsymbol{J}) = \boldsymbol{0}$.
c) (can be difficult) Use Exercises 6.14, 6.15 and the maximality of m_i to show that $\mu_{\boldsymbol{A}}(\boldsymbol{A}) = 0$. Thus a matrix satisfies its minimal equation. Finally show that the degree of any polynomial p such that $p(\boldsymbol{A}) = \boldsymbol{0}$ is at least as large as the degree of the minimal polynomial.
d) Use 2. to show the **Cayley-Hamilton Theorem** which says that a matrix satisfies its characteristic equation $\pi_{\boldsymbol{A}}(\boldsymbol{A}) = \boldsymbol{0}$.

Exercise 6.18 (Cayley Hamilton Theorem (Exam Exercise 1996-3)) Suppose p is a polynomial given by $p(t) := \sum_{j=0}^{r} b_j t^j$, where $b_j \in \mathbb{C}$ and $\boldsymbol{A} \in \mathbb{C}^{n \times n}$. We define the matrix $p(\boldsymbol{A}) \in \mathbb{C}^{n \times n}$ by

$$p(\boldsymbol{A}) := \sum_{j=0}^{r} b_j \boldsymbol{A}^j,$$

where $\boldsymbol{A}^0 := \boldsymbol{I}$. From this it follows that if $p(t) := (t - \alpha_1) \cdots (t - \alpha_r)$ for some $\alpha_0, \ldots, \alpha_r \in \mathbb{C}$ then $p(\boldsymbol{A}) = (\boldsymbol{A} - \alpha_1) \cdots (\boldsymbol{A} - \alpha_r)$. We accept this without proof.

Let $\boldsymbol{U}^* \boldsymbol{A} \boldsymbol{U} = \boldsymbol{T}$, where $\boldsymbol{U}$ is unitary and $\boldsymbol{T}$ upper triangular with the eigenvalues of $\boldsymbol{A}$ on the diagonal.

a) Find the characteristic polynomial $\pi_{\boldsymbol{A}}$ to $\begin{bmatrix} 2 & 1 \\ -1 & 4 \end{bmatrix}$. Show that $\pi(\boldsymbol{A}) = \boldsymbol{0}$.
b) Let now $\boldsymbol{A} \in \mathbb{C}^{n \times n}$ be arbitrary. For any polynomial p show that $p(\boldsymbol{A}) = \boldsymbol{U} p(\boldsymbol{T}) \boldsymbol{U}^*$.
c) Let $n, k \in \mathbb{N}$ with $1 \le k < n$. Let $\boldsymbol{C}, \boldsymbol{D} \in \mathbb{C}^{n \times n}$ be upper triangular. Moreover, $c_{i,j} = 0$ for $i, j \le k$ and $d_{k+1,k+1} = 0$. Define $\boldsymbol{E} := \boldsymbol{C}\boldsymbol{D}$ and show that $e_{i,j} = 0$ for $i, j \le k + 1$.

d) Now let $p := \pi_A$ be the characteristic polynomial of $\boldsymbol{A}$. Show that $p(\boldsymbol{T}) = \boldsymbol{0}$.[3] Then show that $p(\boldsymbol{A}) = \boldsymbol{0}$. (Cayley Hamilton Theorem)

6.6.3 Exercises Sect. 6.3

Exercise 6.19 (Schur Factorization Example) Show that a Schur factorization of $\boldsymbol{A} = \left[\begin{smallmatrix} 1 & 2 \\ 3 & 2 \end{smallmatrix}\right]$ is $\boldsymbol{U}^T\boldsymbol{A}\boldsymbol{U} = \left[\begin{smallmatrix} -1 & -1 \\ 0 & 4 \end{smallmatrix}\right]$, where $\boldsymbol{U} = \frac{1}{\sqrt{2}}\left[\begin{smallmatrix} 1 & 1 \\ -1 & 1 \end{smallmatrix}\right]$.

Exercise 6.20 (Skew-Hermitian Matrix) Suppose $\boldsymbol{C} = \boldsymbol{A} + i\boldsymbol{B}$, where $\boldsymbol{A}, \boldsymbol{B} \in \mathbb{R}^{n\times n}$. Show that $\boldsymbol{C}$ is skew-Hermitian if and only if $\boldsymbol{A}^T = -\boldsymbol{A}$ and $\boldsymbol{B}^T = \boldsymbol{B}$.

Exercise 6.21 (Eigenvalues of a Skew-Hermitian Matrix) Show that any eigenvalue of a skew-Hermitian matrix is purely imaginary.

Exercise 6.22 (Eigenvector Expansion Using Orthogonal Eigenvectors) Show that if the eigenpairs $(\lambda_1, \boldsymbol{u}_1), \ldots, (\lambda_n, \boldsymbol{u}_n)$ of $\boldsymbol{A} \in \mathbb{C}^{n\times n}$ are orthogonal, i.e., $\boldsymbol{u}_j^*\boldsymbol{u}_k = 0$ for $j \neq k$ then the eigenvector expansions of $\boldsymbol{x}$ and $\boldsymbol{A}\boldsymbol{x} \in \mathbb{C}^n$ take the form

$$\boldsymbol{x} = \sum_{j=1}^{n} c_j\boldsymbol{u}_j, \quad \boldsymbol{A}\boldsymbol{x} = \sum_{j=1}^{n} c_j\lambda_j\boldsymbol{u}_j, \text{ where } c_j = \frac{\boldsymbol{u}_j^*\boldsymbol{x}}{\boldsymbol{u}_j^*\boldsymbol{u}_j}. \tag{6.20}$$

Exercise 6.23 (Rayleigh Quotient (Exam Exercise 2015-3))

a) Let $\boldsymbol{A} \in \mathbb{R}^{n\times n}$ be a symmetric matrix. Explain how we can use the spectral theorem for symmetric matrices to show that

$$\lambda_{\min} = \min_{\boldsymbol{x}\neq 0} R(\boldsymbol{x}) = \min_{\|\boldsymbol{x}\|_2=1} R(\boldsymbol{x}),$$

where $\lambda_{\min}$ is the smallest eigenvalue of $\boldsymbol{A}$, and $R(\boldsymbol{x})$ is the Rayleigh quotient given by

$$R(\boldsymbol{x}) := \frac{\boldsymbol{x}^T\boldsymbol{A}\boldsymbol{x}}{\boldsymbol{x}^T\boldsymbol{x}}.$$

[3] Hint: use a suitable factorization of p and use **c**).

b) Let $x, y \in \mathbb{R}^n$ such that $\|x\|_2 = 1$ and $y \neq 0$. Show that

$$R(x - ty) = R(x) - 2t\big(Ax - R(x)x\big)^T y + \mathcal{O}(t^2),$$

where $t > 0$ is small.[4]

c) Based on the characterization given in **a)** above it is tempting to develop an algorithm for computing $\lambda_{\min}$ by approximating the minimum of $R(x)$ over the unit ball

$$B_1 := \{x \in \mathbb{R}^n \mid \|x\|_2 = 1\}.$$

Assume that $x^0 \in B_1$ satisfies $Ax^0 - R(x^0)x^0 \neq 0$, i.e., $(R(x^0), x^0)$ is not an eigenpair for A. Explain how we can find a vector $x^1 \in B_1$ such that $R(x^1) < R(x^0)$.

6.6.4 *Exercises Sect. 6.4*

Exercise 6.24 (Eigenvalue Perturbation for Hermitian Matrices) Show that in Theorem 6.13, if E is symmetric positive semidefinite then $\beta_i \geq \alpha_i$.

Exercise 6.25 (Hoffman-Wielandt) Show that (6.15) does not hold for the matrices $A := \begin{bmatrix} 0 & 0 \\ 0 & 4 \end{bmatrix}$ and $B := \begin{bmatrix} -1 & -1 \\ 1 & 1 \end{bmatrix}$. Why does this not contradict the Hoffman-Wielandt theorem?

Exercise 6.26 (Biorthogonal Expansion) Determine right and left eigenpairs for the matrix $A := \begin{bmatrix} 3 & 1 \\ 2 & 2 \end{bmatrix}$ and the two expansions in (6.16) for any $v \in \mathbb{R}^2$.

Exercise 6.27 (Generalized Rayleigh Quotient) For $A \in \mathbb{C}^{n\times n}$ and any $y, x \in \mathbb{C}^n$ with $y^*x \neq 0$ the quantity $R(y, x) = R_A(y, x) := \frac{y^*Ax}{y^*x}$ is called a **generalized Rayleigh quotient** for A. Show that if (λ, x) is a right eigenpair for A then $R(y, x) = \lambda$ for any y with $y^*x \neq 0$. Also show that if (λ, y) is a left eigenpair for A then $R(y, x) = \lambda$ for any x with $y^*x \neq 0$.

6.7 Review Questions

6.7.1 Does A, A^T and A^* have the same eigenvalues? What about A^*A and AA^*?

6.7.2 Can a matrix with multiple eigenvalues be similar to a diagonal matrix?

---

[4]Hint: Use Taylor's theorem for the function $f(t) = R(x - ty)$.

6.7.3 What is the geometric multiplicity of an eigenvalue? Can it be bigger than the algebraic multiplicity?
6.7.4 What is the Jordan factorization of a matrix?
6.7.5 What are the eigenvalues of a diagonal matrix?
6.7.6 What are the Schur factors of a matrix?
6.7.7 What is a quasi-triangular matrix?
6.7.8 Give some classes of normal matrices. Why are normal matrices important?
6.7.9 State the Courant-Fischer theorem.
6.7.10 State the Hoffman-Wielandt theorem for Hermitian matrices.
6.7.11 What is a left eigenvector of a matrix?

Chapter 7
The Singular Value Decomposition

The singular value decomposition and the reduced form called the singular value factorization are useful both for theory and practice. Some of their applications include solving over-determined equations, principal component analysis in statistics, numerical determination of the rank of a matrix, algorithms used in search engines, and the theory of matrices.

We know from Theorem 6.7 that a square matrix $\boldsymbol{A}$ can be diagonalized by a unitary similarity transformation if and only if it is normal, that is $\boldsymbol{A}^*\boldsymbol{A} = \boldsymbol{A}\boldsymbol{A}^*$. In particular, if $\boldsymbol{A} \in \mathbb{C}^{n\times n}$ is normal then it has a set of orthonormal eigenpairs $(\lambda_1, \boldsymbol{u}_1), \ldots, (\lambda_n, \boldsymbol{u}_n)$. Letting $\boldsymbol{U} := [\boldsymbol{u}_1, \ldots, \boldsymbol{u}_n] \in \mathbb{C}^{n\times n}$ and $\boldsymbol{D} := \operatorname{diag}(\lambda_1, \ldots, \lambda_n)$ we have the spectral decomposition

$$\boldsymbol{A} = \boldsymbol{U}\boldsymbol{D}\boldsymbol{U}^*, \text{ where } \boldsymbol{U}^*\boldsymbol{U} = \boldsymbol{I}. \tag{7.1}$$

The singular value decomposition (SVD) is a decomposition of a matrix in the form $\boldsymbol{A} = \boldsymbol{U}\boldsymbol{\Sigma}\boldsymbol{V}^*$, where $\boldsymbol{U}$ and $\boldsymbol{V}$ are unitary, and $\boldsymbol{\Sigma}$ is a nonnegative diagonal matrix, i.e., $\Sigma_{ij} = 0$ for all $i \neq j$ and $\Sigma_{ii} \geq 0$ for all i. The diagonal elements $\sigma_i := \Sigma_{ii}$ are called **singular values**, while the columns of $\boldsymbol{U}$ and $\boldsymbol{V}$ are called **singular vectors**. To be a singular value decomposition the singular values should be ordered, i.e., $\sigma_i \geq \sigma_{i+1}$ for all i.

Example 7.1 (SVD) The following is a singular value decomposition of a rectangular matrix.

$$\boldsymbol{A} = \frac{1}{15}\begin{bmatrix} 14 & 2 \\ 4 & 22 \\ 16 & 13 \end{bmatrix} = \frac{1}{3}\begin{bmatrix} 1 & 2 & 2 \\ 2 & -2 & 1 \\ 2 & 1 & -2 \end{bmatrix}\begin{bmatrix} 2 & 0 \\ 0 & 1 \\ 0 & 0 \end{bmatrix}\frac{1}{5}\begin{bmatrix} 3 & 4 \\ 4 & -3 \end{bmatrix} = \boldsymbol{U}\boldsymbol{\Sigma}\boldsymbol{V}^*. \tag{7.2}$$

T. Lyche, *Numerical Linear Algebra and Matrix Factorizations*,
Texts in Computational Science and Engineering 22,
https://doi.org/10.1007/978-3-030-36468-7_7

Indeed, $\boldsymbol{U}$ and $\boldsymbol{V}$ are unitary since the columns (the singular vectors) are orthonormal, and $\boldsymbol{\Sigma}$ is a nonnegative diagonal matrix with singular values $\sigma_1 = 2$ and $\sigma_2 = 1$.

7.1 The SVD Always Exists

The singular value decomposition is closely related to the eigenpairs of $\boldsymbol{A}^*\boldsymbol{A}$ and $\boldsymbol{A}\boldsymbol{A}^*$.

7.1.1 The Matrices A^*A, AA^*

To start we show that bases for the four fundamental subspaces $\mathcal{R}(\boldsymbol{A})$, $\mathcal{N}(\boldsymbol{A})$, $\mathcal{R}(\boldsymbol{A}^*)$ and $\mathcal{N}(\boldsymbol{A}^*)$ of a matrix $\boldsymbol{A}$ can be determined from the eigenpairs of $\boldsymbol{A}^*\boldsymbol{A}$ and $\boldsymbol{A}\boldsymbol{A}^*$.

Theorem 7.1 (The Matrices $\boldsymbol{A}^*\boldsymbol{A}$, $\boldsymbol{A}\boldsymbol{A}^*$) *Suppose $m, n \in \mathbb{N}$ and $\boldsymbol{A} \in \mathbb{C}^{m\times n}$.*

1. *The matrices $\boldsymbol{A}^*\boldsymbol{A} \in \mathbb{C}^{n\times n}$ and $\boldsymbol{A}\boldsymbol{A}^* \in \mathbb{C}^{m\times m}$ have the same nonzero eigenvalues with the same algebraic multiplicities. Moreover the extra eigenvalues of the larger matrix are all zero.*
2. *The matrices $\boldsymbol{A}^*\boldsymbol{A}$ and $\boldsymbol{A}\boldsymbol{A}^*$ are Hermitian with nonnegative eigenvalues.*
3. *Let $(\lambda_j, \boldsymbol{v}_j)$ be orthonormal eigenpairs for $\boldsymbol{A}^*\boldsymbol{A}$ with*

$$\lambda_1 \geq \cdots \geq \lambda_r > 0 = \lambda_{r+1} = \cdots = \lambda_n.$$

   Then $\{\boldsymbol{A}\boldsymbol{v}_1, \ldots, \boldsymbol{A}\boldsymbol{v}_r\}$ is an orthogonal basis for the column space $\mathcal{R}(\boldsymbol{A}) := \{\boldsymbol{A}\boldsymbol{y} \in \mathbb{C}^m : \boldsymbol{y} \in \mathbb{C}^n\}$ and $\{\boldsymbol{v}_{r+1}, \ldots, \boldsymbol{v}_n\}$ is an orthonormal basis for the nullspace $\mathcal{N}(\boldsymbol{A}) := \{\boldsymbol{y} \in \mathbb{C}^n : \boldsymbol{A}\boldsymbol{y} = \boldsymbol{0}\}$.
4. *Let $(\lambda_j, \boldsymbol{u}_j)$ be orthonormal eigenpairs for $\boldsymbol{A}\boldsymbol{A}^*$. If $\lambda_j > 0$, $j = 1, \ldots, r$ and $\lambda_j = 0$, $j = r+1, \ldots, m$ then $\{\boldsymbol{A}^*\boldsymbol{u}_1, \ldots, \boldsymbol{A}^*\boldsymbol{u}_r\}$ is an orthogonal basis for the column space $\mathcal{R}(\boldsymbol{A}^*)$ and $\{\boldsymbol{u}_{r+1}, \ldots, \boldsymbol{u}_m\}$ is an orthonormal basis for the nullspace $\mathcal{N}(\boldsymbol{A}^*)$.*
5. *The rank of $\boldsymbol{A}$ equals the number of positive eigenvalues of $\boldsymbol{A}^*\boldsymbol{A}$ and $\boldsymbol{A}\boldsymbol{A}^*$.*

Proof

1. Consider the characteristic polynomials $\pi_{\boldsymbol{A}^*\boldsymbol{A}}$ and $\pi_{\boldsymbol{A}\boldsymbol{A}^*}$. By (6.1) we have

$$\lambda^m \pi_{\boldsymbol{A}^*\boldsymbol{A}}(\lambda) = \lambda^n \pi_{\boldsymbol{A}\boldsymbol{A}^*}(\lambda), \quad \lambda \in \mathbb{C},$$

   and the claim follows.
2. The matrices $\boldsymbol{A}^*\boldsymbol{A}$ and $\boldsymbol{A}\boldsymbol{A}^*$ are Hermitian and positive semidefinite and therefore has nonnegative eigenvalues (cf. Lemmas 4.2 and 4.5). Moreover, if

$A^*Av = \lambda v$ with $v \neq 0$, then

$$\lambda = \frac{v^*A^*Av}{v^*v} = \frac{\|Av\|_2^2}{\|v\|_2^2} \geq 0. \tag{7.3}$$

3. By orthonormality of $v_1, \ldots, v_n$ we have $(Av_j)^*Av_k = v_j^*A^*Av_k = \lambda_k v_j^* v_k = 0$ for $j \neq k$, showing that $Av_1, \ldots, Av_n$ are orthogonal vectors. Moreover, (7.3) implies that $Av_1, \ldots, Av_r$ are nonzero and $Av_j = 0$ for $j = r+1, \ldots, n$. In particular, the elements of $\{Av_1, \ldots, Av_r\}$ and $\{v_{r+1}, \ldots, v_n\}$ are linearly independent vectors in $\mathcal{R}(A)$ and $\mathcal{N}(A)$, respectively. The proof will be complete once it is shown that $\mathcal{R}(A) \subset \text{span}(Av_1, \ldots, Av_r)$ and $\mathcal{N}(A) \subset \text{span}(v_{r+1}, \ldots, v_n)$. Suppose $x \in \mathcal{R}(A)$. Then $x = Ay$ for some $y \in \mathbb{C}^n$, Let $y = \sum_{j=1}^n c_j v_j$ be an eigenvector expansion of y. Since $Av_j = 0$ for $j = r+1, \ldots, n$ we obtain $x = Ay = \sum_{j=1}^n c_j Av_j = \sum_{j=1}^r c_j Av_j \in \text{span}(Av_1, \ldots, Av_r)$. Finally, if $y = \sum_{j=1}^n c_j v_j \in \mathcal{N}(A)$, then we have $Ay = \sum_{j=1}^r c_j Av_j = 0$, and $c_1 = \cdots = c_r = 0$ since $Av_1, \ldots, Av_r$ are linearly independent. But then $y = \sum_{j=r+1}^n c_j v_j \in \text{span}(v_{r+1}, \ldots, v_n)$.
4. Since $AA^* = B^*B$ with $B := A^*$ this follows from part 3 with $A = B$.
5. By part 1 and 2 A^*A and AA^* have the same number r of positive eigenvalues and by part 3 and 4 r is the rank of A.

□

The following theorem shows, in a constructive way, that any matrix has a singular value decomposition.

Theorem 7.2 (Existence of SVD) *Suppose for $m, n, r \in \mathbb{N}$ that $A \in \mathbb{C}^{m\times n}$ has rank r, and that (λ_j, v_j) are orthonormal eigenpairs for A^*A with $\lambda_1 \geq \cdots \geq \lambda_r > 0 = \lambda_{r+1} = \cdots = \lambda_n$. Define*

1. *$V := [v_1, \ldots, v_n] \in \mathbb{C}^{n\times n}$,*
2. *$\Sigma \in \mathbb{R}^{m\times n}$ is a diagonal matrix with diagonal elements $\sigma_j := \sqrt{\lambda_j}$ for $j = 1, \ldots, \min(m, n)$,*
3. *$U := [u_1, \ldots, u_m] \in \mathbb{C}^{m\times m}$, where $u_j = \sigma_j^{-1} Av_j$ for $j = 1, \ldots, r$ and $u_{r+1}, \ldots, u_m$ is any extension of $u_1, \ldots, u_r$ to an orthonormal basis $u_1, \ldots, u_m$ for $\mathbb{C}^m$.*

Then $A = U\Sigma V^*$ is a singular value decomposition of A.

Proof Let U, Σ, V be as in the theorem. The vectors $u_1, \ldots, u_r$ are orthonormal since $Av_1, \ldots, Av_r$ are orthogonal and $\sigma_j = \|Av_j\|_2 > 0$, $j = 1, \ldots, r$ by (7.3). But then U and V are unitary and Σ is a nonnegative diagonal matrix. Moreover,

$$\begin{aligned} U\Sigma &= U[\sigma_1 e_1, \ldots, \sigma_r e_r, 0, \ldots, 0] \\ &= [\sigma_1 u_1, \ldots, \sigma_r u_r, 0, \ldots, 0] \\ &= [Av_1, \ldots, Av_n]. \end{aligned}$$

Thus $\boldsymbol{U\Sigma} = \boldsymbol{AV}$ and since $\boldsymbol{V}$ is square and unitary we find $\boldsymbol{U\Sigma V}^* = \boldsymbol{AVV}^* = \boldsymbol{A}$ and we have an SVD of $\boldsymbol{A}$ with $\sigma_1 \geq \sigma_2 \geq \cdots \geq \sigma_r$. □

Example 7.2 (Find SVD) To derive the SVD in (7.2) where $\boldsymbol{A} = \frac{1}{15}\begin{bmatrix} 14 & 2 \\ 4 & 22 \\ 16 & 13 \end{bmatrix}$, we first compute the eigenpairs of

$$\boldsymbol{B} := \boldsymbol{A}^T\boldsymbol{A} = \frac{1}{25}\begin{bmatrix} 52 & 36 \\ 36 & 73 \end{bmatrix}$$

as

$$\boldsymbol{B}\begin{bmatrix} 3 \\ 4 \end{bmatrix} = 4\begin{bmatrix} 3 \\ 4 \end{bmatrix}, \quad \boldsymbol{B}\begin{bmatrix} 4 \\ -3 \end{bmatrix} = 1\begin{bmatrix} 4 \\ -3 \end{bmatrix}.$$

Thus $\sigma_1 = 2$, $\sigma_2 = 1$, and $\boldsymbol{V} = \frac{1}{5}\begin{bmatrix} 3 & 4 \\ 4 & -3 \end{bmatrix}$. Now $\boldsymbol{u}_1 = \boldsymbol{A}\boldsymbol{v}_1/\sigma_1 = [1, 2, 2]^T/3$, $\boldsymbol{u}_2 = \boldsymbol{A}\boldsymbol{v}_2/\sigma_2 = [2, -2, 1]^T/3$. For an SVD we also need $\boldsymbol{u}_3$ which is any vector of length one orthogonal to $\boldsymbol{u}_1$ and $\boldsymbol{u}_2$. $\boldsymbol{u}_3 = [2, 1, -2]^T/3$ is such a vector and we obtain the singular value decomposition (7.2).

7.2 Further Properties of SVD

We first consider a reduced SVD that is often convenient.

7.2.1 *The Singular Value Factorization*

Suppose $\boldsymbol{A} = \boldsymbol{U\Sigma V}^*$ is a singular value decomposition of $\boldsymbol{A}$ of rank r. Consider the block partitions

$$\begin{aligned}
&\boldsymbol{U} = [\boldsymbol{U}_1, \boldsymbol{U}_2] \in \mathbb{C}^{m\times m}, \quad \boldsymbol{U}_1 := [\boldsymbol{u}_1, \ldots, \boldsymbol{u}_r], \quad \boldsymbol{U}_2 := [\boldsymbol{u}_{r+1}, \ldots, \boldsymbol{u}_m],\\
&\boldsymbol{V} = [\boldsymbol{V}_1, \boldsymbol{V}_2] \in \mathbb{C}^{n\times n}, \quad \boldsymbol{V}_1 := [\boldsymbol{v}_1, \ldots, \boldsymbol{v}_r], \quad \boldsymbol{V}_2 := [\boldsymbol{v}_{r+1}, \ldots, \boldsymbol{v}_n],\\
&\boldsymbol{\Sigma} = \begin{bmatrix} \boldsymbol{\Sigma}_1 & \boldsymbol{0}_{r,n-r} \\ \boldsymbol{0}_{m-r,r} & \boldsymbol{0}_{m-r,n-r} \end{bmatrix} \in \mathbb{R}^{m\times n}, \text{ where } \boldsymbol{\Sigma}_1 := \operatorname{diag}(\sigma_1, \ldots, \sigma_r).
\end{aligned} \tag{7.4}$$

Thus Σ_1 contains the r positive singular values on the diagonal and for $k, l \geq 0$ the symbol $\boldsymbol{0}_{k,l} = [\,]$ denotes the empty matrix if $k = 0$ or $l = 0$, and the zero matrix with k rows and l columns otherwise. We obtain by block multiplication a reduced factorization

$$\boldsymbol{A} = \boldsymbol{U}\boldsymbol{\Sigma}\boldsymbol{V}^* = \boldsymbol{U}_1\boldsymbol{\Sigma}_1\boldsymbol{V}_1^*. \tag{7.5}$$

As an example:

$$\begin{bmatrix}1 & -1\\ 1 & -1\end{bmatrix} = \frac{1}{\sqrt{2}}\begin{bmatrix}1 & 1\\ 1 & -1\end{bmatrix}\begin{bmatrix}2 & 0\\ 0 & 0\end{bmatrix}\frac{1}{\sqrt{2}}\begin{bmatrix}1 & -1\\ 1 & 1\end{bmatrix} = \frac{1}{\sqrt{2}}\begin{bmatrix}1\\ 1\end{bmatrix}\begin{bmatrix}2\end{bmatrix}\frac{1}{\sqrt{2}}\begin{bmatrix}1 & -1\end{bmatrix}.$$

Definition 7.1 (SVF) Let $m, n, r \in \mathbb{N}$ and suppose $\boldsymbol{A} \in \mathbb{C}^{m\times n}$ has r positive singular values, i.e., $\boldsymbol{A}$ has rank r. A **singular value factorization (SVF)** is a factorization of $\boldsymbol{A} \in \mathbb{C}^{m\times n}$ of the form $\boldsymbol{A} = \boldsymbol{U}_1\boldsymbol{\Sigma}_1\boldsymbol{V}_1^*$, where $\boldsymbol{U}_1 \in \mathbb{C}^{m\times r}$ and $\boldsymbol{V}_1 \in \mathbb{C}^{n\times r}$ have orthonormal columns, and $\boldsymbol{\Sigma}_1 \in \mathbb{R}^{r\times r}$ is a diagonal matrix with $\sigma_1 \geq \cdots \geq \sigma_r > 0$.

An SVD and an SVF of a matrix $\boldsymbol{A}$ of rank r are closely related.

1. Let $\boldsymbol{A} = \boldsymbol{U}\boldsymbol{\Sigma}\boldsymbol{V}^*$ be an SVD of $\boldsymbol{A}$. Then $\boldsymbol{A} = \boldsymbol{U}_1\boldsymbol{\Sigma}_1\boldsymbol{V}_1^*$ is an SVF of $\boldsymbol{A}$, where $\boldsymbol{U}_1$, $\boldsymbol{V}_1$ contain the first r columns of $\boldsymbol{U}$, $\boldsymbol{V}$ respectively, and $\Sigma_1 \in \mathbb{R}^{r\times r}$ is a diagonal matrix with the positive singular values on the diagonal.
2. Conversely, suppose $\boldsymbol{A} = \boldsymbol{U}_1\boldsymbol{\Sigma}_1\boldsymbol{V}_1^*$ is a singular value factorization of $\boldsymbol{A}$. Extend $\boldsymbol{U}_1$ and $\boldsymbol{V}_1$ in any way to unitary matrices $\boldsymbol{U} \in \mathbb{C}^{m\times m}$ and $\boldsymbol{V} \in \mathbb{C}^{n\times n}$, and let $\boldsymbol{\Sigma}$ be given by (7.4). Then $\boldsymbol{A} = \boldsymbol{U}\boldsymbol{\Sigma}\boldsymbol{V}^*$ is an SVD of $\boldsymbol{A}$.
3. If $\boldsymbol{A} = [\boldsymbol{u}_1, \ldots, \boldsymbol{u}_r]\operatorname{diag}(\sigma_1, \ldots, \sigma_r)[\boldsymbol{v}_1, \ldots, \boldsymbol{v}_r]^*$ is a singular value factorization of $\boldsymbol{A}$ then

$$\boldsymbol{A} = \sum_{j=1}^{r} \sigma_j \boldsymbol{u}_j \boldsymbol{v}_j^*. \tag{7.6}$$

   This is known as the **outer product form** of the SVD and SVF.
4. We note that a nonsingular square matrix has full rank and only positive singular values. Thus the SVD and SVF are the same for a nonsingular matrix.

Example 7.3 ($r < n < m$) To find the SVF and SVD of

$$\boldsymbol{A} = \begin{bmatrix}1 & 1\\ 1 & 1\\ 0 & 0\end{bmatrix}.$$

we first compute eigenpairs of

$$\boldsymbol{B} := \boldsymbol{A}^T\boldsymbol{A} = \begin{bmatrix}2 & 2\\ 2 & 2\end{bmatrix}$$

as

$$B\begin{bmatrix}1\\1\end{bmatrix}=4\begin{bmatrix}1\\1\end{bmatrix},\quad B\begin{bmatrix}1\\-1\end{bmatrix}=0\begin{bmatrix}1\\-1\end{bmatrix},$$

and we find $\sigma_1 = 2$, $\sigma_2 = 0$, Thus $r = 1$, $m = 3$, $n = 2$ and

$$\boldsymbol{\Sigma}=\begin{bmatrix}\boldsymbol{\Sigma}_1 & 0\\ 0 & 0\\ 0 & 0\end{bmatrix},\quad \boldsymbol{\Sigma}_1=[2],\quad \boldsymbol{V}=\frac{1}{\sqrt{2}}\begin{bmatrix}1 & 1\\ 1 & -1\end{bmatrix}.$$

We find $\boldsymbol{u}_1 = \boldsymbol{A}\boldsymbol{v}_1/\sigma_1 = \boldsymbol{s}_1/\sqrt{2}$, where $\boldsymbol{s}_1 = [1, 1, 0]^T$, and the SVF of $\boldsymbol{A}$ is given by

$$\boldsymbol{A}=\frac{1}{\sqrt{2}}\begin{bmatrix}1\\1\\0\end{bmatrix}[2]\frac{1}{\sqrt{2}}\begin{bmatrix}1 & 1\end{bmatrix}.$$

To find an SVD we need to extend $\boldsymbol{u}_1$ to an orthonormal basis for $\mathbb{R}^3$. We first extend $\boldsymbol{s}_1$ to a basis $\{\boldsymbol{s}_1, \boldsymbol{s}_2, \boldsymbol{s}_3\}$ for $\mathbb{R}^3$, apply the Gram-Schmidt orthogonalization process to $\{\boldsymbol{s}_1, \boldsymbol{s}_2, \boldsymbol{s}_3\}$, and then normalize. Choosing the basis

$$\boldsymbol{s}_1=\begin{bmatrix}1\\1\\0\end{bmatrix},\quad \boldsymbol{s}_2=\begin{bmatrix}0\\1\\0\end{bmatrix},\quad \boldsymbol{s}_3=\begin{bmatrix}0\\0\\1\end{bmatrix},$$

we find from (5.8) $\boldsymbol{w}_1 = \boldsymbol{s}_1$, $\boldsymbol{w}_2 = \boldsymbol{s}_2 - \frac{\boldsymbol{s}_2^T\boldsymbol{w}_1}{\boldsymbol{w}_1^T\boldsymbol{w}_1}\boldsymbol{w}_1 = \begin{bmatrix}-1/2\\1/2\\0\end{bmatrix}$, $\boldsymbol{w}_3 = \boldsymbol{s}_3 - \frac{\boldsymbol{s}_3^T\boldsymbol{w}_1}{\boldsymbol{w}_1^T\boldsymbol{w}_1}\boldsymbol{w}_1 - \frac{\boldsymbol{s}_3^T\boldsymbol{w}_2}{\boldsymbol{w}_2^T\boldsymbol{w}_2}\boldsymbol{w}_2 = \begin{bmatrix}0\\0\\1\end{bmatrix}$. Normalizing the $\boldsymbol{w}_i$'s we obtain $\boldsymbol{u}_1 = \boldsymbol{w}_1/\|\boldsymbol{w}_1\|_2 = [1/\sqrt{2}, 1/\sqrt{2}, 0]^T$, $\boldsymbol{u}_2 = \boldsymbol{w}_2/\|\boldsymbol{w}_2\|_2 = [-1/\sqrt{2}, 1/\sqrt{2}, 0]^T$, and $\boldsymbol{u}_3 = \boldsymbol{s}_3/\|\boldsymbol{s}_3\|_2 = [0, 0, 1]^T$. Therefore, $\boldsymbol{A} = \boldsymbol{U}\boldsymbol{\Sigma}\boldsymbol{V}^T$, is an SVD, where

$$\boldsymbol{U} := \begin{bmatrix}1/\sqrt{2} & -1/\sqrt{2} & 0\\ 1/\sqrt{2} & 1/\sqrt{2} & 0\\ 0 & 0 & 1\end{bmatrix}\in\mathbb{R}^{3,3},\quad \boldsymbol{\Sigma} := \begin{bmatrix}2 & 0\\ 0 & 0\\ 0 & 0\end{bmatrix}\in\mathbb{R}^{3,2},\quad \boldsymbol{V} := \frac{1}{\sqrt{2}}\begin{bmatrix}1 & 1\\ 1 & -1\end{bmatrix}\in\mathbb{R}^{2,2}.$$

The method we used to find the singular value decomposition in the examples and exercises can be suitable for hand calculation with small matrices, but it is not appropriate as a basis for a general purpose numerical method. In particular, the Gram-Schmidt orthogonalization process is not numerically stable, and forming $\boldsymbol{A}^*\boldsymbol{A}$ can lead to extra errors in the computation. Standard computer implementations of the singular value decomposition [16] first reduces $\boldsymbol{A}$ to bidiagonal form and then use an adapted version of the QR algorithm where the matrix $\boldsymbol{A}^*\boldsymbol{A}$ is not formed. The QR algorithm is discussed in Chap. 15.

7.2.2 SVD and the Four Fundamental Subspaces

The singular vectors form orthonormal bases for the four fundamental subspaces $\mathcal{R}(\boldsymbol{A})$, $\mathcal{N}(\boldsymbol{A})$, $\mathcal{R}(\boldsymbol{A}^*)$, and $\mathcal{N}(\boldsymbol{A}^*)$.

Theorem 7.3 (Singular Vectors and Orthonormal Bases) *For positive integers m, n let $\boldsymbol{A} \in \mathbb{C}^{m\times n}$ have rank r and a singular value decomposition $\boldsymbol{A} = [\boldsymbol{u}_1, \dots, \boldsymbol{u}_m]\boldsymbol{\Sigma}[\boldsymbol{v}_1, \dots, \boldsymbol{v}_n]^* = \boldsymbol{U}\boldsymbol{\Sigma}\boldsymbol{V}^*$. Then the singular vectors satisfy*

$$\begin{aligned} \boldsymbol{A}\boldsymbol{v}_i &= \sigma_i \boldsymbol{u}_i,\ i = 1, \dots, r, \quad & \boldsymbol{A}\boldsymbol{v}_i &= 0,\ i = r+1, \dots, n, \\ \boldsymbol{A}^*\boldsymbol{u}_i &= \sigma_i \boldsymbol{v}_i,\ i = 1, \dots, r, \quad & \boldsymbol{A}^*\boldsymbol{u}_i &= 0,\ i = r+1, \dots, m. \end{aligned} \tag{7.7}$$

Moreover,

$$\begin{aligned} &1.\ \{\boldsymbol{u}_1, \dots, \boldsymbol{u}_r\} \text{ is an orthonormal basis for } \mathcal{R}(\boldsymbol{A}), \\ &2.\ \{\boldsymbol{u}_{r+1}, \dots, \boldsymbol{u}_m\} \text{is an orthonormal basis for } \mathcal{N}(\boldsymbol{A}^*), \\ &3.\ \{\boldsymbol{v}_1, \dots, \boldsymbol{v}_r\} \text{ is an orthonormal basis for } \mathcal{R}(\boldsymbol{A}^*), \\ &4.\ \{\boldsymbol{v}_{r+1}, \dots, \boldsymbol{v}_n\} \text{ is an orthonormal basis for } \mathcal{N}(\boldsymbol{A}). \end{aligned} \tag{7.8}$$

Proof If $\boldsymbol{A} = \boldsymbol{U}\boldsymbol{\Sigma}\boldsymbol{V}^*$ then $\boldsymbol{A}\boldsymbol{V} = \boldsymbol{U}\boldsymbol{\Sigma}$, or in terms of the block partition (7.4) $\boldsymbol{A}[\boldsymbol{V}_1, \boldsymbol{V}_2] = [\boldsymbol{U}_1, \boldsymbol{U}_2]\left[\begin{smallmatrix}\Sigma_1 & \boldsymbol{0}\\ \boldsymbol{0} & \boldsymbol{0}\end{smallmatrix}\right]$. But then $\boldsymbol{A}\boldsymbol{V}_1 = \boldsymbol{U}_1\boldsymbol{\Sigma}_1$, $\boldsymbol{A}\boldsymbol{V}_2 = \boldsymbol{0}$, and this implies the first part of (7.7). Taking conjugate transpose of $\boldsymbol{A} = \boldsymbol{U}\boldsymbol{\Sigma}\boldsymbol{V}^*$ gives $\boldsymbol{A}^* = \boldsymbol{V}\boldsymbol{\Sigma}^*\boldsymbol{U}^*$ or $\boldsymbol{A}^*\boldsymbol{U} = \boldsymbol{V}\boldsymbol{\Sigma}^*$. Using the block partition as before we obtain the last part of (7.7).

It follows from Theorem 7.1 that $\{\boldsymbol{A}\boldsymbol{v}_1, \dots, \boldsymbol{A}\boldsymbol{v}_r\}$ is an orthogonal basis for $\mathcal{R}(\boldsymbol{A})$, $\{\boldsymbol{A}^*\boldsymbol{u}_1, \dots, \boldsymbol{A}^*\boldsymbol{u}_r\}$ is an orthogonal basis for $\mathcal{R}(\boldsymbol{A}^*)$, $\{\boldsymbol{v}_{r+1}, \dots, \boldsymbol{v}_m\}$ is an orthonormal basis for $\mathcal{N}(\boldsymbol{A})$ and $\{\boldsymbol{u}_{r+1}, \dots, \boldsymbol{u}_m\}$ is an orthonormal basis for $\mathcal{N}(\boldsymbol{A}^*)$. By (7.7) $\{\boldsymbol{u}_1, \dots, \boldsymbol{u}_r\}$ is an orthonormal basis for $\mathcal{R}(\boldsymbol{A})$ and $\{\boldsymbol{v}_1, \dots, \boldsymbol{v}_r\}$ is an orthonormal basis for $\mathcal{R}(\boldsymbol{A}^*)$. □

7.3 A Geometric Interpretation

The singular value decomposition and factorization give insight into the geometry of a linear transformation. Consider the linear transformation $\boldsymbol{T} : \mathbb{R}^n \to \mathbb{R}^m$ given by $\boldsymbol{T}\boldsymbol{z} := \boldsymbol{A}\boldsymbol{z}$ where $\boldsymbol{A} \in \mathbb{R}^{m\times n}$. Assume that $\operatorname{rank}(\boldsymbol{A}) = n$. In the following theorem we show that the function $\boldsymbol{T}$ maps the unit sphere in $\mathbb{R}^n$ given by $\mathcal{S} := \{\boldsymbol{z} \in \mathbb{R}^n : \|\boldsymbol{z}\|_2 = 1\}$ onto an ellipsoid $\mathcal{E} := \boldsymbol{A}\mathcal{S} = \{\boldsymbol{A}\boldsymbol{z} : \boldsymbol{z} \in \mathcal{S}\}$ in $\mathbb{R}^m$.

Theorem 7.4 (SVF Ellipse) *Suppose $\boldsymbol{A} \in \mathbb{R}^{m\times n}$ has rank $r = n$, and let $\boldsymbol{A} = \boldsymbol{U}_1\Sigma_1\boldsymbol{V}_1^T$ be a singular value factorization of $\boldsymbol{A}$. Then*

$$\mathcal{E} = \boldsymbol{U}_1\tilde{\mathcal{E}} \text{ where } \tilde{\mathcal{E}} := \{\boldsymbol{y} = [y_1, \dots, y_n]^T \in \mathbb{R}^n : \frac{y_1^2}{\sigma_1^2} + \cdots + \frac{y_n^2}{\sigma_n^2} = 1\}.$$

Proof Suppose $z \in \mathcal{S}$. Now $\boldsymbol{A}z = \boldsymbol{U}_1\Sigma_1\boldsymbol{V}_1^T z = \boldsymbol{U}_1\boldsymbol{y}$, where $\boldsymbol{y} := \Sigma_1\boldsymbol{V}_1^T z$. Since rank$(\boldsymbol{A}) = n$ it follows that $\boldsymbol{V}_1 = \boldsymbol{V}$ is square so that $\boldsymbol{V}_1\boldsymbol{V}_1^T = \boldsymbol{I}$. But then $\boldsymbol{V}_1\boldsymbol{\Sigma}_1^{-1}\boldsymbol{y} = z$ and we obtain

$$1 = \|z\|_2^2 = \|\boldsymbol{V}_1\boldsymbol{\Sigma}_1^{-1}\boldsymbol{y}\|_2^2 = \|\boldsymbol{\Sigma}_1^{-1}\boldsymbol{y}\|_2^2 = \frac{y_1^2}{\sigma_1^2} + \cdots + \frac{y_n^2}{\sigma_n^2}.$$

This implies that $\boldsymbol{y} \in \tilde{\mathcal{E}}$. Finally, $\boldsymbol{x} = \boldsymbol{A}z = \boldsymbol{U}_1\Sigma_1\boldsymbol{V}_1^T z = \boldsymbol{U}_1\boldsymbol{y}$, where $\boldsymbol{y} \in \tilde{\mathcal{E}}$ implies that $\mathcal{E} = \boldsymbol{U}_1\tilde{\mathcal{E}}$. □

The equation $1 = \frac{y_1^2}{\sigma_1^2} + \cdots + \frac{y_n^2}{\sigma_n^2}$ describes an ellipsoid in $\mathbb{R}^n$ with semiaxes of length σ_j along the unit vectors $\boldsymbol{e}_j$ for $j = 1, \ldots, n$. Since the orthonormal transformation $\boldsymbol{U}_1\boldsymbol{y} \to \boldsymbol{x}$ preserves length, the image $\mathcal{E} = \boldsymbol{A}\mathcal{S}$ is a rotated ellipsoid with semiaxes along the left singular vectors $\boldsymbol{u}_j = \boldsymbol{U}\boldsymbol{e}_j$, of length σ_j, $j = 1, \ldots, n$. Since $\boldsymbol{A}\boldsymbol{v}_j = \sigma_j\boldsymbol{u}_j$, for $j = 1, \ldots, n$ the right singular vectors defines points in $\mathcal{S}$ that are mapped onto the semiaxes of $\mathcal{E}$.

Example 7.4 (Ellipse) Consider the transformation $\boldsymbol{A} : \mathbb{R}^2 \to \mathbb{R}^2$ given by the matrix

$$\boldsymbol{A} := \frac{1}{25}\begin{bmatrix} 11 & 48 \\ 48 & 39 \end{bmatrix}$$

in Example 7.8. Recall that $\sigma_1 = 3$, $\sigma_2 = 1$, $\boldsymbol{u}_1 = [3, 4]^T/5$ and $\boldsymbol{u}_2 = [-4, 3]^T/5$. The ellipses $y_1^2/\sigma_1^2 + y_2^2/\sigma_2^2 = 1$ and $\mathcal{E} = \boldsymbol{A}\mathcal{S} = \boldsymbol{U}_1\tilde{\mathcal{E}}$ are shown in Fig. 7.1. Since

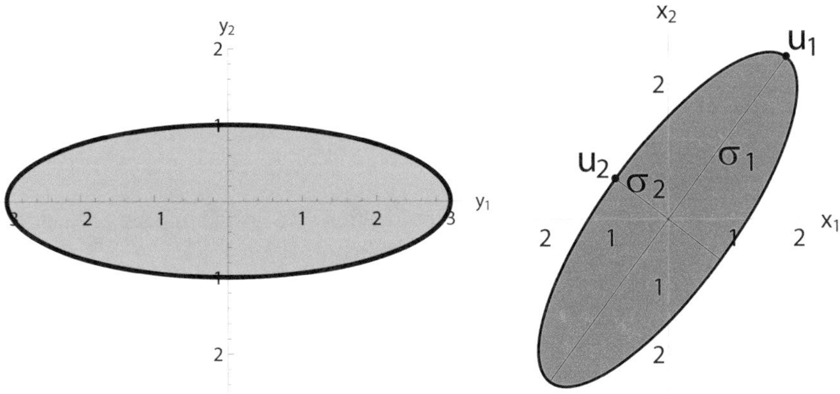

Fig. 7.1 The ellipse $y_1^2/9 + y_2^2 = 1$ (left) and the rotated ellipse $\boldsymbol{A}\mathcal{S}$ (right)

$\boldsymbol{y} = \boldsymbol{U}_1^T \boldsymbol{x} = [3/5x_1 + 4/5x_2, -4/5x_1 + 3/5x_2]^T$, the equation for the ellipse on the right is

$$\frac{(\frac{3}{5}x_1 + \frac{4}{5}x_2)^2}{9} + \frac{(-\frac{4}{5}x_1 + \frac{3}{5}x_2)^2}{1} = 1,$$

7.4 Determining the Rank of a Matrix Numerically

In many elementary linear algebra courses a version of Gaussian elimination, called Gauss-Jordan elimination, is used to determine the rank of a matrix. To carry this out by hand for a large matrix can be a Herculean task and using a computer and floating point arithmetic the result will not be reliable. Entries, which in the final result should have been zero, will have nonzero values because of round-off errors. As an alternative we can use the singular value decomposition to determine rank. Although success is not at all guaranteed, the result will be more reliable than if Gauss-Jordan elimination is used.

By Theorem 7.2 the rank of a matrix is equal to the number of nonzero singular values, and if we have computed the singular values, then all we have to do is to count the nonzero ones. The problem however is the same as for Gaussian elimination. Due to round-off errors none of the computed singular values are likely to be zero.

7.4.1 The Frobenius Norm

This commonly occurring matrix norm will be used here in a discussion of how many of the computed singular values can possibly be considered to be zero. The **Frobenius norm**, of a matrix $\boldsymbol{A} \in \mathbb{C}^{m \times n}$ is defined by

$$\|\boldsymbol{A}\|_F := \Big(\sum_{i=1}^{m}\sum_{j=1}^{n}|a_{ij}|^2\Big)^{1/2}. \tag{7.9}$$

There is a relation between the Frobenius norm of a matrix and its singular values. First we derive some elementary properties of this norm. A systematic study of matrix norms is given in the next chapter.

Lemma 7.1 (Frobenius Norm Properties) *For any $m, n \in \mathbb{N}$ and any matrix $\boldsymbol{A} \in \mathbb{C}^{m \times n}$*

1. $\|\boldsymbol{A}^*\|_F = \|\boldsymbol{A}\|_F$,
2. $\|\boldsymbol{A}\|_F^2 = \sum_{j=1}^n \|\boldsymbol{a}_{:j}\|_2^2$,

3. $\|UA\|_F = \|AV\|_F = \|A\|_F$ *for any unitary matrices* $U \in \mathbb{C}^{m\times m}$ *and* $V \in \mathbb{C}^{n\times n}$,
4. $\|AB\|_F \le \|A\|_F\|B\|_F$ *for any* $B \in \mathbb{C}^{n,k}, \quad k \in \mathbb{N}$,
5. $\|Ax\|_2 \le \|A\|_F\|x\|_2$, *for all* $x \in \mathbb{C}^n$.

Proof

1. $\|A^*\|_F^2 = \sum_{j=1}^n \sum_{i=1}^m |\overline{a}_{ij}|^2 = \sum_{i=1}^m \sum_{j=1}^n |a_{ij}|^2 = \|A\|_F^2$.
2. This follows since the Frobenius norm is the Euclidian norm of a vector, $\|A\|_F := \|\text{vec}(A)\|_2$, where $\text{vec}(A) \in \mathbb{C}^{mn}$ is the vector obtained by stacking the columns $a_{:j}$ of A on top of each other.
3. Recall that if $U^*U = I$ then $\|Ux\|_2 = \|x\|_2$ for all $x \in \mathbb{C}^n$. Applying this to each column $a_{:j}$ of A we find $\|UA\|_F^2 \overset{2.}{=} \sum_{j=1}^n \|Ua_{:j}\|_2^2 = \sum_{j=1}^n \|a_{:j}\|_2^2 \overset{2.}{=} \|A\|_F^2$. Similarly, since $VV^* = I$ we find $\|AV\|_F \overset{1.}{=} \|V^*A^*\|_F = \|A^*\|_F \overset{1.}{=} \|A\|_F$.
4. Using the Cauchy-Schwarz inequality and 2. we obtain

$$\|AB\|_F^2 = \sum_{i=1}^m \sum_{j=1}^k |a_{i:}^* b_{:j}|^2 \le \sum_{i=1}^m \sum_{j=1}^k \|a_{i:}\|_2^2 \|b_{:j}\|_2^2 = \|A\|_F^2 \|B\|_F^2.$$

5. Since $\|v\|_F = \|v\|_2$ for a vector this follows by taking $k = 1$ and $B = x$ in 4.

□

Theorem 7.5 (Frobenius Norm and Singular Values) *We have* $\|A\|_F = \sqrt{\sigma_1^2 + \cdots + \sigma_n^2}$, *where* $\sigma_1, \ldots, \sigma_n$ *are the singular values of* A.

Proof Using Lemma 7.1 we find

$$\|A\|_F \overset{3.}{=} \|U^*AV\|_F = \|\Sigma\|_F = \sqrt{\sigma_1^2 + \cdots + \sigma_n^2}.$$

□

7.4.2 Low Rank Approximation

Suppose $m \ge n \ge 1$ and $A \in \mathbb{C}^{m\times n}$ has a singular value decomposition $A = U \begin{bmatrix} D \\ 0 \end{bmatrix} V^*$, where $D = \text{diag}(\sigma_1, \ldots, \sigma_n)$. We choose $\epsilon > 0$ and let $1 \le r \le n$ be the smallest integer such that $\sigma_{r+1}^2 + \cdots + \sigma_n^2 < \epsilon^2$. Define $A' := U \begin{bmatrix} D' \\ 0 \end{bmatrix} V^*$, where $D' := \text{diag}(\sigma_1, \ldots, \sigma_r, 0, \ldots, 0) \in \mathbb{R}^{n\times n}$. By Lemma 7.1

$$\|A - A'\|_F = \|U \begin{bmatrix} D-D' \\ 0 \end{bmatrix} V^*\|_F = \|\begin{bmatrix} D-D' \\ 0 \end{bmatrix}\|_F = \sqrt{\sigma_{r+1}^2 + \cdots + \sigma_n^2} < \epsilon.$$

Thus, if ϵ is small then $\boldsymbol{A}$ is near a matrix $\boldsymbol{A}'$ of rank r. This can be used to determine rank numerically. We choose an r such that $\sqrt{\sigma_{r+1}^2 + \cdots + \sigma_n^2}$ is "small". Then we postulate that $\operatorname{rank}(\boldsymbol{A}) = r$ since $\boldsymbol{A}$ is close to a matrix of rank r.

The following theorem shows that of all $m \times n$ matrices of rank r, $\boldsymbol{A}'$ is closest to $\boldsymbol{A}$ measured in the Frobenius norm.

Theorem 7.6 (Best Low Rank Approximation) *Suppose* $\boldsymbol{A} \in \mathbb{R}^{m\times n}$ *has singular values* $\sigma_1 \geq \cdots \geq \sigma_n \geq 0$. *For any* $r \leq \operatorname{rank}(\boldsymbol{A})$ *we have*

$$\|\boldsymbol{A} - \boldsymbol{A}'\|_F = \min_{\substack{\boldsymbol{B}\in\mathbb{R}^{m\times n} \\ rank(\boldsymbol{B})=r}} \|\boldsymbol{A} - \boldsymbol{B}\|_F = \sqrt{\sigma_{r+1}^2 + \cdots + \sigma_n^2}.$$

For the proof of this theorem we refer to p. 322 of [16].

7.5 Exercises Chap. 7

7.5.1 Exercises Sect. 7.1

Exercise 7.1 (SVD1) Show that the decomposition

$$\boldsymbol{A} := \begin{bmatrix} 1 & 1 \\ 1 & 1 \end{bmatrix} = \frac{1}{\sqrt{2}} \begin{bmatrix} 1 & 1 \\ 1 & -1 \end{bmatrix} \begin{bmatrix} 2 & 0 \\ 0 & 0 \end{bmatrix} \frac{1}{\sqrt{2}} \begin{bmatrix} 1 & 1 \\ 1 & -1 \end{bmatrix} = \boldsymbol{U}\boldsymbol{D}\boldsymbol{U}^T \tag{7.10}$$

is both a spectral decomposition and a singular value decomposition.

Exercise 7.2 (SVD2) Show that the decomposition

$$\boldsymbol{A} := \begin{bmatrix} 1 & -1 \\ 1 & -1 \end{bmatrix} = \frac{1}{\sqrt{2}} \begin{bmatrix} 1 & 1 \\ 1 & -1 \end{bmatrix} \begin{bmatrix} 2 & 0 \\ 0 & 0 \end{bmatrix} \frac{1}{\sqrt{2}} \begin{bmatrix} 1 & -1 \\ 1 & 1 \end{bmatrix} =: \boldsymbol{U}\boldsymbol{\Sigma}\boldsymbol{V}^T \tag{7.11}$$

is a singular value decomposition. Show that $\boldsymbol{A}$ is defective so it cannot be diagonalized by any similarity transformation.

Exercise 7.3 (SVD Examples) Find the singular value decomposition of the following matrices

a) $\boldsymbol{A} = \begin{bmatrix} 3 \\ 4 \end{bmatrix}$.

b) $\boldsymbol{A} = \begin{bmatrix} 1 & 1 \\ 2 & 2 \\ 2 & 2 \end{bmatrix}$.

Exercise 7.4 (More SVD Examples) Find the singular value decomposition of the following matrices

a) $\boldsymbol{A} = \boldsymbol{e}_1$ the first unit vector in $\mathbb{R}^m$.
b) $\boldsymbol{A} = \boldsymbol{e}_n^T$ the last unit vector in $\mathbb{R}^n$.
c) $\boldsymbol{A} = \begin{bmatrix} -1 & 0 \\ 0 & 3 \end{bmatrix}$.

Exercise 7.5 (Singular Values of a Normal Matrix) Show that

a) the singular values of a normal matrix are the absolute values of its eigenvalues,
b) the singular values of a symmetric positive semidefinite matrix are its eigenvalues.

Exercise 7.6 (The Matrices $\boldsymbol{A}^*\boldsymbol{A}$, $\boldsymbol{A}\boldsymbol{A}^*$ and SVD) Show the following: If $\boldsymbol{A} = \boldsymbol{U}\boldsymbol{\Sigma}\boldsymbol{V}$ is a singular value decomposition of $\boldsymbol{A} \in \mathbb{C}^{m\times n}$ then

a) $\boldsymbol{A}^*\boldsymbol{A} = \boldsymbol{V}\operatorname{diag}(\sigma_1^2, \ldots, \sigma_n^2)\boldsymbol{V}^*$ is a spectral decomposition of $\boldsymbol{A}^*\boldsymbol{A}$.
b) $\boldsymbol{A}\boldsymbol{A}^* = \boldsymbol{U}\operatorname{diag}(\sigma_1^2, \ldots, \sigma_m^2)\boldsymbol{U}^*$ is a spectral decomposition of $\boldsymbol{A}\boldsymbol{A}^*$.
c) The columns of $\boldsymbol{U}$ are orthonormal eigenvectors of $\boldsymbol{A}\boldsymbol{A}^*$.
d) The columns of $\boldsymbol{V}$ are orthonormal eigenvectors of $\boldsymbol{A}^*\boldsymbol{A}$.

Exercise 7.7 (Singular Values (Exam Exercise 2005-2)) Given the statement: "If $\boldsymbol{A} \in \mathbb{R}^{n\times n}$ has singular values $(\sigma_1, \ldots, \sigma_n)$ then $\boldsymbol{A}^2$ has singular values $(\sigma_1^2, \ldots, \sigma_n^2)$". Find a class of matrices for which the statement is true. Show that the statement is not true in general.

7.5.2 *Exercises Sect. 7.2*

Exercise 7.8 (Nonsingular Matrix) Derive the SVF and SVD of the matrix[1] $\boldsymbol{A} = \frac{1}{25}\begin{bmatrix} 11 & 48 \\ 48 & 39 \end{bmatrix}$. Also, using possibly a computer, find its spectral decomposition $\boldsymbol{U}\boldsymbol{D}\boldsymbol{U}^T$. The matrix $\boldsymbol{A}$ is normal, but the spectral decomposition is not an SVD. Why?

Exercise 7.9 (Full Row Rank) Find[2] the SVF and SVD of

$$\boldsymbol{A} := \frac{1}{15}\begin{bmatrix} 14 & 4 & 16 \\ 2 & 22 & 13 \end{bmatrix} \in \mathbb{R}^{2\times 3}.$$

[1]Answer: $\boldsymbol{A} = \frac{1}{5}\begin{bmatrix} 3 & -4 \\ 4 & 3 \end{bmatrix}\begin{bmatrix} 3 & 0 \\ 0 & 1 \end{bmatrix}\frac{1}{5}\begin{bmatrix} 3 & 4 \\ 4 & -3 \end{bmatrix}$.

[2]Hint: Take the transpose of the matrix in (7.2).

Exercise 7.10 (Counting Dimensions of Fundamental Subspaces) Suppose $\boldsymbol{A} \in \mathbb{C}^{m\times n}$. Show using SVD that

a) $\operatorname{rank}(\boldsymbol{A}) = \operatorname{rank}(\boldsymbol{A}^*)$.
b) $\operatorname{rank}(\boldsymbol{A}) + \operatorname{null}(\boldsymbol{A}) = n$,
c) $\operatorname{rank}(\boldsymbol{A}) + \operatorname{null}(\boldsymbol{A}^*) = m$,

where $\operatorname{null}(\boldsymbol{A})$ is defined as the dimension of $\mathcal{N}(\boldsymbol{A})$.

Exercise 7.11 (Rank and Nullity Relations) Use Theorem 7.1 to show that for any $\boldsymbol{A} \in \mathbb{C}^{m\times n}$

a) $\operatorname{rank} \boldsymbol{A} = \operatorname{rank}(\boldsymbol{A}^*\boldsymbol{A}) = \operatorname{rank}(\boldsymbol{A}\boldsymbol{A}^*)$,
b) $\operatorname{null}(\boldsymbol{A}^*\boldsymbol{A}) = \operatorname{null} \boldsymbol{A}$, and $\operatorname{null}(\boldsymbol{A}\boldsymbol{A}^*) = \operatorname{null}(\boldsymbol{A}^*)$.

Exercise 7.12 (Orthonormal Bases Example) Let $\boldsymbol{A}$ and $\boldsymbol{B}$ be as in Example 7.2. Give orthonormal bases for $\mathcal{R}(\boldsymbol{B})$ and $\mathcal{N}(\boldsymbol{B})$.

Exercise 7.13 (Some Spanning Sets) Show for any $\boldsymbol{A} \in \mathbb{C}^{m\times n}$ that $\mathcal{R}(\boldsymbol{A}^*\boldsymbol{A}) = \mathcal{R}(\boldsymbol{V}_1) = \mathcal{R}(\boldsymbol{A}^*)$

Exercise 7.14 (Singular Values and Eigenpair of Composite Matrix) Let $\boldsymbol{A} \in \mathbb{C}^{m\times n}$ with $m \geq n$ have singular values $\sigma_1, \ldots, \sigma_n$, left singular vectors $\boldsymbol{u}_1, \ldots, \boldsymbol{u}_m \in \mathbb{C}^m$, and right singular vectors $\boldsymbol{v}_1, \ldots, \boldsymbol{v}_n \in \mathbb{C}^n$. Show that the matrix

$$\boldsymbol{C} := \begin{bmatrix} \boldsymbol{0} & \boldsymbol{A} \\ \boldsymbol{A}^* & \boldsymbol{0} \end{bmatrix} \in \mathbb{R}^{(m+n)\times(m+n)}$$

has the $n + m$ eigenpairs

$$\{(\sigma_1, \boldsymbol{p}_1), \ldots, (\sigma_n, \boldsymbol{p}_n), (-\sigma_1, \boldsymbol{q}_1), \ldots, (-\sigma_n, \boldsymbol{q}_n), (0, \boldsymbol{r}_{n+1}), \ldots, (0, \boldsymbol{r}_m)\},$$

where

$$\boldsymbol{p}_i = \begin{bmatrix} \boldsymbol{u}_i \\ \boldsymbol{v}_i \end{bmatrix}, \quad \boldsymbol{q}_i = \begin{bmatrix} \boldsymbol{u}_i \\ -\boldsymbol{v}_i \end{bmatrix}, \quad \boldsymbol{r}_j = \begin{bmatrix} \boldsymbol{u}_j \\ \boldsymbol{0} \end{bmatrix}, \text{ for } i = 1, \ldots, n,\ j = n+1, \ldots, m.$$

Exercise 7.15 (Polar Decomposition (Exam Exercise 2011-2)) Given $n \in \mathbb{N}$ and a singular value decomposition $\boldsymbol{A} = \boldsymbol{U}\boldsymbol{\Sigma}\boldsymbol{V}^T$ of a square matrix $\boldsymbol{A} \in \mathbb{R}^{n,n}$, consider the matrices

$$\boldsymbol{Q} := \boldsymbol{U}\boldsymbol{V}^T, \quad \boldsymbol{P} := \boldsymbol{V}\boldsymbol{\Sigma}\boldsymbol{V}^T \tag{7.12}$$

of order n.

a) Show that

$$\boldsymbol{A} = \boldsymbol{Q}\boldsymbol{P} \tag{7.13}$$

and show that $\boldsymbol{Q}$ is orthonormal.

b) Show that $\boldsymbol{P}$ is symmetric positive semidefinite and positive definite if $\boldsymbol{A}$ is nonsingular. The factorization in (7.13) is called a **polar factorization**

c) Use the singular value decomposition of $\boldsymbol{A}$ to give a suitable definition of $\boldsymbol{B} := \sqrt{\boldsymbol{A}^T\boldsymbol{A}}$ so that $\boldsymbol{P} = \boldsymbol{B}$.

For the rest of this problem assume that $\boldsymbol{A}$ is nonsingular. Consider the iterative method

$$\boldsymbol{X}_{k+1} = \frac{1}{2}(\boldsymbol{X}_k + \boldsymbol{X}_k^{-T}), \; k = 0, 1, 2, \dots \text{ with } \boldsymbol{X}_0 = \boldsymbol{A}, \tag{7.14}$$

for finding $\boldsymbol{Q}$.

d) Show that the iteration (7.14) is well defined by showing that $\boldsymbol{X}_k = \boldsymbol{U}\boldsymbol{\Sigma}_k\boldsymbol{V}^T$, for a diagonal matrix Σ_k with positive diagonal elements, $k = 0, 1, 2, \dots$.

e) Show that

$$\boldsymbol{X}_{k+1} - \boldsymbol{Q} = \frac{1}{2}\boldsymbol{X}_k^{-T}(\boldsymbol{X}_k^T - \boldsymbol{Q}^T)(\boldsymbol{X}_k - \boldsymbol{Q}) \tag{7.15}$$

and use (7.15) and the Frobenius norm to show (quadratic convergence to $\boldsymbol{Q}$)

$$\|\boldsymbol{X}_{k+1} - \boldsymbol{Q}\|_F \le \frac{1}{2}\|\boldsymbol{X}_k^{-1}\|_F\|\boldsymbol{X}_k - \boldsymbol{Q}\|_F^2. \tag{7.16}$$

f) Write a MATLAB program `function [Q,P,k] = polardecomp(A,tol,K)` to carry out the iteration in (7.14). The output is approximations $\boldsymbol{Q}$ and $\boldsymbol{P} = \boldsymbol{Q}^T\boldsymbol{A}$ to the polar decomposition $\boldsymbol{A} = \boldsymbol{Q}\boldsymbol{P}$ of $\boldsymbol{A}$ and the number of iterations k such that $\|\boldsymbol{X}_{k+1} - \boldsymbol{X}_k\|_F < tol * \|\boldsymbol{X}_{k+1}\|_F$. Set $k = K + 1$ if convergence is not achieved in K iterations. The Frobenius norm in MATLAB is written `norm(A,'fro')`.

Exercise 7.16 (Underdetermined System (Exam Exercise 2015-1))

a) Let $\boldsymbol{A}$ be the matrix

$$\boldsymbol{A} = \begin{bmatrix} 1 & 2 \\ 0 & 1 \\ -1 & 3 \end{bmatrix}.$$

Compute $\|\boldsymbol{A}\|_1$ and $\|\boldsymbol{A}\|_\infty$.

b) Let $\boldsymbol{B}$ be the matrix

$$\boldsymbol{B} = \begin{bmatrix} 1 & 0 & -1 \\ 1 & 1 & 1 \end{bmatrix}.$$

Find the spaces $\text{span}(\boldsymbol{B}^T)$ and $\ker(\boldsymbol{B})$.

c) Consider the underdetermined linear system

$$
\begin{aligned}
x_1 \qquad - x_3 &= 4, \\
x_1 + x_2 + x_3 &= 12.
\end{aligned}
$$

Find the solution $\boldsymbol{x} \in \mathbb{R}^3$ with $\|\boldsymbol{x}\|_2$ as small as possible.

d) Let $\boldsymbol{A} \in \mathbb{R}^{m\times n}$ be a matrix with linearly independent columns, and $\boldsymbol{b} \in \mathbb{R}^m$ a vector. Assume that we use the Gauss-Seidel method (cf. Chap. 12) to solve the normal equations $\boldsymbol{A}^T\boldsymbol{A}\boldsymbol{x} = \boldsymbol{A}^T\boldsymbol{b}$. Will the method converge? Justify your answer.

7.5.3 Exercises Sect. 7.4

Exercise 7.17 (Rank Example) Consider the singular value decomposition

$$
\boldsymbol{A} := \begin{bmatrix} 0 & 3 & 3 \\ 4 & 1 & -1 \\ 4 & 1 & -1 \\ 0 & 3 & 3 \end{bmatrix} = \begin{bmatrix} \frac{1}{2} & -\frac{1}{2} & -\frac{1}{2} & \frac{1}{2} \\ \frac{1}{2} & \frac{1}{2} & \frac{1}{2} & \frac{1}{2} \\ \frac{1}{2} & \frac{1}{2} & -\frac{1}{2} & -\frac{1}{2} \\ \frac{1}{2} & -\frac{1}{2} & \frac{1}{2} & -\frac{1}{2} \end{bmatrix} \begin{bmatrix} 6 & 0 & 0 \\ 0 & 6 & 0 \\ 0 & 0 & 0 \\ 0 & 0 & 0 \end{bmatrix} \begin{bmatrix} \frac{2}{3} & \frac{2}{3} & \frac{1}{3} \\ \frac{2}{3} & -\frac{1}{3} & -\frac{2}{3} \\ \frac{1}{3} & -\frac{2}{3} & \frac{2}{3} \end{bmatrix}
$$

a) Give orthonormal bases for $\mathcal{R}(\boldsymbol{A})$, $\mathcal{R}(\boldsymbol{A}^T)$, $\mathcal{N}(\boldsymbol{A})$ and $\mathcal{N}(\boldsymbol{A}^T)$.

b) Explain why for all matrices $\boldsymbol{B} \in \mathbb{R}^{4,3}$ of rank one we have $\|\boldsymbol{A} - \boldsymbol{B}\|_F \ge 6$.

c) Give a matrix $\boldsymbol{A}_1$ of rank one such that $\|\boldsymbol{A} - \boldsymbol{A}_1\|_F = 6$.

Exercise 7.18 (Another Rank Example) Let $\boldsymbol{A}$ be the $n \times n$ matrix that for $n = 4$ takes the form

$$
\boldsymbol{A} = \begin{bmatrix} 1 & -1 & -1 & -1 \\ 0 & 1 & -1 & -1 \\ 0 & 0 & 1 & -1 \\ 0 & 0 & 0 & 1 \end{bmatrix}.
$$

Thus $\boldsymbol{A}$ is upper triangular with diagonal elements one and all elements above the diagonal equal to -1. Let $\boldsymbol{B}$ be the matrix obtained from $\boldsymbol{A}$ by changing the $(n, 1)$ element from zero to -2^{2-n}.

a) Show that $\boldsymbol{B}\boldsymbol{x} = \boldsymbol{0}$, where $\boldsymbol{x} := [2^{n-2}, 2^{n-3}, \ldots, 2^0, 1]^T$. Conclude that $\boldsymbol{B}$ is singular, $\det(\boldsymbol{A}) = 1$, and $\|\boldsymbol{A} - \boldsymbol{B}\|_F = 2^{2-n}$. Thus even if $\det(\boldsymbol{A})$ is not small the Frobenius norm of $\boldsymbol{A} - \boldsymbol{B}$ is small for large n, and the matrix $\boldsymbol{A}$ is very close to being singular for large n.

b) Use Theorem 7.6 to show that the smallest singular vale σ_n of $\boldsymbol{A}$ is bounded above by 2^{2-n}.

Exercise 7.19 (Norms, Cholesky and SVD (Exam Exercise 2016-1))

a) Let $\boldsymbol{A}$ be the matrix

$$\boldsymbol{A} = \begin{bmatrix} 3 & 1 \\ 2 & 3 \\ -1 & 5 \end{bmatrix}.$$

Compute $\|\boldsymbol{A}\|_1$, $\|\boldsymbol{A}\|_\infty$ and $\|\boldsymbol{A}\|_F$.

b) Let $\boldsymbol{T}$ be the matrix

$$\boldsymbol{T} = \begin{bmatrix} 2 & -1 \\ -1 & 2 \end{bmatrix}.$$

Show that $\boldsymbol{T}$ is symmetric positive definite, and find the Cholesky factorization $\boldsymbol{T} = \boldsymbol{L}\boldsymbol{L}^T$ of $\boldsymbol{T}$.

c) Let $\boldsymbol{A} = \boldsymbol{U}\boldsymbol{\Sigma}\boldsymbol{V}^*$ be a singular value decomposition of the $m \times n$-matrix $\boldsymbol{A}$ with $m \geq n$, and let $\boldsymbol{A}' = \sum_{i=1}^{r} \sigma_i \boldsymbol{u}_i \boldsymbol{v}_i^*$, where $1 \leq r \leq n$, σ_i are the singular values of $\boldsymbol{A}$, and where $\boldsymbol{u}_i$, $\boldsymbol{v}_i$ are the columns of $\boldsymbol{U}$ and $\boldsymbol{V}$. Prove that

$$\|\boldsymbol{A} - \boldsymbol{A}'\|_F^2 = \sigma_{r+1}^2 + \cdots + \sigma_n^2.$$

7.6 Review Questions

7.6.1 Consider an SVD and an SVF of a matrix $\boldsymbol{A}$.

- What are the singular values of $\boldsymbol{A}$?
- how is the SVD defined?
- how can we find an SVF if we know an SVD?
- how can we find an SVD if we know an SVF?
- what are the relations between the singular vectors?
- which singular vectors form bases for $\mathcal{R}(\boldsymbol{A})$ and $\mathcal{N}(\boldsymbol{A}^*)$?

7.6.2 How are the Frobenius norm and singular values related?

Part III
Matrix Norms and Least Squares

We introduce vector and matrix norms and use them to study how sensitive the solution of a linear system is to perturbation in the data. This leads to the important concept of condition number.

In the second chapter in this part we consider solving linear systems in the least squares sense. We give examples, the basic theory, discuss numerical methods and perturbation theory. Singular values and the important concept of generalized inverses play a central role in our presentation.

Chapter 8
Matrix Norms and Perturbation Theory for Linear Systems

Norms are used to measure the size of vector and matrices.

8.1 Vector Norms

Definition 8.1 (Vector Norm) A **(vector) norm** in a real (resp. complex) vector space $\mathcal{V}$ is a function $\|\cdot\| : \mathcal{V} \to \mathbb{R}$ that satisfies for all $\boldsymbol{x}, \boldsymbol{y}$ in $\mathcal{V}$ and all a in $\mathbb{R}$ (resp. $\mathbb{C}$)

1. $\|\boldsymbol{x}\| \geq 0$ with equality if and only if $\boldsymbol{x} = \boldsymbol{0}$. (positivity)
2. $\|a\boldsymbol{x}\| = |a|\,\|\boldsymbol{x}\|$. (homogeneity)
3. $\|\boldsymbol{x} + \boldsymbol{y}\| \leq \|\boldsymbol{x}\| + \|\boldsymbol{y}\|$. (subadditivity)

The triple $(\mathcal{V}, \mathbb{R}, \|\cdot\|)$ (resp. $(\mathcal{V}, \mathbb{C}, \|\cdot\|)$) is called a **normed vector space** and the inequality 3. is called the **triangle inequality**.

In this book the vector space will be one of $\mathbb{R}^n$, $\mathbb{C}^n$ or one of the matrix spaces $\mathbb{R}^{m\times n}$, or $\mathbb{C}^{m\times n}$. Vector addition is defined by element wise addition and scalar multiplication is defined by multiplying every element by the scalar.

We encountered norms associated with any inner product in $\mathbb{R}^n$ or $\mathbb{C}^n$ in Chap. 5. That these inner product norms are really norms was shown in Theorem 5.2. In this book we will use the following family of vector norms on $\mathcal{V} = \mathbb{C}^n$ and $\mathcal{V} = \mathbb{R}^n$.

T. Lyche, *Numerical Linear Algebra and Matrix Factorizations*,
Texts in Computational Science and Engineering 22,
https://doi.org/10.1007/978-3-030-36468-7_8

Definition 8.2 (Vector p-Norms) We define for $p \geq 1$ and $\boldsymbol{x} \in \mathbb{R}^n$ or $\boldsymbol{x} \in \mathbb{C}^n$ the p-**norms** by

$$\|\boldsymbol{x}\|_p := \Big(\sum_{j=1}^n |x_j|^p\Big)^{1/p}, \tag{8.1}$$

$$\|\boldsymbol{x}\|_\infty := \max_{1\leq j\leq n} |x_j|. \tag{8.2}$$

The most important cases are $p = 1, 2, \infty$:

1. $\|\boldsymbol{x}\|_1 := \sum_{j=1}^n |x_j|$, **(the one-norm** or l_1**-norm)**
2. $\|\boldsymbol{x}\|_2 := \sqrt{\sum_{j=1}^n |x_j|^2}$, **(the two-norm**, l_2**-norm**, or **Euclidian norm)**
3. $\|\boldsymbol{x}\|_\infty := \max_{1\leq j\leq n} |x_j|$, **(the infinity-norm**, l_∞**-norm**, or **max norm)**

Some remarks are in order.

1. In Sect. 8.4, we show that the p-norms are vector norms for $1 \leq p \leq \infty$.
2. The triangle inequality $\|\boldsymbol{x} + \boldsymbol{y}\|_p \leq \|\boldsymbol{x}\|_p + \|\boldsymbol{y}\|_p$ is called **Minkowski's inequality**.
3. To prove it one first establishes **Hölder's inequality**

$$\sum_{j=1}^n |x_j y_j| \leq \|\boldsymbol{x}\|_p \|\boldsymbol{y}\|_q, \quad \frac{1}{p} + \frac{1}{q} = 1, \quad \boldsymbol{x}, \boldsymbol{y} \in \mathbb{C}^n. \tag{8.3}$$

The relation $\frac{1}{p} + \frac{1}{q} = 1$ means that if $p = 1$ then $q = \infty$ and vice versa. The Hölder's inequality is the same as the Cauchy-Schwarz inequality (cf. Theorem 5.1) for the Euclidian norm $p = 2$.

4. The infinity norm is related to the other p-norms by

$$\lim_{p\to\infty} \|\boldsymbol{x}\|_p = \|\boldsymbol{x}\|_\infty \text{ for all } \boldsymbol{x} \in \mathbb{C}^n. \tag{8.4}$$

5. The equation (8.4) clearly holds for $\boldsymbol{x} = \boldsymbol{0}$. For $\boldsymbol{x} \neq \boldsymbol{0}$ we write

$$\|\boldsymbol{x}\|_p := \|\boldsymbol{x}\|_\infty \Big(\sum_{j=1}^n \big(\frac{|x_j|}{\|\boldsymbol{x}\|_\infty}\big)^p\Big)^{1/p}.$$

Now each term in the sum is not greater than one and at least one term is equal to one, and we obtain

$$\|\boldsymbol{x}\|_\infty \leq \|\boldsymbol{x}\|_p \leq n^{1/p} \|\boldsymbol{x}\|_\infty, \quad p \geq 1. \tag{8.5}$$

Since $\lim_{p\to\infty} n^{1/p} = 1$ for any fixed $n \in \mathbb{N}$, we see that (8.4) follows.

6. In Exercise 8.28 we show the following generalization of inequality (8.5)

$$\|\boldsymbol{x}\|_{p'} \le \|\boldsymbol{x}\|_p \le n^{1/p-1/p'}\|\boldsymbol{x}\|_{p'}, \quad \boldsymbol{x} \in \mathbb{C}^n, \quad 1 \le p \le p' \le \infty. \tag{8.6}$$

We return now to the general vector norm case.

Definition 8.3 (Equivalent Norms) We say that two norms $\|\cdot\|$ and $\|\cdot\|'$ on $\mathcal{V}$ are **equivalent** if there are positive constants m and M such that for all vectors $\boldsymbol{x} \in \mathcal{V}$ we have

$$m\|\boldsymbol{x}\|' \le \|\boldsymbol{x}\| \le M\|\boldsymbol{x}\|'. \tag{8.7}$$

By (8.5) the p- and ∞-norms are equivalent for any $p \ge 1$. This result is generalized in the following theorem.

Theorem 8.1 (Basic Properties of Vector Norms) *The following holds for a normed vector space* $(\mathcal{V}, \mathbb{C}, \|\cdot\|)$.

1. $\|\boldsymbol{x} - \boldsymbol{y}\| \ge |\,\|\boldsymbol{x}\| - \|\boldsymbol{y}\|\,|$, *for all* $\boldsymbol{x}, \boldsymbol{y} \in \mathbb{C}^n$ *(inverse triangle inequality).*
2. *The vector norm is a continuous function* $\mathcal{V} \to \mathbb{R}$.
3. *All vector norms on* $\mathcal{V}$ *are equivalent provided* $\mathcal{V}$ *is finite dimensional.*

Proof

1. Since $\|\boldsymbol{x}\| = \|\boldsymbol{x} - \boldsymbol{y} + \boldsymbol{y}\| \le \|\boldsymbol{x} - \boldsymbol{y}\| + \|\boldsymbol{y}\|$ we obtain $\|\boldsymbol{x} - \boldsymbol{y}\| \ge \|\boldsymbol{x}\| - \|\boldsymbol{y}\|$. By symmetry $\|\boldsymbol{x} - \boldsymbol{y}\| = \|\boldsymbol{y} - \boldsymbol{x}\| \ge \|\boldsymbol{y}\| - \|\boldsymbol{x}\|$ and we obtain the inverse triangle inequality.
2. This follows from the inverse triangle inequality.
3. The following proof can be skipped by those who do not have the necessary background in advanced calculus. Define the $\|\cdot\|'$ unit sphere

$$\mathcal{S} := \{\boldsymbol{y} \in \mathcal{V} : \|\boldsymbol{y}\|' = 1\}.$$

The set $\mathcal{S}$ is a closed and bounded set and the function $f : \mathcal{S} \to \mathbb{R}$ given by $f(\boldsymbol{y}) = \|\boldsymbol{y}\|$ is continuous by what we just showed. Therefore f attains its minimum and maximum value on $\mathcal{S}$. Thus, there are positive constants m and M such that

$$m \le \|\boldsymbol{y}\| \le M, \quad \boldsymbol{y} \in \mathcal{S}. \tag{8.8}$$

For any $\boldsymbol{x} \in \mathcal{V}$ we have $\boldsymbol{y} := \boldsymbol{x}/\|\boldsymbol{x}\|' \in \mathcal{S}$, and (8.7) follows if we apply (8.8) to these $\boldsymbol{y}$.

□

8.2 Matrix Norms

For simplicity we consider only norms on the vector space $(\mathbb{C}^{m\times n}, \mathbb{C})$. All results also holds for $(\mathbb{R}^{m\times n}, \mathbb{R})$. A matrix norm $\| \ \| : \mathbb{C}^{m\times n}, \to \mathbb{R}$ is simply a vector norm on $\mathbb{C}^{m\times n}$. Thus 1., 2. and 3. in Definition 8.1 holds, where we replace $\boldsymbol{x}$ and $\boldsymbol{y}$ by $m \times n$ matrices $\boldsymbol{A}$ and $\boldsymbol{B}$, respectively. The Frobenius norm

$$\|\boldsymbol{A}\|_F := \Big(\sum_{i=1}^m \sum_{j=1}^n |a_{ij}|^2\Big)^{1/2}$$

is a matrix norm. Indeed, writing all elements in $\boldsymbol{A}$ in a string of length mn we see that the Frobenius norm is the Euclidian norm on the space $\mathbb{C}^{mn}$.

Adapting Theorem 8.1 to the matrix situation gives

Theorem 8.2 (Matrix Norm Equivalence) *All matrix norms on $\mathbb{C}^{m\times n}$ are equivalent. Thus, if $\|\cdot\|$ and $\|\cdot\|'$ are two matrix norms on $\mathbb{C}^{m\times n}$ then there are positive constants μ and M such that*

$$\mu\|\boldsymbol{A}\| \le \|\boldsymbol{A}\|' \le M\|\boldsymbol{A}\|$$

holds for all $\boldsymbol{A} \in \mathbb{C}^{m\times n}$. Moreover, a matrix norm is a continuous function.

Any vector norm $\|\cdot\|_V$ on $\mathbb{C}^{mn}$ defines a matrix norm on $\mathbb{C}^{m\times n}$ given by $\|\boldsymbol{A}\| := \|\mathrm{vec}(\boldsymbol{A})\|_V$, where $\mathrm{vec}(\boldsymbol{A}) \in \mathbb{C}^{mn}$ is the vector obtained by stacking the columns of $\boldsymbol{A}$ on top of each other. In particular, to the p vector norms for $p = 1, 2, \infty$, we have the corresponding **sum norm**, **Frobenius norm**, and **max norm** defined by

$$\|\boldsymbol{A}\|_S := \sum_{i=1}^m \sum_{j=1}^n |a_{ij}|, \quad \|\boldsymbol{A}\|_F := \Big(\sum_{i=1}^m \sum_{j=1}^n |a_{ij}|^2\Big)^{1/2}, \quad \|\boldsymbol{A}\|_M := \max_{i,j} |a_{ij}|. \tag{8.9}$$

Of these norms the Frobenius norm is the most useful. Some of its properties were derived in Lemma 7.1 and Theorem 7.5.

8.2.1 *Consistent and Subordinate Matrix Norms*

Since matrices can be multiplied it is useful to have an analogue of subadditivity for matrix multiplication. For square matrices the product $\boldsymbol{AB}$ is defined in a fixed space $\mathbb{C}^{n\times n}$, while in the rectangular case matrix multiplication combines matrices in different spaces. The following definition captures this distinction.

Definition 8.4 (Consistent Matrix Norms) A matrix norm is called **consistent on** $\mathbb{C}^{n\times n}$ if

4. $\|\boldsymbol{A}\boldsymbol{B}\| \le \|\boldsymbol{A}\|\,\|\boldsymbol{B}\|$ (submultiplicativity)

holds for all $\boldsymbol{A}, \boldsymbol{B} \in \mathbb{C}^{n\times n}$. A matrix norm is **consistent** if it is defined on $\mathbb{C}^{m\times n}$ for all $m, n \in \mathbb{N}$, and 4. holds for all matrices $\boldsymbol{A}, \boldsymbol{B}$ for which the product $\boldsymbol{A}\boldsymbol{B}$ is defined.

Clearly the Frobenius norm is defined for all $m, n \in \mathbb{N}$. From Lemma 7.1 it follows that the Frobenius norm is consistent.

For a consistent matrix norm on $\mathbb{C}^{n\times n}$ we have the inequality

$$\|\boldsymbol{A}^k\| \le \|\boldsymbol{A}\|^k \text{ for } \boldsymbol{A} \in \mathbb{C}^{n\times n} \text{ and } k \in \mathbb{N}. \tag{8.10}$$

When working with norms one often has to bound the vector norm of a matrix times a vector by the norm of the matrix times the norm of the vector. This leads to the following definition.

Definition 8.5 (Subordinate Matrix Norms) Suppose $m, n \in \mathbb{N}$ are given, let $\|\ \|$ on $\mathbb{C}^m$ and $\|\ \|_\beta$ on $\mathbb{C}^n$ be vector norms, and let $\|\ \|$ be a matrix norm on $\mathbb{C}^{m\times n}$. We say that the matrix norm $\|\ \|$ is **subordinate** to the vector norms $\|\ \|$ and $\|\ \|_\beta$ if $\|\boldsymbol{A}\boldsymbol{x}\| \le \|\boldsymbol{A}\|\,\|\boldsymbol{x}\|_\beta$ for all $\boldsymbol{A} \in \mathbb{C}^{m\times n}$ and all $\boldsymbol{x} \in \mathbb{C}^n$.

By Lemma 7.1 we have $\|\boldsymbol{A}\boldsymbol{x}\|_2 \le \|\boldsymbol{A}\|_F\|\boldsymbol{x}\|_2$, for all $\boldsymbol{x} \in \mathbb{C}^n$. Thus the Frobenius norm is subordinate to the Euclidian vector norm.

For consistent matrix norms we have

Proposition 8.1 *For $m, n \in \mathbb{N}$, $\boldsymbol{A} \in \mathbb{C}^{m\times n}$, all $\boldsymbol{x} \in \mathbb{C}^n$ and any consistent matrix norm $\|\ \|$*

$$\|\boldsymbol{A}\boldsymbol{x}\| \le \|\boldsymbol{A}\|\,\|\boldsymbol{x}\|, \tag{8.11}$$

i.e., a consistent matrix norm is subordinate to itself. Moreover, the matrix power bound (8.10) *holds for all square matrices $\boldsymbol{A} \in \mathbb{C}^{n\times n}$.*

Proof Since a consistent matrix norm is defined on $\mathbb{C}^{m\times n}$ for all $m, n \in \mathbb{N}$ the consistency implies that (8.11) holds for $\boldsymbol{A} \in \mathbb{C}^{m\times n}$ and $\boldsymbol{B} := \boldsymbol{x} \in \mathbb{C}^{n\times 1}$. The last statement also follows immediately from the consistency. □

8.2.2 *Operator Norms*

Corresponding to vector norms on $\mathbb{C}^n$ and $\mathbb{C}^m$ there is an induced matrix norm on $\mathbb{C}^{m\times n}$ which we call the **operator norm**. It is possible to consider one vector norm on $\mathbb{C}^m$ and another vector norm on $\mathbb{C}^n$, but we treat only the case of one vector norm defined on $\mathbb{C}^n$ for all $n \in \mathbb{N}$.[1]

Definition 8.6 (Operator Norm) Let $\| \ \|$ be a vector norm defined on $\mathbb{C}^n$ for all $n \in \mathbb{N}$. For given $m, n \in \mathbb{N}$ and $\boldsymbol{A} \in \mathbb{C}^{m \times n}$ we define

$$\|\boldsymbol{A}\| := \max_{x \neq 0} \frac{\|\boldsymbol{A}\boldsymbol{x}\|}{\|\boldsymbol{x}\|}. \tag{8.12}$$

We call this the **operator norm** corresponding to the vector norm $\| \ \|$.

With a risk of confusion we use the same symbol for the operator norm and the corresponding vector norm. Before we show that the operator norm is a matrix norm we make some observations.

1. It is enough to take the max over subsets of $\mathbb{C}^n$. For example

$$\|\boldsymbol{A}\| = \max_{\|\boldsymbol{x}\|=1} \|\boldsymbol{A}\boldsymbol{x}\|. \tag{8.13}$$

The set

$$\mathcal{S} := \{x \in \mathbb{C}^n : \|\boldsymbol{x}\| = 1\} \tag{8.14}$$

is the unit sphere in $\mathbb{C}^n$ with respect to the vector norm $\| \ \|$. It is enough to take the max over this unit sphere since

$$\max_{x \neq 0} \frac{\|\boldsymbol{A}\boldsymbol{x}\|}{\|\boldsymbol{x}\|} = \max_{x \neq 0} \left\| \boldsymbol{A}\Big(\frac{\boldsymbol{x}}{\|\boldsymbol{x}\|}\Big) \right\| = \max_{\|\boldsymbol{y}\|=1} \|\boldsymbol{A}\boldsymbol{y}\|.$$

2. The operator norm is subordinate to the corresponding vector norm. Thus,

$$\|\boldsymbol{A}\boldsymbol{x}\| \le \|\boldsymbol{A}\|\|\boldsymbol{x}\| \text{ for all } \boldsymbol{A} \in \mathbb{C}^{m \times n} \text{ and } \boldsymbol{x} \in \mathbb{C}^n. \tag{8.15}$$

3. We can use max instead of sup in (8.12). This follows by the following compactness argument. The unit sphere $\mathcal{S}$ given by (8.14) is bounded. It is also finite dimensional and closed, and hence compact. Moreover, since the vector norm $\| \ \| : \mathcal{S} \to \mathbb{R}$ is a continuous function, it follows that the function $f : \mathcal{S} \to \mathbb{R}$ given by $f(\boldsymbol{x}) = \|\boldsymbol{A}\boldsymbol{x}\|$ is continuous. But then f attains its max and min and we have

$$\|\boldsymbol{A}\| = \|\boldsymbol{A}\boldsymbol{x}^*\| \text{ for some } \boldsymbol{x}^* \in \mathcal{S}. \tag{8.16}$$

[1]In the case of one vector norm $\| \ \|$ on $\mathbb{C}^m$ and another vector norm $\| \ \|_\beta$ on $\mathbb{C}^n$ we would define $\|\boldsymbol{A}\| := \max_{\boldsymbol{x} \neq 0} \frac{\|\boldsymbol{A}\boldsymbol{x}\|}{\|\boldsymbol{x}\|_\beta}$.

Lemma 8.1 (The Operator Norm Is a Consistent Matrix Norm) *If $\| \ \|$ is vector norm defined on $\mathbb{C}^n$ for all $n \in \mathbb{N}$, then the operator norm given by (8.12) is a consistent matrix norm. Moreover, $\|\boldsymbol{I}\| = 1$.*

Proof We use (8.13). In 2. and 3. below we take the max over the unit sphere $\mathcal{S}$ given by (8.14).

1. Nonnegativity is obvious. If $\|\boldsymbol{A}\| = 0$ then $\|\boldsymbol{A}\boldsymbol{y}\| = 0$ for each $\boldsymbol{y} \in \mathbb{C}^n$. In particular, each column $\boldsymbol{A}\boldsymbol{e}_j$ in $\boldsymbol{A}$ is zero. Hence $\boldsymbol{A} = 0$.
2. $\|c\boldsymbol{A}\| = \max_{\boldsymbol{x}} \|c\boldsymbol{A}\boldsymbol{x}\| = \max_{\boldsymbol{x}} |c| \, \|\boldsymbol{A}\boldsymbol{x}\| = |c| \, \|\boldsymbol{A}\|$.
3. $\|\boldsymbol{A} + \boldsymbol{B}\| = \max_{\boldsymbol{x}} \|(\boldsymbol{A} + \boldsymbol{B})\boldsymbol{x}\| \le \max_{\boldsymbol{x}} \|\boldsymbol{A}\boldsymbol{x}\| + \max_{\boldsymbol{x}} \|\boldsymbol{B}\boldsymbol{x}\| = \|\boldsymbol{A}\| + \|\boldsymbol{B}\|$.
4. $\|\boldsymbol{A}\boldsymbol{B}\| = \max_{\boldsymbol{x} \neq \boldsymbol{0}} \frac{\|\boldsymbol{A}\boldsymbol{B}\boldsymbol{x}\|}{\|\boldsymbol{x}\|} = \max_{\boldsymbol{B}\boldsymbol{x} \neq \boldsymbol{0}} \frac{\|\boldsymbol{A}\boldsymbol{B}\boldsymbol{x}\|}{\|\boldsymbol{x}\|} = \max_{\boldsymbol{B}\boldsymbol{x} \neq \boldsymbol{0}} \frac{\|\boldsymbol{A}\boldsymbol{B}\boldsymbol{x}\|}{\|\boldsymbol{B}\boldsymbol{x}\|} \frac{\|\boldsymbol{B}\boldsymbol{x}\|}{\|\boldsymbol{x}\|}$
   $\le \max_{\boldsymbol{y} \neq \boldsymbol{0}} \frac{\|\boldsymbol{A}\boldsymbol{y}\|}{\|\boldsymbol{y}\|} \max_{\boldsymbol{x} \neq \boldsymbol{0}} \frac{\|\boldsymbol{B}\boldsymbol{x}\|}{\|\boldsymbol{x}\|} = \|\boldsymbol{A}\| \, \|\boldsymbol{B}\|$.

That $\|\boldsymbol{I}\| = 1$ for any operator norm follows immediately from the definition. □

Since $\|\boldsymbol{I}\|_F = \sqrt{n}$, we see that the Frobenius norm is not an operator norm for $n > 1$.

8.2.3 *The Operator p-Norms*

Recall that the p or ℓ_p vector norms (8.1) are given by

$$\|\boldsymbol{x}\|_p := \Big(\sum_{j=1}^{n} |x_j|^p\Big)^{1/p}, \ p \ge 1, \quad \|\boldsymbol{x}\|_\infty := \max_{1 \le j \le n} |x_j|.$$

The operator norms $\| \ \|_p$ defined from these p-vector norms are used quite frequently for $p = 1, 2, \infty$. We define for any $1 \le p \le \infty$

$$\|\boldsymbol{A}\|_p := \max_{\boldsymbol{x} \neq 0} \frac{\|\boldsymbol{A}\boldsymbol{x}\|_p}{\|\boldsymbol{x}\|_p} = \max_{\|\boldsymbol{y}\|_p = 1} \|\boldsymbol{A}\boldsymbol{y}\|_p. \tag{8.17}$$

For $p = 1, 2, \infty$ we have explicit expressions for these norms.

Theorem 8.3 (One-Two-Inf-Norms) *For $\mathbf{A} \in \mathbb{C}^{m \times n}$ we have*

$$\begin{aligned} \|\boldsymbol{A}\|_1 &:= \max_{1 \le j \le n} \|\boldsymbol{A}\boldsymbol{e}_j\|_1 = \max_{1 \le j \le n} \sum_{k=1}^{m} |a_{k,j}|, && \textit{(max column sum)} \\ \|\boldsymbol{A}\|_2 &:= \sigma_1, && \textit{(largest singular value of } \boldsymbol{A}\textit{)} \\ \|\boldsymbol{A}\|_\infty &= \max_{1 \le k \le m} \|\boldsymbol{e}_k^T \boldsymbol{A}\|_1 = \max_{1 \le k \le m} \sum_{j=1}^{n} |a_{k,j}|. && \textit{(max row sum)} \end{aligned} \tag{8.18}$$

The **two-norm** $\|A\|_2$ is also called the **spectral norm** of A.

Proof We proceed as follows:

(a) We derive a constant K_p such that $\|Ax\|_p \le K_p$ for any $x \in \mathbb{C}^n$ with $\|x\|_p = 1$.
(b) We give an extremal vector $y_e \in \mathbb{C}^n$ with $\|y_e\|_p = 1$ so that $\|Ay_e\|_p = K_p$.

It then follows from (8.17) that $\|A\|_p = \|Ay_e\|_p = K_p$.

2-norm: Let $A = U\Sigma V^*$ be a singular value decomposition of A, define $K_2 = \sigma_1$, $c := V^*x$, and $y_e = v_1$ the singular vector corresponding to σ_1. Then $x = Vc$, $\|c\|_2 = \|x\|_2 = 1$, and using (7.7) in **(b)** we find

(a) $\|Ax\|_2^2 = \|U\Sigma V^*x\|_2^2 = \|\Sigma c\|_2^2 = \sum_{j=1}^n \sigma_j^2 |c_j|^2 \le \sigma_1^2 \sum_{j=1}^n |c_j|^2 = \sigma_1^2$.
(b) $\|Av_1\|_2 = \|\sigma_1 u_1\|_2 = \sigma_1$.

1-norm: Define K_1, c and y_e by $K_1 := \|Ae_c\|_1 = \max_{1\le j\le n}\|Ae_j\|_1$ and $y_e := e_c$, a unit vector. Then $\|y_e\|_1 = 1$ and we obtain

(a)

$$\|Ax\|_1 = \sum_{k=1}^m \Big|\sum_{j=1}^n a_{kj}x_j\Big| \le \sum_{k=1}^m\sum_{j=1}^n |a_{kj}||x_j| = \sum_{j=1}^n \Big(\sum_{k=1}^m |a_{kj}|\Big)|x_j| \le K_1.$$

(b) $\|Ay_e\|_1 = K_1$.

∞-norm: Define K_∞, r and y_e by $K_\infty := \|e_r^T A\|_1 = \max_{1\le k\le m}\|e_k^T A\|_1$ and $y_e := [e^{-i\theta_1}, \dots, e^{-i\theta_n}]^T$, where $a_{rj} = |a_{rj}|e^{i\theta_j}$ for $j = 1, \dots, n$.

(a)

$$\|Ax\|_\infty = \max_{1\le k\le m}\Big|\sum_{j=1}^n a_{kj}x_j\Big| \le \max_{1\le k\le m}\sum_{j=1}^n |a_{kj}||x_j| \le K_\infty.$$

(b) $\|Ay^*\|_\infty = \max_{1\le k\le m}\Big|\sum_{j=1}^n a_{kj}e^{-i\theta_j}\Big| = K_\infty$.

The last equality is correct because $\Big|\sum_{j=1}^n a_{kj}e^{-i\theta_j}\Big| \le \sum_{j=1}^n |a_{kj}| \le K_\infty$ with equality for $k = r$.

□

Example 8.1 (Comparing One-Two-Inf-Norms) The largest singular value of the matrix $A := \frac{1}{15}\begin{bmatrix} 14 & 4 & 16 \\ 2 & 22 & 13 \end{bmatrix}$, is $\sigma_1 = 2$ (cf. Example 7.9). We find

$$\|A\|_1 = \frac{29}{15}, \quad \|A\|_2 = 2, \quad \|A\|_\infty = \frac{37}{15}, \quad \|A\|_F = \sqrt{5}.$$

The values of these norms do not differ by much.

In some cases the spectral norm is equal to an eigenvalue of the matrix.

Theorem 8.4 (Spectral Norm) *Suppose* $A \in \mathbb{C}^{n\times n}$ *has singular values* $\sigma_1 \geq \sigma_2 \geq \cdots \geq \sigma_n$ *and eigenvalues* $|\lambda_1| \geq |\lambda_2| \geq \cdots \geq |\lambda_n|$. *Then*

$$\|A\|_2 = \sigma_1 \text{ and } \|A^{-1}\|_2 = \frac{1}{\sigma_n}, \tag{8.19}$$

$$\|A\|_2 = \lambda_1 \text{ and } \|A^{-1}\|_2 = \frac{1}{\lambda_n}, \quad \text{if } A \text{ is positive definite}, \tag{8.20}$$

$$\|A\|_2 = |\lambda_1| \text{ and } \|A^{-1}\|_2 = \frac{1}{|\lambda_n|}, \quad \text{if } A \text{ is normal}. \tag{8.21}$$

For the norms of A^{-1} *we assume that* A *is nonsingular.*

Proof Since $1/\sigma_n$ is the largest singular value of A^{-1}, (8.19) follows. By Exercise 7.5 the singular values of a positive definite matrix (normal matrix) are equal to the eigenvalues (absolute value of the eigenvalues). This implies (8.20) and (8.21). □

The following result is sometimes useful.

Theorem 8.5 (Spectral Norm Bound) *For any* $A \in \mathbb{C}^{m\times n}$ *we have* $\|A\|_2^2 \leq \|A\|_1\|A\|_\infty$.

Proof Let (σ^2, v) be an eigenpair for A^*A corresponding to the largest singular value σ of A. Then

$$\|A\|_2^2\|v\|_1 = \sigma^2\|v\|_1 = \|\sigma^2 v\|_1 = \|A^*Av\|_1 \leq \|A^*\|_1\|A\|_1\|v\|_1.$$

Observing that $\|A^*\|_1 = \|A\|_\infty$ by Theorem 8.3 and canceling $\|v\|_1$ proves the result. □

8.2.4 Unitary Invariant Matrix Norms

Definition 8.7 (Unitary Invariant Norm) A matrix norm $\|\ \|$ on $\mathbb{C}^{m\times n}$ is called **unitary invariant** if $\|UAV\| = \|A\|$ for any $A \in \mathbb{C}^{m\times n}$ and any unitary matrices $U \in \mathbb{C}^{m\times m}$ and $V \in \mathbb{C}^{n\times n}$.

When a unitary invariant matrix norm is used, the size of a perturbation is not increased by a unitary transformation. Thus if U and V are unitary then $U(A + E)V = UAV + F$, where $\|F\| = \|E\|$.

It follows from Lemma 7.1 that the Frobenius norm is unitary invariant. We show here that this also holds for the spectral norm.

Theorem 8.6 (Unitary Invariant Norms) *The Frobenius norm and the spectral norm are unitary invariant. Moreover,*

$$\|A^*\|_F = \|A\|_F \text{ and } \|A^*\|_2 = \|A\|_2.$$

Proof The results for the Frobenius norm follow from Lemma 7.1. Suppose $A \in \mathbb{C}^{m\times n}$ and let $U \in \mathbb{C}^{m\times m}$ and $V \in \mathbb{C}^{n\times n}$ be unitary. Since the 2-vector norm is unitary invariant we obtain

$$\|UA\|_2 = \max_{\|x\|_2=1} \|UAx\|_2 = \max_{\|x\|_2=1} \|Ax\|_2 = \|A\|_2.$$

Now A and A^* have the same nonzero singular values, and it follows from Theorem 8.3 that $\|A^*\|_2 = \|A\|_2$. Moreover V^* is unitary. Using these facts we find

$$\|AV\|_2 = \|(AV)^*\|_2 = \|V^*A^*\|_2 = \|A^*\|_2 = \|A\|_2.$$

□

It can be shown that the spectral norm is the only unitary invariant operator norm, see [10] p. 357.

8.2.5 *Absolute and Monotone Norms*

A vector norm on $\mathbb{C}^n$ is an **absolute norm** if $\|x\| = \|\,|x|\,\|$ for all $x \in \mathbb{C}^n$. Here $|x| := [|x_1|, \ldots, |x_n|]^T$, the absolute values of the components of x. Clearly the vector p norms are absolute norms. We state without proof (see Theorem 5.5.10 of [10]) that a vector norm on $\mathbb{C}^n$ is an absolute norm if and only if it is a **monotone norm**, i.e.,

$$|x_i| \le |y_i|,\ i = 1, \ldots, n \Longrightarrow \|x\| \le \|y\|, \text{ for all } x, y \in \mathbb{C}^n.$$

Absolute and monotone matrix norms are defined as for vector norms.

The study of matrix norms will be continued in Chap. 12.

8.3 The Condition Number with Respect to Inversion

Consider the system of two linear equations

$$\begin{aligned} x_1 + \phantom{(1-10^{-16})}x_2 &= 20 \\ x_1 + (1-10^{-16})x_2 &= 20 - 10^{-15} \end{aligned}$$

whose exact solution is $x_1 = x_2 = 10$. If we replace the second equation by

$$x_1 + (1 + 10^{-16})x_2 = 20 - 10^{-15},$$

the exact solution changes to $x_1 = 30$, $x_2 = -10$. Here a small change in one of the coefficients, from $1 - 10^{-16}$ to $1 + 10^{-16}$, changed the exact solution by a large amount.

A mathematical problem in which the solution is very sensitive to changes in the data is called **ill-conditioned.** Such problems can be difficult to solve on a computer.

In this section we consider what effect a small change (perturbation) in the data $\boldsymbol{A}$,$\boldsymbol{b}$ has on the inverse of $\boldsymbol{A}$ and on the solution $\boldsymbol{x}$ of a linear system $\boldsymbol{A}\boldsymbol{x} = \boldsymbol{b}$. To measure this we use vector and matrix norms. In this section $\| \; \|$ will denote a vector norm on $\mathbb{C}^n$ and also a matrix norm on $\mathbb{C}^{n\times n}$. We assume that the matrix norm is consistent on $\mathbb{C}^{n\times n}$ and subordinate to the vector norm. Thus, for any $\boldsymbol{A}, \boldsymbol{B} \in \mathbb{C}^{n\times n}$ and any $\boldsymbol{x} \in \mathbb{C}^n$ we have

$$\|\boldsymbol{A}\boldsymbol{B}\| \le \|\boldsymbol{A}\| \, \|\boldsymbol{B}\| \text{ and } \|\boldsymbol{A}\boldsymbol{x}\| \le \|\boldsymbol{A}\| \, \|\boldsymbol{x}\|.$$

Recall that this holds if the matrix norm is the operator norm corresponding to the given vector norm. It also holds for the Frobenius matrix norm and the Euclidian vector norm. This follows from Lemma 7.1. We recall that if $\boldsymbol{I} \in \mathbb{R}^{n\times n}$ then $\|\boldsymbol{I}\| = 1$ for an operator norm, while $\|\boldsymbol{I}\|_F = \sqrt{n}$.

8.3.1 *Perturbation of the Right Hand Side in a Linear Systems*

Suppose $\boldsymbol{x}$, $\boldsymbol{y}$ solve $\boldsymbol{A}\boldsymbol{x} = \boldsymbol{b}$ and $(\boldsymbol{A}+\boldsymbol{E})\boldsymbol{y} = \boldsymbol{b}+\boldsymbol{e}$, respectively. where $\boldsymbol{A}, \boldsymbol{A}+\boldsymbol{E} \in \mathbb{C}^{n\times n}$ are nonsingular and $\boldsymbol{b}, \boldsymbol{e} \in \mathbb{C}^n$. How large can $\boldsymbol{y}$–$\boldsymbol{x}$ be? The difference $\|\boldsymbol{y}-\boldsymbol{x}\|$ measures the **absolute error** in $\boldsymbol{y}$ as an approximation to $\boldsymbol{x}$, while $\|\boldsymbol{y}-\boldsymbol{x}\|/\|\boldsymbol{x}\|$ and $\|\boldsymbol{y}-\boldsymbol{x}\|/\|\boldsymbol{y}\|$ are measures for the **relative error**.

We consider first the simpler case of a perturbation in the right-hand side $\boldsymbol{b}$.

Theorem 8.7 (Perturbation in the Right-Hand Side) *Suppose $\boldsymbol{A} \in \mathbb{C}^{n\times n}$ is nonsingular, $\boldsymbol{b}, \boldsymbol{e} \in \mathbb{C}^n$, $\boldsymbol{b} \neq \boldsymbol{0}$ and $\boldsymbol{A}\boldsymbol{x} = \boldsymbol{b}$, $\boldsymbol{A}\boldsymbol{y} = \boldsymbol{b}+\boldsymbol{e}$. Then*

$$\frac{1}{K(\boldsymbol{A})}\frac{\|\boldsymbol{e}\|}{\|\boldsymbol{b}\|} \le \frac{\|\boldsymbol{y}-\boldsymbol{x}\|}{\|\boldsymbol{x}\|} \le K(\boldsymbol{A})\frac{\|\boldsymbol{e}\|}{\|\boldsymbol{b}\|}, \tag{8.22}$$

where $K(\boldsymbol{A}) = \|\boldsymbol{A}\| \, \|\boldsymbol{A}^{-1}\|$ is the condition number of $\boldsymbol{A}$.

Proof Subtracting $\boldsymbol{A}\boldsymbol{x} = \boldsymbol{b}$ from $\boldsymbol{A}\boldsymbol{y} = \boldsymbol{b}+\boldsymbol{e}$ we have $\boldsymbol{A}(\boldsymbol{y}-\boldsymbol{x}) = \boldsymbol{e}$ or $\boldsymbol{y}-\boldsymbol{x} = \boldsymbol{A}^{-1}\boldsymbol{e}$. Combining $\|\boldsymbol{y}-\boldsymbol{x}\| = \|\boldsymbol{A}^{-1}\boldsymbol{e}\| \le \|\boldsymbol{A}^{-1}\| \, \|\boldsymbol{e}\|$ and $\|\boldsymbol{b}\| = \|\boldsymbol{A}\boldsymbol{x}\| \le \|\boldsymbol{A}\| \, \|\boldsymbol{x}\|$ we obtain the upper bound in (8.22). Combining $\|\boldsymbol{e}\| \le \|\boldsymbol{A}\| \, \|\boldsymbol{y}-\boldsymbol{x}\|$ and $\|\boldsymbol{x}\| \le \|\boldsymbol{A}^{-1}\| \, \|\boldsymbol{b}\|$ we obtain the lower bound. □

Consider (8.22). $\|\boldsymbol{e}\|/\|\boldsymbol{b}\|$ is a measure of the size of the perturbation $\boldsymbol{e}$ relative to the size of $\boldsymbol{b}$. The upper bound says that $\|\boldsymbol{y}-\boldsymbol{x}\|/\|\boldsymbol{x}\|$ in the worst case can be $K(\boldsymbol{A})$ times as large as $\|\boldsymbol{e}\|/\|\boldsymbol{b}\|$.

The bounds in (8.22) depends on $K(\boldsymbol{A})$. This number is called the **condition number with respect to inversion of a matrix**, or just the condition number of $\boldsymbol{A}$, if it is clear from the context that we are talking about inverting a matrix. The condition number depends on the matrix $\boldsymbol{A}$ and on the norm used. If $K(\boldsymbol{A})$ is large, $\boldsymbol{A}$ is called **ill-conditioned** (with respect to inversion). If $K(\boldsymbol{A})$ is small, $\boldsymbol{A}$ is called **well-conditioned** (with respect to inversion). We always have $K(\boldsymbol{A}) \geq 1$. For since $\|\boldsymbol{x}\| = \|\boldsymbol{I}\boldsymbol{x}\| \leq \|\boldsymbol{I}\|\|\boldsymbol{x}\|$ for any $\boldsymbol{x}$ we have $\|\boldsymbol{I}\| \geq 1$ and therefore $\|\boldsymbol{A}\|\,\|\boldsymbol{A}^{-1}\| \geq \|\boldsymbol{A}\boldsymbol{A}^{-1}\| = \|\boldsymbol{I}\| \geq 1$.

Since all matrix norms are equivalent, the dependence of $K(\boldsymbol{A})$ on the norm chosen is less important than the dependence on $\boldsymbol{A}$. Example 8.1 provided an illustration of this. See also Exercise 8.19. Sometimes one chooses the spectral norm when discussing properties of the condition number, and the ℓ_1, ℓ_∞, or Frobenius norm when one wishes to compute it or estimate it.

Suppose we have computed an approximate solution $\boldsymbol{y}$ to $\boldsymbol{A}\boldsymbol{x} = \boldsymbol{b}$. The vector $\boldsymbol{r}(\boldsymbol{y}) := \boldsymbol{A}\boldsymbol{y} - \boldsymbol{b}$ is called the **residual vector**, or just the residual. We can bound $\boldsymbol{x}-\boldsymbol{y}$ in terms of $\boldsymbol{r}$.

Theorem 8.8 (Perturbation and Residual) *Suppose $\boldsymbol{A} \in \mathbb{C}^{n\times n}$, $\boldsymbol{b} \in \mathbb{C}^n$, $\boldsymbol{A}$ is nonsingular and $\boldsymbol{b} \neq \boldsymbol{0}$. Let $\boldsymbol{r}(\boldsymbol{y}) = \boldsymbol{A}\boldsymbol{y} - \boldsymbol{b}$ for $\boldsymbol{y} \in \mathbb{C}^n$. If $\boldsymbol{A}\boldsymbol{x} = \boldsymbol{b}$ then*

$$\frac{1}{K(\boldsymbol{A})}\frac{\|\boldsymbol{r}(\boldsymbol{y})\|}{\|\boldsymbol{b}\|} \leq \frac{\|\boldsymbol{y}-\boldsymbol{x}\|}{\|\boldsymbol{x}\|} \leq K(\boldsymbol{A})\frac{\|\boldsymbol{r}(\boldsymbol{y})\|}{\|\boldsymbol{b}\|}. \tag{8.23}$$

Proof We simply take $\boldsymbol{e} = \boldsymbol{r}(\boldsymbol{y})$ in Theorem 8.7. □

Consider next a perturbation in the coefficient matrix in a linear system. Suppose $\boldsymbol{A}, \boldsymbol{E} \in \mathbb{C}^{n\times n}$ with $\boldsymbol{A}$, $\boldsymbol{A}+\boldsymbol{E}$ nonsingular. We like to compare the solution $\boldsymbol{x}$ and $\boldsymbol{y}$ of the systems $\boldsymbol{A}\boldsymbol{x} = \boldsymbol{b}$ and $(\boldsymbol{A}+\boldsymbol{E})\boldsymbol{y} = \boldsymbol{b}$.

Theorem 8.9 (Perturbation in Matrix) *Suppose $\boldsymbol{A}, \boldsymbol{E} \in \mathbb{C}^{n\times n}$, $\boldsymbol{b} \in \mathbb{C}^n$ with $\boldsymbol{A}$ nonsingular and $\boldsymbol{b} \neq \boldsymbol{0}$. If $r := \|\boldsymbol{A}^{-1}\boldsymbol{E}\| < 1$ then $\boldsymbol{A}+\boldsymbol{E}$ is nonsingular. If $\boldsymbol{A}\boldsymbol{x} = \boldsymbol{b}$ and $(\boldsymbol{A}+\boldsymbol{E})\boldsymbol{y} = \boldsymbol{b}$ then*

$$\frac{\|\boldsymbol{y}-\boldsymbol{x}\|}{\|\boldsymbol{y}\|} \leq r \leq K(\boldsymbol{A})\frac{\|\boldsymbol{E}\|}{\|\boldsymbol{A}\|}, \tag{8.24}$$

$$\frac{\|\boldsymbol{y}-\boldsymbol{x}\|}{\|\boldsymbol{x}\|} \leq \frac{r}{1-r} \leq \frac{K(\boldsymbol{A})}{1-r}\frac{\|\boldsymbol{E}\|}{\|\boldsymbol{A}\|}. \tag{8.25}$$

Proof We show $\boldsymbol{A}+\boldsymbol{E}$ singular implies $r \geq 1$. Suppose $\boldsymbol{A}+\boldsymbol{E}$ is singular. Then $(\boldsymbol{A}+\boldsymbol{E})\boldsymbol{x} = \boldsymbol{0}$ for some nonzero $\boldsymbol{x} \in \mathbb{C}^n$. Multiplying by $\boldsymbol{A}^{-1}$ it follows that $(\boldsymbol{I}+\boldsymbol{A}^{-1}\boldsymbol{E})\boldsymbol{x} = \boldsymbol{0}$ and this implies that $\|\boldsymbol{x}\| = \|\boldsymbol{A}^{-1}\boldsymbol{E}\boldsymbol{x}\| \leq r\|\boldsymbol{x}\|$. But then $r \geq 1$.

Subtracting $(\boldsymbol{A}+\boldsymbol{E})\boldsymbol{y} = \boldsymbol{b}$ from $\boldsymbol{A}\boldsymbol{x} = \boldsymbol{b}$ gives $\boldsymbol{A}(\boldsymbol{x}-\boldsymbol{y}) = \boldsymbol{E}\boldsymbol{y}$ or $\boldsymbol{x}-\boldsymbol{y} = \boldsymbol{A}^{-1}\boldsymbol{E}\boldsymbol{y}$. Taking norms and dividing by $\|\boldsymbol{y}\|$ proves (8.24). Solving

$\boldsymbol{x} - \boldsymbol{y} = \boldsymbol{A}^{-1}\boldsymbol{E}\boldsymbol{y}$ for $\boldsymbol{y}$ we obtain $\boldsymbol{y} = (\boldsymbol{I} + \boldsymbol{A}^{-1}\boldsymbol{E})^{-1}\boldsymbol{x}$. By Theorem 12.14 we have $\|\boldsymbol{y}\| \le \|(\boldsymbol{I} + \boldsymbol{A}^{-1}\boldsymbol{E})^{-1}\|\|\boldsymbol{x}\| \le \frac{\|\boldsymbol{x}\|}{1-r}$ and (8.24) implies $\|\boldsymbol{y} - \boldsymbol{x}\| \le r\|\boldsymbol{y}\| \le \frac{r}{1-r}\|\boldsymbol{x}\| \le \frac{K(\boldsymbol{A})}{1-r}\frac{\|\boldsymbol{E}\|}{\|\boldsymbol{A}\|}\|\boldsymbol{x}\|$. Dividing by $\|\boldsymbol{x}\|$ gives (8.25). □

In Theorem 8.9 we gave bounds for the relative error in $\boldsymbol{x}$ as an approximation to $\boldsymbol{y}$ and the relative error in $\boldsymbol{y}$ as an approximation to $\boldsymbol{x}$. $\|\boldsymbol{E}\|/\|\boldsymbol{A}\|$ is a measure for the size of the perturbation $\boldsymbol{E}$ in $\boldsymbol{A}$ relative to the size of $\boldsymbol{A}$. The condition number again plays a crucial role. $\|\boldsymbol{y}-\boldsymbol{x}\|/\|\boldsymbol{y}\|$ can be as large as $K(\boldsymbol{A})$ times $\|\boldsymbol{E}\|/\|\boldsymbol{A}\|$. It can be shown that the upper bound can be attained for any $\boldsymbol{A}$ and any $\boldsymbol{b}$. In deriving the upper bound we used the inequality $\|\boldsymbol{A}^{-1}\boldsymbol{E}\boldsymbol{y}\| \le \|\boldsymbol{A}^{-1}\|\,\|\boldsymbol{E}\|\,\|\boldsymbol{y}\|$. For a more or less random perturbation $\boldsymbol{E}$ this is not a severe overestimate for $\|\boldsymbol{A}^{-1}\boldsymbol{E}\boldsymbol{y}\|$. In the situation where $\boldsymbol{E}$ is due to round-off errors (8.24) can give a fairly realistic estimate for $\|\boldsymbol{y} - \boldsymbol{x}\|/\|\boldsymbol{y}\|$.

The following explicit expressions for the 2-norm condition number follow from Theorem 8.4.

Theorem 8.10 (Spectral Condition Number) *Suppose $\boldsymbol{A} \in \mathbb{C}^{n\times n}$ is nonsingular with singular values $\sigma_1 \ge \sigma_2 \ge \cdots \ge \sigma_n > 0$ and eigenvalues $|\lambda_1| \ge |\lambda_2| \ge \cdots \ge |\lambda_n| > 0$. Then*

$$K_2(\boldsymbol{A}) = \begin{cases} \lambda_1/\lambda_n, & \textit{if } \boldsymbol{A} \textit{ is positive definite}, \\ |\lambda_1|/|\lambda_n|, & \textit{if } \boldsymbol{A} \textit{ is normal}, \\ \sigma_1/\sigma_n, & \textit{in general}. \end{cases} \tag{8.26}$$

It follows that $\boldsymbol{A}$ is ill-conditioned with respect to inversion if and only if σ_1/σ_n is large, or λ_1/λ_n is large when $\boldsymbol{A}$ is positive definite.

If $\boldsymbol{A}$ is well-conditioned, (8.23) says that $\|\boldsymbol{y} - \boldsymbol{x}\|/\|\boldsymbol{x}\| \approx \|\boldsymbol{r}(\boldsymbol{y})\|/\|\boldsymbol{b}\|$. In other words, the accuracy in $\boldsymbol{y}$ is about the same order of magnitude as the residual as long as $\|\boldsymbol{b}\| \approx 1$. If $\boldsymbol{A}$ is ill-conditioned, anything can happen. We can for example have an accurate solution even if the residual is large.

8.3.2 *Perturbation of a Square Matrix*

Suppose $\boldsymbol{A}$ is nonsingular and $\boldsymbol{E}$ a perturbation of $\boldsymbol{A}$. We expect $\boldsymbol{B} := \boldsymbol{A} + \boldsymbol{E}$ to be nonsingular when $\boldsymbol{E}$ is small relative to $\boldsymbol{A}$. But how small is small? It is also useful to have bounds on $\|\boldsymbol{B}^{-1}\|$ in terms of $\|\boldsymbol{A}^{-1}\|$ and the difference $\|\boldsymbol{B}^{-1} - \boldsymbol{A}^{-1}\|$. We consider the relative errors $\|\boldsymbol{B}^{-1}\|/\|\boldsymbol{A}^{-1}\|$, $\|\boldsymbol{B}^{-1} - \boldsymbol{A}^{-1}\|/\|\boldsymbol{B}^{-1}\|$ and $\|\boldsymbol{B}^{-1} - \boldsymbol{A}^{-1}\|/\|\boldsymbol{A}^{-1}\|$.

Theorem 8.11 (Perturbation of Inverse Matrix) *Suppose $A \in \mathbb{C}^{n\times n}$ is nonsingular and let $B := A + E \in \mathbb{C}^{n\times n}$ be nonsingular. For any consistent matrix norm $\| \ \|$ we have*

$$\frac{\|B^{-1} - A^{-1}\|}{\|B^{-1}\|} \le \|A^{-1}E\| \le K(A)\frac{\|E\|}{\|A\|}, \tag{8.27}$$

where $K(A) := \|A\|\,\|A^{-1}\|$. If $r := \|A^{-1}E\| < 1$ then B is nonsingular and

$$\frac{1}{1+r} \le \frac{\|B^{-1}\|}{\|A^{-1}\|} \le \frac{1}{1-r}. \tag{8.28}$$

We also have

$$\frac{\|B^{-1} - A^{-1}\|}{\|A^{-1}\|} \le \frac{r}{1-r} \le \frac{K(A)}{1-r}\frac{\|E\|}{\|A\|}. \tag{8.29}$$

We can replace $\|A^{-1}E\|$ by $\|EA^{-1}\|$ everywhere.

Proof That B is nonsingular if $r < 1$ follows from Theorem 8.9. We have $-E = A - B = A(B^{-1} - A^{-1})B = B(B^{-1} - A^{-1})A$ so that

$$B^{-1} - A^{-1} = -A^{-1}EB^{-1} = -B^{-1}EA^{-1}. \tag{8.30}$$

Therefore, if B is nonsingular then by (8.30)

$$\|B^{-1} - A^{-1}\| \le \|A^{-1}E\|\|B^{-1}\| \le K(A)\frac{\|E\|}{\|A\|}\|B^{-1}\|.$$

Dividing through by $\|B^{-1}\|$ gives the upper bounds in (8.27). Next, (8.30) implies

$$\|B^{-1}\| \le \|A^{-1}\| + \|A^{-1}EB^{-1}\| \le \|A^{-1}\| + r\|B^{-1}\|.$$

Solving for $\|B^{-1}\|$ and dividing by $\|A^{-1}\|$ we obtain the upper bound in (8.28). Similarly we obtain the lower bound in (8.28) from $\|A^{-1}\| \le \|B^{-1}\| + r\|B^{-1}\|$.

The bound in (8.29) follows by multiplying (8.27) by $\|B^{-1}\|/\|A^{-1}\|$ and using (8.28).

That we can replace $\|A^{-1}E\|$ by $\|EA^{-1}\|$ everywhere follows from (8.30). □

8.4 Proof That the p-Norms Are Norms

We want to show

Theorem 8.12 (The p Vector Norms Are Norms) *Let for $1 \le p \le \infty$ and $\boldsymbol{x} \in \mathbb{C}^n$*

$$\|\boldsymbol{x}\|_p := \Big(\sum_{j=1}^{n} |x_j|^p\Big)^{1/p}, \qquad \|\boldsymbol{x}\|_\infty := \max_{1\le j\le n} |x_j|.$$

Then for all $1 \le p \le \infty$, $\boldsymbol{x}, \boldsymbol{y} \in \mathbb{C}^n$ and all $a \in \mathbb{C}$

1. $\|\boldsymbol{x}\|_p \ge 0$ *with equality if and only if* $\boldsymbol{x} = \boldsymbol{0}$. *(positivity)*
2. $\|a\boldsymbol{x}\|_p = |a|\,\|\boldsymbol{x}\|_p$. *(homogeneity)*
3. $\|\boldsymbol{x}+\boldsymbol{y}\|_p \le \|\boldsymbol{x}\|_p + \|\boldsymbol{y}\|_p$. *(subadditivity)*

Positivity and homogeneity follows immediately. To show the subadditivity we need some elementary properties of convex functions.

Definition 8.8 (Convex Function) Let $I \subset \mathbb{R}$ be an interval. A function $f : I \to \mathbb{R}$ is convex if

$$f\big((1-\lambda)x_1 + \lambda x_2\big) \le (1-\lambda)f(x_1) + \lambda f(x_2) \tag{8.31}$$

for all $x_1, x_2 \in I$ with $x_1 < x_2$ and all $\lambda \in [0, 1]$. The sum $\sum_{j=1}^{n} \lambda_j x_j$ is called a **convex combination** of $x_1, \dots, x_n$ if $\lambda_j \ge 0$ for $j = 1, \dots, n$ and $\sum_{j=1}^{n} \lambda_j = 1$.

The convexity condition is illustrated in Fig. 8.1.

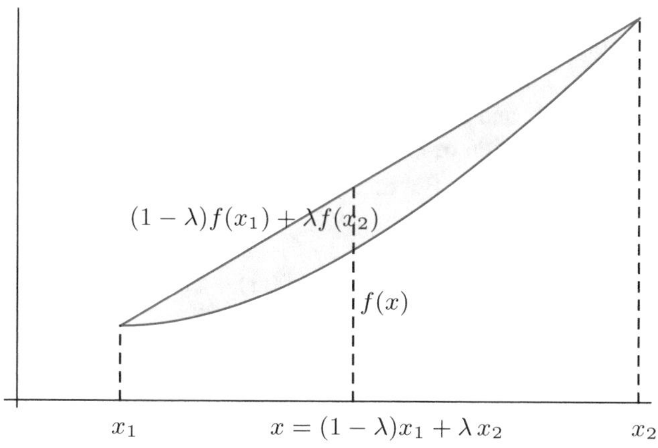

Fig. 8.1 A convex function

Lemma 8.2 (A Sufficient Condition for Convexity) *If* $f \in C^2[a,b]$ *and* $f''(x) \geq 0$ *for* $x \in [a,b]$ *then* f *is convex.*

Proof We recall the formula for linear interpolation with remainder, (cf a book on numerical methods) For any $a \leq x_1 \leq x \leq x_2 \leq b$ there is a $c \in [x_1, x_2]$ such that

$$\begin{aligned} f(x) &= \frac{x_2 - x}{x_2 - x_1} f(x_1) + \frac{x - x_1}{x_2 - x_1} f(x_2) + (x - x_1)(x - x_2) f''(c)/2 \\ &= (1 - \lambda) f(x_1) + \lambda f(x_2) + (x_2 - x_1)^2 \lambda(\lambda - 1) f''(c)/2, \quad \lambda := \frac{x - x_1}{x_2 - x_1}. \end{aligned}$$

Since $\lambda \in [0, 1]$ we have $f(x) \leq (1 - \lambda) f(x_1) + \lambda f(x_2)$. Moreover,

$$x = \frac{x_2 - x}{x_2 - x_1} x_1 + \frac{x - x_1}{x_2 - x_1} x_2 = (1 - \lambda) x_1 + \lambda x_2$$

so that (8.31) holds, and f is convex. □

The following inequality is elementary, but can be used to prove many nontrivial inequalities.

Theorem 8.13 (Jensen's Inequality) *Suppose* $I \in \mathbb{R}$ *is an interval and* $f : I \to \mathbb{R}$ *is convex. Then for all* $n \in \mathbb{N}$, *all* $\lambda_1, \ldots, \lambda_n$ *with* $\lambda_j \geq 0$ *for* $j = 1, \ldots, n$ *and* $\sum_{j=1}^{n} \lambda_j = 1$, *and all* $z_1, \ldots, z_n \in I$

$$f\Big(\sum_{j=1}^{n} \lambda_j z_j\Big) \leq \sum_{j=1}^{n} \lambda_j f(z_j).$$

Proof We use induction on n. The result is trivial for $n = 1$. Let $n \geq 2$, assume the inequality holds for $n - 1$, and let λ_j, z_j for $j = 1, \ldots, n$ be given as in the theorem. Since $n \geq 2$ we have $\lambda_i < 1$ for at least one i so assume without loss of generality that $\lambda_1 < 1$, and define $u := \sum_{j=2}^{n} \frac{\lambda_j}{1-\lambda_1} z_j$. Since $\sum_{j=2}^{n} \lambda_j = 1 - \lambda_1$ this is a convex combination of $n - 1$ terms and the induction hypothesis implies that $f(u) \leq \sum_{j=2}^{n} \frac{\lambda_j}{1-\lambda_1} f(z_j)$. But then by the convexity of f

$$f\Big(\sum_{j=1}^{n} \lambda_j z_j\Big) = f(\lambda_1 z_1 + (1 - \lambda_1) u) \leq \lambda_1 f(z_1) + (1 - \lambda_1) f(u) \leq \sum_{j=1}^{n} \lambda_j f(z_j)$$

and the inequality holds for n. □

Corollary 8.1 (Weighted Geometric/Arithmetic Mean Inequality) *Suppose $\sum_{j=1}^{n} \lambda_j a_j$ is a convex combination of nonnegative numbers $a_1, \ldots, a_n$. Then*

$$a_1^{\lambda_1} a_2^{\lambda_2} \cdots a_n^{\lambda_n} \le \sum_{j=1}^{n} \lambda_j a_j, \tag{8.32}$$

where $0^0 := 0$.

Proof The result is trivial if one or more of the a_j's are zero so assume $a_j > 0$ for all j. Consider the function $f : (0, \infty) \to \mathbb{R}$ given by $f(x) = -\log x$. Since $f''(x) = 1/x^2 > 0$ for $x \in (0, \infty)$, it follows from Lemma 8.2 that this function is convex. By Jensen's inequality

$$-\log\Big(\sum_{j=1}^{n} \lambda_j a_j\Big) \le -\sum_{j=1}^{n} \lambda_j \log(a_j) = -\log\big(a_1^{\lambda_1} \cdots a_n^{\lambda_n}\big)$$

or $\log\big(a_1^{\lambda_1} \cdots a_n^{\lambda_n}\big) \le \log\big(\sum_{j=1}^{n} \lambda_j a_j\big)$. The inequality follows since $\exp(\log x) = x$ for $x > 0$ and the exponential function is monotone increasing. □

Taking $\lambda_j = \frac{1}{n}$ for all j in (8.32) we obtain the classical **geometric/arithmetic mean inequality**

$$(a_1 a_2 \cdots a_n)^{\frac{1}{n}} \le \frac{1}{n} \sum_{j=1}^{n} a_j. \tag{8.33}$$

Corollary 8.2 (Hölder's Inequality) *For $\boldsymbol{x}, \boldsymbol{y} \in \mathbb{C}^n$ and $1 \le p \le \infty$*

$$\sum_{j=1}^{n} |x_j y_j| \le \|\boldsymbol{x}\|_p \|\boldsymbol{y}\|_q, \text{ where } \frac{1}{p} + \frac{1}{q} = 1.$$

Proof We leave the proof for $p = 1$ and $p = \infty$ as an exercise so assume $1 < p < \infty$. For any $a, b \ge 0$ the weighted arithmetic/geometric mean inequality implies that

$$a^{\frac{1}{p}} b^{\frac{1}{q}} \le \frac{1}{p} a + \frac{1}{q} b, \text{ where } \frac{1}{p} + \frac{1}{q} = 1. \tag{8.34}$$

If $\boldsymbol{x} = \boldsymbol{0}$ or $\boldsymbol{y} = \boldsymbol{0}$ there is nothing to prove so assume that both $\boldsymbol{x}$ and $\boldsymbol{y}$ are nonzero. Using 8.34 on each term in the middle sum we obtain

$$\frac{1}{\|\boldsymbol{x}\|_p \|\boldsymbol{y}\|_q} \sum_{j=1}^{n} |x_j y_j| = \sum_{j=1}^{n} \left(\frac{|x_j|^p}{\|\boldsymbol{x}\|_p^p}\right)^{\frac{1}{p}} \left(\frac{|y_j|^q}{\|\boldsymbol{y}\|_q^q}\right)^{\frac{1}{q}} \le \sum_{j=1}^{n} \left(\frac{1}{p} \frac{|x_j|^p}{\|\boldsymbol{x}\|_p^p} + \frac{1}{q} \frac{|y_j|^q}{\|\boldsymbol{y}\|_q^q}\right) = 1$$

and the proof of the inequality is complete. □

Corollary 8.3 (Minkowski's Inequality) *For $\boldsymbol{x}, \boldsymbol{y} \in \mathbb{C}^n$ and $1 \le p \le \infty$*

$$\|\boldsymbol{x} + \boldsymbol{y}\|_p \le \|\boldsymbol{x}\|_p + \|\boldsymbol{y}\|_p.$$

Proof We leave the proof for $p = 1$ and $p = \infty$ as an exercise so assume $1 < p < \infty$. We write

$$\|\boldsymbol{x} + \boldsymbol{y}\|_p^p = \sum_{j=1}^n |x_j + y_j|^p \le \sum_{j=1}^n |x_j||x_j + y_j|^{p-1} + \sum_{j=1}^n |y_j||x_j + y_j|^{p-1}.$$

We apply Hölder's inequality with exponent p and q to each sum. In view of the relation $(p-1)q = p$ the result is

$$\|\boldsymbol{x} + \boldsymbol{y}\|_p^p \le \|\boldsymbol{x}\|_p \|\boldsymbol{x} + \boldsymbol{y}\|_p^{p/q} + \|\boldsymbol{y}\|_p \|\boldsymbol{x} + \boldsymbol{y}\|_p^{p/q} = (\|\boldsymbol{x}\|_p + \|\boldsymbol{y}\|_p)\|\boldsymbol{x} + \boldsymbol{y}\|_p^{p-1},$$

and canceling the common factor, the inequality follows. □

8.4.1 p-Norms and Inner Product Norms

It is possible to characterize the p-norms that are derived from an inner product. We start with the following identity.

Theorem 8.14 (Parallelogram Identity) *For all $\boldsymbol{x}, \boldsymbol{y}$ in a real or complex inner product space*

$$\|\boldsymbol{x} + \boldsymbol{y}\|^2 + \|\boldsymbol{x} - \boldsymbol{y}\|^2 = 2\|\boldsymbol{x}\|^2 + 2\|\boldsymbol{y}\|^2, \tag{8.35}$$

where $\| \ \|$ is the inner product norm in the space.

Proof We set $a = \pm 1$ in (5.5) and add the two equations. □

Theorem 8.15 (When Is a Norm an Inner Product Norm?) *To a given norm on a real or complex vector space $\mathcal{V}$ there exists an inner product on $\mathcal{V}$ such that $\langle \boldsymbol{x}, \boldsymbol{x} \rangle = \|\boldsymbol{x}\|^2$ if and only if the parallelogram identity* (8.35) *holds for all $\boldsymbol{x}, \boldsymbol{y} \in \mathcal{V}$.*

Proof If $\langle \boldsymbol{x}, \boldsymbol{x} \rangle = \|\boldsymbol{x}\|^2$ then

$$\|\boldsymbol{x} + \boldsymbol{y}\|^2 + \|\boldsymbol{x} - \boldsymbol{y}\|^2 = \langle \boldsymbol{x} + \boldsymbol{y}, \boldsymbol{x} + \boldsymbol{y} \rangle + \langle \boldsymbol{x} - \boldsymbol{y}, \boldsymbol{x} - \boldsymbol{y} \rangle = 2\|\boldsymbol{x}\|^2 + 2\|\boldsymbol{y}\|^2$$

and the parallelogram identity holds. For the converse we prove the real case and leave the complex case as an exercise. Suppose (8.35) holds for all $\boldsymbol{x}, \boldsymbol{y}$ in the real vector space $\mathcal{V}$. We show that

$$\langle \boldsymbol{x}, \boldsymbol{y} \rangle := \frac{1}{4}\big(\|\boldsymbol{x} + \boldsymbol{y}\|^2 - \|\boldsymbol{x} - \boldsymbol{y}\|^2\big), \quad \boldsymbol{x}, \boldsymbol{y} \in \mathcal{V} \tag{8.36}$$

defines an inner product on $\mathcal{V}$. Clearly 1. and 2. in Definition 5.1 hold. The hard part is to show 3. We need to show that

$$\langle \boldsymbol{x}, \boldsymbol{z}\rangle + \langle \boldsymbol{y}, \boldsymbol{z}\rangle = \langle \boldsymbol{x}+\boldsymbol{y}, \boldsymbol{z}\rangle, \quad \boldsymbol{x}, \boldsymbol{y}, \boldsymbol{z} \in \mathcal{V}, \tag{8.37}$$

$$\langle a\boldsymbol{x}, \boldsymbol{y}\rangle = a\langle \boldsymbol{x}, \boldsymbol{y}\rangle, \quad a \in \mathbb{R}, \quad \boldsymbol{x}, \boldsymbol{y} \in \mathcal{V}. \tag{8.38}$$

Now

$$\begin{aligned}
4\langle \boldsymbol{x}, \boldsymbol{z}\rangle + 4\langle \boldsymbol{y}, \boldsymbol{z}\rangle &\overset{(8.36)}{=} \|\boldsymbol{x}+\boldsymbol{z}\|^2 - \|\boldsymbol{x}-\boldsymbol{z}\|^2 + \|\boldsymbol{y}+\boldsymbol{z}\|^2 - \|\boldsymbol{y}-\boldsymbol{z}\|^2 \\
&= \|\big(\boldsymbol{z}+\frac{\boldsymbol{x}+\boldsymbol{y}}{2}\big) + \frac{\boldsymbol{x}-\boldsymbol{y}}{2}\|^2 - \|\big(\boldsymbol{z}-\frac{\boldsymbol{x}+\boldsymbol{y}}{2}\big) + \frac{\boldsymbol{y}-\boldsymbol{x}}{2}\|^2 \\
&+ \|\big(\boldsymbol{z}+\frac{\boldsymbol{x}+\boldsymbol{y}}{2}\big) - \frac{\boldsymbol{x}-\boldsymbol{y}}{2}\|^2 - \|\big(\boldsymbol{z}-\frac{\boldsymbol{x}+\boldsymbol{y}}{2}\big) - \frac{\boldsymbol{y}-\boldsymbol{x}}{2}\|^2 \\
&\overset{(8.35)}{=} 2\|\boldsymbol{z}+\frac{\boldsymbol{x}+\boldsymbol{y}}{2}\|^2 + 2\|\frac{\boldsymbol{x}-\boldsymbol{y}}{2}\|^2 - 2\|\boldsymbol{z}-\frac{\boldsymbol{x}+\boldsymbol{y}}{2}\|^2 - 2\|\frac{\boldsymbol{y}-\boldsymbol{x}}{2}\|^2 \\
&\overset{(8.36)}{=} 8\langle \frac{\boldsymbol{x}+\boldsymbol{y}}{2}, \boldsymbol{z}\rangle,
\end{aligned}$$

or

$$\langle \boldsymbol{x}, \boldsymbol{z}\rangle + \langle \boldsymbol{y}, \boldsymbol{z}\rangle = 2\langle \frac{\boldsymbol{x}+\boldsymbol{y}}{2}, \boldsymbol{z}\rangle, \quad \boldsymbol{x}, \boldsymbol{y}, \boldsymbol{z} \in \mathcal{V}.$$

In particular, since $\boldsymbol{y} = \boldsymbol{0}$ implies $\langle \boldsymbol{y}, \boldsymbol{z}\rangle = 0$ we obtain $\langle \boldsymbol{x}, \boldsymbol{z}\rangle = 2\langle \frac{\boldsymbol{x}}{2}, \boldsymbol{z}\rangle$ for all $\boldsymbol{x}, \boldsymbol{z} \in \mathcal{V}$. This means that $2\langle \frac{\boldsymbol{x}+\boldsymbol{y}}{2}, \boldsymbol{z}\rangle = \langle \boldsymbol{x}+\boldsymbol{y}, \boldsymbol{z}\rangle$ for all $\boldsymbol{x}, \boldsymbol{y}, \boldsymbol{z} \in \mathcal{V}$ and (8.37) follows.

We first show (8.38) when $a = n$ is a positive integer. By induction

$$\langle n\boldsymbol{x}, \boldsymbol{y}\rangle = \langle (n-1)\boldsymbol{x} + \boldsymbol{x}, \boldsymbol{y}\rangle \overset{(8.37)}{=} \langle (n-1)\boldsymbol{x}, \boldsymbol{y}\rangle + \langle \boldsymbol{x}, \boldsymbol{y}\rangle = n\langle \boldsymbol{x}, \boldsymbol{y}\rangle. \tag{8.39}$$

If $m, n \in \mathbb{N}$ then

$$m^2\langle \frac{n}{m}\boldsymbol{x}, \boldsymbol{y}\rangle \overset{(8.39)}{=} m\langle n\boldsymbol{x}, \boldsymbol{y}\rangle \overset{(8.39)}{=} mn\langle \boldsymbol{x}, \boldsymbol{y}\rangle,$$

implying that (8.38) holds for positive rational numbers

$$\langle \frac{n}{m}\boldsymbol{x}, \boldsymbol{y}\rangle = \frac{n}{m}\langle \boldsymbol{x}, \boldsymbol{y}\rangle.$$

Now if $a > 0$ there is a sequence $\{a_n\}$ of positive rational numbers converging to a. For each n

$$a_n\langle \boldsymbol{x}, \boldsymbol{y}\rangle = \langle a_n\boldsymbol{x}, \boldsymbol{y}\rangle \overset{(8.36)}{=} \frac{1}{4}\big(\|a_n\boldsymbol{x}+\boldsymbol{y}\|^2 - \|a_n\boldsymbol{x}-\boldsymbol{y}\|^2\big).$$

Taking limits and using continuity of norms we obtain $a\langle \boldsymbol{x}, \boldsymbol{y}\rangle = \langle a\boldsymbol{x}, \boldsymbol{y}\rangle$. This also holds for $a = 0$. Finally, if $a < 0$ then $(-a) > 0$ and from what we just showed

$$(-a)\langle \boldsymbol{x}, \boldsymbol{y}\rangle = \langle (-a)\boldsymbol{x}, \boldsymbol{y}\rangle \overset{(8.36)}{=} \frac{1}{4}\big(\|-a\boldsymbol{x} + \boldsymbol{y}\|^2 - \|-a\boldsymbol{x} - \boldsymbol{y}\|^2\big) = -\langle a\boldsymbol{x}, \boldsymbol{y}\rangle,$$

so (8.38) also holds for negative a. □

Corollary 8.4 (Are the p-Norms Inner Product Norms?) *For the p vector norms on $\mathcal{V} = \mathbb{R}^n$ or $\mathcal{V} = \mathbb{C}^n$, $1 \le p \le \infty$, $n \ge 2$, there is an inner product on $\mathcal{V}$ such that $\langle \boldsymbol{x}, \boldsymbol{x}\rangle = \|\boldsymbol{x}\|_p^2$ for all $\boldsymbol{x} \in \mathcal{V}$ if and only if $p = 2$.*

Proof For $p = 2$ the p-norm is the Euclidian norm which corresponds to the standard inner product. If $p \ne 2$ then the parallelogram identity (8.35) does not hold for say $\boldsymbol{x} := \boldsymbol{e}_1$ and $\boldsymbol{y} := \boldsymbol{e}_2$. □

8.5 Exercises Chap. 8

8.5.1 Exercises Sect. 8.1

Exercise 8.1 (An A-Norm Inequality (Exam Exercise 1982-4)) Given a symmetric positive definite matrix $\boldsymbol{A} \in \mathbb{R}^{n\times n}$ with eigenvalues $0 < \lambda_n \le \cdots \le \lambda_1$. Show that

$$\|\boldsymbol{x}\|_A \le \|\boldsymbol{y}\|_A \implies \|\boldsymbol{x}\|_2 \le \sqrt{\frac{\lambda_1}{\lambda_n}}\|\boldsymbol{y}\|_2,$$

where

$$\|\boldsymbol{x}\|_A := \sqrt{\boldsymbol{x}^T\boldsymbol{A}\boldsymbol{x}}, \quad \boldsymbol{x} \in \mathbb{R}^n.$$

Exercise 8.2 (A Orthogonal Bases (Exam Exercise 1995-4)) Let $\boldsymbol{A} \in \mathbb{R}^{n\times n}$ be a symmetric and positive definite matrix and assume $\boldsymbol{b}_1, \ldots, \boldsymbol{b}_n$ is a basis for $\mathbb{R}^n$. We define $\boldsymbol{B}_k := [\boldsymbol{b}_1, \ldots, \boldsymbol{b}_k] \in \mathbb{R}^{n\times k}$ for $k = 1, \ldots, n$. We consider in this exercise the inner product $\langle \cdot, \cdot\rangle$ defined by $\langle \boldsymbol{x}, \boldsymbol{y}\rangle := \boldsymbol{x}^T\boldsymbol{A}\boldsymbol{y}$ for $\boldsymbol{x}, \boldsymbol{y} \in \mathbb{R}^n$ and the corresponding norm $\|\boldsymbol{x}\|_A := \langle \boldsymbol{x}, \boldsymbol{x}\rangle^{1/2}$. We define $\tilde{\boldsymbol{b}}_1 := \boldsymbol{b}_1$ and

$$\tilde{\boldsymbol{b}}_k := \boldsymbol{b}_k - \boldsymbol{B}_{k-1}\big(\boldsymbol{B}_{k-1}^T\boldsymbol{A}\boldsymbol{B}_{k-1}\big)^{-1}\boldsymbol{B}_{k-1}^T\boldsymbol{A}\boldsymbol{b}_k, \quad k = 2, \ldots, n.$$

a) Show that $\boldsymbol{B}_k^T\boldsymbol{A}\boldsymbol{B}_k$ is positive definite for $k = 1, \ldots, n$.
b) Show that for $k = 2, \ldots, n$ we have (i) $\langle \tilde{\boldsymbol{b}}_k, \boldsymbol{b}_j\rangle = 0$ for $j = 1, \ldots, k-1$ and (ii) $\tilde{\boldsymbol{b}}_k - \boldsymbol{b}_k \in \operatorname{span}(\boldsymbol{b}_1, \ldots, \boldsymbol{b}_{k-1})$.

c) Explain why $\tilde{\boldsymbol{b}}_1, \ldots, \tilde{\boldsymbol{b}}_n$ is a basis for $\mathbb{R}^n$ which in addition is $\boldsymbol{A}$-orthogonal, i.e., $\langle \tilde{\boldsymbol{b}}_i, \tilde{\boldsymbol{b}}_j \rangle = 0$ for all $i, j \le n, i \ne j$.

d) Define $\tilde{\boldsymbol{B}}_n := [\tilde{\boldsymbol{b}}_1, \ldots, \tilde{\boldsymbol{b}}_n]$. Show that there is an upper triangular matrix $\boldsymbol{T} \in \mathbb{R}^{n\times n}$ with ones on the diagonal and satisfies $\boldsymbol{B}_n = \tilde{\boldsymbol{B}}_n \boldsymbol{T}$.

e) Assume that the matrix $\boldsymbol{T}$ in d) is such that $|t_{ij}| \le \frac{1}{2}$ for all $i, j \le n, i \ne j$. Assume also that $\|\tilde{\boldsymbol{b}}_k\|_A^2 \le 2\|\tilde{\boldsymbol{b}}_{k+1}\|_A^2$ for $k = 1, \ldots, n-1$ and that $\det(\boldsymbol{B}_n) = 1$. Show that then[2]

$$\|\boldsymbol{b}_1\|_A \|\boldsymbol{b}_2\|_A \cdots \|\boldsymbol{b}_n\|_A \le 2^{n(n-1)/4}\sqrt{\det(\boldsymbol{A})}.$$

8.5.2 Exercises Sect. 8.2

Exercise 8.3 (Consistency of Sum Norm?) Show that the sum norm is consistent.

Exercise 8.4 (Consistency of Max Norm?) Show that the max norm is not consistent by considering $\left[\begin{smallmatrix} 1 & 1 \\ 1 & 1 \end{smallmatrix}\right]$.

Exercise 8.5 (Consistency of Modified Max Norm)

a) Show that the norm

$$\|\boldsymbol{A}\| := \sqrt{mn}\|\boldsymbol{A}\|_M, \quad \boldsymbol{A} \in \mathbb{C}^{m\times n}$$

is a consistent matrix norm.

b) Show that the constant $\sqrt{mn}$ can be replaced by m and by n.

Exercise 8.6 (What Is the Sum Norm Subordinate to?) Show that the sum norm is subordinate to the l_1-norm.

Exercise 8.7 (What Is the Max Norm Subordinate to?)

a) Show that the max norm is subordinate to the ∞ and 1 norm, i.e., $\|\boldsymbol{Ax}\|_\infty \le \|\boldsymbol{A}\|_M \|\boldsymbol{x}\|_1$ holds for all $\boldsymbol{A} \in \mathbb{C}^{m\times n}$ and all $\boldsymbol{x} \in \mathbb{C}^n$.

b) Show that if $\|\boldsymbol{A}\|_M = |a_{kl}|$, then $\|\boldsymbol{Ae}_l\|_\infty = \|\boldsymbol{A}\|_M \|\boldsymbol{e}_l\|_1$.

c) Show that $\|\boldsymbol{A}\|_M = \max_{\boldsymbol{x}\ne\boldsymbol{0}} \frac{\|\boldsymbol{Ax}\|_\infty}{\|\boldsymbol{x}\|_1}$.

Exercise 8.8 (Spectral Norm) Let $m, n \in \mathbb{N}$ and $\boldsymbol{A} \in \mathbb{C}^{m\times n}$. Show that

$$\|\boldsymbol{A}\|_2 = \max_{\|\boldsymbol{x}\|_2 = \|\boldsymbol{y}\|_2 = 1} |\boldsymbol{y}^* \boldsymbol{Ax}|.$$

---

[2]Hint: Show that $\|\tilde{\boldsymbol{b}}_1\|_A^2 \cdots \|\tilde{\boldsymbol{b}}_n\|_A^2 = \det(\boldsymbol{A})$.

Exercise 8.9 (Spectral Norm of the Inverse) Suppose $\boldsymbol{A} \in \mathbb{C}^{n\times n}$ is nonsingular. Show that $\|\boldsymbol{A}\boldsymbol{x}\|_2 \geq \sigma_n$ for all $\boldsymbol{x} \in \mathbb{C}^n$ with $\|\boldsymbol{x}\|_2 = 1$. Show that

$$\|\boldsymbol{A}^{-1}\|_2 = \max_{\boldsymbol{x}\neq\boldsymbol{0}} \frac{\|\boldsymbol{x}\|_2}{\|\boldsymbol{A}\boldsymbol{x}\|_2}.$$

Exercise 8.10 (p-Norm Example) Let

$$\boldsymbol{A} = \begin{bmatrix} 2 & -1 \\ -1 & 2 \end{bmatrix}.$$

Compute $\|\boldsymbol{A}\|_p$ and $\|\boldsymbol{A}^{-1}\|_p$ for $p = 1, 2, \infty$.

Exercise 8.11 (Unitary Invariance of the Spectral Norm) Show that $\|\boldsymbol{V}\boldsymbol{A}\|_2 = \|\boldsymbol{A}\|_2$ holds even for a rectangular $\boldsymbol{V}$ as long as $\boldsymbol{V}^*\boldsymbol{V} = \boldsymbol{I}$.

Exercise 8.12 ($\|\boldsymbol{A}\boldsymbol{U}\|_2$ Rectangular $\boldsymbol{A}$) Find $\boldsymbol{A} \in \mathbb{R}^{2\times 2}$ and $\boldsymbol{U} \in \mathbb{R}^{2\times 1}$ with $\boldsymbol{U}^T\boldsymbol{U} = \boldsymbol{I}$ such that $\|\boldsymbol{A}\boldsymbol{U}\|_2 < \|\boldsymbol{A}\|_2$. Thus, in general, $\|\boldsymbol{A}\boldsymbol{U}\|_2 = \|\boldsymbol{A}\|_2$ does not hold for a rectangular $\boldsymbol{U}$ even if $\boldsymbol{U}^*\boldsymbol{U} = \boldsymbol{I}$.

Exercise 8.13 (p-Norm of Diagonal Matrix) Show that $\|\boldsymbol{A}\|_p = \rho(\boldsymbol{A}) := \max|\lambda_i|$ (the largest eigenvalue of $\boldsymbol{A}$), $1 \leq p \leq \infty$, when $\boldsymbol{A}$ is a diagonal matrix.

Exercise 8.14 (Spectral Norm of a Column Vector) A vector $\boldsymbol{a} \in \mathbb{C}^m$ can also be considered as a matrix $\boldsymbol{A} \in \mathbb{C}^{m,1}$.

a) Show that the spectral matrix norm (2-norm) of $\boldsymbol{A}$ equals the Euclidean vector norm of $\boldsymbol{a}$.
b) Show that $\|\boldsymbol{A}\|_p = \|\boldsymbol{a}\|_p$ for $1 \leq p \leq \infty$.

Exercise 8.15 (Norm of Absolute Value Matrix) If $\boldsymbol{A} \in \mathbb{C}^{m\times n}$ has elements a_{ij}, let $|\boldsymbol{A}| \in \mathbb{R}^{m\times n}$ be the matrix with elements $|a_{ij}|$.

a) Compute $|\boldsymbol{A}|$ if $\boldsymbol{A} = \begin{bmatrix} 1+i & -2 \\ 1 & 1-i \end{bmatrix}$, $i = \sqrt{-1}$.
b) Show that for any $\boldsymbol{A} \in \mathbb{C}^{m\times n}$ $\|\boldsymbol{A}\|_F = \|\,|\boldsymbol{A}|\,\|_F$, $\|\boldsymbol{A}\|_p = \|\,|\boldsymbol{A}|\,\|_p$ for $p = 1, \infty$.
c) Show that for any $\boldsymbol{A} \in \mathbb{C}^{m\times n}$ $\|\boldsymbol{A}\|_2 \leq \|\,|\boldsymbol{A}|\,\|_2$.
d) Find a real symmetric 2×2 matrix $\boldsymbol{A}$ such that $\|\boldsymbol{A}\|_2 < \|\,|\boldsymbol{A}|\,\|_2$.

Exercise 8.16 (An Iterative Method (Exam Exercise 2017-3)) Assume that $\boldsymbol{A} \in \mathbb{C}^{n\times n}$ is non-singular and nondefective (the eigenvectors of $\boldsymbol{A}$ form a basis for $\mathbb{C}^n$). We wish to solve $\boldsymbol{A}\boldsymbol{x} = \boldsymbol{b}$. Assume that we have a list of the eigenvalues $\{\lambda_1, \lambda_2, \ldots, \lambda_m\}$, in no particular order. We have that $m \leq n$, since some of the eigenvalues may have multiplicity larger than one. Given $\boldsymbol{x}_0 \in \mathbb{C}^n$, and $k \geq 0$, we define the sequence $\{\boldsymbol{x}_k\}_{k=0}^{m-1}$ by

$$\boldsymbol{x}_{k+1} = \boldsymbol{x}_k + \frac{1}{\lambda_{k+1}} \boldsymbol{r}_k, \text{ where } \boldsymbol{r}_k = \boldsymbol{b} - \boldsymbol{A}\boldsymbol{x}_k.$$

a) Let the coefficients c_{ik} be defined by

$$\boldsymbol{r}_k = \sum_{i=1}^{n} c_{ik} \boldsymbol{u}_i,$$

where $\{(\sigma_i, \boldsymbol{u}_i)\}_{i=1}^n$ are the eigenpairs of $\boldsymbol{A}$. Show that

$$c_{i,k+1} = \begin{cases} 0 & \text{if } \sigma_i = \lambda_{k+1}, \\ c_{i,k}\left(1 - \frac{\sigma_i}{\lambda_{k+1}}\right) & \text{otherwise.} \end{cases}$$

b) Show that for some $l \le m$, we have that $\boldsymbol{x}_l = \boldsymbol{x}_{l+1} = \cdots = \boldsymbol{x}_m = \boldsymbol{x}$, where $\boldsymbol{A}\boldsymbol{x} = \boldsymbol{b}$.

c) Consider this iteration for the $n \times n$ matrix $\boldsymbol{T} = \operatorname{tridiag}(c, d, c)$, where d and c are positive real numbers and $d > 2c$. The eigenvalues of $\boldsymbol{T}$ are

$$\lambda_j = d + 2c\cos\left(\frac{j\pi}{n+1}\right), \qquad j = 1, \ldots, n.$$

What is the operation count for solving $\boldsymbol{T}\boldsymbol{x} = \boldsymbol{b}$ using the iterative algorithm above?

d) Let now $\boldsymbol{B}$ be a symmetric $n \times n$ matrix which is zero on the "tridiagonal", i.e., $b_{ij} = 0$ if $|i - j| \le 1$. Set $\boldsymbol{A} = \boldsymbol{T} + \boldsymbol{B}$, where $\boldsymbol{T}$ is the tridiagonal matrix above. We wish to solve $\boldsymbol{A}\boldsymbol{x} = \boldsymbol{b}$ by the iterative scheme

$$\boldsymbol{T}\boldsymbol{x}_{k+1} = \boldsymbol{b} - \boldsymbol{B}\boldsymbol{x}_k. \tag{8.40}$$

Recall that if $\boldsymbol{E} \in \mathbb{R}^{n \times n}$ has eigenvalues $\lambda_1, \ldots, \lambda_n$ then $\rho(\boldsymbol{E}) := \max_i |\lambda_i|$ is the spectral radius of $\boldsymbol{E}$. Show that $\rho(\boldsymbol{T}^{-1}\boldsymbol{B}) \le \rho(\boldsymbol{T}^{-1})\rho(\boldsymbol{B})$.

e) Show that the iteration (8.40) will converge if[3]

$$\min\left\{\max_i \sum_{j=1}^{n} |b_{ij}|, \max_j \sum_{i=1}^{n} |b_{ij}|\right\} < d - 2c.$$

[3]Hint: use Gershgorin's theorem.

8.5.3 Exercises Sect. 8.3

Exercise 8.17 (Perturbed Linear Equation (Exam Exercise 1981-2)) Given the systems $\boldsymbol{A}\boldsymbol{x} = \boldsymbol{b}$, $\boldsymbol{A}\boldsymbol{y} = \boldsymbol{b} + \boldsymbol{e}$, where

$$\boldsymbol{A} := \begin{bmatrix} 1.1 & 1 \\ 1 & 1 \end{bmatrix}, \quad \boldsymbol{b} := \begin{bmatrix} b_1 \\ b_2 \end{bmatrix} = \begin{bmatrix} 2.1 \\ 2.0 \end{bmatrix}, \quad \boldsymbol{e} := \begin{bmatrix} e_1 \\ e_2 \end{bmatrix}, \quad \|\boldsymbol{e}\|_2 = 0.1.$$

We define $\delta := \|\boldsymbol{x} - \boldsymbol{y}\|_2 / \|\boldsymbol{x}\|_2$.

a) Determine $K_2(\boldsymbol{A}) = \|\boldsymbol{A}\|_2 \|\boldsymbol{A}^{-1}\|_2$. Give an upper bound and a positive lower bound for δ without computing $\boldsymbol{x}$ and $\boldsymbol{y}$.

b) Suppose as before that $b_2 = 2.0$ and $\|\boldsymbol{e}\|_2 = 0.1$. Determine b_1 and $\boldsymbol{e}$ which maximize δ.

Exercise 8.18 (Sharpness of Perturbation Bounds) The upper and lower bounds for $\|\boldsymbol{y} - \boldsymbol{x}\|/\|\boldsymbol{x}\|$ given by (8.22) can be attained for any matrix $\boldsymbol{A}$, but only for special choices of $\boldsymbol{b}$. Suppose $\boldsymbol{y}_A$ and $\boldsymbol{y}_{A^{-1}}$ are vectors with $\|\boldsymbol{y}_A\| = \|\boldsymbol{y}_{A^{-1}}\| = 1$ and $\|\boldsymbol{A}\| = \|\boldsymbol{A}\boldsymbol{y}_A\|$ and $\|\boldsymbol{A}^{-1}\| = \|\boldsymbol{A}^{-1}\boldsymbol{y}_{A^{-1}}\|$.

a) Show that the upper bound in (8.22) is attained if $\boldsymbol{b} = \boldsymbol{A}\boldsymbol{y}_A$ and $\boldsymbol{e} = \boldsymbol{y}_{A^{-1}}$.

b) Show that the lower bound is attained if $\boldsymbol{b} = \boldsymbol{y}_{A^{-1}}$ and $\boldsymbol{e} = \boldsymbol{A}\boldsymbol{y}_A$.

Exercise 8.19 (Condition Number of 2. Derivative Matrix) In this exercise we will show that for $m \ge 1$

$$\frac{4}{\pi^2}(m+1)^2 - 2/3 < \operatorname{cond}_p(\boldsymbol{T}) \le \frac{1}{2}(m+1)^2, \quad p = 1, 2, \infty, \tag{8.41}$$

where $\boldsymbol{T} := \operatorname{tridiag}(-1, 2, -1) \in \mathbb{R}^{m \times m}$ and $\operatorname{cond}_p(\boldsymbol{T}) := \|\boldsymbol{T}\|_p \|\boldsymbol{T}^{-1}\|_p$ is the p-norm condition number of $\boldsymbol{T}$. The p matrix norm is given by (8.17). You will need the explicit inverse of $\boldsymbol{T}$ given by (2.39) and the eigenvalues given in Lemma 2.2. As usual we define $h := 1/(m+1)$.

a) Show that for $m \ge 3$

$$\operatorname{cond}_1(\boldsymbol{T}) = \operatorname{cond}_\infty(\boldsymbol{T}) = \frac{1}{2}\begin{cases} h^{-2}, & m \text{ odd}, \\ h^{-2} - 1, & m \text{ even}. \end{cases} \tag{8.42}$$

and that $\operatorname{cond}_1(\boldsymbol{T}) = \operatorname{cond}_\infty(\boldsymbol{T}) = 3$ for $m = 2$.

b) Show that for $p = 2$ and $m \ge 1$ we have

$$\operatorname{cond}_2(\boldsymbol{T}) = \cot^2\left(\frac{\pi h}{2}\right) = 1/\tan^2\left(\frac{\pi h}{2}\right).$$

c) Show the bounds

$$\frac{4}{\pi^2}h^{-2} - \frac{2}{3} < \mathrm{cond}_2(\boldsymbol{T}) < \frac{4}{\pi^2}h^{-2}. \tag{8.43}$$

Hint: For the upper bound use the inequality $\tan x > x$ valid for $0 < x < \pi/2$. For the lower bound we use (without proof) the inequality $\cot^2 x > \frac{1}{x^2} - \frac{2}{3}$ for $x > 0$.

d) Show (8.41).

Exercise 8.20 (Perturbation of the Identity Matrix) Let $\boldsymbol{E}$ be a square matrix.

a) Show that if $\boldsymbol{I} - \boldsymbol{E}$ is nonsingular then

$$\frac{\|(\boldsymbol{I} - \boldsymbol{E})^{-1} - \boldsymbol{I}\|}{\|(\boldsymbol{I} - \boldsymbol{E})^{-1}\|} \le \|\boldsymbol{E}\|$$

b) If $\|\boldsymbol{E}\| < 1$ then $(\boldsymbol{I} - \boldsymbol{E})^{-1}$ is nonsingular by exists and

$$\frac{1}{1 + \|\boldsymbol{E}\|} \le \|(\boldsymbol{I} - \boldsymbol{E})^{-1}\| \le \frac{1}{1 - \|\boldsymbol{E}\|}$$

Show the lower bound. Show the upper bound if $\|\boldsymbol{I}\| = 1$. In general for a consistent matrix norm (i.e., the Frobenius norm) the upper bound follows from Theorem 12.14 using Neumann series.

c) Show that if $\|\boldsymbol{E}\| < 1$ then

$$\|(\boldsymbol{I} - \boldsymbol{E})^{-1} - \boldsymbol{I}\| \le \frac{\|\boldsymbol{E}\|}{1 - \|\boldsymbol{E}\|}.$$

Exercise 8.21 (Lower Bounds in (8.27) **and** (8.29)**)**

a) Solve for $\boldsymbol{E}$ in (8.30) and show that

$$K(\boldsymbol{B})^{-1}\frac{\|\boldsymbol{E}\|}{\|\boldsymbol{A}\|} \le \frac{\|\boldsymbol{B}^{-1} - \boldsymbol{A}^{-1}\|}{\|\boldsymbol{B}^{-1}\|}.$$

b) Show using **a)** and (8.28) that

$$\frac{K(\boldsymbol{B})^{-1}}{1 + r}\frac{\|\boldsymbol{E}\|}{\|\boldsymbol{A}\|} \le \frac{\|\boldsymbol{B}^{-1} - \boldsymbol{A}^{-1}\|}{\|\boldsymbol{A}^{-1}\|}.$$

Exercise 8.22 (Periodic Spline Interpolation (Exam Exercise 1993-2)) Let the components of $\boldsymbol{x} = [x_0, \ldots, x_n]^T \in \mathbb{R}^{n+1}$ define a partition of the interval $[a, b]$,

$$a = x_0 < x_1 < \cdots < x_n = b,$$

and given a dataset $\boldsymbol{y} := [y_0, \ldots, y_n]^T \in \mathbb{R}^{n+1}$, where we assume $y_0 = y_n$. The periodic cubic spline interpolation problem is defined by finding a cubic spline function g satisfying the conditions

$$g(x_i) = y_i, \qquad i = 0, 1, \ldots, n,$$
$$g'(a) = g'(b), \quad g''(a) = g''(b).$$

(Recall that g is a cubic polynomial on each interval (x_{i-1}, x_i), for $i = 1, \ldots, n$ with smoothness $C^2[a, b]$.)

We define $s_i := g'(x_i)$, $i = 0, \ldots, n$. It can be shown that the vector $\boldsymbol{s} := [s_1, \ldots, s_n]^T$ is determined from a linear system

$$\boldsymbol{A}\boldsymbol{s} = \boldsymbol{b}, \tag{8.44}$$

where $\boldsymbol{b} \in \mathbb{R}^n$ is a given vector determined by $\boldsymbol{x}$ and $\boldsymbol{y}$. The matrix $\boldsymbol{A} \in \mathbb{R}^{n \times n}$ is given by

$$\boldsymbol{A} := \begin{bmatrix} 2 & \mu_1 & 0 & \cdots & 0 & \lambda_1 \\ \lambda_2 & 2 & \mu_2 & \ddots & & 0 \\ 0 & \ddots & \ddots & \ddots & \ddots & \vdots \\ \vdots & \ddots & \ddots & \ddots & \ddots & 0 \\ 0 & & \ddots & \lambda_{n-1} & 2 & \mu_{n-1} \\ \mu_n & 0 & \cdots & 0 & \lambda_n & 2 \end{bmatrix},$$

where

$$\lambda_i := \frac{h_i}{h_{i-1} + h_i}, \qquad \mu_i := \frac{h_{i-1}}{h_{i-1} + h_i}, \qquad , i = 1, \ldots, n,$$

and

$$h_i = x_{i+1} - x_i, \quad i = 0, \ldots, n-1, \text{ and } h_n = h_0.$$

You shall not argue or prove the system (8.44). Throughout this exercise we assume that

$$\frac{1}{2} \le \frac{h_i}{h_{i-1}} \le 2, \quad i = 1, \ldots, n.$$

a) Show that

$$\|A\|_\infty = 3 \quad \text{and that} \quad \|A\|_1 \le \frac{10}{3}.$$

b) Show that $\|A^{-1}\|_\infty \le 1$.
c) Show that $\|A^{-1}\|_1 \le \frac{3}{2}$.
d) Let s and b be as in (8.44), where we assume $b \ne 0$. Let $e \in \mathbb{R}^n$ be such that $\|e\|_p/\|b\|_p \le 0.01$. Suppose $\hat{s}$ satisfies

$$A\hat{s} = b + e.$$

Give estimates for

$$\frac{\|\hat{s} - s\|_\infty}{\|s\|_\infty} \quad \text{and} \quad \frac{\|\hat{s} - s\|_1}{\|s\|_1}.$$

Exercise 8.23 (LSQ MATLAB Program (Exam Exercise 2013-4)) Suppose $A \in \mathbb{R}^{m\times n}, b \in \mathbb{R}^m$, where A has rank n and let $A = U\Sigma V^T$ be a singular value factorization of A. Thus $U \in \mathbb{R}^{m\times n}$ and $\Sigma, V \in \mathbb{R}^{n\times n}$. Write a MATLAB `function [x,K]=lsq(A,b)` that uses the singular value factorization of A to calculate a least squares solution $x = V\Sigma^{-1}U^T b$ to the system $Ax = b$ and the spectral (2-norm) condition number of A. The MATLAB command `[U,Sigma,V]=svd(A,0)` computes the singular value factorization of A.

8.5.4 Exercises Sect. 8.4

Exercise 8.24 (When Is a Complex Norm an Inner Product Norm?) Given a vector norm in a complex vector space $\mathcal{V}$, and suppose (8.35) holds for all $x, y \in \mathcal{V}$. Show that

$$\langle x, y\rangle := \frac{1}{4}\left(\|x+y\|^2 - \|x-y\|^2 + i\|x+iy\|^2 - i\|x-iy\|^2\right), \tag{8.45}$$

defines an inner product on $\mathcal{V}$, where $i = \sqrt{-1}$. The identity (8.45) is called the **polarization identity**.[4]

Exercise 8.25 (p Norm for $p = 1$ and $p = \infty$) Show that $\|\cdot\|_p$ is a vector norm in $\mathbb{R}^n$ for $p = 1, p = \infty$.

[4]Hint: We have $\langle x, y\rangle = s(x, y) + is(x, iy)$, where $s(x, y) := \frac{1}{4}\left(\|x+y\|^2 - \|x-y\|^2\right)$.

Exercise 8.26 (The p-Norm Unit Sphere) The set

$$S_p = \{\boldsymbol{x} \in \mathbb{R}^n : \|\boldsymbol{x}\|_p = 1\}$$

is called the unit sphere in $\mathbb{R}^n$ with respect to p. Draw S_p for $p = 1, 2, \infty$ for $n = 2$.

Exercise 8.27 (Sharpness of p-Norm Inequality) For $p \geq 1$, and any $\boldsymbol{x} \in \mathbb{C}^n$ we have $\|\boldsymbol{x}\|_\infty \leq \|\boldsymbol{x}\|_p \leq n^{1/p}\|\boldsymbol{x}\|_\infty$ (cf. (8.5)).

Produce a vector $\boldsymbol{x}_l$ such that $\|\boldsymbol{x}_l\|_\infty = \|\boldsymbol{x}_l\|_p$ and another vector $\boldsymbol{x}_u$ such that $\|\boldsymbol{x}_u\|_p = n^{1/p}\|\boldsymbol{x}_u\|_\infty$. Thus, these inequalities are sharp.

Exercise 8.28 (p-Norm Inequalities for Arbitrary p) If $1 \leq q \leq p \leq \infty$ then

$$\|\boldsymbol{x}\|_p \leq \|\boldsymbol{x}\|_q \leq n^{1/q-1/p}\|\boldsymbol{x}\|_p, \quad \boldsymbol{x} \in \mathbb{C}^n.$$

Hint: For the rightmost inequality use Jensen's inequality Cf. Theorem 8.13 with $f(z) = z^{p/q}$ and $z_i = |x_i|^q$. For the left inequality consider first $y_i = x_i/\|\boldsymbol{x}\|_\infty$, $i = 1, 2, \ldots, n$.

8.6 Review Questions

8.6.1

- What is a consistent matrix norm?
- what is a subordinate matrix norm?
- is an operator norm consistent?
- why is the Frobenius norm not an operator norm?
- what is the spectral norm of a matrix?
- how do we compute $\|\boldsymbol{A}\|_\infty$?
- what is the spectral condition number of a symmetric positive definite matrix?

8.6.2 Does there exist a vector norm $\|\ \|$ such that $\|\boldsymbol{A}\boldsymbol{x}\| \leq \|\boldsymbol{A}\|_F\|\boldsymbol{x}\|$ for all $\boldsymbol{A} \in \mathbb{C}^{n\times n}$, $\boldsymbol{x} \in \mathbb{C}^n$, $m, n \in \mathbb{N}$?

8.6.3 Why is $\|\boldsymbol{A}\|_2 \leq \|\boldsymbol{A}\|_F$ for any matrix $\boldsymbol{A}$?

8.6.4 What is the spectral norm of the inverse of a normal matrix?

Chapter 9
Least Squares

Consider the linear system $\boldsymbol{A}\boldsymbol{x} = \boldsymbol{b}$ of m equations in n unknowns. It is overdetermined, if $m > n$, square, if $m = n$, and underdetermined, if $m < n$. In either case the system can only be solved approximately if $\boldsymbol{b} \notin \mathcal{R}(\boldsymbol{A})$, the column space of $\boldsymbol{A}$. One way to solve $\boldsymbol{A}\boldsymbol{x} = \boldsymbol{b}$ approximately is to select a vector norm $\|\cdot\|$, say a p-norm, and look for $\boldsymbol{x} \in \mathbb{C}^n$ which minimizes $\|\boldsymbol{A}\boldsymbol{x} - \boldsymbol{b}\|$. The use of the one and ∞ norm can be formulated as linear programming problems, while the Euclidian norm leads to a linear system and has applications in statistics. Only this norm is considered here.

Definition 9.1 (Least Squares Problem (LSQ)) Suppose $m, n \in \mathbb{N}$, $\boldsymbol{A} \in \mathbb{C}^{m\times n}$ and $\boldsymbol{b} \in \mathbb{C}^m$. To find $\boldsymbol{x} \in \mathbb{C}^n$ that minimizes $E : \mathbb{C}^n \to \mathbb{R}$ given by

$$E(\boldsymbol{x}) := \|\boldsymbol{A}\boldsymbol{x} - \boldsymbol{b}\|_2^2,$$

is called the **least squares problem**. A minimizer $\boldsymbol{x}$ is called a **least squares solution**.

Since the square root function is monotone, minimizing $E(\boldsymbol{x})$ or $\sqrt{E(\boldsymbol{x})}$ is equivalent.

Example 9.1 (Average) Consider an overdetermined linear system of 3 equations in one unknown

$$\begin{aligned} x_1 &= 1 \\ x_1 &= 1, \\ x_1 &= 2 \end{aligned} \qquad \boldsymbol{A} = \begin{bmatrix} 1 \\ 1 \\ 1 \end{bmatrix}, \qquad \boldsymbol{x} = [x_1], \qquad \boldsymbol{b} = \begin{bmatrix} 1 \\ 1 \\ 2 \end{bmatrix}.$$

T. Lyche, *Numerical Linear Algebra and Matrix Factorizations*,
Texts in Computational Science and Engineering 22,
https://doi.org/10.1007/978-3-030-36468-7_9

To solve this as a least squares problem we compute

$$\|\boldsymbol{A}\boldsymbol{x}-\boldsymbol{b}\|_2^2 = (x_1-1)^2+(x_1-1)^2+(x_1-2)^2 = 3x_1^2-8x_1+6.$$

Setting the first derivative with respect to x_1 equal to zero we obtain $6x_1 - 8 = 0$ or $x_1 = 4/3$, the average of b_1, b_2, b_3. The second derivative is positive and $x_1 = 4/3$ is a global minimum.

We will show below the following results, valid for any $m, n \in \mathbb{N}$, $\boldsymbol{A} \in \mathbb{C}^{m\times n}$ and $\boldsymbol{b} \in \mathbb{C}^n$.

Theorem 9.1 (Existence) *The least squares problem always has a solution.*

Theorem 9.2 (Uniqueness) *The solution of the least squares problem is unique if and only if $\boldsymbol{A}$ has linearly independent columns.*

Theorem 9.3 (Characterization) *$\boldsymbol{x} \in \mathbb{C}^n$ is a solution of the least squares problem if and only if $\boldsymbol{A}^*\boldsymbol{A}\boldsymbol{x} = \boldsymbol{A}^*\boldsymbol{b}$.*

The linear system $\boldsymbol{A}^*\boldsymbol{A}\boldsymbol{x} = \boldsymbol{A}^*\boldsymbol{b}$ is known as the **normal equations**. By Lemma 4.2 the coefficient matrix $\boldsymbol{A}^*\boldsymbol{A}$ is symmetric and positive semidefinite, and it is positive definite if and only if $\boldsymbol{A}$ has linearly independent columns. This is the same condition which guarantees that the least squares problem has a unique solution.

9.1 Examples

Example 9.2 (Linear Regression) We want to fit a straight line $p(t) = x_1 + x_2 t$ to $m \geq 2$ given data $(t_k, y_k) \in \mathbb{R}^2$, $k = 1, \ldots, m$. This is part of the linear regression process in statistics. We obtain the linear system

$$\boldsymbol{A}\boldsymbol{x} = \begin{bmatrix} p(t_1) \\ \vdots \\ p(t_m) \end{bmatrix} = \begin{bmatrix} 1 & t_1 \\ \vdots & \vdots \\ 1 & t_m \end{bmatrix} \begin{bmatrix} x_1 \\ x_2 \end{bmatrix} = \begin{bmatrix} y_1 \\ \vdots \\ y_m \end{bmatrix} = \boldsymbol{b}.$$

This is square for $m = 2$ and overdetermined for $m > 2$. The matrix $\boldsymbol{A}$ has linearly independent columns if and only if the set $\{t_1, \ldots, t_m\}$ of sites contains at least two distinct elements. For if say $t_i \neq t_j$ then

$$c_1 \begin{bmatrix} 1 \\ \vdots \\ 1 \end{bmatrix} + c_2 \begin{bmatrix} t_1 \\ \vdots \\ t_m \end{bmatrix} = \begin{bmatrix} 0 \\ \vdots \\ 0 \end{bmatrix} \Longrightarrow \begin{bmatrix} 1 & t_i \\ 1 & t_j \end{bmatrix} \begin{bmatrix} c_1 \\ c_2 \end{bmatrix} = \begin{bmatrix} 0 \\ 0 \end{bmatrix} \Longrightarrow c_1 = c_2 = 0.$$

Conversely, if $t_1 = \cdots = t_m$ then the columns of $\boldsymbol{A}$ are linearly dependent. The normal equations are

$$\boldsymbol{A}^*\boldsymbol{A}\boldsymbol{x} = \begin{bmatrix} 1 & \cdots & 1 \\ t_1 & \cdots & t_m \end{bmatrix} \begin{bmatrix} 1 & t_1 \\ \vdots & \\ 1 & t_m \end{bmatrix} \begin{bmatrix} x_1 \\ x_2 \end{bmatrix} = \begin{bmatrix} m & \sum t_k \\ \sum t_k & \sum t_k^2 \end{bmatrix} \begin{bmatrix} x_1 \\ x_2 \end{bmatrix},$$

$$= \begin{bmatrix} 1 & \cdots & 1 \\ t_1 & \cdots & t_m \end{bmatrix} \begin{bmatrix} y_1 \\ \vdots \\ y_m \end{bmatrix} = \begin{bmatrix} \sum y_k \\ \sum t_k y_k \end{bmatrix} = \boldsymbol{A}^*\boldsymbol{b},$$

where k ranges from 1 to m in the sums. By what we showed the coefficient matrix is positive semidefinite and positive definite if we have at least two distinct cites. If $m = 2$ and $t_1 \neq t_2$ then both systems $\boldsymbol{A}\boldsymbol{x} = \boldsymbol{b}$ and $\boldsymbol{A}^*\boldsymbol{A}\boldsymbol{x} = \boldsymbol{A}^*\boldsymbol{b}$ are square, and p is the linear interpolant to the data. Indeed, p is linear and $p(t_k) = y_k$, $k = 1, 2$.

With the data

t	1.0	2.0	3.0	4.0
y	3.1	1.8	1.0	0.1

the normal equations become $\begin{bmatrix} 4 & 10 \\ 10 & 30 \end{bmatrix} \begin{bmatrix} x_1 \\ x_2 \end{bmatrix} = \begin{bmatrix} 6 \\ 10.1 \end{bmatrix}$. The data and the least squares polynomial $p(t) = x_1 + x_2 t = 3.95 - 0.98t$ are shown in Fig. 9.1.

Example 9.3 (Input/Output Model) Suppose we have a simple input/output model. To every input $\boldsymbol{u} \in \mathbb{R}^n$ we obtain an output $y \in \mathbb{R}$. Assuming we have a linear relation

$$y = \boldsymbol{u}^*\boldsymbol{x} = \sum_{i=1}^{n} u_i x_i,$$

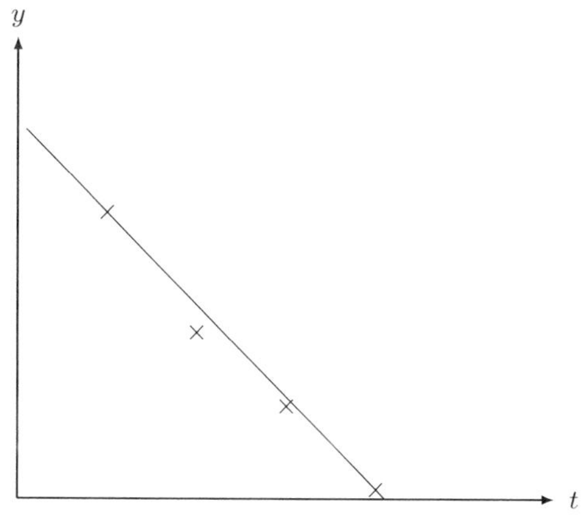

Fig. 9.1 A least squares fit to data

between $\boldsymbol{u}$ and y, how can we determine $\boldsymbol{x}$?

Performing $m \geq n$ experiments we obtain a table of values

$$\begin{array}{c|c|c|c|c} \boldsymbol{u} & \boldsymbol{u}_1 & \boldsymbol{u}_2 & \cdots & \boldsymbol{u}_m \\ \hline y & y_1 & y_2 & \cdots & y_m \end{array}.$$

We would like to find $\boldsymbol{x}$ such that

$$\boldsymbol{A}\boldsymbol{x} = \begin{bmatrix} \boldsymbol{u}_1^* \\ \boldsymbol{u}_2^* \\ \vdots \\ \boldsymbol{u}_m^* \end{bmatrix} \boldsymbol{x} = \begin{bmatrix} y_1 \\ y_2 \\ \vdots \\ y_m \end{bmatrix} = \boldsymbol{b}.$$

We can estimate $\boldsymbol{x}$ by solving the least squares problem $\min\|\boldsymbol{A}\boldsymbol{x} - \boldsymbol{b}\|_2^2$.

9.1.1 Curve Fitting

Given

- size: $1 \leq n \leq m$,
- sites: $\mathcal{S} := \{t_1, t_2, \ldots, t_m\} \subset [a, b]$,
- y-values: $\boldsymbol{y} = [y_1, y_2, \ldots, y_m]^* \in \mathbb{R}^m$,
- functions: $\phi_j : [a, b] \to \mathbb{R}$, $j = 1, \ldots, n$.

Find a function (curve fit) $p : [a, b] \to \mathbb{R}$ given by $p := \sum_{j=1}^n x_j \phi_j$ such that $p(t_k) \approx y_k$ for $k = 1, \ldots, m$.

A solution to the curve fitting problem is found by finding an approximate solution to the following overdetermined set of linear equations

$$\boldsymbol{A}\boldsymbol{x} = \begin{bmatrix} p(t_1) \\ \vdots \\ p(t_m) \end{bmatrix} = \begin{bmatrix} \phi_1(t_1) & \cdots & \phi_n(t_1) \\ \vdots & & \vdots \\ \phi_1(t_m) & \cdots & \phi_n(t_m) \end{bmatrix} \begin{bmatrix} x_1 \\ \vdots \\ x_n \end{bmatrix} = \begin{bmatrix} y_1 \\ \vdots \\ y_m \end{bmatrix} =: \boldsymbol{b}. \tag{9.1}$$

We propose to find $\boldsymbol{x} \in \mathbb{R}^n$ as a solution of the corresponding least squares problem given by

$$E(\boldsymbol{x}) := \|\boldsymbol{A}\boldsymbol{x} - \boldsymbol{b}\|_2^2 = \sum_{k=1}^m \Big(\sum_{j=1}^n x_j \phi_j(t_k) - y_k \Big)^2. \tag{9.2}$$

Typical examples of functions ϕ_j are polynomials, trigonometric functions, exponential functions, or splines.

In (9.2) one can also include **weights** $w_k > 0$ for $k = 1, \ldots, m$ and minimize

$$E(\boldsymbol{x}) := \sum_{k=1}^{m} w_k \Big(\sum_{j=1}^{n} x_j \phi_j(t_k) - y_k \Big)^2.$$

If y_k is an accurate observation, we can choose a large weight w_k. This will force $p(t_k) - y_k$ to be small. Similarly, a small w_k will allow $p(t_k) - y_k$ to be large. If an estimate for the standard deviation δy_k in y_k is known for each k, we can choose $w_k = 1/(\delta y_k)^2$, $k = 1, 2, \ldots, m$. For simplicity we will assume in the following that $w_k = 1$ for all k.

Lemma 9.1 (Curve Fitting) *Let $\boldsymbol{A}$ be given by* (9.1). *The matrix $\boldsymbol{A}^*\boldsymbol{A}$ is symmetric positive definite if and only if*

$$p(t_k) := \sum_{j=1}^{n} x_j \phi_j(t_k) = 0, \quad k = 1, \ldots, m \Rightarrow x_1 = \cdots = x_n = 0. \tag{9.3}$$

Proof By Lemma 4.2 $\boldsymbol{A}^*\boldsymbol{A}$ is positive definite if and only if $\boldsymbol{A}$ has linearly independent columns. Since $(\boldsymbol{A}\boldsymbol{x})_k = \sum_{j=1}^{n} x_j \phi_j(t_k)$, $k = 1, \ldots, m$ this is equivalent to (9.3). □

Example 9.4 (Ill Conditioning and the Hilbert Matrix) The normal equations can be extremely ill-conditioned. Consider the curve fitting problem using the polynomials $\phi_j(t) := t^{j-1}$, for $j = 1, \ldots, n$ and equidistant sites $t_k = (k-1)/(m-1)$ for $k = 1, \ldots, m$. The normal equations are $\boldsymbol{B}_n \boldsymbol{x} = \boldsymbol{c}_n$, where for $n = 3$

$$\boldsymbol{B}_3 \boldsymbol{x} := \begin{bmatrix} m & \sum t_k & \sum t_k^2 \\ \sum t_k & \sum t_k^2 & \sum t_k^3 \\ \sum t_k^2 & \sum t_k^3 & \sum t_k^4 \end{bmatrix} \begin{bmatrix} x_1 \\ x_2 \\ x_3 \end{bmatrix} = \begin{bmatrix} \sum y_k \\ \sum t_k y_k \\ \sum t_k^2 y_k \end{bmatrix}.$$

$\boldsymbol{B}_n$ is positive definite if at least n of the t's are distinct. However $\boldsymbol{B}_n$ is extremely ill-conditioned even for moderate n. Indeed, $\frac{1}{m}\boldsymbol{B}_n \approx \boldsymbol{H}_n$, where $\boldsymbol{H}_n \in \mathbb{R}^{n \times n}$ is the **Hilbert Matrix** with i, j element $1/(i + j - 1)$. Thus, for $n = 3$

$$\boldsymbol{H}_3 = \begin{bmatrix} 1 & \frac{1}{2} & \frac{1}{3} \\ \frac{1}{2} & \frac{1}{3} & \frac{1}{4} \\ \frac{1}{3} & \frac{1}{4} & \frac{1}{5} \end{bmatrix}.$$

The elements of $\frac{1}{m}\boldsymbol{B}_n$ are Riemann sums approximations to the elements of $\boldsymbol{H}_n$. In fact, if $\boldsymbol{B}_n = [b_{i,j}]_{i,j=1}^n$ then

$$\begin{aligned}\frac{1}{m}b_{i,j} &= \frac{1}{m}\sum_{k=1}^{m} t_k^{i+j-2} = \frac{1}{m}\sum_{k=1}^{m}\left(\frac{k-1}{m-1}\right)^{i+j-2} \\ &\approx \int_0^1 x^{i+j-2}dx = \frac{1}{i+j-1} = h_{i,j}.\end{aligned}$$

The elements of $\boldsymbol{H}_n^{-1}$ are determined in Exercise 1.13. We find $K_1(\boldsymbol{H}_6) := \|\boldsymbol{H}_6\|_1\|\boldsymbol{H}_6^{-1}\|_1 \approx 3\cdot 10^7$. It appears that $\frac{1}{m}\boldsymbol{B}_n$ and hence $\boldsymbol{B}_n$ is ill-conditioned for moderate n at least if m is large. The cure for this problem is to use a different basis for polynomials. Orthogonal polynomials are an excellent choice. Another possibility is to use the shifted power basis $(t-\tilde{t})^{j-1}$, $j = 1,\ldots,n$, for a suitable $\tilde{t}$.

9.2 Geometric Least Squares Theory

The least squares problem can be studied as a quadratic minimization problem. In the real case we have

$$E(\boldsymbol{x}) := \|\boldsymbol{A}\boldsymbol{x}-\boldsymbol{b}\|_2^2 = (\boldsymbol{A}\boldsymbol{x}-\boldsymbol{b})^*(\boldsymbol{A}\boldsymbol{x}-\boldsymbol{b}) = \boldsymbol{x}^*\boldsymbol{A}^*\boldsymbol{A}\boldsymbol{x} - 2\boldsymbol{x}^*\boldsymbol{A}^*\boldsymbol{b} + \boldsymbol{b}^*\boldsymbol{b}.$$

Minimization of a quadratic function like $E(\boldsymbol{x})$ will be considered in Chap. 13. Here we consider a geometric approach based on orthogonal sums of subspaces, cf. Sect. 5.1.3.

With the usual inner product $\langle \boldsymbol{x}, \boldsymbol{y}\rangle = \boldsymbol{y}^*\boldsymbol{x}$, orthogonal sums and projections we can prove the existence, uniqueness and characterization theorems for least squares problems. For $\boldsymbol{A} \in \mathbb{C}^{m\times n}$ we consider the column space $\mathcal{S} := \mathcal{R}(\boldsymbol{A})$ of $\boldsymbol{A}$ and the null space $\mathcal{T} := \mathcal{N}(\boldsymbol{A}^*)$ of $\boldsymbol{A}^*$. These are subspaces of $\mathbb{C}^m$ and by Theorem 7.3 we have the orthogonal sum

$$\mathbb{C}^m = \mathcal{R}(\boldsymbol{A})\overset{\perp}{\oplus}\mathcal{N}(\boldsymbol{A}^*). \tag{9.4}$$

Proof of Theorem 9.1 It follows from (9.4) that any $\boldsymbol{b} \in \mathbb{C}^m$ can be decomposed uniquely as $\boldsymbol{b} = \boldsymbol{b}_1 + \boldsymbol{b}_2$, where $\boldsymbol{b}_1$ is the orthogonal projection of $\boldsymbol{b}$ into $\mathcal{R}(\boldsymbol{A})$ and $\boldsymbol{b}_2$ is the orthogonal projection of $\boldsymbol{b}$ into $\mathcal{N}(\boldsymbol{A}^*)$. Suppose $\boldsymbol{x} \in \mathbb{C}^n$. Clearly $\boldsymbol{A}\boldsymbol{x}-\boldsymbol{b}_1 \in \mathcal{R}(\boldsymbol{A})$ since it is a subspace and $\boldsymbol{b}_2 \in \mathcal{N}(\boldsymbol{A}^*)$. But then $\langle \boldsymbol{A}\boldsymbol{x}-\boldsymbol{b}_1, \boldsymbol{b}_2\rangle = 0$ and by Pythagoras

$$\|\boldsymbol{A}\boldsymbol{x}-\boldsymbol{b}\|_2^2 = \|(\boldsymbol{A}\boldsymbol{x}-\boldsymbol{b}_1)-\boldsymbol{b}_2\|_2^2 = \|\boldsymbol{A}\boldsymbol{x}-\boldsymbol{b}_1\|_2^2 + \|\boldsymbol{b}_2\|_2^2 \geq \|\boldsymbol{b}_2\|_2^2$$

with equality if and only if $Ax = b_1$. It follows that the set of all least squares solutions is

$$\{x \in \mathbb{C}^n : Ax = b_1\}. \tag{9.5}$$

This set is nonempty since $b_1 \in \mathcal{R}(A)$. □

Proof of Theorem 9.2 The set (9.5) contains exactly one element if and only if A has linearly independent columns. □

Proof of Theorem 9.3 If x solves the least squares problem then $Ax - b_1 = 0$ and it follows that $A^*(Ax - b) = A^*(Ax - b_1) = 0$ since $b_2 \in \mathcal{N}(A^*)$. This shows that the normal equations hold. Conversely, if $A^*Ax = A^*b$ then $A^*b_2 = 0$ implies that $A^*(Ax - b_1) = 0$. But then $Ax - b_1 \in \mathcal{R}(A) \cap \mathcal{N}(A^*)$ showing that $Ax - b_1 = 0$, and x is a least squares solution. □

9.3 Numerical Solution

We assume that $m \geq n$, $A \in \mathbb{C}^{m \times n}$, $b \in \mathbb{C}^m$. We consider numerical methods based on normal equations, QR factorization, or Singular Value Factorization. For more see [2]. We discuss the first two approaches in this section. Another possibility is to use an iterative method like the conjugate gradient method (cf. Exercise 13.10).

9.3.1 Normal Equations

We assume that $\operatorname{rank}(A) = n$, i.e., A has linearly independent columns. The coefficient matrix $B := A^*A$ in the normal equations is positive definite, and we can solve these equations using the Cholesky factorization of B. Consider forming the normal equations. We can use either a column oriented (inner product)- or a row oriented (outer product) approach.

1. inner product: $(A^*A)_{i,j} = \sum_{k=1}^{m} \overline{a}_{k,i} a_{k,j}$, $i, j = 1, \ldots, n$,
   $(A^*b)_i = \sum_{k=1}^{m} \overline{a}_{k,i} b_k$, $i = 1, \ldots, n$,
2. outer product: $A^*A = \sum_{k=1}^{m} \begin{bmatrix} \overline{a}_{k,1} \\ \vdots \\ \overline{a}_{k,n} \end{bmatrix} [a_{k1} \cdots a_{kn}]$, $A^*b = \sum_{k=1}^{m} \begin{bmatrix} \overline{a}_{k,1} \\ \vdots \\ \overline{a}_{k,n} \end{bmatrix} b_k$.

The outer product form is suitable for large problems since it uses only one pass through the data importing one row of A at a time from some separate storage.

Consider the number of operations to find the least squares solution for real data. We need $2m$ arithmetic operations for each inner product. Since B is symmetric we only need to compute $n(n+1)/2$ such inner products. It follows that B can be computed in approximately mn^2 arithmetic operations. In conclusion the number of operations are mn^2 to find $\boldsymbol{B}$, $2mn$ to find $\boldsymbol{c} := \boldsymbol{A}^*\boldsymbol{b}$, $n^3/3$ to find $\boldsymbol{L}$ such that $\boldsymbol{B} = \boldsymbol{L}\boldsymbol{L}^*$, n^2 to solve $\boldsymbol{L}\boldsymbol{y} = \boldsymbol{c}$ and n^2 to solve $\boldsymbol{L}^*\boldsymbol{x} = \boldsymbol{y}$. If $m \approx n$ it takes $\frac{4}{3}n^3 = 2G_n$ arithmetic operations. If m is much bigger than n the number of operations is approximately mn^2, the work to compute $\boldsymbol{B}$.

Conditioning of $\boldsymbol{A}$ can be a problem with the normal equation approach. We have

Theorem 9.4 (Spectral Condition Number of A^*A) *Suppose $1 \le n \le m$ and that $\boldsymbol{A} \in \mathbb{C}^{m\times n}$ has linearly independent columns. Then*

$$K_2(\boldsymbol{A}^*\boldsymbol{A}) := \|\boldsymbol{A}^*\boldsymbol{A}\|_2\|(\boldsymbol{A}^*\boldsymbol{A})^{-1}\|_2 = \frac{\lambda_1}{\lambda_n} = \frac{\sigma_1^2}{\sigma_n^2} = K_2(\boldsymbol{A})^2, \tag{9.6}$$

where $\lambda_1 \ge \cdots \ge \lambda_n > 0$ are the eigenvalues of $\boldsymbol{A}^*\boldsymbol{A}$, and $\sigma_1 \ge \cdots \ge \sigma_n > 0$ are the singular values of $\boldsymbol{A}$.

Proof Since $\boldsymbol{A}^*\boldsymbol{A}$ is Hermitian it follows from Theorem 8.10 that $K_2(\boldsymbol{A}) = \frac{\sigma_1}{\sigma_n}$ and $K_2(\boldsymbol{A}^*\boldsymbol{A}) = \frac{\lambda_1}{\lambda_n}$. But $\lambda_i = \sigma_i^2$ by Theorem 7.2 and the proof is complete. □

It follows from Theorem 9.4 that the 2-norm condition number of $\boldsymbol{B} := \boldsymbol{A}^*\boldsymbol{A}$ is the square of the condition number of $\boldsymbol{A}$ and therefore can be quite large even if $\boldsymbol{A}$ is only mildly ill-conditioned. Another difficulty which can be encountered is that the computed $\boldsymbol{A}^*\boldsymbol{A}$ might not be positive definite. See Problem 9.21 for an example.

9.3.2 QR Factorization

The QR factorization can be used to solve the least squares problem. We assume that $\text{rank}(\boldsymbol{A}) = n$, i.e., $\boldsymbol{A}$ has linearly independent columns. Suppose $\boldsymbol{A} = \boldsymbol{Q}_1\boldsymbol{R}_1$ is a QR factorization of $\boldsymbol{A}$. Since $\boldsymbol{Q}_1 \in \mathbb{C}^{m\times n}$ has orthonormal columns we find

$$\boldsymbol{A}^*\boldsymbol{A} = \boldsymbol{R}_1^*\boldsymbol{Q}_1^*\boldsymbol{Q}_1\boldsymbol{R}_1 = \boldsymbol{R}_1^*\boldsymbol{R}_1, \quad \boldsymbol{A}^*\boldsymbol{b} = \boldsymbol{R}_1^*\boldsymbol{Q}_1^*\boldsymbol{b}.$$

Since $\boldsymbol{A}$ has rank n the matrix $\boldsymbol{R}_1^*$ is nonsingular and can be canceled. Thus

$$\boldsymbol{A}^*\boldsymbol{A}\boldsymbol{x} = \boldsymbol{A}^*\boldsymbol{b} \implies \boldsymbol{R}_1\boldsymbol{x} = \boldsymbol{c}_1, \quad \boldsymbol{c}_1 := \boldsymbol{Q}_1^*\boldsymbol{b}.$$

We can use Householder transformations or Givens rotations to find $\boldsymbol{R}_1$ and $\boldsymbol{c}_1$. Consider using the Householder triangulation algorithm Algorithm 5.2. We find $\boldsymbol{R} = \boldsymbol{Q}^*\boldsymbol{A}$ and $\boldsymbol{c} = \boldsymbol{Q}^*\boldsymbol{b}$, where $\boldsymbol{A} = \boldsymbol{Q}\boldsymbol{R}$ is the QR decomposition of $\boldsymbol{A}$. The matrices $\boldsymbol{R}_1$ and $\boldsymbol{c}_1$ are located in the first n rows of $\boldsymbol{R}$ and $\boldsymbol{c}$. Using also Algorithm 3.2 we have the following method to solve the full rank least squares problem.

1. `[R,c]=housetriang(A,b)`.
2. `x=rbacksolve(R(1:n,1:n),c(1:n),n)`.

Example 9.5 (Solution Using QR Factorization) Consider the least squares problem with

$$A = \begin{bmatrix} 1 & 3 & 1 \\ 1 & 3 & 7 \\ 1 & -1 & -4 \\ 1 & -1 & 2 \end{bmatrix} \text{ and } \boldsymbol{b} = \begin{bmatrix} 1 \\ 1 \\ 1 \\ 1 \end{bmatrix}.$$

This is the matrix in Example 5.2. The least squares solution $\boldsymbol{x}$ is found by solving the system

$$\begin{bmatrix} 2 & 2 & 3 \\ 0 & 4 & 5 \\ 0 & 0 & 6 \end{bmatrix} \begin{bmatrix} x_1 \\ x_2 \\ x_3 \end{bmatrix} = \frac{1}{2} \begin{bmatrix} 1 & 1 & 1 & 1 \\ 1 & 1 & -1 & -1 \\ -1 & 1 & -1 & 1 \end{bmatrix} \times \begin{bmatrix} 1 \\ 1 \\ 1 \\ 1 \end{bmatrix} = \begin{bmatrix} 2 \\ 0 \\ 0 \end{bmatrix},$$

and we find $\boldsymbol{x} = [1, 0, 0]^*$.

Using Householder triangulation is a useful alternative to normal equations for solving full rank least squares problems. It can even be extended to rank deficient problems, see [2]. The 2 norm condition number for the system $\boldsymbol{R}_1\boldsymbol{x} = \boldsymbol{c}_1$ is $K_2(\boldsymbol{R}_1) = K_2(\boldsymbol{Q}_1\boldsymbol{R}_1) = K_2(\boldsymbol{A})$, and as discussed in the previous section this is the square root of $K_2(\boldsymbol{A}^*\boldsymbol{A})$, the condition number for the normal equations. Thus if $\boldsymbol{A}$ is mildly ill-conditioned the normal equations can be quite ill-conditioned and solving the normal equations can give inaccurate results. On the other hand Algorithm 5.2 is quite stable.

But using Householder transformations requires more work. The leading term in the number of arithmetic operations in Algorithm 5.2 is approximately $2mn^2 - 2n^3/3$, (cf. (5.16)) while the number of arithmetic operations needed to form the normal equations, taking advantage of symmetry is approximately mn^2. Thus for m much larger than n using Householder triangulation requires twice as many arithmetic operations as the approach based on the normal equations. Also, Householder triangulation have problems taking advantage of the structure in sparse problems.

Using MATLAB a least squares solution can be found using `x=A\b` if $\boldsymbol{A}$ has full rank.For rank deficient problems the function `x=lscov(A,b)` finds a least squares solution with a maximal number of zeros in $\boldsymbol{x}$.

9.3.3 Singular Value Decomposition, Generalized Inverses and Least Squares

Further insight into the least squares problem can be obtained by considering a singular value decomposition of $\boldsymbol{A}$ and the corresponding singular value factorization.

If $\boldsymbol{A}$ has rank r then

$$\boldsymbol{A} = \boldsymbol{U}\boldsymbol{\Sigma}\boldsymbol{V}^* = [\boldsymbol{U}_1, \boldsymbol{U}_2]\begin{bmatrix}\Sigma_1 & \mathbf{0}\\ \mathbf{0} & \mathbf{0}\end{bmatrix}\begin{bmatrix}\boldsymbol{V}_1^*\\ \boldsymbol{V}_2^*\end{bmatrix} = \boldsymbol{U}_1\boldsymbol{\Sigma}_1\boldsymbol{V}_1^*, \quad \Sigma_1 = \operatorname{diag}(\sigma_1, \ldots, \sigma_r), \tag{9.7}$$

where

$$\begin{aligned}
\boldsymbol{U}_1 &= [\boldsymbol{u}_1, \ldots, \boldsymbol{u}_r],\ \boldsymbol{U}_2 = [\boldsymbol{u}_{r+1}, \ldots, \boldsymbol{u}_m], \quad \boldsymbol{U}_1^*\boldsymbol{U}_1 = \boldsymbol{I},\ \boldsymbol{U}_2^*\boldsymbol{U}_2 = \boldsymbol{I},\\
\boldsymbol{V}_1 &= [\boldsymbol{v}_1, \ldots, \boldsymbol{v}_r],\ \boldsymbol{V}_2 = [\boldsymbol{v}_{r+1}, \ldots, \boldsymbol{v}_n], \quad \boldsymbol{V}_1^*\boldsymbol{V}_1 = \boldsymbol{I},\ \boldsymbol{V}_2^*\boldsymbol{V}_2 = \boldsymbol{I}
\end{aligned}$$

and $\sigma_1 \geq \cdots \geq \sigma_r > 0$. We recall (cf. Theorem 7.3)

- the set of columns of $\boldsymbol{U}_1$ is an orthonormal basis for $\mathcal{R}(\boldsymbol{A})$,
- the set of columns of $\boldsymbol{U}_2$ is an orthonormal basis for $\mathcal{N}(\boldsymbol{A}^*)$,
- the set of columns of $\boldsymbol{V}_1$ is an orthonormal basis for $\mathcal{R}(\boldsymbol{A}^*)$,
- the set of columns of $\boldsymbol{V}_2$ is an orthonormal basis for $\mathcal{N}(\boldsymbol{A})$.

The concept of the inverse of a matrix can be generalized to any rectangular matrix.

Theorem 9.5 (The Generalized Inverse) *For any $m, n \in \mathbb{N}$ and any $\boldsymbol{A} \in \mathbb{C}^{m\times n}$ there is a unique matrix $\boldsymbol{A}^\dagger \in \mathbb{C}^{n\times m}$ such that*

$$\boldsymbol{A}\boldsymbol{A}^\dagger\boldsymbol{A} = \boldsymbol{A},\ \boldsymbol{A}^\dagger\boldsymbol{A}\boldsymbol{A}^\dagger = \boldsymbol{A}^\dagger,\ (\boldsymbol{A}^\dagger\boldsymbol{A})^* = \boldsymbol{A}^\dagger\boldsymbol{A},\ (\boldsymbol{A}\boldsymbol{A}^\dagger)^* = \boldsymbol{A}\boldsymbol{A}^\dagger. \tag{9.8}$$

If $\boldsymbol{U}_1\boldsymbol{\Sigma}_1\boldsymbol{V}_1^*$ is a singular value factorization of $\boldsymbol{A}$ then

$$\boldsymbol{A}^\dagger = \boldsymbol{V}_1\boldsymbol{\Sigma}_1^{-1}\boldsymbol{U}_1^*. \tag{9.9}$$

Proof For existence we show that the matrices

$$\boldsymbol{A} = \boldsymbol{U}_1\boldsymbol{\Sigma}_1\boldsymbol{V}_1^*, \quad \boldsymbol{A}^\dagger := \boldsymbol{V}_1\boldsymbol{\Sigma}_1^{-1}\boldsymbol{U}_1^*$$

satisfies (9.8). Since $\boldsymbol{U}_1$ and $\boldsymbol{V}_1$ have orthonormal columns we find

$$\begin{aligned}
\boldsymbol{A}^\dagger\boldsymbol{A} &= \boldsymbol{V}_1\boldsymbol{\Sigma}_1^{-1}\boldsymbol{U}_1^*\boldsymbol{U}_1\boldsymbol{\Sigma}_1\boldsymbol{V}_1^* = \boldsymbol{V}_1\boldsymbol{V}_1^*,\\
\boldsymbol{A}\boldsymbol{A}^\dagger &= \boldsymbol{U}_1\boldsymbol{\Sigma}_1\boldsymbol{V}_1^*\boldsymbol{V}_1\boldsymbol{\Sigma}_1^{-1}\boldsymbol{U}_1^* = \boldsymbol{U}_1\boldsymbol{U}_1^*.
\end{aligned} \tag{9.10}$$

A similar calculation shows that $(\boldsymbol{A}^\dagger\boldsymbol{A})^* = \boldsymbol{V}_1\boldsymbol{V}_1^*$ and $(\boldsymbol{A}\boldsymbol{A}^\dagger)^* = \boldsymbol{U}_1\boldsymbol{U}_1^*$ showing that $\boldsymbol{A}^\dagger\boldsymbol{A}$ and $\boldsymbol{A}\boldsymbol{A}^\dagger$ are Hermitian. Moreover, by (9.10)

$$\begin{aligned}
\boldsymbol{A}\boldsymbol{A}^\dagger\boldsymbol{A} &= \boldsymbol{U}_1\boldsymbol{\Sigma}_1\boldsymbol{V}_1^*\boldsymbol{V}_1\boldsymbol{V}_1^* = \boldsymbol{U}_1\boldsymbol{\Sigma}_1\boldsymbol{V}_1^* = \boldsymbol{A}\\
\boldsymbol{A}^\dagger\boldsymbol{A}\boldsymbol{A}^\dagger &= \boldsymbol{V}_1\boldsymbol{\Sigma}_1^{-1}\boldsymbol{U}_1^*\boldsymbol{U}_1\boldsymbol{U}_1^* = \boldsymbol{V}_1\Sigma^{-1}\boldsymbol{U}_1^* = \boldsymbol{A}^\dagger.
\end{aligned}$$

Thus (9.8) follows.

That there is only one matrix $\boldsymbol{A}^\dagger \in \mathbb{C}^{n\times m}$ satisfying (9.8) is shown in Exercise 9.5. □

The matrix $\boldsymbol{A}^\dagger$ is called the **generalized inverse** of $\boldsymbol{A}$. We note that

1. If $\boldsymbol{A}$ is square and nonsingular then $\boldsymbol{A}^{-1}$ satisfies (9.8) so that $\boldsymbol{A}^{-1} = \boldsymbol{A}^\dagger$. Indeed, $\boldsymbol{A}^{-1}\boldsymbol{A} = \boldsymbol{A}\boldsymbol{A}^{-1} = \boldsymbol{I}$ implies that $\boldsymbol{A}^{-1}\boldsymbol{A}$ and $\boldsymbol{A}\boldsymbol{A}^{-1}$ are Hermitian. Moreover, $\boldsymbol{A}\boldsymbol{A}^{-1}\boldsymbol{A} = \boldsymbol{A}$, $\boldsymbol{A}^{-1}\boldsymbol{A}\boldsymbol{A}^{-1} = \boldsymbol{A}^{-1}$. By uniqueness $\boldsymbol{A}^{-1} = \boldsymbol{A}^\dagger$.
2. We show in Exercise 9.7 that if $\boldsymbol{A}$ has linearly independent columns then

$$\boldsymbol{A}^\dagger = (\boldsymbol{A}^*\boldsymbol{A})^{-1}\boldsymbol{A}^*. \tag{9.11}$$

For further properties and examples of the generalized inverse see the exercises.

Orthogonal projections can be expressed in terms of generalized inverses and singular vectors.

Theorem 9.6 (Orthogonal Projections) *Given $m, n \in \mathbb{N}$, $\boldsymbol{A} \in \mathbb{C}^{m\times n}$ of rank r, and let $\mathcal{S}$ be one of the subspaces $\mathcal{R}(\boldsymbol{A})$, $\mathcal{N}(\boldsymbol{A}^*)$. The orthogonal projection of $\boldsymbol{v} \in \mathbb{C}^m$ into $\mathcal{S}$ can be written as a matrix $\boldsymbol{P}_\mathcal{S}$ times the vector $\boldsymbol{v}$ in the form $\boldsymbol{P}_\mathcal{S}\boldsymbol{v}$, where*

$$\begin{aligned} \boldsymbol{P}_{\mathcal{R}(\boldsymbol{A})} &= \boldsymbol{A}\boldsymbol{A}^\dagger = \boldsymbol{U}_1\boldsymbol{U}_1^* = \sum_{j=1}^{r} \boldsymbol{u}_j\boldsymbol{u}_j^* \in \mathbb{C}^{m\times m}, \\ \boldsymbol{P}_{\mathcal{N}(\boldsymbol{A}^*)} &= \boldsymbol{I} - \boldsymbol{A}\boldsymbol{A}^\dagger = \boldsymbol{U}_2\boldsymbol{U}_2^* = \sum_{j=r+1}^{m} \boldsymbol{u}_j\boldsymbol{u}_j^* \in \mathbb{C}^{m\times m}. \end{aligned} \tag{9.12}$$

where $\boldsymbol{A}^\dagger$ is the generalized inverse of $\boldsymbol{A}$, and $\boldsymbol{A} = \boldsymbol{U}\boldsymbol{\Sigma}\boldsymbol{V}^* \in \mathbb{C}^{m\times n}$ is a singular value decomposition of $\boldsymbol{A}$ (cf. (9.7)).

Proof By block multiplication we have for any $\boldsymbol{v} \in \mathbb{C}^m$

$$\boldsymbol{v} = \boldsymbol{U}\boldsymbol{U}^*\boldsymbol{v} = [\boldsymbol{U}_1, \boldsymbol{U}_2\begin{bmatrix}\boldsymbol{U}_1^*\\ \boldsymbol{U}_2^*\end{bmatrix}\boldsymbol{v} = \boldsymbol{s} + \boldsymbol{t},$$

where $\boldsymbol{s} = \boldsymbol{U}_1\boldsymbol{U}_1^*\boldsymbol{v} \in \mathcal{R}(\boldsymbol{A})$ and $\boldsymbol{t} = \boldsymbol{U}_2\boldsymbol{U}_2^*\boldsymbol{v} \in \mathcal{N}(\boldsymbol{A}^*)$. By uniqueness and (9.10) we obtain the first equation in (9.12). Since $\boldsymbol{v} = (\boldsymbol{U}_1\boldsymbol{U}_1^* + \boldsymbol{U}_2\boldsymbol{U}_2^*)\boldsymbol{v}$ for any $\boldsymbol{v} \in \mathbb{C}^m$ we have $\boldsymbol{U}_1\boldsymbol{U}_1^* + \boldsymbol{U}_2\boldsymbol{U}_2^* = \boldsymbol{I}$, and hence $\boldsymbol{U}_2\boldsymbol{U}_2^* = \boldsymbol{I} - \boldsymbol{U}_1\boldsymbol{U}_1^* = \boldsymbol{I} - \boldsymbol{A}\boldsymbol{A}^\dagger$ and the second equation in (9.12) follows. □

Corollary 9.1 (LSQ Characterization Using Generalized Inverse) $\boldsymbol{x} \in \mathbb{C}^n$ *solves the least squares problem $\min_x \|\boldsymbol{A}\boldsymbol{x} - b\|_2^2$ if and only if $\boldsymbol{x} = \boldsymbol{A}^\dagger\boldsymbol{b} + \boldsymbol{z}$, where $\boldsymbol{A}^\dagger$ is the generalized inverse of $\boldsymbol{A}$ and $\boldsymbol{z} \in \mathcal{N}(\boldsymbol{A})$.*

Proof It follows from Theorem 9.6 that $\boldsymbol{b}_1 := \boldsymbol{A}\boldsymbol{A}^\dagger\boldsymbol{b}$ is the orthogonal projection of $\boldsymbol{b} \in \mathbb{C}^m$ into $\mathcal{R}(\boldsymbol{A})$. Moreover, (9.5) implies that $\boldsymbol{x}$ is a least squares solution if and only if $\boldsymbol{A}\boldsymbol{x} = \boldsymbol{b}_1$.

Let $\boldsymbol{x}$ be a least squares solution, i.e., $\boldsymbol{A}\boldsymbol{x} = \boldsymbol{b}_1$. If $\boldsymbol{z} := \boldsymbol{x} - \boldsymbol{A}^\dagger\boldsymbol{b}$ then $\boldsymbol{A}\boldsymbol{z} = \boldsymbol{A}\boldsymbol{x} - \boldsymbol{A}\boldsymbol{A}^\dagger\boldsymbol{b} = \boldsymbol{b}_1 - \boldsymbol{b}_1 = \boldsymbol{0}$ and $\boldsymbol{x} = \boldsymbol{A}^\dagger\boldsymbol{b} + \boldsymbol{z}$.

Conversely, if $\boldsymbol{x} = \boldsymbol{A}^\dagger\boldsymbol{b} + \boldsymbol{z}$ with $\boldsymbol{A}\boldsymbol{z} = \boldsymbol{0}$ then $\boldsymbol{A}\boldsymbol{x} = \boldsymbol{A}(\boldsymbol{A}^\dagger\boldsymbol{b} + \boldsymbol{z}) = \boldsymbol{b}_1$ and $\boldsymbol{x}$ is a least squares solution. □

The least squares solution $\boldsymbol{A}^\dagger\boldsymbol{b}$ has an interesting property.

Theorem 9.7 (Minimal Norm Solution) *The least squares solution with minimal Euclidian norm is* $\boldsymbol{x} = \boldsymbol{A}^\dagger\boldsymbol{b}$ *corresponding to* $\boldsymbol{z} = \boldsymbol{0}$.

Proof Consider a singular value decomposition of $\boldsymbol{A}$ using the notation in (9.7). Suppose $\boldsymbol{x} = \boldsymbol{A}^\dagger\boldsymbol{b} + \boldsymbol{z}$, with $\boldsymbol{z} \in \mathcal{N}(\boldsymbol{A})$. Since the columns of $\boldsymbol{V}_2$ form a basis for $\mathcal{N}(\boldsymbol{A})$ we have $\boldsymbol{z} = \boldsymbol{V}_2\boldsymbol{y}$ for some $\boldsymbol{y}$. Moreover, $\boldsymbol{V}_2^*\boldsymbol{V}_1 = \boldsymbol{0}$ since $\boldsymbol{V}$ has orthonormal columns. But then $\boldsymbol{z}^*\boldsymbol{A}^\dagger\boldsymbol{b} = \boldsymbol{y}^*\boldsymbol{V}_2^*\boldsymbol{V}_1\boldsymbol{\Sigma}_1^{-1}\boldsymbol{U}_1^*\boldsymbol{b} = \boldsymbol{0}$. Thus $\boldsymbol{z}$ and $\boldsymbol{A}^\dagger\boldsymbol{b}$ are orthogonal so that by Pythagoras $\|\boldsymbol{x}\|_2^2 = \|\boldsymbol{A}^\dagger\boldsymbol{b} + \boldsymbol{z}\|_2^2 = \|\boldsymbol{A}^\dagger\boldsymbol{b}\|_2^2 + \|\boldsymbol{z}\|_2^2 \geq \|\boldsymbol{A}^\dagger\boldsymbol{b}\|_2^2$ with equality for $\boldsymbol{z} = \boldsymbol{0}$. □

Example 9.6 (Rank Deficient Least Squares Solution) Consider the least squares problem with $\boldsymbol{A} = \begin{bmatrix} 1 & 1 \\ 1 & 1 \end{bmatrix}$ and $\boldsymbol{b} = [1, 1]^*$. The singular value factorization, $\boldsymbol{A}^\dagger$ and $\boldsymbol{A}^\dagger\boldsymbol{b}$ are given by

$$\boldsymbol{A} := \frac{1}{\sqrt{2}}\begin{bmatrix} 1 \\ 1 \end{bmatrix}[2]\frac{1}{\sqrt{2}}\begin{bmatrix} 1 & 1 \end{bmatrix}, \quad \boldsymbol{A}^\dagger = \frac{1}{\sqrt{2}}\begin{bmatrix} 1 \\ 1 \end{bmatrix}\begin{bmatrix} \frac{1}{2} \end{bmatrix}\frac{1}{\sqrt{2}}\begin{bmatrix} 1 & 1 \end{bmatrix} = \frac{1}{4}\boldsymbol{A}, \quad \boldsymbol{A}^\dagger\boldsymbol{b} = \begin{bmatrix} 1/2 \\ 1/2 \end{bmatrix}.$$

Using Corollary 9.1 we find the general solution $[1/2, 1/2] + [a, -a]$ for any $a \in \mathbb{C}$. The MATLAB function `lscov` gives the solution $[1, 0]^*$ corresponding to $a = 1/2$, while the minimal norm solution is $[1/2, 1/2]$ obtained for $a = 0$.

9.4 Perturbation Theory for Least Squares

In this section we consider what effect small changes in the data $\boldsymbol{A}, \boldsymbol{b}$ have on the solution $\boldsymbol{x}$ of the least squares problem $\min\|\boldsymbol{A}\boldsymbol{x} - \boldsymbol{b}\|_2$.

If $\boldsymbol{A}$ has linearly independent columns then we can write the least squares solution $\boldsymbol{x}$ (the solution of $\boldsymbol{A}^*\boldsymbol{A}\boldsymbol{x} = \boldsymbol{A}^*\boldsymbol{b}$) as (cf. Exercise 9.7)

$$\boldsymbol{x} = \boldsymbol{A}^\dagger\boldsymbol{b} = \boldsymbol{A}^\dagger\boldsymbol{b}_1, \quad \boldsymbol{A}^\dagger = (\boldsymbol{A}^*\boldsymbol{A})^{-1}\boldsymbol{A}^*,$$

where $\boldsymbol{b}_1$ is the orthogonal projection of $\boldsymbol{b}$ into the column space $\mathcal{R}(\boldsymbol{A})$.

9.4.1 Perturbing the Right Hand Side

Let us now consider the effect of a perturbation in $\boldsymbol{b}$ on $\boldsymbol{x}$.

Theorem 9.8 (Perturbing the Right Hand Side) *Suppose $\boldsymbol{A} \in \mathbb{C}^{m\times n}$ has linearly independent columns, and let $\boldsymbol{b}, \boldsymbol{e} \in \mathbb{C}^m$. Let $\boldsymbol{x}, \boldsymbol{y} \in \mathbb{C}^n$ be the solutions of* $\min\|\boldsymbol{Ax} - \boldsymbol{b}\|_2$ *and* $\min\|\boldsymbol{Ay} - \boldsymbol{b} - \boldsymbol{e}\|_2$. *Finally, let $\boldsymbol{b}_1$, $\boldsymbol{e}_1$ be the orthogonal projections of $\boldsymbol{b}$ and $\boldsymbol{e}$ into $\mathcal{R}(\boldsymbol{A})$. If $\boldsymbol{b}_1 \neq \boldsymbol{0}$, we have for any operator norm*

$$\frac{1}{K(\boldsymbol{A})}\frac{\|\boldsymbol{e}_1\|}{\|\boldsymbol{b}_1\|} \le \frac{\|\boldsymbol{y}-\boldsymbol{x}\|}{\|\boldsymbol{x}\|} \le K(\boldsymbol{A})\frac{\|\boldsymbol{e}_1\|}{\|\boldsymbol{b}_1\|}, \qquad K(\boldsymbol{A}) = \|\boldsymbol{A}\|\|\boldsymbol{A}^\dagger\|. \tag{9.13}$$

Proof Subtracting $\boldsymbol{x} = \boldsymbol{A}^\dagger\boldsymbol{b}_1$ from $\boldsymbol{y} = \boldsymbol{A}^\dagger\boldsymbol{b}_1 + \boldsymbol{A}^\dagger\boldsymbol{e}_1$ we have $\boldsymbol{y} - \boldsymbol{x} = \boldsymbol{A}^\dagger\boldsymbol{e}_1$. Thus $\|\boldsymbol{y}-\boldsymbol{x}\| = \|\boldsymbol{A}^\dagger\boldsymbol{e}_1\| \le \|\boldsymbol{A}^\dagger\|\|\boldsymbol{e}_1\|$. Moreover, $\|\boldsymbol{b}_1\| = \|\boldsymbol{Ax}\| \le \|\boldsymbol{A}\|\|\boldsymbol{x}\|$. Therefore $\|\boldsymbol{y}-\boldsymbol{x}\|/\|\boldsymbol{x}\| \le \|\boldsymbol{A}\|\|\boldsymbol{A}^\dagger\|\|\boldsymbol{e}_1\|/\|\boldsymbol{b}_1\|$ proving the rightmost inequality. From $\boldsymbol{A}(\boldsymbol{x}-\boldsymbol{y}) = \boldsymbol{e}_1$ and $\boldsymbol{x} = \boldsymbol{A}^\dagger\boldsymbol{b}_1$ we obtain the leftmost inequality. □

Equation (9.13) is analogous to the bound (8.22) for linear systems. We see that the number $K(\boldsymbol{A}) = \|\boldsymbol{A}\|\|\boldsymbol{A}^\dagger\|$ generalizes the condition number $\|\boldsymbol{A}\|\|\boldsymbol{A}^{-1}\|$ for a square matrix. The main difference between (9.13) and (8.22) is however that $\|\boldsymbol{e}\|/\|\boldsymbol{b}\|$ in (8.22) has been replaced by $\|\boldsymbol{e}_1\|/\|\boldsymbol{b}_1\|$, the orthogonal projections of $\boldsymbol{e}$ and $\boldsymbol{b}$ into $\mathcal{R}(\boldsymbol{A})$. If $\boldsymbol{b}$ lies almost entirely in $\mathcal{N}(\boldsymbol{A}^*)$, i.e. $\|\boldsymbol{b}\|/\|\boldsymbol{b}_1\|$ is large, then $\|\boldsymbol{e}_1\|/\|\boldsymbol{b}_1\|$ can be much larger than $\|\boldsymbol{e}\|/\|\boldsymbol{b}\|$. This is illustrated in Fig. 9.2. If $\boldsymbol{b}$ is almost orthogonal to $\mathcal{R}(\boldsymbol{A})$, then $\|\boldsymbol{e}_1\|/\|\boldsymbol{b}_1\|$ will normally be much larger than $\|\boldsymbol{e}\|/\|\boldsymbol{b}\|$.

Example 9.7 (Perturbing the Right Hand Side) Suppose

$$\boldsymbol{A} = \begin{bmatrix} 1 & 1 \\ 0 & 1 \\ 0 & 0 \end{bmatrix}, \quad \boldsymbol{b} = \begin{bmatrix} 10^{-4} \\ 0 \\ 1 \end{bmatrix}, \quad \boldsymbol{e} = \begin{bmatrix} 10^{-6} \\ 0 \\ 0 \end{bmatrix}.$$

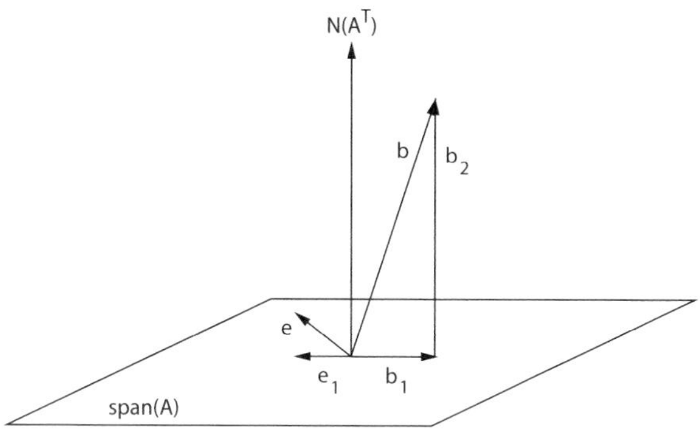

Fig. 9.2 Graphical interpretation of the bounds in Theorem 9.8

For this example we can compute $K(\boldsymbol{A})$ by finding $\boldsymbol{A}^\dagger$ explicitly. Indeed,

$$\boldsymbol{A}^*\boldsymbol{A} = \begin{bmatrix} 1 & 1 \\ 1 & 2 \end{bmatrix}, \quad (\boldsymbol{A}^*\boldsymbol{A})^{-1} = \begin{bmatrix} 2 & -1 \\ -1 & 1 \end{bmatrix}, \quad \boldsymbol{A}^\dagger = (\boldsymbol{A}^*\boldsymbol{A})^{-1}\boldsymbol{A}^* = \begin{bmatrix} 1 & -1 & 0 \\ 0 & 1 & 0 \end{bmatrix}.$$

Thus $K_\infty(\boldsymbol{A}) = \|\boldsymbol{A}\|_\infty \|\boldsymbol{A}^\dagger\|_\infty = 2 \cdot 2 = 4$ is quite small.

Consider now the projections $\boldsymbol{b}_1$ and $\boldsymbol{e}_1$. We find $\boldsymbol{A}\boldsymbol{A}^\dagger = \begin{bmatrix} 1 & 0 & 0 \\ 0 & 1 & 0 \\ 0 & 0 & 0 \end{bmatrix}$. Hence

$$\boldsymbol{b}_1 = \boldsymbol{A}\boldsymbol{A}^\dagger \boldsymbol{b} = [10^{-4}, 0, 0]^*, \quad \text{and} \quad \boldsymbol{e}_1 = \boldsymbol{A}\boldsymbol{A}^\dagger \boldsymbol{e} = [10^{-6}, 0, 0]^*.$$

Thus $\|\boldsymbol{e}_1\|_\infty / \|\boldsymbol{b}_1\|_\infty = 10^{-2}$ and (9.13) takes the form

$$\frac{1}{4}10^{-2} \leq \frac{\|\boldsymbol{y} - \boldsymbol{x}\|_\infty}{\|\boldsymbol{x}\|_\infty} \leq 4 \cdot 10^{-2}. \tag{9.14}$$

To verify the bounds we compute the solutions as $\boldsymbol{x} = \boldsymbol{A}^\dagger \boldsymbol{b} = [10^{-4}, 0]^*$ and $\boldsymbol{y} = \boldsymbol{A}^\dagger(\boldsymbol{b} + \boldsymbol{e}) = [10^{-4} + 10^{-6}, 0]^*$. Hence

$$\frac{\|\boldsymbol{x} - \boldsymbol{y}\|_\infty}{\|\boldsymbol{x}\|_\infty} = \frac{10^{-6}}{10^{-4}} = 10^{-2},$$

in agreement with (9.14)

For each $\boldsymbol{A}$ we can find $\boldsymbol{b}$ and $\boldsymbol{e}$ so that we have equality in the upper bound in (9.13). The lower bound is best possible in a similar way.

9.4.2 Perturbing the Matrix

The analysis of the effects of a perturbation $\boldsymbol{E}$ in $\boldsymbol{A}$ is quite difficult. The following result is stated without proof, see [12, p. 51]. For other estimates see [2] and [19].

Theorem 9.9 (Perturbing the Matrix) *Suppose $\boldsymbol{A}, \boldsymbol{E} \in \mathbb{C}^{m \times n}$, $m > n$, where $\boldsymbol{A}$ has linearly independent columns and $\alpha := 1 - \|\boldsymbol{E}\|_2 \|\boldsymbol{A}^\dagger\|_2 > 0$. Then $\boldsymbol{A} + \boldsymbol{E}$ has linearly independent columns. Let $\boldsymbol{b} = \boldsymbol{b}_1 + \boldsymbol{b}_2 \in \mathbb{C}^m$ where $\boldsymbol{b}_1$ and $\boldsymbol{b}_2$ are the orthogonal projections into $\mathcal{R}(\boldsymbol{A})$ and $\mathcal{N}(\boldsymbol{A}^*)$ respectively. Suppose $\boldsymbol{b}_1 \neq \boldsymbol{0}$. Let $\boldsymbol{x}$ and $\boldsymbol{y}$ be the solutions of* $\min\|\boldsymbol{A}\boldsymbol{x} - \boldsymbol{b}\|_2$ *and* $\min\|(\boldsymbol{A} + \boldsymbol{E})\boldsymbol{y} - \boldsymbol{b}\|_2$. *Then*

$$\rho = \frac{\|\boldsymbol{x} - \boldsymbol{y}\|_2}{\|\boldsymbol{x}\|_2} \leq \frac{1}{\alpha} K(1 + \beta K)\frac{\|\boldsymbol{E}\|_2}{\|\boldsymbol{A}\|_2}, \quad \beta = \frac{\|\boldsymbol{b}_2\|_2}{\|\boldsymbol{b}_1\|_2}, \quad K = \|\boldsymbol{A}\|_2 \|\boldsymbol{A}^\dagger\|_2. \tag{9.15}$$

Equation (9.15) says that the relative error in $\boldsymbol{y}$ as an approximation to $\boldsymbol{x}$ can be at most $K(1+\beta K)/\alpha$ times as large as the size $\|\boldsymbol{E}\|_2 / \|\boldsymbol{A}\|_2$ of the relative perturbation in $\boldsymbol{A}$. β will be small if $\boldsymbol{b}$ lies almost entirely in $\mathcal{R}(\boldsymbol{A})$, and we have approximately $\rho \le \frac{1}{\alpha} K \|\boldsymbol{E}\|_2/\|\boldsymbol{A}\|_2$. This corresponds to the estimate (8.25) for linear systems. If β is not small, the term $\frac{1}{\alpha} K^2 \beta \|\boldsymbol{E}\|_2/\|\boldsymbol{A}\|_2$ will dominate. In other words, the condition number is roughly $K(\boldsymbol{A})$ if β is small and $K(\boldsymbol{A})^2\beta$ if β is not small. Note that β is large if $\boldsymbol{b}$ is almost orthogonal to $\mathcal{R}(\boldsymbol{A})$ and that $\boldsymbol{b}_2 = \boldsymbol{b} - \boldsymbol{A}\boldsymbol{x}$ is the residual of $\boldsymbol{x}$.

9.5 Perturbation Theory for Singular Values

In this section we consider what effect a small change in the matrix $\boldsymbol{A}$ has on the singular values.

9.5.1 *The Minmax Theorem for Singular Values and the Hoffman-Wielandt Theorem*

We have a minmax and maxmin characterization for singular values.

Theorem 9.10 (The Courant-Fischer Theorem for Singular Values) *Suppose $\boldsymbol{A} \in \mathbb{C}^{m\times n}$ has singular values $\sigma_1, \sigma_2, \ldots, \sigma_n$ ordered so that $\sigma_1 \ge \cdots \ge \sigma_n$. Then for $k = 1, \ldots, n$*

$$\sigma_k = \min_{\dim(\mathcal{S})=n-k+1} \max_{\substack{\boldsymbol{x}\in\mathcal{S}\\ \boldsymbol{x}\neq\boldsymbol{0}}} \frac{\|\boldsymbol{A}\boldsymbol{x}\|_2}{\|\boldsymbol{x}\|_2} = \max_{\dim(\mathcal{S})=k} \min_{\substack{\boldsymbol{x}\in\mathcal{S}\\ \boldsymbol{x}\neq\boldsymbol{0}}} \frac{\|\boldsymbol{A}\boldsymbol{x}\|_2}{\|\boldsymbol{x}\|_2}. \tag{9.16}$$

Proof We have

$$\frac{\|\boldsymbol{A}\boldsymbol{x}\|_2^2}{\|\boldsymbol{x}\|_2^2} = \frac{(\boldsymbol{A}\boldsymbol{x})^*(\boldsymbol{A}\boldsymbol{x})}{\boldsymbol{x}^*\boldsymbol{x}} = \frac{\boldsymbol{x}^*\boldsymbol{A}^*\boldsymbol{A}\boldsymbol{x}}{\boldsymbol{x}^*\boldsymbol{x}} = R_{\boldsymbol{A}^*\boldsymbol{A}}(\boldsymbol{x}),$$

the Rayleigh quotient of $\boldsymbol{A}^*\boldsymbol{A}$. Since the singular values of $\boldsymbol{A}$ are the nonnegative square roots of the eigenvalues of $\boldsymbol{A}^*\boldsymbol{A}$, the results follow from the Courant-Fischer Theorem for eigenvalues, see Theorem 6.1. □

By taking $k = 1$ and $k = n$ in (9.16) we obtain for any $\boldsymbol{A} \in \mathbb{C}^{m\times n}$

$$\sigma_1 = \max_{\substack{\boldsymbol{x}\in\mathbb{C}^n\\ \boldsymbol{x}\neq\boldsymbol{0}}} \frac{\|\boldsymbol{A}\boldsymbol{x}\|_2}{\|\boldsymbol{x}\|_2}, \quad \sigma_n = \min_{\substack{\boldsymbol{x}\in\mathbb{C}^n\\ \boldsymbol{x}\neq\boldsymbol{0}}} \frac{\|\boldsymbol{A}\boldsymbol{x}\|_2}{\|\boldsymbol{x}\|_2}. \tag{9.17}$$

This follows since the only subspace of $\mathbb{C}^n$ of dimension n is $\mathbb{C}^n$ itself.

Using Theorem 9.10 we obtain the following result.

Theorem 9.11 (Perturbation of Singular Values) *Let $\boldsymbol{A}, \boldsymbol{B} \in \mathbb{R}^{m\times n}$ be rectangular matrices with singular values $\alpha_1 \ge \alpha_2 \ge \cdots \ge \alpha_n$ and $\beta_1 \ge \beta_2 \ge \cdots \ge \beta_n$. Then*

$$|\alpha_j - \beta_j| \le \|\boldsymbol{A} - \boldsymbol{B}\|_2, \text{ for } j = 1, 2, \ldots, n. \tag{9.18}$$

Proof Fix j and let $\mathcal{S}$ be the $n-j+1$ dimensional subspace for which the minimum in Theorem 9.10 is obtained for $\boldsymbol{B}$. Then

$$\alpha_j \le \max_{\substack{x\in\mathcal{S}\\ x\ne 0}} \frac{\|(\boldsymbol{B}+(\boldsymbol{A}-\boldsymbol{B}))\boldsymbol{x}\|_2}{\|\boldsymbol{x}\|_2} \le \max_{\substack{x\in\mathcal{S}\\ x\ne 0}} \frac{\|\boldsymbol{B}\boldsymbol{x}\|_2}{\|\boldsymbol{x}\|_2} + \max_{\substack{x\in\mathcal{S}\\ x\ne 0}} \frac{\|(\boldsymbol{A}-\boldsymbol{B})\boldsymbol{x}\|_2}{\|\boldsymbol{x}\|_2}$$

$$\le \beta_j + \|\boldsymbol{A}-\boldsymbol{B}\|_2.$$

By symmetry we obtain $\beta_j \le \alpha_j + \|\boldsymbol{A}-\boldsymbol{B}\|_2$ and the proof is complete. □

The following result is an analogue of Theorem 8.11.

Theorem 9.12 (Generalized Inverse When Perturbing the Matrix) *Let $\boldsymbol{A}, \boldsymbol{E} \in \mathbb{R}^{m\times n}$ have singular values $\alpha_1 \ge \cdots \ge \alpha_n$ and $\epsilon_1 \ge \cdots \ge \epsilon_n$. If* $\operatorname{rank}(\boldsymbol{A}+\boldsymbol{E}) \le \operatorname{rank}(\boldsymbol{A}) = r$ *and* $\|\boldsymbol{A}^\dagger\|_2\|\boldsymbol{E}\|_2 < 1$ *then*

1. $\operatorname{rank}(\boldsymbol{A}+\boldsymbol{E}) = \operatorname{rank}(\boldsymbol{A})$,
2. $\|(\boldsymbol{A}+\boldsymbol{E})^\dagger\|_2 \le \frac{\|\boldsymbol{A}^\dagger\|_2}{1-\|\boldsymbol{A}^\dagger\|_2\|\boldsymbol{E}\|_2} = \frac{1}{\alpha_r-\epsilon_1}$.

Proof Suppose $\boldsymbol{A}$ has rank r and let $\boldsymbol{B} := \boldsymbol{A}+\boldsymbol{E}$ have singular values $\beta_1 \ge \cdots \ge \beta_n$. In terms of singular values the inequality $\|\boldsymbol{A}^\dagger\|_2\|\boldsymbol{E}\|_2 < 1$ can be written $\epsilon_1/\alpha_r < 1$ or $\alpha_r > \epsilon_1$. By Theorem 9.11 we have $\alpha_r - \beta_r \le \|\boldsymbol{E}\|_2 = \epsilon_1$, which implies $\beta_r \ge \alpha_r - \epsilon_1 > 0$, and this shows that $\operatorname{rank}(\boldsymbol{A}+\boldsymbol{E}) \ge r$. Thus 1. follows. To prove 2., the inequality $\beta_r \ge \alpha_r - \epsilon_1$ implies that

$$\|(\boldsymbol{A}+\boldsymbol{E})^\dagger\|_2 = \frac{1}{\beta_r} \le \frac{1}{\alpha_r-\epsilon_1} = \frac{1/\alpha_r}{1-\epsilon_1/\alpha_r} = \frac{\|\boldsymbol{A}^\dagger\|_2}{1-\|\boldsymbol{A}^\dagger\|_2\|\boldsymbol{E}\|_2}.$$

□

The Hoffman-Wielandt Theorem, see Theorem 6.14, for eigenvalues of Hermitian matrices can be written

$$\sum_{j=1}^n |\mu_j - \lambda_j|^2 \le \|\boldsymbol{A}-\boldsymbol{B}\|_F^2 := \sum_{i=1}^n\sum_{j=1}^n |a_{ij}-b_{ij}|^2, \tag{9.19}$$

where $\boldsymbol{A}, \boldsymbol{B} \in \mathbb{C}^{n\times n}$ are both Hermitian matrices with eigenvalues $\lambda_1 \ge \cdots \ge \lambda_n$ and $\mu_1 \ge \cdots \ge \mu_n$, respectively.

For singular values we have a similar result.

Theorem 9.13 (Hoffman-Wielandt Theorem for Singular Values) *For any $m, n \in \mathbb{N}$ and $\boldsymbol{A}, \boldsymbol{B} \in \mathbb{C}^{m\times n}$ we have*

$$\sum_{j=1}^{n} |\beta_j - \alpha_j|^2 \le \|\boldsymbol{A} - \boldsymbol{B}\|_F^2. \tag{9.20}$$

where $\alpha_1 \ge \cdots \ge \alpha_n$ and $\beta_1 \ge \cdots \ge \beta_n$ are the singular values of $\boldsymbol{A}$ and $\boldsymbol{B}$, respectively.

Proof We apply the Hoffman-Wielandt Theorem for eigenvalues to the Hermitian matrices

$$\boldsymbol{C} := \begin{bmatrix} \boldsymbol{0} & \boldsymbol{A} \\ \boldsymbol{A}^* & \boldsymbol{0} \end{bmatrix} \text{ and } \boldsymbol{D} := \begin{bmatrix} \boldsymbol{0} & \boldsymbol{B} \\ \boldsymbol{B}^* & \boldsymbol{0} \end{bmatrix} \in \mathbb{C}^{(m+n)\times(m+n)}.$$

If $\boldsymbol{C}$ and $\boldsymbol{D}$ have eigenvalues $\lambda_1 \ge \cdots \ge \lambda_{m+n}$ and $\mu_1 \ge \cdots \ge \mu_{m+n}$, respectively then

$$\sum_{j=1}^{m+n} |\lambda_j - \mu_j|^2 \le \|\boldsymbol{C} - \boldsymbol{D}\|_F^2. \tag{9.21}$$

Suppose $\boldsymbol{A}$ has rank r and SVD $[\boldsymbol{u}_1, \ldots, \boldsymbol{u}_m]\boldsymbol{\Sigma}[\boldsymbol{v}_1, \ldots, \boldsymbol{v}_n]^*$. We use (7.7) and determine the eigenpairs of $\boldsymbol{C}$ as follows.

$$\begin{bmatrix} \boldsymbol{0} & \boldsymbol{A} \\ \boldsymbol{A}^* & \boldsymbol{0} \end{bmatrix}\begin{bmatrix} \boldsymbol{u}_i \\ \boldsymbol{v}_i \end{bmatrix} = \begin{bmatrix} \boldsymbol{A}\boldsymbol{v}_i \\ \boldsymbol{A}^*\boldsymbol{u}_i \end{bmatrix} = \begin{bmatrix} \alpha_i\boldsymbol{u}_i \\ \alpha_i\boldsymbol{v}_i \end{bmatrix} = \alpha_i\begin{bmatrix} \boldsymbol{u}_i \\ \boldsymbol{v}_i \end{bmatrix}, \quad i = 1, \ldots, r,$$

$$\begin{bmatrix} \boldsymbol{0} & \boldsymbol{A} \\ \boldsymbol{A}^* & \boldsymbol{0} \end{bmatrix}\begin{bmatrix} \boldsymbol{u}_i \\ -\boldsymbol{v}_i \end{bmatrix} = \begin{bmatrix} -\boldsymbol{A}\boldsymbol{v}_i \\ \boldsymbol{A}^*\boldsymbol{u}_i \end{bmatrix} = \begin{bmatrix} -\alpha_i\boldsymbol{u}_i \\ \alpha_i\boldsymbol{v}_i \end{bmatrix} = -\alpha_i\begin{bmatrix} \boldsymbol{u}_i \\ -\boldsymbol{v}_i \end{bmatrix}, \quad i = 1, \ldots, r,$$

$$\begin{bmatrix} \boldsymbol{0} & \boldsymbol{A} \\ \boldsymbol{A}^* & \boldsymbol{0} \end{bmatrix}\begin{bmatrix} \boldsymbol{u}_i \\ \boldsymbol{0} \end{bmatrix} = \begin{bmatrix} \boldsymbol{0} \\ \boldsymbol{A}^*\boldsymbol{u}_i \end{bmatrix} = \begin{bmatrix} \boldsymbol{0} \\ \boldsymbol{0} \end{bmatrix} = 0\begin{bmatrix} \boldsymbol{u}_i \\ \boldsymbol{0} \end{bmatrix}, \quad i = r+1, \ldots, m,$$

$$\begin{bmatrix} \boldsymbol{0} & \boldsymbol{A} \\ \boldsymbol{A}^* & \boldsymbol{0} \end{bmatrix}\begin{bmatrix} \boldsymbol{0} \\ \boldsymbol{v}_i \end{bmatrix} = \begin{bmatrix} \boldsymbol{A}\boldsymbol{v}_i \\ \boldsymbol{0} \end{bmatrix} = \begin{bmatrix} \boldsymbol{0} \\ \boldsymbol{0} \end{bmatrix} = 0\begin{bmatrix} \boldsymbol{0} \\ \boldsymbol{v}_i \end{bmatrix}, \quad i = r+1, \ldots, n.$$

Thus $\boldsymbol{C}$ has the $2r$ eigenvalues $\alpha_1, -\alpha_1, \ldots, \alpha_r, -\alpha_r$ and $m + n - 2r$ additional zero eigenvalues. Similarly, if $\boldsymbol{B}$ has rank s then $\boldsymbol{D}$ has the $2s$ eigenvalues $\beta_1, -\beta_1, \ldots, \beta_s, -\beta_s$ and $m + n - 2s$ additional zero eigenvalues. Let

$$t := \max(r, s).$$

Then

$$\lambda_1 \ge \cdots \ge \lambda_{m+n} = \alpha_1 \ge \cdots \ge \alpha_t \ge 0 = \cdots = 0 \ge -\alpha_t \ge \cdots \ge -\alpha_1,$$

$$\mu_1 \ge \cdots \ge \mu_{m+n} = \beta_1 \ge \cdots \ge \beta_t \ge 0 = \cdots = 0 \ge -\beta_t \ge \cdots \ge -\beta_1.$$

We find

$$\sum_{j=1}^{m+n} |\lambda_j - \mu_j|^2 = \sum_{i=1}^{t} |\alpha_i - \beta_i|^2 + \sum_{i=1}^{t} |-\alpha_i + \beta_i|^2 = 2\sum_{i=1}^{t} |\alpha_i - \beta_i|^2$$

and

$$\|\boldsymbol{C}-\boldsymbol{D}\|_F^2 = \|\begin{bmatrix} \boldsymbol{0} & \boldsymbol{A}-\boldsymbol{B} \\ \boldsymbol{A}^* - \boldsymbol{B}^* & \boldsymbol{0} \end{bmatrix}\|_F^2 = \|\boldsymbol{B}-\boldsymbol{A}\|_F^2 + \|(\boldsymbol{B}-\boldsymbol{A})^*\|_F^2 = 2\|\boldsymbol{B}-\boldsymbol{A}\|_F^2.$$

But then (9.21) implies $\sum_{i=1}^{t} |\alpha_i - \beta_i|^2 \leq \|\boldsymbol{B}-\boldsymbol{A}\|_F^2$. Since $t \leq n$ and $\alpha_i = \beta_i = 0$ for $i = t+1, \ldots, n$ we obtain (9.20). □

Because of Theorem 9.11 and the Hoffman-Wielandt Theorem for singular values, Theorem 9.13 we will say that the singular values of a matrix are **well conditioned**. Changing the Frobenius norm or the spectral norm of a matrix by small amount only changes the singular values by a small amount.

9.6 Exercises Chap. 9

9.6.1 Exercises Sect. 9.1

Exercise 9.1 (Fitting a Circle to Points) In this problem we derive an algorithm to fit a circle $(t - c_1)^2 + (y - c_2)^2 = r^2$ to $m \geq 3$ given points $(t_i, y_i)_{i=1}^m$ in the (t, y)-plane. We obtain the overdetermined system

$$(t_i - c_1)^2 + (y_i - c_2)^2 = r^2, \; i = 1, \ldots, m, \tag{9.22}$$

of m equations in the three unknowns c_1, c_2 and r. This system is nonlinear, but it can be solved from the linear system

$$t_i x_1 + y_i x_2 + x_3 = t_i^2 + y_i^2, \; i = 1, \ldots, m, \tag{9.23}$$

and then setting $c_1 = x_1/2$, $c_2 = x_2/2$ and $r^2 = c_1^2 + c_2^2 + x_3$.

a) Derive (9.23) from (9.22). Explain how we can find c_1, c_2, r once $[x_1, x_2, x_3]$ is determined.
b) Formulate (9.23) as a linear least squares problem for suitable $\boldsymbol{A}$ and $\boldsymbol{b}$.
c) Does the matrix $\boldsymbol{A}$ in **b)** have linearly independent columns?
d) Use (9.23) to find the circle passing through the three points $(1, 4)$, $(3, 2)$, $(1, 0)$.

Exercise 9.2 (Least Square Fit (Exam Exercise 2018-1))

a) Let $\boldsymbol{A}$ be the matrix $\begin{bmatrix} \sqrt{2} & \sqrt{2} \\ 0 & \sqrt{3} \end{bmatrix}$. Find the singular values of $\boldsymbol{A}$, and compute $\|\boldsymbol{A}\|_2$.

b) Consider the matrix $\boldsymbol{A} = \begin{bmatrix} 3 & \alpha \\ \alpha & 1 \end{bmatrix}$, where α is a real number. For which values of α is $\boldsymbol{A}$ positive definite?

c) We would like to fit the points $\boldsymbol{p}_1 = (0, 1)$, $\boldsymbol{p}_2 = (1, 0)$, $\boldsymbol{p}_3 = (2, 1)$ to a straight line in the plane. Find a line $p(x) = mx + b$ which minimizes

$$\sum_{i=1}^{3} \|p(x_i) - y_i\|^2,$$

where $\boldsymbol{p}_i = (x_i, y_i)$. Is this solution unique?

9.6.2 Exercises Sect. 9.2

Exercise 9.3 (A Least Squares Problem (Exam Exercise 1983-2)) Suppose $\boldsymbol{A} \in \mathbb{R}^{m\times n}$ and let $\boldsymbol{I} \in \mathbb{R}^{n\times n}$ be the identity matrix. We define $F : \mathbb{R}^n \to \mathbb{R}$ by

$$F(\boldsymbol{x}) := \|\boldsymbol{A}\boldsymbol{x} - \boldsymbol{b}\|_2^2 + \|\boldsymbol{x}\|_2^2.$$

a) Show that the matrix $\boldsymbol{B} := \boldsymbol{I} + \boldsymbol{A}^T\boldsymbol{A}$ is symmetric and positive definite.

b) Show that

$$F(\boldsymbol{x}) = \boldsymbol{x}^T\boldsymbol{B}\boldsymbol{x} - 2\boldsymbol{c}^T\boldsymbol{x} + \boldsymbol{b}^T\boldsymbol{b}, \quad \text{where} \quad \boldsymbol{c} = \boldsymbol{A}^T\boldsymbol{b}.$$

c) Show that to every $\boldsymbol{b} \in \mathbb{R}^m$ there is a unique $\boldsymbol{x}$ which minimizes F. Moreover, $\boldsymbol{x}$ is the unique solution of the linear system $(\boldsymbol{I} + \boldsymbol{A}^T\boldsymbol{A})\boldsymbol{x} = \boldsymbol{A}^T\boldsymbol{b}$.

Exercise 9.4 (Weighted Least Squares (Exam Exercise 1977-2)) For $m \ge n$ we are given $\boldsymbol{A} \in \mathbb{R}^{m\times n}$ with linearly independent columns, $\boldsymbol{b} \in \mathbb{R}^m$, and $\boldsymbol{D} := \mathrm{diag}(d_1, d_2, \ldots, d_m) \in \mathbb{R}^{m\times m}$, where $d_i > 0$, $i = 1, 2, \ldots, m$. We want to minimize

$$\|\boldsymbol{r}(\boldsymbol{x})\|_D^2 := \sum_{i=1}^{m} r_i(x)^2 d_i, \quad \boldsymbol{x} \in \mathbb{R}^n, \tag{9.24}$$

where $r_i = r_i(\boldsymbol{x})$, $i = 1, 2, \ldots, m$ are the components of the vector

$$\boldsymbol{r} = \boldsymbol{r}(\boldsymbol{x}) = \boldsymbol{b} - \boldsymbol{A}\boldsymbol{x}.$$

a) Show that $\|\boldsymbol{r}(\boldsymbol{x})\|_D^2$ in (9.24) obtains a unique minimum when $\boldsymbol{x} = \boldsymbol{x}_{min}$ is the solution of the system

$$\boldsymbol{A}^T\boldsymbol{D}\boldsymbol{A}\boldsymbol{x} = \boldsymbol{A}^T\boldsymbol{D}\boldsymbol{b}.$$

b) Show that

$$K_2(\boldsymbol{A}^T\boldsymbol{D}\boldsymbol{A}) \leq K_2(\boldsymbol{A}^T\boldsymbol{A})K_2(\boldsymbol{D}),$$

where for any nonsingular matrix $K_2(\boldsymbol{B}) := \|\boldsymbol{B}\|_2\|\boldsymbol{B}^{-1}\|_2$.

9.6.3 Exercises Sect. 9.3

Exercise 9.5 (Uniqueness of Generalized Inverse) Given $\boldsymbol{A} \in \mathbb{C}^{m\times n}$, and suppose $\boldsymbol{B}, \boldsymbol{C} \in \mathbb{C}^{n\times m}$ satisfy

$$\begin{array}{rcl} \boldsymbol{ABA} = \boldsymbol{A} & (1) & \boldsymbol{ACA} = \boldsymbol{A}, \\ \boldsymbol{BAB} = \boldsymbol{B} & (2) & \boldsymbol{CAC} = \boldsymbol{C}, \\ (\boldsymbol{AB})^* = \boldsymbol{AB} & (3) & (\boldsymbol{AC})^* = \boldsymbol{AC}, \\ (\boldsymbol{BA})^* = \boldsymbol{BA} & (4) & (\boldsymbol{CA})^* = \boldsymbol{CA}. \end{array}$$

Verify the following proof that $\boldsymbol{B} = \boldsymbol{C}$.

$$\begin{aligned} \boldsymbol{B} &= (\boldsymbol{BA})\boldsymbol{B} = (\boldsymbol{A}^*)\boldsymbol{B}^*\boldsymbol{B} = (\boldsymbol{A}^*\boldsymbol{C}^*)\boldsymbol{A}^*\boldsymbol{B}^*\boldsymbol{B} = \boldsymbol{CA}(\boldsymbol{A}^*\boldsymbol{B}^*)\boldsymbol{B} \\ &= \boldsymbol{CA}(\boldsymbol{BAB}) = (\boldsymbol{C})\boldsymbol{AB} = \boldsymbol{C}(\boldsymbol{AC})\boldsymbol{AB} = \boldsymbol{CC}^*\boldsymbol{A}^*(\boldsymbol{AB}) \\ &= \boldsymbol{CC}^*(\boldsymbol{A}^*\boldsymbol{B}^*\boldsymbol{A}^*) = \boldsymbol{C}(\boldsymbol{C}^*\boldsymbol{A}^*) = \boldsymbol{CAC} = \boldsymbol{C}. \end{aligned}$$

Exercise 9.6 (Verify That a Matrix Is a Generalized Inverse) Show that the generalized inverse of $\boldsymbol{A} = \begin{bmatrix} 1 & 1 \\ 1 & 1 \\ 0 & 0 \end{bmatrix}$ is $\boldsymbol{A}^\dagger = \frac{1}{4}\begin{bmatrix} 1 & 1 & 0 \\ 1 & 1 & 0 \end{bmatrix}$ without using the singular value decomposition of $\boldsymbol{A}$.

Exercise 9.7 (Linearly Independent Columns and Generalized Inverse) Suppose $\boldsymbol{A} \in \mathbb{C}^{m\times n}$ has linearly independent columns. Show that $\boldsymbol{A}^*\boldsymbol{A}$ is nonsingular and $\boldsymbol{A}^\dagger = (\boldsymbol{A}^*\boldsymbol{A})^{-1}\boldsymbol{A}^*$. If $\boldsymbol{A}$ has linearly independent rows, then show that $\boldsymbol{A}\boldsymbol{A}^*$ is nonsingular and $\boldsymbol{A}^\dagger = \boldsymbol{A}^*(\boldsymbol{A}\boldsymbol{A}^*)^{-1}$.

Exercise 9.8 (More Orthogonal Projections) Given $m, n \in \mathbb{N}$, $\boldsymbol{A} \in \mathbb{C}^{m\times n}$ of rank r, and let $\mathcal{S}$ be one of the subspaces $\mathcal{R}(\boldsymbol{A}^*)$, $\mathcal{N}(\boldsymbol{A})$. Show that the orthogonal projection of $\boldsymbol{v} \in \mathbb{C}^n$ into $\mathcal{S}$ can be written as a matrix $\boldsymbol{P}_{\mathcal{S}}$ times the vector $\boldsymbol{v}$ in the form $\boldsymbol{P}_{\mathcal{S}}\boldsymbol{v}$, where

$$\begin{aligned} \boldsymbol{P}_{\mathcal{R}(\boldsymbol{A}^*)} &= \boldsymbol{A}^\dagger\boldsymbol{A} = \boldsymbol{V}_1\boldsymbol{V}_1^* = \sum_{j=1}^{r} \boldsymbol{v}_j\boldsymbol{v}_j^* \in \mathbb{C}^{n\times n} \\ \boldsymbol{P}_{\mathcal{N}(\boldsymbol{A})} &= \boldsymbol{I} - \boldsymbol{A}^\dagger\boldsymbol{A} = \boldsymbol{V}_2\boldsymbol{V}_2^* = \sum_{j=r+1}^{n} \boldsymbol{v}_j\boldsymbol{v}_j^* \in \mathbb{C}^{n\times n}. \end{aligned} \tag{9.25}$$

where $\boldsymbol{A}^\dagger$ is the generalized inverse of $\boldsymbol{A}$ and $\boldsymbol{A} = \boldsymbol{U\Sigma V}^* \in \mathbb{C}^{m\times n}$, is a singular value decomposition of $\boldsymbol{A}$ (cf. (9.7)). Thus (9.12) and (9.25) give the orthogonal projections into the 4 fundamental subspaces. Hint: by Theorem 7.3 we have the orthogonal sum $\mathbb{C}^n = \mathcal{R}(\boldsymbol{A}^*)\overset{\perp}{\oplus}\mathcal{N}(\boldsymbol{A})$.

Exercise 9.9 (The Generalized Inverse of a Vector) Show that $\boldsymbol{u}^\dagger = (\boldsymbol{u}^*\boldsymbol{u})^{-1}\boldsymbol{u}^*$ if $\boldsymbol{u} \in \mathbb{C}^{n,1}$ is nonzero.

Exercise 9.10 (The Generalized Inverse of an Outer Product) If $\boldsymbol{A} = \boldsymbol{u}\boldsymbol{v}^*$ where $\boldsymbol{u} \in \mathbb{C}^m$, $\boldsymbol{v} \in \mathbb{C}^n$ are nonzero, show that

$$\boldsymbol{A}^\dagger = \frac{1}{\alpha}\boldsymbol{A}^*, \qquad \alpha = \|\boldsymbol{u}\|_2^2\|\boldsymbol{v}\|_2^2.$$

Exercise 9.11 (The Generalized Inverse of a Diagonal Matrix) Show that $\mathrm{diag}(\lambda_1, \ldots, \lambda_n)^\dagger = \mathrm{diag}(\lambda_1^\dagger, \ldots, \lambda_n^\dagger)$ where

$$\lambda_i^\dagger = \begin{cases} 1/\lambda_i, & \lambda_i \neq 0 \\ 0 & \lambda_i = 0. \end{cases}$$

Exercise 9.12 (Properties of the Generalized Inverse) Suppose $\boldsymbol{A} \in \mathbb{C}^{m\times n}$. Show that

a) $(\boldsymbol{A}^*)^\dagger = (\boldsymbol{A}^\dagger)^*$.
b) $(\boldsymbol{A}^\dagger)^\dagger = \boldsymbol{A}$.
c) $(\alpha\boldsymbol{A})^\dagger = \frac{1}{\alpha}\boldsymbol{A}^\dagger$, $\alpha \neq 0$.

Exercise 9.13 (The Generalized Inverse of a Product) Suppose $k, m, n \in \mathbb{N}$, $\boldsymbol{A} \in \mathbb{C}^{m\times n}$, $\boldsymbol{B} \in \mathbb{C}^{n\times k}$. Suppose $\boldsymbol{A}$ has linearly independent columns and $\boldsymbol{B}$ has linearly independent rows.

a) Show that $(\boldsymbol{AB})^\dagger = \boldsymbol{B}^\dagger\boldsymbol{A}^\dagger$.
Hint: Let $\boldsymbol{E} = \boldsymbol{AB}$, $\boldsymbol{F} = \boldsymbol{B}^\dagger\boldsymbol{A}^\dagger$. Show by using $\boldsymbol{A}^\dagger\boldsymbol{A} = \boldsymbol{B}\boldsymbol{B}^\dagger = \boldsymbol{I}$ that $\boldsymbol{F}$ is the generalized inverse of $\boldsymbol{E}$.
b) Find $\boldsymbol{A} \in \mathbb{R}^{1,2}$, $\boldsymbol{B} \in \mathbb{R}^{2,1}$ such that $(\boldsymbol{AB})^\dagger \neq \boldsymbol{B}^\dagger\boldsymbol{A}^\dagger$.

Exercise 9.14 (The Generalized Inverse of the Conjugate Transpose) Show that $\boldsymbol{A}^* = \boldsymbol{A}^\dagger$ if and only if all singular values of $\boldsymbol{A}$ are either zero or one.

Exercise 9.15 (Linearly Independent Columns) Show that if $\boldsymbol{A}$ has rank n then $\boldsymbol{A}(\boldsymbol{A}^*\boldsymbol{A})^{-1}\boldsymbol{A}^*\boldsymbol{b}$ is the projection of $\boldsymbol{b}$ into $\mathcal{R}(\boldsymbol{A})$. (Cf. Exercise 9.8.)

Exercise 9.16 (Analysis of the General Linear System) Consider the linear system $\boldsymbol{A}\boldsymbol{x} = \boldsymbol{b}$ where $\boldsymbol{A} \in \mathbb{C}^{n\times n}$ has rank $r > 0$ and $\boldsymbol{b} \in \mathbb{C}^n$. Let

$$\boldsymbol{U}^*\boldsymbol{A}\boldsymbol{V} = \begin{bmatrix} \boldsymbol{\Sigma}_1 & \boldsymbol{0} \\ \boldsymbol{0} & \boldsymbol{0} \end{bmatrix}$$

represent the singular value decomposition of $\boldsymbol{A}$.

a) Let $\boldsymbol{c} = [c_1, \ldots, c_n]^* = \boldsymbol{U}^*\boldsymbol{b}$ and $\boldsymbol{y} = [y_1, \ldots, y_n]^* = \boldsymbol{V}^*\boldsymbol{x}$. Show that $\boldsymbol{A}\boldsymbol{x} = \boldsymbol{b}$ if and only if

$$\begin{bmatrix} \boldsymbol{\Sigma}_1 & \boldsymbol{0} \\ \boldsymbol{0} & \boldsymbol{0} \end{bmatrix} \boldsymbol{y} = \boldsymbol{c}.$$

b) Show that $\boldsymbol{A}\boldsymbol{x} = \boldsymbol{b}$ has a solution $\boldsymbol{x}$ if and only if $c_{r+1} = \cdots = c_n = 0$.
c) Deduce that a linear system $\boldsymbol{A}\boldsymbol{x} = \boldsymbol{b}$ has either no solution, one solution or infinitely many solutions.

Exercise 9.17 (Fredholm's Alternative) For any $\boldsymbol{A} \in \mathbb{C}^{m\times n}$, $\boldsymbol{b} \in \mathbb{C}^n$ show that one and only one of the following systems has a solution

$$(1) \quad \boldsymbol{A}\boldsymbol{x} = \boldsymbol{b}, \qquad (2) \quad \boldsymbol{A}^*\boldsymbol{y} = \boldsymbol{0},\ \boldsymbol{y}^*\boldsymbol{b} \neq 0.$$

In other words either $\boldsymbol{b} \in \mathcal{R}(\boldsymbol{A})$, or we can find $\boldsymbol{y} \in \mathcal{N}(\boldsymbol{A}^*)$ such that $\boldsymbol{y}^*\boldsymbol{b} \neq 0$. This is called **Fredholm's alternative**.

Exercise 9.18 (SVD (Exam Exercise 2017-2)) Let $\boldsymbol{A} \in \mathbb{C}^{m\times n}$, with $m \geq n$, be a matrix on the form

$$\boldsymbol{A} = \begin{bmatrix} \boldsymbol{B} \\ \boldsymbol{C} \end{bmatrix}$$

where $\boldsymbol{B}$ is a non-singular $n \times n$ matrix and $\boldsymbol{C}$ is in $\mathbb{C}^{(m-n)\times n}$. Let $\boldsymbol{A}^\dagger$ denote the pseudoinverse of $\boldsymbol{A}$. Show that $\|\boldsymbol{A}^\dagger\|_2 \leq \|\boldsymbol{B}^{-1}\|_2$.

9.6.4 Exercises Sect. 9.4

Exercise 9.19 (Condition Number) Let

$$\boldsymbol{A} = \begin{bmatrix} 1 & 2 \\ 1 & 1 \\ 1 & 1 \end{bmatrix}, \quad \boldsymbol{b} = \begin{bmatrix} b_1 \\ b_2 \\ b_3 \end{bmatrix}.$$

a) Determine the projections $\boldsymbol{b}_1$ and $\boldsymbol{b}_2$ of $\boldsymbol{b}$ on $\mathcal{R}(\boldsymbol{A})$ and $\mathcal{N}(\boldsymbol{A}^*)$.
b) Compute $K(\boldsymbol{A}) = \|\boldsymbol{A}\|_2 \|\boldsymbol{A}^\dagger\|_2$.

Exercise 9.20 (Equality in Perturbation Bound) Let $\boldsymbol{A} \in \mathbb{C}^{m \times n}$. Suppose $\boldsymbol{y}_A$ and $\boldsymbol{y}_{A^\dagger}$ are vectors with $\|\boldsymbol{y}_A\| = \|\boldsymbol{y}_{A^\dagger}\| = 1$ and $\|\boldsymbol{A}\| = \|\boldsymbol{A}\boldsymbol{y}_A\|$ and $\|\boldsymbol{A}^\dagger\| = \|\boldsymbol{A}^\dagger \boldsymbol{y}_{A^\dagger}\|$.

a) Show that we have equality to the right in (9.13) if $\boldsymbol{b} = \boldsymbol{A}\boldsymbol{y}_A$, $\boldsymbol{e}_1 = \boldsymbol{y}_{A^\dagger}$.
b) Show that we have equality to the left if we switch $\boldsymbol{b}$ and $\boldsymbol{e}_1$ in **a)**.
c) Let $\boldsymbol{A}$ be as in Example 9.7. Find extremal $\boldsymbol{b}$ and $\boldsymbol{e}$ when the l_∞ norm is used.

This generalizes the sharpness results in Exercise 8.18. For if $m = n$ and $\boldsymbol{A}$ is nonsingular then $\boldsymbol{A}^\dagger = \boldsymbol{A}^{-1}$ and $\boldsymbol{e}_1 = \boldsymbol{e}$.

Exercise 9.21 (Problem Using Normal Equations) Consider the least squares problems where

$$\boldsymbol{A} = \begin{bmatrix} 1 & 1 \\ 1 & 1 \\ 1 & 1+\epsilon \end{bmatrix}, \quad \boldsymbol{b} = \begin{bmatrix} 2 \\ 3 \\ 2 \end{bmatrix}, \quad \epsilon \in \mathbb{R}.$$

a) Find the normal equations and the exact least squares solution.
b) Suppose ϵ is small and we replace the $(2, 2)$ entry $3+2\epsilon+\epsilon^2$ in $\boldsymbol{A}^*\boldsymbol{A}$ by $3+2\epsilon$. (This will be done in a computer if $\epsilon < \sqrt{u}$, u being the round-off unit). For example, if $u = 10^{-16}$ then $\sqrt{u} = 10^{-8}$. Solve $\boldsymbol{A}^*\boldsymbol{A}\boldsymbol{x} = \boldsymbol{A}^*\boldsymbol{b}$ for $\boldsymbol{x}$ and compare with the $\boldsymbol{x}$ found in **a)**. (We will get a much more accurate result using the QR factorization or the singular value decomposition on this problem).

9.6.5 Exercises Sect. 9.5

Exercise 9.22 (Singular Values Perturbation (Exam Exercise 1980-2)) Let $\boldsymbol{A}(\epsilon) \in \mathbb{R}^{n \times n}$ be bidiagonal with $a_{i,j} = 0$ for $i, j = 1, \ldots, n$ and $j \neq i, i+1$. Moreover, for some $1 \leq k \leq n-1$ we have $a_{k,k+1} = \epsilon \in \mathbb{R}$. Show that

$$|\sigma_i(\epsilon) - \sigma_i(0)| \leq |\epsilon|, \quad i = 1, \ldots, n,$$

where $\sigma_i(\epsilon)$, $i = 1, \ldots, n$ are the singular values of $\boldsymbol{A}(\epsilon)$.

9.7 Review Questions

9.7.1 Do the normal equations always have a solution?
9.7.2 When is the least squares solution unique?
9.7.3 Express the general least squares solution in terms of the generalized inverse.
9.7.4 Consider perturbing the right-hand side in a linear equation and a least squares problem. What is the main difference in the perturbation inequalities?
9.7.5 Why does one often prefer using QR factorization instead of normal equations for solving least squares problems.
9.7.6 What is an orthogonal sum?
9.7.7 How is an orthogonal projection defined?

Part IV
Kronecker Products and Fourier Transforms

We give an introduction to Kronecker products of matrices and the fast Fourier transform. We illustrate the usefulness by giving a fast method for solving the 2 dimensional discrete Poison Equation based on the fast Fourier transform.

Chapter 10
The Kronecker Product

Matrices arising from 2D and 3D problems sometimes have a Kronecker product structure. Identifying a Kronecker structure can be very rewarding since it simplifies the derivation of properties of such matrices.

10.1 The 2D Poisson Problem

Let $\Omega := (0,1)^2 = \{(x,y) : 0 < x, y < 1\}$ be the open unit square with boundary $\partial\Omega$. Consider the problem

$$- \Delta u := -\frac{\partial^2 u}{\partial x^2} - \frac{\partial^2 u}{\partial y^2} = f \text{ on } \Omega, \tag{10.1}$$

$$u := 0 \text{ on } \partial\Omega.$$

Here the function f is given and continuous on Ω, and we seek a function $u = u(x,y)$ such that (10.1) holds and which is zero on $\partial\Omega$.

Let m be a positive integer. We solve the problem numerically by finding approximations $v_{j,k} \approx u(jh,kh)$ on a grid of points given by

$$\overline{\Omega}_h := \{(jh,kh) : \; j,k = 0,1,\ldots,m+1\}, \quad \text{where} \quad h = 1/(m+1).$$

The points $\Omega_h := \{(jh,kh) : \; j,k = 1,\ldots,m\}$ are called the interior points, while $\overline{\Omega}_h \setminus \Omega_h$ are the boundary points. The solution is zero at the boundary points. Using the difference approximation from Chap. 2 for the second derivative we obtain the

T. Lyche, *Numerical Linear Algebra and Matrix Factorizations*,
Texts in Computational Science and Engineering 22,
https://doi.org/10.1007/978-3-030-36468-7_10

following approximations for the partial derivatives

$$\frac{\partial^2 u(jh,kh)}{\partial x^2} \approx \frac{v_{j-1,k} - 2v_{j,k} + v_{j+1,k}}{h^2}, \quad \frac{\partial^2 u(jh,kh)}{\partial y^2} \approx \frac{v_{j,k-1} - 2v_{j,k} + v_{j,k+1}}{h^2}.$$

Inserting this in (10.1) we get the following discrete analog of (10.1)

$$-\Delta_h v_{j,k} = f_{j,k}, \quad (jh,kh) \in \Omega_h, \tag{10.2}$$

$$v_{j,k} = 0, \quad (jh,kh) \in \partial\Omega_h, \tag{10.3}$$

where $f_{j,k} := f(jh,kh)$ and

$$-\Delta_h v_{j,k} := \frac{-v_{j-1,k} + 2v_{j,k} - v_{j+1,k}}{h^2} + \frac{-v_{j,k-1} + 2v_{j,k} - v_{j,k+1}}{h^2}. \tag{10.4}$$

Let us take a closer look at (10.2). It consists of $n := m^2$ linear equations. Since the values at the boundary points are known, the unknowns are the n numbers $v_{j,k}$ at the interior points. These linear equations can be written as a matrix equation in the form

$$\boldsymbol{TV} + \boldsymbol{VT} = h^2\boldsymbol{F} \quad \text{with} \quad h = 1/(m+1), \tag{10.5}$$

where $\boldsymbol{T} = \text{tridiag}(-1, 2, -1) \in \mathbb{R}^{m\times m}$ is the second derivative matrix given by (2.27) and

$$\boldsymbol{V} := \begin{bmatrix} v_{1,1} & \cdots & v_{1,m} \\ \vdots & & \vdots \\ v_{m,1} & \cdots & v_{m,m} \end{bmatrix} \in \mathbb{R}^{m\times m}, \quad \boldsymbol{F} := \begin{bmatrix} f_{1,1} & \cdots & f_{1,m} \\ \vdots & & \vdots \\ f_{m,1} & \cdots & f_{m,m} \end{bmatrix} \in \mathbb{R}^{m\times m}. \tag{10.6}$$

Indeed, the (j,k) element in $\boldsymbol{TV} + \boldsymbol{VT}$ is given by

$$\sum_{i=1}^{m} \boldsymbol{T}_{j,i} v_{i,k} + \sum_{i=1}^{m} v_{j,i} \boldsymbol{T}_{i,k} = -v_{j-1,k} + 2v_{j,k} - v_{j+1,k} - v_{j,k-1} + 2v_{j,k} - v_{j,k+1},$$

and if we divide by h^2 this is precisely the left hand side of (10.2).

To write (10.5) in standard form $\boldsymbol{Ax} = \boldsymbol{b}$ we need to order the unknowns $v_{j,k}$ in some way. The following operation of **vectorization** of a matrix gives one possible ordering.

Definition 10.1 (vec Operation) For any $\boldsymbol{B} \in \mathbb{R}^{m\times n}$ we define the vector

$$vec(\boldsymbol{B}) := [b_{11}, \ldots, b_{m1}, b_{12}, \ldots, b_{m2}, \ldots, b_{1n}, \ldots, b_{mn}]^T \in \mathbb{R}^{mn}$$

by stacking the columns of $\boldsymbol{B}$ on top of each other.

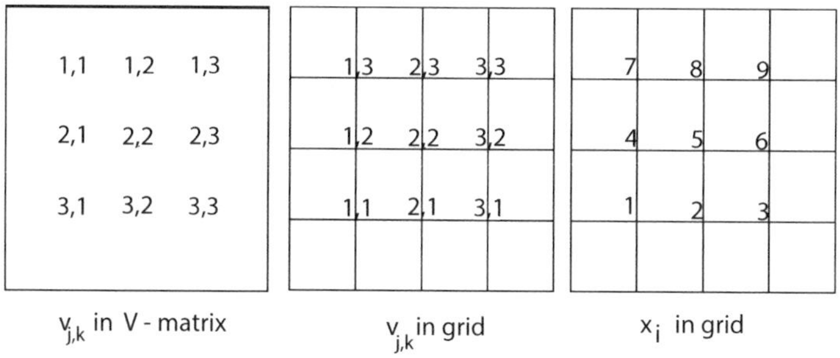

Fig. 10.1 Numbering of grid points

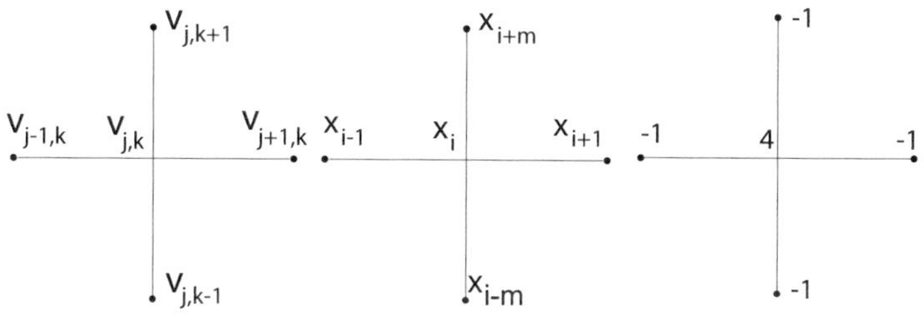

Fig. 10.2 The 5-point stencil

Let $\boldsymbol{x} := vec(\boldsymbol{V}) \in \mathbb{R}^n$, where $n = m^2$. Note that forming $\boldsymbol{x}$ by stacking the columns of $\boldsymbol{V}$ on top of each other means an ordering of the grid points. For $m = 3$ this is illustrated in Fig. 10.1. We call this the **natural ordering**. The elements in (10.2) defines a 5-point stencil, as shown in Fig. 10.2.

To find the matrix $\boldsymbol{A}$ we note that for values of j, k where the 5-point stencil does not touch the boundary, (10.2) implies

$$4x_i - x_{i-1} - x_{i+1} - x_{i-m} - x_{i+m} = b_i,$$

where $x_i = v_{j,k}$ and $b_i = h^2 f_{j,k}$. This must be modified close to the boundary. We then obtain the linear system

$$\boldsymbol{A}\boldsymbol{x} = \boldsymbol{b}, \quad \boldsymbol{A} \in \mathbb{R}^{n \times n}, \quad b \in \mathbb{R}^n, \quad n = m^2, \tag{10.7}$$

where $\boldsymbol{x} = vec(\boldsymbol{V})$, $\boldsymbol{b} = h^2 vec(\boldsymbol{F})$ and $\boldsymbol{A}$ is the **Poisson matrix** given by

$$\begin{aligned} a_{ii} &= 4, & & i = 1, \dots, n, \\ a_{i+1,i} = a_{i,i+1} &= -1, & & i = 1, \dots, n-1, \ i \neq m, 2m, \dots, (m-1)m, \\ a_{i+m,i} = a_{i,i+m} &= -1, & & i = 1, \dots, n-m, \\ a_{ij} &= 0, & & \text{otherwise}. \end{aligned} \tag{10.8}$$

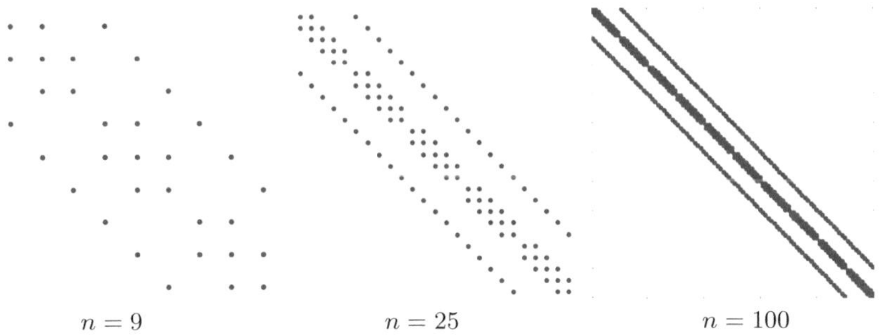

Fig. 10.3 Band structure of the 2D test matrix

For $m = 3$ we have the following matrix

$$\boldsymbol{A} = \begin{bmatrix} 4 & -1 & 0 & -1 & 0 & 0 & 0 & 0 & 0 \\ -1 & 4 & -1 & 0 & -1 & 0 & 0 & 0 & 0 \\ 0 & -1 & 4 & 0 & 0 & -1 & 0 & 0 & 0 \\ -1 & 0 & 0 & 4 & -1 & 0 & -1 & 0 & 0 \\ 0 & -1 & 0 & -1 & 4 & -1 & 0 & -1 & 0 \\ 0 & 0 & -1 & 0 & -1 & 4 & 0 & 0 & -1 \\ 0 & 0 & 0 & -1 & 0 & 0 & 4 & -1 & 0 \\ 0 & 0 & 0 & 0 & -1 & 0 & -1 & 4 & -1 \\ 0 & 0 & 0 & 0 & 0 & -1 & 0 & -1 & 4 \end{bmatrix}.$$

The bands of the weakly diagonally dominant matrix $\boldsymbol{A}$ are illustrated in Fig. 10.3.

10.1.1 The Test Matrices

In Sect. 2.4 we encountered the 1-dimensional test matrix $\boldsymbol{T}_1 \in \mathbb{R}^{m \times m}$ defined for any real numbers a, d by

$$\boldsymbol{T}_1 := \text{tridiag}(a, d, a). \tag{10.9}$$

The (2-dimensional) Poisson matrix is a special case of the matrix $\boldsymbol{T}_2 = [a_{ij}] \in \mathbb{R}^{n \times n}$ with elements

$$\begin{aligned} a_{ii} &= 2d, \ i = 1, \dots, n, \\ a_{i,i+1} = a_{i+1,i} &= a, \quad i = 1, \dots, n-1, \quad i \neq m, 2m, \dots, (m-1)m, \\ a_{i,i+m} = a_{i+m,i} &= a, \quad i = 1, \dots, n-m, \\ a_{ij} &= 0, \quad \text{otherwise}, \end{aligned} \tag{10.10}$$

and where a, d are real numbers. We will refer to this matrix as simply the **2D test matrix**. For $m = 3$ the 2D test matrix looks as follows

$$T_2 = \left[\begin{array}{ccc|ccc|ccc} 2d & a & 0 & a & 0 & 0 & 0 & 0 & 0 \\ a & 2d & a & 0 & a & 0 & 0 & 0 & 0 \\ 0 & a & 2d & 0 & 0 & a & 0 & 0 & 0 \\ \hline a & 0 & 0 & 2d & a & 0 & a & 0 & 0 \\ 0 & a & 0 & a & 2d & a & 0 & a & 0 \\ 0 & 0 & a & 0 & a & 2d & 0 & 0 & a \\ \hline 0 & 0 & 0 & a & 0 & 0 & 2d & a & 0 \\ 0 & 0 & 0 & 0 & a & 0 & a & 2d & a \\ 0 & 0 & 0 & 0 & 0 & a & 0 & a & 2d \end{array}\right]. \tag{10.11}$$

The partition into 3×3 sub matrices shows that T_2 is block tridiagonal.

Properties of T_2 can be derived from properties of T_1 by using properties of the Kronecker product.

10.2 The Kronecker Product

Definition 10.2 (Kronecker Product) For any positive integers p, q, r, s we define the **Kronecker product** of two matrices $A \in \mathbb{R}^{p \times q}$ and $B \in \mathbb{R}^{r \times s}$ as a matrix $C \in \mathbb{R}^{pr \times qs}$ given in block form as

$$C = \begin{bmatrix} Ab_{1,1} & Ab_{1,2} & \cdots & Ab_{1,s} \\ Ab_{2,1} & Ab_{2,2} & \cdots & Ab_{2,s} \\ \vdots & \vdots & \ddots & \vdots \\ Ab_{r,1} & Ab_{r,2} & \cdots & Ab_{r,s} \end{bmatrix}.$$

We denote the Kronecker product of A and B by $C = A \otimes B$.

This definition of the Kronecker product is known more precisely as the **left Kronecker product**. In the literature one often finds the **right Kronecker product** which in our notation is given by $B \otimes A$.

The Kronecker product $u \otimes v = \left[u^T v_1, \dots, u^T v_r\right]^T$ of two column vectors $u \in \mathbb{R}^p$ and $v \in \mathbb{R}^r$ is a column vector of length $p \cdot r$.

The test matrix T_2 can be written as a sum of Kronecker products. Indeed, if $m = 3$ then

$$T_1 = \begin{bmatrix} d & a & 0 \\ a & d & a \\ 0 & a & d \end{bmatrix}, \quad I = \begin{bmatrix} 1 & 0 & 0 \\ 0 & 1 & 0 \\ 0 & 0 & 1 \end{bmatrix}$$

and

$$\boldsymbol{T}_1 \otimes \boldsymbol{I} + \boldsymbol{I} \otimes \boldsymbol{T}_1 = \begin{bmatrix} \boldsymbol{T}_1 & \boldsymbol{0} & \boldsymbol{0} \\ \boldsymbol{0} & \boldsymbol{T}_1 & \boldsymbol{0} \\ \boldsymbol{0} & \boldsymbol{0} & \boldsymbol{T}_1 \end{bmatrix} + \begin{bmatrix} d\boldsymbol{I} & a\boldsymbol{I} & \boldsymbol{0} \\ a\boldsymbol{I} & d\boldsymbol{I} & a\boldsymbol{I} \\ \boldsymbol{0} & a\boldsymbol{I} & d\boldsymbol{I} \end{bmatrix} = \boldsymbol{T}_2$$

given by (10.11). This formula holds for any integer $m \geq 2$

$$\boldsymbol{T}_2 = \boldsymbol{T}_1 \otimes \boldsymbol{I} + \boldsymbol{I} \otimes \boldsymbol{T}_1, \quad \boldsymbol{T}_1, \boldsymbol{I} \in \mathbb{R}^{m \times m}, \quad \boldsymbol{T}_2 \in \mathbb{R}^{(m^2) \times (m^2)}. \tag{10.12}$$

The sum of two Kronecker products involving the identity matrix is worthy of a special name.

Definition 10.3 (Kronecker Sum) For positive integers r, s, k, let $\boldsymbol{A} \in \mathbb{R}^{r \times r}$, $\boldsymbol{B} \in \mathbb{R}^{s \times s}$, and $\boldsymbol{I}_k$ be the identity matrix of order k. The sum $\boldsymbol{A} \otimes \boldsymbol{I}_s + \boldsymbol{I}_r \otimes \boldsymbol{B}$ is known as the **Kronecker sum** of $\boldsymbol{A}$ and $\boldsymbol{B}$.

In other words, the 2D test matrix $\boldsymbol{T}_2$ is the Kronecker sum involving the 1D test matrix $\boldsymbol{T}_1$.

The following simple arithmetic rules hold for Kronecker products. For scalars λ, μ and matrices $\boldsymbol{A}, \boldsymbol{A}_1, \boldsymbol{A}_2, \boldsymbol{B}, \boldsymbol{B}_1, \boldsymbol{B}_2, \boldsymbol{C}$ of dimensions such that the operations are defined, we have

$$\begin{aligned} (\lambda \boldsymbol{A}) \otimes (\mu \boldsymbol{B}) &= \lambda\mu(\boldsymbol{A} \otimes \boldsymbol{B}), \\ (\boldsymbol{A}_1 + \boldsymbol{A}_2) \otimes \boldsymbol{B} &= \boldsymbol{A}_1 \otimes \boldsymbol{B} + \boldsymbol{A}_2 \otimes \boldsymbol{B}, \\ \boldsymbol{A} \otimes (\boldsymbol{B}_1 + \boldsymbol{B}_2) &= \boldsymbol{A} \otimes \boldsymbol{B}_1 + \boldsymbol{A} \otimes \boldsymbol{B}_2, \\ (\boldsymbol{A} \otimes \boldsymbol{B}) \otimes \boldsymbol{C} &= \boldsymbol{A} \otimes (\boldsymbol{B} \otimes \boldsymbol{C}). \end{aligned} \tag{10.13}$$

Note however that in general we have $\boldsymbol{A} \otimes \boldsymbol{B} \neq \boldsymbol{B} \otimes \boldsymbol{A}$, but it can be shown that there are permutation matrices $\boldsymbol{P}, \boldsymbol{Q}$ such that $\boldsymbol{B} \otimes \boldsymbol{A} = \boldsymbol{P}(\boldsymbol{A} \otimes \boldsymbol{B})\boldsymbol{Q}$, see [9].

The following **mixed product rule** is an essential tool for dealing with Kronecker products and sums.

Lemma 10.1 (Mixed Product Rule) *Suppose $\boldsymbol{A}, \boldsymbol{B}, \boldsymbol{C}, \boldsymbol{D}$ are rectangular matrices with dimensions so that the products $\boldsymbol{AC}$ and $\boldsymbol{BD}$ are well defined. Then the product $(\boldsymbol{A} \otimes \boldsymbol{B})(\boldsymbol{C} \otimes \boldsymbol{D})$ is defined and*

$$(\boldsymbol{A} \otimes \boldsymbol{B})(\boldsymbol{C} \otimes \boldsymbol{D}) = (\boldsymbol{AC}) \otimes (\boldsymbol{BD}). \tag{10.14}$$

Proof If $\boldsymbol{B} \in \mathbb{R}^{r,t}$ and $\boldsymbol{D} \in \mathbb{R}^{t,s}$ for some integers r, s, t, then

$$(\boldsymbol{A} \otimes \boldsymbol{B})(\boldsymbol{C} \otimes \boldsymbol{D}) = \begin{bmatrix} \boldsymbol{A}b_{1,1} & \cdots & \boldsymbol{A}b_{1,t} \\ \vdots & & \vdots \\ \boldsymbol{A}b_{r,1} & \cdots & \boldsymbol{A}b_{r,t} \end{bmatrix} \begin{bmatrix} \boldsymbol{C}d_{1,1} & \cdots & \boldsymbol{C}d_{1,s} \\ \vdots & & \vdots \\ \boldsymbol{C}d_{t,1} & \cdots & \boldsymbol{C}d_{t,s} \end{bmatrix}.$$

Thus for all i, j

$$((\boldsymbol{A} \otimes \boldsymbol{B})(\boldsymbol{C} \otimes \boldsymbol{D}))_{i,j} = \boldsymbol{AC} \sum_{k=1}^{t} b_{i,k} d_{k,j} = (\boldsymbol{AC})(\boldsymbol{BD})_{i,j} = ((\boldsymbol{AC}) \otimes (\boldsymbol{BD}))_{i,j}.$$

□

Using the mixed product rule we obtain the following properties of Kronecker products and sums.

Theorem 10.1 (Properties of Kronecker Products) *Suppose for $r, s \in \mathbb{N}$ that $\boldsymbol{A} \in \mathbb{R}^{r \times r}$ and $\boldsymbol{B} \in \mathbb{R}^{s \times s}$ are square matrices with eigenpairs $(\lambda_i, \boldsymbol{u}_i)$ $i = 1, \ldots, r$ and $(\mu_j, \boldsymbol{v}_j)$, $j = 1, \ldots, s$. Moreover, let $\boldsymbol{F}, \boldsymbol{V} \in \mathbb{R}^{r \times s}$. Then*

1. *$(\boldsymbol{A} \otimes \boldsymbol{B})^T = \boldsymbol{A}^T \otimes \boldsymbol{B}^T$, (this also holds for rectangular matrices).*
2. *If $\boldsymbol{A}$ and $\boldsymbol{B}$ are nonsingular then $\boldsymbol{A} \otimes \boldsymbol{B}$ is nonsingular. with $(\boldsymbol{A} \otimes \boldsymbol{B})^{-1} = \boldsymbol{A}^{-1} \otimes \boldsymbol{B}^{-1}$.*
3. *If $\boldsymbol{A}$ and $\boldsymbol{B}$ are symmetric then $\boldsymbol{A} \otimes \boldsymbol{B}$ and $\boldsymbol{A} \otimes \boldsymbol{I}_s + \boldsymbol{I}_r \otimes \boldsymbol{B}$ are symmetric.*
4. *$(\boldsymbol{A} \otimes \boldsymbol{B})(\boldsymbol{u}_i \otimes \boldsymbol{v}_j) = \lambda_i \mu_j (\boldsymbol{u}_i \otimes \boldsymbol{v}_j)$, $\quad i = 1, \ldots, r, \quad j = 1, \ldots, s,$*
5. *$(\boldsymbol{A} \otimes \boldsymbol{I}_s + \boldsymbol{I}_r \otimes \boldsymbol{B})(\boldsymbol{u}_i \otimes \boldsymbol{v}_j) = (\lambda_i + \mu_j)(\boldsymbol{u}_i \otimes \boldsymbol{v}_j)$, $\quad i = 1, \ldots, r, \quad j = 1, \ldots, s.$,*
6. *If one of $\boldsymbol{A}$, $\boldsymbol{B}$ is positive definite and the other is positive semidefinite then $\boldsymbol{A} \otimes \boldsymbol{I} + \boldsymbol{I} \otimes \boldsymbol{B}$ is positive definite.*
7. *$\boldsymbol{AVB}^T = \boldsymbol{F} \quad \Leftrightarrow \quad (\boldsymbol{A} \otimes \boldsymbol{B}) \operatorname{vec}(\boldsymbol{V}) = \operatorname{vec}(\boldsymbol{F})$,*
8. *$\boldsymbol{AV} + \boldsymbol{VB}^T = \boldsymbol{F} \quad \Leftrightarrow \quad (\boldsymbol{A} \otimes \boldsymbol{I}_s + \boldsymbol{I}_r \otimes \boldsymbol{B}) \operatorname{vec}(\boldsymbol{V}) = \operatorname{vec}(\boldsymbol{F})$.*

Before giving the simple proofs of this theorem we present some comments.

1. The transpose (or the inverse) of an ordinary matrix product equals the transpose (or the inverse) of the matrices in reverse order. For Kronecker products the order is kept.
2. The eigenvalues of the Kronecker product (sum) are the product (sum) of the eigenvalues of the factors. The eigenvectors are the Kronecker products of the eigenvectors of the factors. In particular, the eigenvalues of the test matrix $\boldsymbol{T}_2$ are sums of eigenvalues of $\boldsymbol{T}_1$.
3. Since we already know that $\boldsymbol{T} = \operatorname{tridiag}(-1, 2, -1)$ is positive definite the 2D Poisson matrix $\boldsymbol{A} = \boldsymbol{T} \otimes \boldsymbol{I} + \boldsymbol{I} \otimes \boldsymbol{T}$ is also positive definite.
4. The system $\boldsymbol{AVB}^T = \boldsymbol{F}$ in part 7 can be solved by first finding $\boldsymbol{W}$ from $\boldsymbol{AW} = \boldsymbol{F}$, and then finding $\boldsymbol{V}$ from $\boldsymbol{BV}^T = \boldsymbol{W}^T$. This is preferable to solving the much larger linear system $(\boldsymbol{A} \otimes \boldsymbol{B}) \operatorname{vec}(\boldsymbol{V}) = \operatorname{vec}(\boldsymbol{F})$.
5. A fast way to solve the 2D Poisson problem in the form $\boldsymbol{TV} + \boldsymbol{VT} = \boldsymbol{F}$ will be considered in the next chapter.

Proof of Theorem 10.1

1. Exercise.
2. By the mixed product rule $(\boldsymbol{A} \otimes \boldsymbol{B})(\boldsymbol{A}^{-1} \otimes \boldsymbol{B}^{-1}) = (\boldsymbol{A}\boldsymbol{A}^{-1}) \otimes (\boldsymbol{B}\boldsymbol{B}^{-1}) = \boldsymbol{I}_r \otimes \boldsymbol{I}_s = \boldsymbol{I}_{rs}$.Thus $(\boldsymbol{A} \otimes \boldsymbol{B})$ is nonsingular with the indicated inverse.
3. By 1, $(\boldsymbol{A} \otimes \boldsymbol{B})^T = \boldsymbol{A}^T \otimes \boldsymbol{B}^T = \boldsymbol{A} \otimes \boldsymbol{B}$. Moreover, since then $\boldsymbol{A} \otimes \boldsymbol{I}$ and $\boldsymbol{I} \otimes \boldsymbol{B}$ are symmetric, their sum is symmetric.
4. $(\boldsymbol{A} \otimes \boldsymbol{B})(\boldsymbol{u}_i \otimes \boldsymbol{v}_j) = (\boldsymbol{A}\boldsymbol{u}_i) \otimes (\boldsymbol{B}\boldsymbol{v}_j) = (\lambda_i \boldsymbol{u}_i) \otimes (\mu_j \boldsymbol{v}_j) = (\lambda_i \mu_j)(\boldsymbol{u}_i \otimes \boldsymbol{v}_j)$, for all i, j, where we used the mixed product rule.
5. $(\boldsymbol{A} \otimes \boldsymbol{I}_s)(\boldsymbol{u}_i \otimes \boldsymbol{v}_j) = \lambda_i (\boldsymbol{u}_i \otimes \boldsymbol{v}_j)$, and $(\boldsymbol{I}_r \otimes \boldsymbol{B})(\boldsymbol{u}_i \otimes \boldsymbol{v}_j) = \mu_j (\boldsymbol{u}_i \otimes \boldsymbol{v}_j)$. The result now follows by summing these relations.
6. By 1, $\boldsymbol{A} \otimes \boldsymbol{I} + \boldsymbol{I} \otimes \boldsymbol{B}$ is symmetric. Moreover, the eigenvalues $\lambda_i + \mu_j$ are positive since for all i, j, both λ_i and μ_j are nonnegative and one of them is positive. It follows that $\boldsymbol{A} \otimes \boldsymbol{I} + \boldsymbol{I} \otimes \boldsymbol{B}$ is positive definite.
7. We partition $\boldsymbol{V}$, $\boldsymbol{F}$, and $\boldsymbol{B}^T$ by columns as $\boldsymbol{V} = [\boldsymbol{v}_1, \ldots, \boldsymbol{v}_s]$, $\boldsymbol{F} = [\boldsymbol{f}_1, \ldots, \boldsymbol{f}_s]$ and $\boldsymbol{B}^T = [\boldsymbol{b}_1, \ldots, \boldsymbol{b}_s]$. Then we have

$$(\boldsymbol{A} \otimes \boldsymbol{B}) \operatorname{vec}(\boldsymbol{V}) = \operatorname{vec}(\boldsymbol{F})$$

$$\Leftrightarrow \quad \begin{bmatrix} \boldsymbol{A}b_{11} & \cdots & \boldsymbol{A}b_{1s} \\ \vdots & & \vdots \\ \boldsymbol{A}b_{s1} & \cdots & \boldsymbol{A}b_{ss} \end{bmatrix} \begin{bmatrix} \boldsymbol{v}_1 \\ \vdots \\ \boldsymbol{v}_s \end{bmatrix} = \begin{bmatrix} \boldsymbol{f}_1 \\ \vdots \\ \boldsymbol{f}_s \end{bmatrix}$$

$$\Leftrightarrow \quad \boldsymbol{A}\Big[\sum_j b_{1j}\boldsymbol{v}_j, \ldots, \sum_j b_{sj}\boldsymbol{v}_j\Big] = [\boldsymbol{f}_1, \ldots, \boldsymbol{f}_s]$$

$$\Leftrightarrow \quad \boldsymbol{A}[\boldsymbol{V}\boldsymbol{b}_1, \ldots, \boldsymbol{V}\boldsymbol{b}_s] = \boldsymbol{F} \quad \Leftrightarrow \quad \boldsymbol{A}\boldsymbol{V}\boldsymbol{B}^T = \boldsymbol{F}.$$

8. From the proof of 7. we have $(\boldsymbol{A} \otimes \boldsymbol{I}_s) \operatorname{vec}(\boldsymbol{V}) = \boldsymbol{A}\boldsymbol{V}\boldsymbol{I}_s^T$ and $(\boldsymbol{I}_r \otimes \boldsymbol{B}) \operatorname{vec}(\boldsymbol{V}) = \boldsymbol{I}_r\boldsymbol{V}\boldsymbol{B}^T$. But then

$$(\boldsymbol{A} \otimes \boldsymbol{I}_s + \boldsymbol{I}_r \otimes \boldsymbol{B}) \operatorname{vec}(\boldsymbol{V}) = \operatorname{vec}(\boldsymbol{F})$$

$$\Leftrightarrow \quad (\boldsymbol{A}\boldsymbol{V}\boldsymbol{I}_s^T + \boldsymbol{I}_r\boldsymbol{V}\boldsymbol{B}^T) = \boldsymbol{F} \quad \Leftrightarrow \quad \boldsymbol{A}\boldsymbol{V} + \boldsymbol{V}\boldsymbol{B}^T = \boldsymbol{F}.$$

For more on Kronecker products see [9]. □

10.3 Properties of the 2D Test Matrices

Using Theorem 10.1 we can derive properties of the 2D test matrix $\boldsymbol{T}_2$ from those of $\boldsymbol{T}_1$. Recall (cf. Lemma 2.2) that $\boldsymbol{T}_1\boldsymbol{s}_j = \lambda_j \boldsymbol{s}_j$ for $j = 1, \ldots, m$, where

$$\lambda_j = d + 2a\cos(j\pi h), \quad h := \frac{1}{m+1}, \tag{10.15}$$

$$s_j = [\sin(j\pi h), \sin(2j\pi h), \ldots, \sin(mj\pi h)]^T. \tag{10.16}$$

Moreover, the eigenvalues are distinct and the eigenvectors are orthogonal

$$s_j^T s_k = \frac{m+1}{2}\delta_{j,k} = \frac{1}{2h}\delta_{j,k}, \quad j,k = 1,\ldots,m. \tag{10.17}$$

Theorem 10.2 (Eigenpairs of 2D Test Matrix) *For fixed $m \geq 2$ let $\boldsymbol{T}_2$ be the matrix given by (10.10) and let $h = 1/(m+1)$. Then*

$$\boldsymbol{T}_2(s_j \otimes s_k) = (\lambda_j + \lambda_k)(s_j \otimes s_k) \quad j,k = 1,\ldots,m, \tag{10.18}$$

where (λ_j, s_j) are the eigenpairs of $\boldsymbol{T}_1$ given by (10.15) and (10.16). The eigenvectors $s_j \otimes s_k$ are orthogonal

$$(s_j \otimes s_k)^T(s_p \otimes s_q) = \frac{1}{4h^2}\delta_{j,p}\delta_{k,q}, \quad j,k,p,q = 1,\ldots,m, \tag{10.19}$$

and $\boldsymbol{T}_2$ is positive definite if $d > 0$ and $d \geq 2|a|$.

Proof Equation (10.18) follows from Part 5. of Theorem 10.1. Using the transpose rule, the mixed product rule and (2.32) we find for $j,k,p,q = 1,\ldots,m$

$$\left(s_j \otimes s_k\right)^T\left(s_p \otimes s_q\right) = \left(s_j^T \otimes s_k^T\right)\left(s_p \otimes s_q\right) = \left(s_j^T s_p\right) \otimes \left(s_k^T s_q\right) = \frac{1}{4h^2}\delta_{j,p}\delta_{k,q}$$

and (10.19) follows. Since $\boldsymbol{T}_2$ is symmetric, $\boldsymbol{T}_2$ is positive definite if the λ_j given by (10.15) are positive. But this is true whenever $d > 0$ and $d \geq 2|a|$ (cf. Exercise 10.5). □

Corollary 10.1 *The spectral condition number of the discrete Poisson matrix $\boldsymbol{A} \in \mathbb{R}^{m^2 \times m^2}$ given by (10.8) is given by*

$$\|\boldsymbol{A}\|_2\|\boldsymbol{A}^{-1}\|_2 = \frac{\cos^2 w}{\sin^2 w}, \quad w := \frac{\pi}{2(m+1)}. \tag{10.20}$$

Proof Recall that by (10.15) with $d = 2$, $a = -1$, and (10.18), the eigenvalues $\lambda_{j,k}$ of $\boldsymbol{A}$ are

$$\lambda_{j,k} = 4 - 2\cos(2jw) - 2\cos(2kw) = 4\sin^2(jw) + 4\sin^2(kw), \quad j,k = 1,\ldots,m.$$

Using trigonometric formulas, it follows that the largest and smallest eigenvalue of $\boldsymbol{A}$, are given by

$$\lambda_{max} = 8\cos^2 w, \quad \lambda_{min} = 8\sin^2 w.$$

Since $d > 0$ and $d \geq 2|a|$ it follows that $\boldsymbol{A}$ is positive definite. By (8.26) we have $\|\boldsymbol{A}\|_2\|\boldsymbol{A}^{-1}\|_2 = \frac{\lambda_{max}}{\lambda_{min}}$ and (10.20) follows. □

10.4 Exercises Chap. 10

10.4.1 Exercises Sects. 10.1, 10.2

Exercise 10.1 (4×4 **Poisson Matrix)** Write down the Poisson matrix for $m = 2$ and show that it is strictly diagonally dominant.

Exercise 10.2 (Properties of Kronecker Products) Prove (10.13).

Exercise 10.3 (Eigenpairs of Kronecker Products (Exam Exercise 2008-3)) Let $\boldsymbol{A}, \boldsymbol{B} \in \mathbb{R}^{n\times n}$. Show that the eigenvalues of the Kronecker product $\boldsymbol{A} \otimes \boldsymbol{B}$ are products of the eigenvalues of $\boldsymbol{A}$ and $\boldsymbol{B}$ and that the eigenvectors of $\boldsymbol{A} \otimes \boldsymbol{B}$ are Kronecker products of the eigenvectors of $\boldsymbol{A}$ and $\boldsymbol{B}$.

10.4.2 Exercises Sect. 10.3

Exercise 10.4 (2. Derivative Matrix Is Positive Definite) Write down the eigenvalues of $\boldsymbol{T} = \text{tridiag}(-1, 2, -1)$ using (10.15) and conclude that $\boldsymbol{T}$ is symmetric positive definite.

Exercise 10.5 (1D Test Matrix Is Positive Definite?) Show that the matrix $\boldsymbol{T}_1$ is symmetric positive definite if $d > 0$ and $d \geq 2|a|$.

Exercise 10.6 (Eigenvalues for 2D Test Matrix of Order 4) For $m = 2$ the matrix (10.10) is given by

$$\boldsymbol{A} = \begin{bmatrix} 2d & a & a & 0 \\ a & 2d & 0 & a \\ a & 0 & 2d & a \\ 0 & a & a & 2d \end{bmatrix}.$$

Show that $\lambda = 2a + 2d$ is an eigenvalue corresponding to the eigenvector $\boldsymbol{x} = [1, 1, 1, 1]^T$. Verify that apart from a scaling of the eigenvector this agrees with (10.15) and (10.16) for $j = k = 1$ and $m = 2$.

Exercise 10.7 (Nine Point Scheme for Poisson Problem) Consider the following 9 point difference approximation to the Poisson problem $-\Delta u = f$, $u = 0$ on the boundary of the unit square (cf. (10.1))

$$\begin{aligned}
&(a)\ -(\Box_h v)_{j,k} = (\mu f)_{j,k} && j,k = 1,\ldots,m\\
&(b)\qquad 0 = v_{0,k} = v_{m+1,k} = v_{j,0} = v_{j,m+1}, && j,k = 0,1,\ldots,m+1,\\
&(c)\ -(\Box_h v)_{j,k} = [20v_{j,k} - 4v_{j-1,k} - 4v_{j,k-1} - 4v_{j+1,k} - 4v_{j,k+1}\\
&\qquad - v_{j-1,k-1} - v_{j+1,k-1} - v_{j-1,k+1} - v_{j+1,k+1}]/(6h^2),\\
&(d)\quad (\mu f)_{j,k} = [8f_{j,k} + f_{j-1,k} + f_{j,k-1} + f_{j+1,k} + f_{j,k+1}]/12.
\end{aligned} \tag{10.21}$$

a) Write down the 4-by-4 system we obtain for $m = 2$.
b) Find $v_{j,k}$ for $j,k = 1,2$, if $f(x,y) = 2\pi^2 \sin(\pi x)\sin(\pi y)$ and $m = 2$. Answer: $v_{j,k} = 5\pi^2/66$.

It can be shown that (10.21) defines an $O(h^4)$ approximation to (10.1).

Exercise 10.8 (Matrix Equation for Nine Point Scheme) Consider the nine point difference approximation to (10.1) given by (10.21) in Problem 10.7.

a) Show that (10.21) is equivalent to the matrix equation

$$\boldsymbol{T}\boldsymbol{V} + \boldsymbol{V}\boldsymbol{T} - \frac{1}{6}\boldsymbol{T}\boldsymbol{V}\boldsymbol{T} = h^2\mu\boldsymbol{F}. \tag{10.22}$$

Here $\mu\boldsymbol{F}$ has elements $(\mu f)_{j,k}$ given by (10.21d) and $\boldsymbol{T} = \text{tridiag}(-1,2,-1)$.
b) Show that the standard form of the matrix equation (10.22) is $\boldsymbol{A}\boldsymbol{x} = \boldsymbol{b}$, where $\boldsymbol{A} = \boldsymbol{T}\otimes\boldsymbol{I} + \boldsymbol{I}\otimes\boldsymbol{T} - \frac{1}{6}\boldsymbol{T}\otimes\boldsymbol{T}$, $\boldsymbol{x} = \text{vec}(\boldsymbol{V})$, and $\boldsymbol{b} = h^2\text{vec}(\mu\boldsymbol{F})$.

Exercise 10.9 (Biharmonic Equation) Consider the biharmonic equation

$$\begin{aligned}
&\Delta^2 u(s,t) := \Delta\big(\Delta u(s,t)\big) = f(s,t) \quad (s,t)\in\Omega,\\
&u(s,t) = 0, \quad \Delta u(s,t) = 0 \qquad (s,t)\in\partial\Omega.
\end{aligned} \tag{10.23}$$

Here Ω is the open unit square. The condition $\Delta u = 0$ is called the *Navier boundary condition*. Moreover, $\Delta^2 u = u_{xxxx} + 2u_{xxyy} + u_{yyyy}$.

a) Let $v = -\Delta u$. Show that (10.23) can be written as a system

$$\begin{aligned}
-\Delta v(s,t) &= f(s,t) && (s,t)\in\Omega\\
-\Delta u(s,t) &= v(s,t) && (s,t)\in\Omega\\
u(s,t) &= v(s,t) = 0 && (s,t)\in\partial\Omega.
\end{aligned} \tag{10.24}$$

b) Discretizing, using (10.4), with $\boldsymbol{T} = \text{tridiag}(-1,2,-1)\in\mathbb{R}^{m\times m}$, $h = 1/(m+1)$, and $\boldsymbol{F} = \big(f(jh,kh)\big)_{j,k=1}^m$ we get two matrix equations

$$\boldsymbol{T}\boldsymbol{V} + \boldsymbol{V}\boldsymbol{T} = h^2\boldsymbol{F}, \qquad \boldsymbol{T}\boldsymbol{U} + \boldsymbol{U}\boldsymbol{T} = h^2\boldsymbol{V}.$$

Show that

$$(\boldsymbol{T}\otimes\boldsymbol{I}+\boldsymbol{I}\otimes\boldsymbol{T})\mathrm{vec}(\boldsymbol{V})=h^2\mathrm{vec}(\boldsymbol{F}),\quad (\boldsymbol{T}\otimes\boldsymbol{I}+\boldsymbol{I}\otimes\boldsymbol{T})\mathrm{vec}(\boldsymbol{U})=h^2\mathrm{vec}(\boldsymbol{V}).$$

and hence $\boldsymbol{A}=(\boldsymbol{T}\otimes\boldsymbol{I}+\boldsymbol{I}\otimes\boldsymbol{T})^2$ is the matrix for the standard form of the discrete biharmonic equation.

c) Show that with $n=m^2$ the vector form and standard form of the systems in b) can be written

$$\boldsymbol{T}^2\boldsymbol{U}+2\boldsymbol{T}\boldsymbol{U}\boldsymbol{T}+\boldsymbol{U}\boldsymbol{T}^2=h^4\boldsymbol{F}\quad\text{and}\quad \boldsymbol{A}\boldsymbol{x}=\boldsymbol{b}, \tag{10.25}$$

where $\boldsymbol{A}=\boldsymbol{T}^2\otimes\boldsymbol{I}+2\boldsymbol{T}\otimes\boldsymbol{T}+\boldsymbol{I}\otimes\boldsymbol{T}^2\in\mathbb{R}^{n\times n}$, $\boldsymbol{x}=\mathrm{vec}(\boldsymbol{U})$, and $\boldsymbol{b}=h^4\,\mathrm{vec}(\boldsymbol{F})$.

d) Determine the eigenvalues and eigenvectors of the matrix $\boldsymbol{A}$ in **c)** and show that it is positive definite. Also determine the bandwidth of $\boldsymbol{A}$.

e) Suppose we want to solve the standard form equation $\boldsymbol{A}\boldsymbol{x}=\boldsymbol{b}$. We have two representations for the matrix $\boldsymbol{A}$, the product one in **b)** and the one in **c)**. Which one would you prefer for the basis of an algorithm? Why?

10.5 Review Questions

10.5.1 Consider the Poisson matrix.

- Write this matrix as a Kronecker sum,
- how are its eigenvalues and eigenvectors related to the second derivative matrix?
- is it symmetric? positive definite?

10.5.2 What are the eigenpairs of $\boldsymbol{T}_1 :=$ tridiagonal(a,d,a)?

10.5.3 What are the inverse and transpose of a Kronecker product?

10.5.4

- give an economical general way to solve the linear system $(\boldsymbol{A}\otimes\boldsymbol{B})\,\mathrm{vec}(\boldsymbol{V})=\mathrm{vec}(\boldsymbol{F})$?
- Same for $(\boldsymbol{A}\otimes\boldsymbol{I}_s+\boldsymbol{I}_r\otimes\boldsymbol{B})\,\mathrm{vec}(\boldsymbol{V})=\mathrm{vec}(\boldsymbol{F})$.

Chapter 11
Fast Direct Solution of a Large Linear System

11.1 Algorithms for a Banded Positive Definite System

In this chapter we present a fast method for solving $\boldsymbol{Ax} = \boldsymbol{b}$, where $\boldsymbol{A}$ is the Poisson matrix (10.8). Thus, for $n = 9$

$$\boldsymbol{A} = \left[\begin{array}{ccc|ccc|ccc} 4 & -1 & 0 & -1 & 0 & 0 & 0 & 0 & 0 \\ -1 & 4 & -1 & 0 & -1 & 0 & 0 & 0 & 0 \\ 0 & -1 & 4 & 0 & 0 & -1 & 0 & 0 & 0 \\ \hline -1 & 0 & 0 & 4 & -1 & 0 & -1 & 0 & 0 \\ 0 & -1 & 0 & -1 & 4 & -1 & 0 & -1 & 0 \\ 0 & 0 & -1 & 0 & -1 & 4 & 0 & 0 & -1 \\ \hline 0 & 0 & 0 & -1 & 0 & 0 & 4 & -1 & 0 \\ 0 & 0 & 0 & 0 & -1 & 0 & -1 & 4 & -1 \\ 0 & 0 & 0 & 0 & 0 & -1 & 0 & -1 & 4 \end{array}\right]$$

$$= \begin{bmatrix} \boldsymbol{T} + 2\boldsymbol{I} & -\boldsymbol{I} & \boldsymbol{0} \\ -\boldsymbol{I} & \boldsymbol{T} + 2\boldsymbol{I} & -\boldsymbol{I} \\ \boldsymbol{0} & -\boldsymbol{I} & \boldsymbol{T} + 2\boldsymbol{I} \end{bmatrix},$$

where $\boldsymbol{T} = \text{tridiag}(-1, 2, -1)$. For the matrix $\boldsymbol{A}$ we know by now that

1. It is positive definite.
2. It is banded.
3. It is block-tridiagonal.
4. We know the eigenvalues and eigenvectors of $\boldsymbol{A}$.
5. The eigenvectors are orthogonal.

T. Lyche, *Numerical Linear Algebra and Matrix Factorizations*,
Texts in Computational Science and Engineering 22,
https://doi.org/10.1007/978-3-030-36468-7_11

11.1.1 Cholesky Factorization

Since $\boldsymbol{A}$ is positive definite we can use the Cholesky factorization $\boldsymbol{A} = \boldsymbol{L}\boldsymbol{L}^*$, with $\boldsymbol{L}$ lower triangular, to solve $\boldsymbol{A}\boldsymbol{x} = \boldsymbol{b}$. Since $\boldsymbol{A}$ and $\boldsymbol{L}$ has the same bandwidth $d = \sqrt{n}$ the complexity of this factorization is $O(nd^2) = O(n^2)$, cf. Algorithm 4.2. We need to store $\boldsymbol{A}$, and this can be done in sparse form.

The nonzero elements in $\boldsymbol{L}$ are shown in Fig. 11.1 for $n = 100$. Note that most of the zeros between the diagonals in $\boldsymbol{A}$ have become nonzero in $\boldsymbol{L}$. This is known as **fill-inn**.

11.1.2 Block LU Factorization of a Block Tridiagonal Matrix

The Poisson matrix has a block tridiagonal structure. Consider finding the block LU factorization of a block tridiagonal matrix. We are looking for a factorization of the form

$$\begin{bmatrix} \boldsymbol{D}_1 & \boldsymbol{C}_1 & & & \\ \boldsymbol{A}_1 & \boldsymbol{D}_2 & \boldsymbol{C}_2 & & \\ & \ddots & \ddots & \ddots & \\ & & \boldsymbol{A}_{m-2} & \boldsymbol{D}_{m-1} & \boldsymbol{C}_{m-1} \\ & & & \boldsymbol{A}_{m-1} & \boldsymbol{D}_m \end{bmatrix} = \begin{bmatrix} \boldsymbol{I} & & & \\ \boldsymbol{L}_1 & \boldsymbol{I} & & \\ & \ddots & \ddots & \\ & & \boldsymbol{L}_{m-1} & \boldsymbol{I} \end{bmatrix} \begin{bmatrix} \boldsymbol{U}_1 & \boldsymbol{C}_1 & & \\ & \ddots & \ddots & \\ & & \boldsymbol{U}_{m-1} & \boldsymbol{C}_{m-1} \\ & & & \boldsymbol{U}_m \end{bmatrix}. \tag{11.1}$$

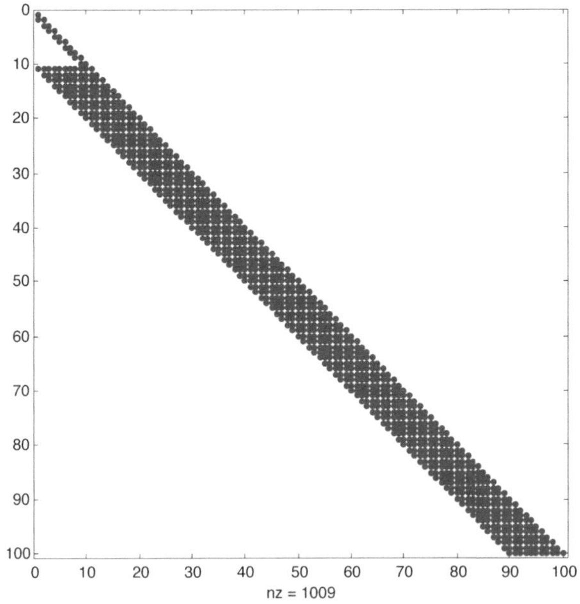

Fig. 11.1 Fill-inn in the Cholesky factor of the Poisson matrix ($n = 100$)

Here $\boldsymbol{D}_1, \ldots, \boldsymbol{D}_m$ and $\boldsymbol{U}_1, \ldots, \boldsymbol{U}_m$ are square matrices while $\boldsymbol{A}_1, \ldots, \boldsymbol{A}_{m-1}, \boldsymbol{L}_1, \ldots, \boldsymbol{L}_{m-1}$ and $\boldsymbol{C}_1, \ldots, \boldsymbol{C}_{m-1}$ can be rectangular.

Using block multiplication the formulas (2.16) generalize to

$$\boldsymbol{U}_1 = \boldsymbol{D}_1, \quad \boldsymbol{L}_k = \boldsymbol{A}_k \boldsymbol{U}_k^{-1}, \quad \boldsymbol{U}_{k+1} = \boldsymbol{D}_{k+1} - \boldsymbol{L}_k \boldsymbol{C}_k, \quad k = 1, 2, \ldots, m-1. \tag{11.2}$$

To solve the system $\boldsymbol{A}\boldsymbol{x} = \boldsymbol{b}$ we partition $\boldsymbol{b}$ conformally with $\boldsymbol{A}$ in the form $\boldsymbol{b}^T = [\boldsymbol{b}_1^T, \ldots, \boldsymbol{b}_m^T]$. The formulas for solving $\boldsymbol{L}\boldsymbol{y} = \boldsymbol{b}$ and $\boldsymbol{U}\boldsymbol{x} = \boldsymbol{y}$ are as follows:

$$\begin{aligned} &\boldsymbol{y}_1 = \boldsymbol{b}_1, && \boldsymbol{y}_k = \boldsymbol{b}_k - \boldsymbol{L}_{k-1}\boldsymbol{y}_{k-1}, && k = 2, 3, \ldots, m, \\ &\boldsymbol{x}_m = \boldsymbol{U}_m^{-1}\boldsymbol{y}_m, && \boldsymbol{x}_k = \boldsymbol{U}_k^{-1}(\boldsymbol{y}_k - \boldsymbol{C}_k\boldsymbol{x}_{k+1}), && k = m-1, \ldots, 2, 1. \end{aligned} \tag{11.3}$$

The solution is then $\boldsymbol{x}^T = [\boldsymbol{x}_1^T, \ldots, \boldsymbol{x}_m^T]$. To find $\boldsymbol{L}_k$ in (11.2) we solve the linear systems $\boldsymbol{L}_k \boldsymbol{U}_k = \boldsymbol{A}_k$. Similarly we need to solve a linear system to find $\boldsymbol{x}_k$ in (11.3).

The number of arithmetic operations using block factorizations is $O(n^2)$, asymptotically the same as for Cholesky factorization. However we only need to store the $m \times m$ blocks, and using matrix operations can be advantageous.

11.1.3 Other Methods

Other methods include

- Iterative methods, (we study this in Chaps. 12 and 13),
- multigrid. See [5],
- fast solvers based on diagonalization and the fast Fourier transform. See Sects. 11.2, 11.3.

11.2 A Fast Poisson Solver Based on Diagonalization

The algorithm we now derive will only require $O(n^{3/2})$ arithmetic operations and we only need to work with matrices of order m. Using the fast Fourier transform the number of arithmetic operations can be reduced further to $O(n \log n)$.

To start we recall that $\boldsymbol{A}\boldsymbol{x} = \boldsymbol{b}$ can be written as a matrix equation in the form (cf. (10.5))

$$\boldsymbol{T}\boldsymbol{V} + \boldsymbol{V}\boldsymbol{T} = h^2 \boldsymbol{F} \quad \text{with} \quad h = 1/(m+1),$$

where $\boldsymbol{T} = \text{tridiag}(-1, 2, -1) \in \mathbb{R}^{m \times m}$ is the second derivative matrix, $\boldsymbol{V} = (v_{j,k}) \in \mathbb{R}^{m \times m}$ are the unknowns, and $\boldsymbol{F} = (f_{j,k}) = (f(jh, kh)) \in \mathbb{R}^{m \times m}$ contains function values.

Recall (cf. Lemma 2.2) that the eigenpairs of $\boldsymbol{T}$ are given by

$$\begin{aligned}\boldsymbol{T}\boldsymbol{s}_j &= \lambda_j \boldsymbol{s}_j, \quad j = 1, \ldots, m,\\ \boldsymbol{s}_j &= [\sin(j\pi h), \sin(2j\pi h), \ldots, \sin(mj\pi h)]^T,\\ \lambda_j &= 2 - 2\cos(j\pi h) = 4\sin^2(j\pi h/2), \quad h = 1/(m+1),\\ \boldsymbol{s}_j^T \boldsymbol{s}_k &= \delta_{j,k}/(2h) \text{ for all } j, k.\end{aligned}$$

Let

$$\boldsymbol{S} := [\boldsymbol{s}_1, \ldots, \boldsymbol{s}_m] = \left[\sin(jk\pi h)\right]_{j,k=1}^m \in \mathbb{R}^{m\times m}, \quad \boldsymbol{D} = \operatorname{diag}(\lambda_1, \ldots, \lambda_m). \tag{11.4}$$

Then

$$\boldsymbol{T}\boldsymbol{S} = [\boldsymbol{T}\boldsymbol{s}_1, \ldots, \boldsymbol{T}\boldsymbol{s}_m] = [\lambda_1 \boldsymbol{s}_1, \ldots, \lambda_m \boldsymbol{s}_m] = \boldsymbol{S}\boldsymbol{D}, \quad \boldsymbol{S}^2 = \boldsymbol{S}^T\boldsymbol{S} = \frac{1}{2h}\boldsymbol{I}.$$

Define $\boldsymbol{X} \in \mathbb{R}^{m\times m}$ by $\boldsymbol{V} = \boldsymbol{S}\boldsymbol{X}\boldsymbol{S}$, where $\boldsymbol{V}$ is the solution of $\boldsymbol{T}\boldsymbol{V} + \boldsymbol{V}\boldsymbol{T} = h^2\boldsymbol{F}$. Then

$$\begin{aligned} && \boldsymbol{T}\boldsymbol{V} + \boldsymbol{V}\boldsymbol{T} &= h^2\boldsymbol{F}\\ &\overset{\boldsymbol{V}=\boldsymbol{S}\boldsymbol{X}\boldsymbol{S}}{\iff} & \boldsymbol{T}\boldsymbol{S}\boldsymbol{X}\boldsymbol{S} + \boldsymbol{S}\boldsymbol{X}\boldsymbol{S}\boldsymbol{T} &= h^2\boldsymbol{F}\\ &\overset{\boldsymbol{S}(\,)\boldsymbol{S}}{\iff} & \boldsymbol{S}\boldsymbol{T}\boldsymbol{S}\boldsymbol{X}\boldsymbol{S}^2 + \boldsymbol{S}^2\boldsymbol{X}\boldsymbol{S}\boldsymbol{T}\boldsymbol{S} &= h^2\boldsymbol{S}\boldsymbol{F}\boldsymbol{S} = h^2\boldsymbol{G}\\ &\overset{\boldsymbol{T}\boldsymbol{S}=\boldsymbol{S}\boldsymbol{D}}{\iff} & \boldsymbol{S}^2\boldsymbol{D}\boldsymbol{X}\boldsymbol{S}^2 + \boldsymbol{S}^2\boldsymbol{X}\boldsymbol{S}^2\boldsymbol{D} &= h^2\boldsymbol{G}\\ &\overset{\boldsymbol{S}^2=\boldsymbol{I}/(2h)}{\iff} & \boldsymbol{D}\boldsymbol{X} + \boldsymbol{X}\boldsymbol{D} &= 4h^4\boldsymbol{G}.\end{aligned}$$

Since $\boldsymbol{D}$ is diagonal, the equation $\boldsymbol{D}\boldsymbol{X} + \boldsymbol{X}\boldsymbol{D} = 4h^4\boldsymbol{G}$, is easy to solve. For the j, k element we find

$$(\boldsymbol{D}\boldsymbol{X} + \boldsymbol{X}\boldsymbol{D})_{j,k} = \sum_{\ell=1}^m d_{j,\ell}x_{\ell,k} + \sum_{\ell=1}^m x_{j,\ell}d_{\ell,k} = \lambda_j x_{j,k} + \lambda_k x_{j,k}$$

so that for all j, k

$$x_{j,k} = 4h^4 g_{j,k}/(\lambda_j + \lambda_k) = h^4 g_{j,k}/(\sigma_j + \sigma_k), \quad \sigma_j := \lambda_j/4 = \sin^2(j\pi h/2).$$

Thus to find $\boldsymbol{V}$ we compute

1. $\boldsymbol{G} = \boldsymbol{SFS}$,
2. $x_{j,k} = h^4 g_{j,k}/(\sigma_j + \sigma_k), \quad j,k = 1,\dots,m$,
3. $\boldsymbol{V} = \boldsymbol{SXS}$.

We can compute $\boldsymbol{X}, \boldsymbol{S}$ and the σ's without using loops. Using outer products, element by element division, and raising a matrix element by element to a power we find

$$\boldsymbol{X} = h^4\boldsymbol{G}/\boldsymbol{M}, \text{ where } \boldsymbol{M} := \begin{bmatrix}\sigma_1\\ \vdots\\ \sigma_m\end{bmatrix}[\,1, \dots, 1\,] + \begin{bmatrix}1\\ \vdots\\ 1\end{bmatrix}[\,\sigma_1. \dots .\sigma_m\,],$$

$$\boldsymbol{S} = \sin\Big(\pi h\begin{bmatrix}1\\ 2\\ \vdots\\ m\end{bmatrix}[\,1\ 2 \dots m\,]\Big), \quad \boldsymbol{\sigma} = \sin\Big(\frac{\pi h}{2}\begin{bmatrix}1\\ 2\\ \vdots\\ m\end{bmatrix}\Big)\wedge 2.$$

We now get the following algorithm to solve numerically the Poisson problem $-\Delta u = f$ on $\Omega = (0,1)^2$ and $u = 0$ on $\partial\Omega$ using the 5-point scheme, i.e., let $m \in \mathbb{N}$, $h = 1/(m+1)$, and $\boldsymbol{F} = (f(jh,kh)) \in \mathbb{R}^{m\times m}$. We compute $\boldsymbol{V} \in \mathbb{R}^{(m+2)\times(m+2)}$ using diagonalization of $\boldsymbol{T} = \text{tridiag}(-1,2,-1) \in \mathbb{R}^{m\times m}$.

```
function V=fastpoisson(F)
%function V=fastpoisson(F)
m=length(F); h=1/(m+1); hv=pi*h*(1:m)';
sigma=sin(hv/2).^2;
S=sin(hv*(1:m));
G=S*F*S;
X=h^4*G./(sigma*ones(1,m)+ ones(m,1)*sigma');
V=zeros(m+2,m+2);
V(2:m+1,2:m+1)=S*X*S;
end
```

Listing 11.1 fastpoisson

The formulas are fully vectorized. Since the 6th line in Algorithm 11.1 only requires $O(m^3)$ arithmetic operations, the complexity of this algorithm is for large m determined by 4 m-by-m matrix multiplications and is given by $O(4 \times 2m^3) = O(8n^{3/2})$.[1] The method is very fast and will be used as a preconditioner for a more complicated problem in Chap. 13. In 2012 it took about 0.2 seconds on a laptop to find the 10^6 unknowns $v_{j,k}$ on a 1000×1000 grid.

[1] It is possible to compute $\boldsymbol{V}$ using only two matrix multiplications and hence reduce the complexity to $O(4n^{3/2})$. This is detailed in Problem 11.4.

11.3 A Fast Poisson Solver Based on the Discrete Sine and Fourier Transforms

In Algorithm 11.1 we need to compute the product of the sine matrix $\boldsymbol{S} \in \mathbb{R}^{m\times m}$ given by (11.4) and a matrix $\boldsymbol{A} \in \mathbb{R}^{m\times m}$. Since the matrices are m-by-m this will normally require $O(m^3)$ operations. In this section we show that it is possible to calculate the products $\boldsymbol{SA}$ and $\boldsymbol{AS}$ in $O(m^2 \log_2 m)$ operations.

We need to discuss certain transforms known as the **discrete sine transform**, the **discrete Fourier transform** and the **fast Fourier transform**. In addition we have the **discrete cosine transform** which will not be discussed here. These transforms are of independent interest. They have applications to signal processing and image analysis, and are often used when one is dealing with discrete samples of data on a computer.

11.3.1 The Discrete Sine Transform (DST)

Given $\boldsymbol{v} = [v_1, \ldots, v_m]^T \in \mathbb{R}^m$ we say that the vector $\boldsymbol{w} = [w_1, \ldots, w_m]^T$ given by

$$w_j = \sum_{k=1}^{m} \sin\left(\frac{jk\pi}{m+1}\right) v_k, \quad j = 1, \ldots, m$$

is the **discrete sine transform** (DST) of $\boldsymbol{v}$. In matrix form we can write the DST as the matrix times vector $\boldsymbol{w} = \boldsymbol{Sv}$, where $\boldsymbol{S}$ is the sine matrix given by (11.4). We can then identify the matrix $\boldsymbol{B} = \boldsymbol{SA}$ as the DST of $\boldsymbol{A} \in \mathbb{R}^{m,n}$, i.e. as the DST of the columns of $\boldsymbol{A}$. The product $\boldsymbol{B} = \boldsymbol{AS}$ can also be interpreted as a DST. Indeed, since $\boldsymbol{S}$ is symmetric we have $\boldsymbol{B} = (\boldsymbol{SA}^T)^T$ which means that $\boldsymbol{B}$ is the transpose of the DST of the rows of $\boldsymbol{A}$. It follows that we can compute the unknowns $\boldsymbol{V}$ in Algorithm 11.1 by carrying out discrete sine transforms on 4 m-by-m matrices in addition to the computation of $\boldsymbol{X}$.

11.3.2 The Discrete Fourier Transform (DFT)

The fast computation of the DST is based on its relation to the discrete Fourier transform (DFT) and the fact that the DFT can be computed by a technique known as the fast Fourier transform (FFT). To define the DFT let for $N \in \mathbb{N}$

$$\omega_N = \exp^{-2\pi i/N} = \cos(2\pi/N) - i\sin(2\pi/N), \tag{11.5}$$

where $i = \sqrt{-1}$ is the imaginary unit. Given $\boldsymbol{y} = [y_1, \ldots, y_N]^T \in \mathbb{R}^N$ we say that $\boldsymbol{z} = [z_1, \ldots, z_N]^T$ given by

$$z_{j+1} = \sum_{k=0}^{N-1} \omega_N^{jk} y_{k+1}, \quad j = 0, \ldots, N-1,$$

is the **discrete Fourier transform** (DFT) of $\boldsymbol{y}$. We can write this as a matrix times vector product $\boldsymbol{z} = \boldsymbol{F}_N \boldsymbol{y}$, where the **Fourier matrix** $\boldsymbol{F}_N \in \mathbb{C}^{N\times N}$ has elements ω_N^{jk}, $j, k = 0, 1, \ldots, N-1$. For a matrix we say that $\boldsymbol{B} = \boldsymbol{F}_N \boldsymbol{A}$ is the DFT of $\boldsymbol{A}$.

As an example, since

$$\omega_4 = \exp^{-2\pi i/4} = \cos(\pi/2) - i\sin(\pi/2) = -i$$

we find $\omega_4^2 = (-i)^2 = -1$, $\omega_4^3 = (-i)(-1) = i$, $\omega_4^4 = (-1)^2 = 1$, $\omega_4^6 = i^2 = -1$, $\omega_4^9 = i^3 = -i$, and so

$$\boldsymbol{F}_4 = \begin{bmatrix} 1 & 1 & 1 & 1 \\ 1 & \omega_4 & \omega_4^2 & \omega_4^3 \\ 1 & \omega_4^2 & \omega_4^4 & \omega_4^6 \\ 1 & \omega_4^3 & \omega_4^6 & \omega_4^9 \end{bmatrix} = \begin{bmatrix} 1 & 1 & 1 & 1 \\ 1 & -i & -1 & i \\ 1 & -1 & 1 & -1 \\ 1 & i & -1 & -i \end{bmatrix}. \tag{11.6}$$

The following lemma shows how the discrete sine transform of order m can be computed from the discrete Fourier transform of order $2m + 2$. We recall that for any complex number w

$$\sin w = \frac{e^{iw} - e^{-iw}}{2i}.$$

Lemma 11.1 (Sine Transform as Fourier Transform) *Given a positive integer m and a vector $\boldsymbol{x} \in \mathbb{R}^m$. Component k of $\boldsymbol{S}\boldsymbol{x}$ is equal to $i/2$ times component $k+1$ of $\boldsymbol{F}_{2m+2}\boldsymbol{z}$ where*

$$\boldsymbol{z}^T = [0, \boldsymbol{x}^T, 0, -\boldsymbol{x}_B^T] \in \mathbb{R}^{2m+2}, \quad \boldsymbol{x}_B^T := [x_m, \ldots, x_2, x_1].$$

In symbols

$$(\boldsymbol{S}\boldsymbol{x})_k = \frac{i}{2}(\boldsymbol{F}_{2m+2}\boldsymbol{z})_{k+1}, \quad k = 1, \ldots, m.$$

Proof Let $\omega = \omega_{2m+2} = e^{-2\pi i/(2m+2)} = e^{-\pi i/(m+1)}$. We note that

$$\omega^{jk} = e^{-\pi ijk/(m+1)}, \quad \omega^{(2m+2-j)k} = e^{-2\pi i} e^{\pi ijk/(m+1)} = e^{\pi ijk/(m+1)}.$$

Component $k+1$ of $\boldsymbol{F}_{2m+2}\boldsymbol{z}$ is then given by

$$\begin{aligned}(\boldsymbol{F}_{2m+2}\boldsymbol{z})_{k+1} &= \sum_{j=0}^{2m-1} \omega^{jk} z_{j+1} = \sum_{j=1}^{m} x_j \omega^{jk} - \sum_{j=1}^{m} x_j \omega^{(2m+2-j)k} \\ &= \sum_{j=1}^{m} x_j \big(e^{-\pi ijk/(m+1)} - e^{\pi ijk/(m+1)}\big) \\ &= -2i \sum_{j=1}^{m} x_j \sin\left(\frac{jk\pi}{m+1}\right) = -2i(\boldsymbol{S}_m\boldsymbol{x})_k.\end{aligned}$$

Dividing both sides by $-2i$ and noting $-1/(2i) = -i/(2i^2) = i/2$, proves the lemma. □

It follows that we can compute the DST of length m by extracting m components from the DFT of length $N = 2m+2$.

11.3.3 The Fast Fourier Transform (FFT)

From a linear algebra viewpoint the fast Fourier transform is a quick way to compute the matrix- vector product $\boldsymbol{F}_N\boldsymbol{y}$. Suppose N is even. The key to the FFT is a connection between $\boldsymbol{F}_N$ and $\boldsymbol{F}_{N/2}$ which makes it possible to compute the FFT of order N as two FFT's of order $N/2$. By repeating this process we can reduce the number of arithmetic operations to compute a DFT from $O(N^2)$ to $O(N\log_2 N)$.

Suppose N is even. The connection between $\boldsymbol{F}_N$ and $\boldsymbol{F}_{N/2}$ involves a permutation matrix $\boldsymbol{P}_N \in \mathbb{R}^{N\times N}$ given by

$$\boldsymbol{P}_N = [\boldsymbol{e}_1, \boldsymbol{e}_3, \ldots, \boldsymbol{e}_{N-1}, \boldsymbol{e}_2, \boldsymbol{e}_4, \ldots, \boldsymbol{e}_N],$$

where the $\boldsymbol{e}_k = (\delta_{j,k})$ are unit vectors. If $\boldsymbol{A}$ is a matrix with N columns $[\boldsymbol{a}_1, \ldots, \boldsymbol{a}_N]$ then

$$\boldsymbol{A}\boldsymbol{P}_N = [\boldsymbol{a}_1, \boldsymbol{a}_3, \ldots, \boldsymbol{a}_{N-1}, \boldsymbol{a}_2, \boldsymbol{a}_4, \ldots, \boldsymbol{a}_N],$$

i.e. post multiplying $\boldsymbol{A}$ by $\boldsymbol{P}_N$ permutes the columns of $\boldsymbol{A}$ so that all the odd-indexed columns are followed by all the even-indexed columns. For example we have from (11.6)

$$\boldsymbol{P}_4 = [\boldsymbol{e}_1\ \boldsymbol{e}_3\ \boldsymbol{e}_2\ \boldsymbol{e}_4] = \begin{bmatrix} 1&0&0&0\\0&0&1&0\\0&1&0&0\\0&0&0&1\end{bmatrix} \quad \boldsymbol{F}_4\boldsymbol{P}_4 = \left[\begin{array}{cc|cc} 1&1&1&1\\ 1&-1&-i&i\\ \hline 1&1&-1&-1\\ 1&-1&i&-i\end{array}\right],$$

where we have indicated a certain block structure of $\boldsymbol{F}_4\boldsymbol{P}_4$. These blocks can be related to the 2-by-2 matrix $\boldsymbol{F}_2$. We define the diagonal scaling matrix $\boldsymbol{D}_2$ by

$$\boldsymbol{D}_2 = \operatorname{diag}(1, \omega_4) = \begin{bmatrix} 1 & 0 \\ 0 & -i \end{bmatrix}.$$

Since $\omega_2 = \exp^{-2\pi i/2} = -1$ we find

$$\boldsymbol{F}_2 = \begin{bmatrix} 1 & 1 \\ 1 & -1 \end{bmatrix}, \quad \boldsymbol{D}_2\boldsymbol{F}_2 = \begin{bmatrix} 1 & 1 \\ -i & i \end{bmatrix},$$

and we see that

$$\boldsymbol{F}_4\boldsymbol{P}_4 = \left[\begin{array}{c|c} \boldsymbol{F}_2 & \boldsymbol{D}_2\boldsymbol{F}_2 \\ \hline \boldsymbol{F}_2 & -\boldsymbol{D}_2\boldsymbol{F}_2 \end{array}\right].$$

This result holds in general.

Theorem 11.1 (Fast Fourier Transform) *If $N = 2m$ is even then*

$$\boldsymbol{F}_{2m}\boldsymbol{P}_{2m} = \left[\begin{array}{c|c} \boldsymbol{F}_m & \boldsymbol{D}_m\boldsymbol{F}_m \\ \hline \boldsymbol{F}_m & -\boldsymbol{D}_m\boldsymbol{F}_m \end{array}\right], \tag{11.7}$$

where

$$\boldsymbol{D}_m = \operatorname{diag}(1, \omega_N, \omega_N^2, \ldots, \omega_N^{m-1}). \tag{11.8}$$

Proof Fix integers p, q with $1 \le p, q \le m$ and set $j := p - 1$ and $k := q - 1$. Since

$$\omega_m^m = 1,\ \omega_{2m}^{2k} = \omega_m^k,\ \omega_{2m}^m = -1,\ (\boldsymbol{F}_m)_{p,q} = \omega_m^{jk},\ (\boldsymbol{D}_m\boldsymbol{F}_m)_{p,q} = \omega_{2m}^j\omega_m^{jk},$$

we find by considering elements in the four sub-blocks in turn

$$\begin{aligned}
(\boldsymbol{F}_{2m}\boldsymbol{P}_{2m})_{p,q} &= \omega_{2m}^{j(2k)} &&= \omega_m^{jk}, \\
(\boldsymbol{F}_{2m}\boldsymbol{P}_{2m})_{p+m,q} &= \omega_{2m}^{(j+m)(2k)} &&= \omega_m^{(j+m)k} &&= \omega_m^{jk}, \\
(\boldsymbol{F}_{2m}\boldsymbol{P}_{2m})_{p,q+m} &= \omega_{2m}^{j(2k+1)} &&= \omega_{2m}^{j}\omega_m^{jk}, \\
(\boldsymbol{F}_{2m}\boldsymbol{P}_{2m})_{p+m,q+m} &= \omega_{2m}^{(j+m)(2k+1)} &&= \omega_{2m}^{j+m}\omega_m^{(j+m)k} &&= -\omega_{2m}^{j}\omega_m^{jk}.
\end{aligned}$$

It follows that the four m-by-m blocks of $\boldsymbol{F}_{2m}\boldsymbol{P}_{2m}$ have the required structure. □

Using Theorem 11.1 we can carry out the DFT as a block multiplication. Let $\boldsymbol{y} \in \mathbb{R}^{2m}$ and set $\boldsymbol{w} = \boldsymbol{P}_{2m}^T\boldsymbol{y} = [\boldsymbol{w}_1^T, \boldsymbol{w}_2^T]^T$, where

$$\boldsymbol{w}_1^T = [y_1, y_3, \ldots, y_{2m-1}], \quad \boldsymbol{w}_2^T = [y_2, y_4, \ldots, y_{2m}].$$

Then

$$\begin{aligned} F_{2m}y &= F_{2m}P_{2m}P_{2m}^T y = F_{2m}P_{2m}w \\ &= \left[\begin{array}{c|c} F_m & D_m F_m \\ \hline F_m & -D_m F_m \end{array}\right] \begin{bmatrix} w_1 \\ w_2 \end{bmatrix} = \begin{bmatrix} q_1 + q_2 \\ q_1 - q_2 \end{bmatrix}, \end{aligned}$$

where

$$q_1 = F_m w_1, \quad \text{and} \quad q_2 = D_m(F_m w_2).$$

In order to compute $F_{2m}y$ we need to compute $F_m w_1$ and $F_m w_2$. Thus, by combining two FFT's of order m we obtain an FFT of order $2m$. If $n = 2^k$ then this process can be applied recursively as in the following MATLAB function:

```
function z=fftrec(y)
%function z=fftrec(y)
y=y(:);
n=length(y);
if n==1
   z=y;
else
   q1=fftrec(y(1:2:n-1));
   q2=exp(-2*pi*1i/n).^(0:n/2-1)'.*fftrec(y(2:2:n));
   z=[q1+q2; q1-q2];
end
```

Listing 11.2 fftrec

Statement 3 is included so that the input $y \in \mathbb{R}^n$ can be either a row or column vector, while the output z is a column vector.

Such a recursive version of FFT is useful for testing purposes, but is much too slow for large problems. A challenge for FFT code writers is to develop nonrecursive versions and also to handle efficiently the case where N is not a power of two. We refer to [14] for further details.

The complexity of the FFT is given by $\gamma N \log_2 N$ for some constant γ independent of N. To show this for the special case when N is a power of two let x_k be the complexity (the number of arithmetic operations) when $N = 2^k$. Since we need two FFT's of order $N/2 = 2^{k-1}$ and a multiplication with the diagonal matrix $D_{N/2}$, it is reasonable to assume that $x_k = 2x_{k-1} + \gamma 2^k$ for some constant γ independent of k. Since $x_0 = 0$ we obtain by induction on k that $x_k = \gamma k 2^k$. Indeed, this holds for $k = 0$ and if $x_{k-1} = \gamma(k-1)2^{k-1}$ then $x_k = 2x_{k-1} + \gamma 2^k = 2\gamma(k-1)2^{k-1} + \gamma 2^k = \gamma k 2^k$. Reasonable implementations of FFT typically have $\gamma \approx 5$, see [14].

The efficiency improvement using the FFT to compute the DFT is spectacular for large N. The direct multiplication $F_N y$ requires $O(8n^2)$ arithmetic operations since complex arithmetic is involved. Assuming that the FFT uses $5N \log_2 N$ arithmetic operations we find for $N = 2^{20} \approx 10^6$ the ratio

$$\frac{8N^2}{5N \log_2 N} \approx 84000.$$

Thus if the FFT takes one second of computing time and the computing time is proportional to the number of arithmetic operations then the direct multiplication would take something like 84000 seconds or 23 hours.

11.3.4 A Poisson Solver Based on the FFT

We now have all the ingredients to compute the matrix products $\boldsymbol{SA}$ and $\boldsymbol{AS}$ using FFT's of order $2m + 2$ where m is the order of $\boldsymbol{S}$ and $\boldsymbol{A}$. This can then be used for quick computation of the exact solution $\boldsymbol{V}$ of the discrete Poisson problem in Algorithm 11.1. We first compute $\boldsymbol{H} = \boldsymbol{SF}$ using Lemma 11.1 and m FFT's, one for each of the m columns of $\boldsymbol{F}$. We then compute $\boldsymbol{G} = \boldsymbol{HS}$ by m FFT's, one for each of the rows of $\boldsymbol{H}$. After $\boldsymbol{X}$ is determined we compute $\boldsymbol{Z} = \boldsymbol{SX}$ and $\boldsymbol{V} = \boldsymbol{ZS}$ by another $2m$ FFT's. In total the work amounts to $4m$ FFT's of order $2m+2$. Since one FFT requires $O(\gamma(2m + 2)\log_2(2m + 2))$ arithmetic operations the $4m$ FFT's amount to

$$8\gamma m(m + 1)\log_2(2m + 2) \approx 8\gamma m^2 \log_2 m = 4\gamma n \log_2 n,$$

where $n = m^2$ is the size of the linear system $\boldsymbol{Ax} = \boldsymbol{b}$ we would be solving if Cholesky factorization was used. This should be compared to the $O(8n^{3/2})$ arithmetic operations used in Algorithm 11.1 requiring 4 straightforward matrix multiplications with $\boldsymbol{S}$. What is faster will depend heavily on the programming of the FFT and the size of the problem. We refer to [14] for other efficient ways to implement the DST.

11.4 Exercises Chap. 11

11.4.1 Exercises Sect. 11.3

Exercise 11.1 (Fourier Matrix) Show that the Fourier matrix $\boldsymbol{F}_4$ is symmetric, but not Hermitian.

Exercise 11.2 (Sine Transform as Fourier Transform) Verify Lemma 11.1 directly when $m = 1$.

Exercise 11.3 (Explicit Solution of the Discrete Poisson Equation) Show that the exact solution of the discrete Poisson equation (10.5) can be written $V = (v_{i,j})_{i,j=1}^m$, where

$$v_{ij} = \frac{1}{(m+1)^4} \sum_{p=1}^{m} \sum_{r=1}^{m} \sum_{k=1}^{m} \sum_{l=1}^{m} \frac{\sin\left(\frac{ip\pi}{m+1}\right)\sin\left(\frac{jr\pi}{m+1}\right)\sin\left(\frac{kp\pi}{m+1}\right)\sin\left(\frac{lr\pi}{m+1}\right)}{\left[\sin\left(\frac{p\pi}{2(m+1)}\right)\right]^2 + \left[\sin\left(\frac{r\pi}{2(m+1)}\right)\right]^2} f_{k,l}.$$

Exercise 11.4 (Improved Version of Algorithm 11.1) Algorithm 11.1 involves multiplying a matrix by S four times. In this problem we show that it is enough to multiply by S two times. We achieve this by diagonalizing only the second T in $TV + VT = h^2 F$. Let $D = \mathrm{diag}(\lambda_1, \ldots, \lambda_m)$, where $\lambda_j = 4\sin^2(j\pi h/2)$, $j = 1, \ldots, m$.

a) Show that

$$TX + XD = C, \text{ where } X = VS, \text{ and } C = h^2 FS.$$

b) Show that

$$(T + \lambda_j I)x_j = c_j \quad j = 1, \ldots, m, \tag{11.9}$$

where $X = [x_1, \ldots, x_m]$ and $C = [c_1, \ldots, c_m]$. Thus we can find X by solving m linear systems, one for each of the columns of X. Recall that a tridiagonal $m \times m$ system can be solved by Algorithms 2.1 and 2.2 in $8m - 7$ arithmetic operations. Give an algorithm to find X which only requires $O(\delta m^2)$ arithmetic operations for some constant δ independent of m.

c) Describe a method to compute V which only requires $O(4m^3) = O(4n^{3/2})$ arithmetic operations.

d) Describe a method based on the fast Fourier transform which requires $O(2\gamma n \log_2 n)$ where γ is the same constant as mentioned at the end of the last section.

Exercise 11.5 (Fast Solution of 9 Point Scheme) Consider the equation

$$TV + VT - \frac{1}{6}TVT = h^2 \mu F,$$

that was derived in Exercise 10.8 for the 9-point scheme. Define the matrix X by $V = SXS = (x_{j,k})$ where V is the solution of (10.22). Show that

$$DX + XD - \frac{1}{6}DXD = 4h^4 G, \text{ where } G = S\mu FS,$$

where $\boldsymbol{D} = \operatorname{diag}(\lambda_1, \ldots, \lambda_m)$, with $\lambda_j = 4\sin^2(j\pi h/2)$, $j = 1, \ldots, m$, and that

$$x_{j,k} = \frac{h^4 g_{j,k}}{\sigma_j + \sigma_k - \frac{2}{3}\sigma_j\sigma_k}, \quad \text{where } \sigma_j = \sin^2\big((j\pi h)/2\big) \text{ for } j, k = 1, 2, \ldots, m.$$

Show that $\sigma_j + \sigma_k - \frac{2}{3}\sigma_j\sigma_k > 0$ for $j, k = 1, 2, \ldots, m$. Conclude that the matrix $\boldsymbol{A}$ in Exercise 10.8 b) is symmetric positive definite and that (10.21) always has a solution $\boldsymbol{V}$.

Exercise 11.6 (Algorithm for Fast Solution of 9 Point Scheme) Derive an algorithm for solving (10.21) which for large m requires essentially the same number of operations as in Algorithm 11.1. (We assume that $\mu\boldsymbol{F}$ already has been formed).

Exercise 11.7 (Fast Solution of Biharmonic Equation) For the biharmonic problem we derived in Exercise 10.9 the equation

$$\boldsymbol{T}^2\boldsymbol{U} + 2\boldsymbol{T}\boldsymbol{U}\boldsymbol{T} + \boldsymbol{U}\boldsymbol{T}^2 = h^4\boldsymbol{F}.$$

Define the matrix $\boldsymbol{X} = (x_{j,k})$ by $\boldsymbol{U} = \boldsymbol{S}\boldsymbol{X}\boldsymbol{S}$ where $\boldsymbol{U}$ is the solution of (10.25). Show that

$$\boldsymbol{D}^2\boldsymbol{X} + 2\boldsymbol{D}\boldsymbol{X}\boldsymbol{D} + \boldsymbol{X}\boldsymbol{D}^2 = 4h^6\boldsymbol{G}, \quad \text{where } \boldsymbol{G} = \boldsymbol{S}\boldsymbol{F}\boldsymbol{S},$$

and that

$$x_{j,k} = \frac{h^6 g_{j,k}}{4(\sigma_j + \sigma_k)^2}, \quad \text{where } \sigma_j = \sin^2((j\pi h)/2) \text{ for } j, k = 1, 2, \ldots, m.$$

Exercise 11.8 (Algorithm for Fast Solution of Biharmonic Equation) Use Exercise 11.7 to derive an algorithm

```
function U=simplefastbiharmonic(F)
```

which requires only $O(\delta n^{3/2})$ operations to find $\boldsymbol{U}$ in Problem 10.9. Here δ is some constant independent of n.

Exercise 11.9 (Check Algorithm for Fast Solution of Biharmonic Equation) In Exercise 11.8 compute the solution $\boldsymbol{U}$ corresponding to $\boldsymbol{F}$ = `ones(m,m)`. For some small m's check that you get the same solution obtained by solving the standard form $\boldsymbol{A}\boldsymbol{x} = \boldsymbol{b}$ in (10.25). You can use $\boldsymbol{x} = \boldsymbol{A}\backslash\boldsymbol{b}$ for solving $\boldsymbol{A}\boldsymbol{x} = \boldsymbol{b}$. Use `F(:)` to vectorize a matrix and `reshape(x,m,m)` to turn a vector $\boldsymbol{x} \in \mathbb{R}^{m^2}$ into an $m \times m$ matrix. Use the MATLAB command `surf(U)` for plotting U for, say, $m = 50$. Compare the result with Exercise 11.8 by plotting the difference between both matrices.

Exercise 11.10 (Fast Solution of Biharmonic Equation Using 9 Point Rule) Repeat Exercises 10.9, 11.8 and 11.9 using the nine point rule (10.21) to solve the system (10.24).

11.5 Review Questions

11.5.1 Consider the Poisson matrix.

- What is the bandwidth of its Cholesky factor?
- approximately how many arithmetic operations does it take to find the Cholesky factor?
- same question for block LU,
- same question for the fast Poisson solver with and without FFT.

11.5.2 What is the discrete sine transform and discrete Fourier transform of a vector?

Part V
Iterative Methods for Large Linear Systems

Gaussian elimination, LU and Cholesky factorization are direct methods. In absence of rounding errors they are used to find the exact solution of a linear system using a finite number of arithmetic operations. In an iterative method we start with an approximation $\boldsymbol{x}_0$ to the exact solution $\boldsymbol{x}$ and then compute a sequence $\{\boldsymbol{x}_k\}$ such that hopefully $\boldsymbol{x}_k \to \boldsymbol{x}$. Iterative methods are mainly used for large sparse systems, i.e., where many of the elements in the coefficient matrix are zero. The main advantages of iterative methods are reduced storage requirements and ease of implementation. In an iterative method the main work in each iteration is a matrix times vector multiplication, an operation which often does not need storing the matrix, not even in sparse form.

In this part we consider the iterative methods of Jacobi, Gauss-Seidel, successive over relaxation (SOR), steepest descent and conjugate gradients.

Chapter 12
The Classical Iterative Methods

In this chapter we consider the classical iterative methods of Richardson, Jacobi, Gauss-Seidel and an accelerated version of Gauss-Seidel's method called successive overrelaxation (SOR). David Young developed in his thesis a beautiful theory describing the convergence rate of SOR, see [22].

We give the main points of this theory specialized to the discrete Poisson matrix. With a careful choice of an acceleration parameter the amount of work using SOR on the discrete Poisson problem is the same as for the fast Poisson solver without FFT (cf. Algorithm 11.1). Moreover, SOR is not restricted to constant coefficient methods on a rectangle. However, to obtain fast convergence using SOR it is necessary to have a good estimate for an acceleration parameter.

For convergence we need to study convergence of powers of matrices. In this chapter we only use matrix norms which are consistent on $\mathbb{C}^{n\times n}$ and subordinate to a vector norm on $\mathbb{C}^n$, (cf. Definitions 8.4 and 8.5).

12.1 Classical Iterative Methods; Component Form

We start with an example showing how a linear system can be solved using an iterative method.

Example 12.1 (Iterative Methods on a Special 2×2 *Matrix)* Solving for the diagonal elements, the linear system $\left[\begin{smallmatrix} 2 & -1 \\ -1 & 2 \end{smallmatrix}\right]\left[\begin{smallmatrix} y \\ z \end{smallmatrix}\right] = \left[\begin{smallmatrix} 1 \\ 1 \end{smallmatrix}\right]$ can be written in component form as $y = (z+1)/2$ and $z = (y+1)/2$. Starting with y_0, z_0 we generate two sequences $\{y_k\}$ and $\{z_k\}$ using the difference equations $y_{k+1} = (z_k + 1)/2$ and $z_{k+1} = (y_k + 1)/2$. This is an example of Jacobi's method. If $y_0 = z_0 = 0$ then we find $y_1 = z_1 = 1/2$ and in general $y_k = z_k = 1 - 2^{-k}$ for $k = 0, 1, 2, 3, \ldots$. The iteration converges to the exact solution $[1, 1]^T$, and the error is halved in each iteration.

T. Lyche, *Numerical Linear Algebra and Matrix Factorizations*,
Texts in Computational Science and Engineering 22,
https://doi.org/10.1007/978-3-030-36468-7_12

We can improve the convergence rate by using the most current approximation in each iteration. This leads to Gauss-Seidel's method: $y_{k+1} = (z_k + 1)/2$ and $z_{k+1} = (y_{k+1} + 1)/2$. If $y_0 = z_0 = 0$ then we find $y_1 = 1/2$, $z_1 = 3/4$, $y_2 = 7/8$, $z_2 = 15/16$, and in general $y_k = 1 - 2 \cdot 4^{-k}$ and $z_k = 1 - 4^{-k}$ for $k = 1, 2, 3, \ldots$. The error is now reduced by a factor 4 in each iteration.

Consider the general case. Suppose $\boldsymbol{A} \in \mathbb{C}^{n \times n}$ is nonsingular and $\boldsymbol{b} \in \mathbb{C}^n$. Suppose we know an approximation $\boldsymbol{x}_k = [\boldsymbol{x}_k(1), \ldots, \boldsymbol{x}_k(n)]^T$ to the exact solution $\boldsymbol{x}$ of $\boldsymbol{A}\boldsymbol{x} = \boldsymbol{b}$. We need to assume that the rows are ordered so that $\boldsymbol{A}$ has nonzero diagonal elements. Solving the ith equation of $\boldsymbol{A}\boldsymbol{x} = \boldsymbol{b}$ for $\boldsymbol{x}(i)$, we obtain a **fixed-point form** of $\boldsymbol{A}\boldsymbol{x} = \boldsymbol{b}$

$$\boldsymbol{x}(i) = \Big(- \sum_{j=1}^{i-1} a_{ij}\boldsymbol{x}(j) - \sum_{j=i+1}^{n} a_{ij}\boldsymbol{x}(j) + b_i \Big)/a_{ii}, \quad i = 1, 2, \ldots, n. \tag{12.1}$$

1. In **Jacobi's method (J method)** we substitute $\boldsymbol{x}_k$ into the right hand side of (12.1) and compute a new approximation by

$$\boldsymbol{x}_{k+1}(i) = \Big(- \sum_{j=1}^{i-1} a_{ij}\boldsymbol{x}_k(j) - \sum_{j=i+1}^{n} a_{ij}\boldsymbol{x}_k(j) + b_i \Big)/a_{ii}, \text{ for } i = 1, 2, \ldots, n. \tag{12.2}$$

2. **Gauss-Seidel's method (GS method)** is a modification of Jacobi's method, where we use the new $\boldsymbol{x}_{k+1}(i)$ immediately after it has been computed.

$$\boldsymbol{x}_{k+1}(i) = \Big(- \sum_{j=1}^{i-1} a_{ij}\boldsymbol{x}_{k+1}(j) - \sum_{j=i+1}^{n} a_{ij}\boldsymbol{x}_k(j) + b_i \Big)/a_{ii}, \text{ for } i = 1, 2, \ldots, n. \tag{12.3}$$

3. The **Successive overrelaxation method (SOR method)** is obtained by introducing an acceleration parameter $0 < \omega < 2$ in the GS method. We write $\boldsymbol{x}(i) = \omega\boldsymbol{x}(i) + (1 - \omega)\boldsymbol{x}(i)$ and this leads to the method

$$\boldsymbol{x}_{k+1}(i) = \omega\Big(- \sum_{j=1}^{i-1} a_{ij}\boldsymbol{x}_{k+1}(j) - \sum_{j=i+1}^{n} a_{ij}\boldsymbol{x}_k(j) + b_i \Big)/a_{ii} + (1 - \omega)\boldsymbol{x}_k(i). \tag{12.4}$$

The SOR method reduces to the Gauss-Seidel method for $\omega = 1$. Denoting the right hand side of (12.3) by $\boldsymbol{x}_{k+1}^{gs}$ we can write (12.4) as $\boldsymbol{x}_{k+1} = \omega\boldsymbol{x}_{k+1}^{gs} + (1 - \omega)\boldsymbol{x}_k$, and we see that $\boldsymbol{x}_{k+1}$ is located on the straight line passing through the two points $\boldsymbol{x}_{k+1}^{gs}$ and $\boldsymbol{x}_k$. The restriction $0 < \omega < 2$ is necessary for convergence (cf. Theorem 12.6). Normally, the best results are obtained for the relaxation parameter ω in the range $1 \le \omega < 2$ and then $\boldsymbol{x}_{k+1}$ is computed by linear extrapolation, i.e., it is not located between $\boldsymbol{x}_{k+1}^{gs}$ and $\boldsymbol{x}_k$.

4. We mention also briefly the symmetric successive overrelaxation method **SSOR**. One iteration in SSOR consists of two SOR sweeps. A forward SOR sweep (12.4), computing an approximation denoted $\boldsymbol{x}_{k+1/2}$ instead of $\boldsymbol{x}_{k+1}$, is followed by a backward SOR sweep computing

$$\boldsymbol{x}_{k+1}(i) = \omega\Big(- \sum_{j=1}^{i-1} a_{ij}\boldsymbol{x}_{k+1/2}(j) - \sum_{j=i+1}^{n} a_{ij}\boldsymbol{x}_{k+1}(j) + b_i\Big)/a_{ii} + (1-\omega)\boldsymbol{x}_{k+1/2}(i) \tag{12.5}$$

in the order $i = n, n-1, \ldots 1$. The method is slower and more complicated than the SOR method. Its main use is as a symmetric preconditioner. For if $\boldsymbol{A}$ is symmetric then SSOR combines the two SOR steps in such a way that the resulting iteration matrix is similar to a symmetric matrix. We will not discuss this method any further here and refer to Sect. 13.6 for an alternative example of a preconditioner.

We will refer to the J, GS and SOR methods as the **classical (iterative) methods**.

12.1.1 The Discrete Poisson System

Consider the classical methods applied to the discrete Poisson matrix $\boldsymbol{A} \in \mathbb{R}^{n\times n}$ given by (10.8). Let $n = m^2$ and set $h = 1/(m+1)$. In component form the linear system $\boldsymbol{A}\boldsymbol{x} = \boldsymbol{b}$ can be written (cf. (10.4))

$$4\boldsymbol{v}(i,j) - \boldsymbol{v}(i-1,j) - \boldsymbol{v}(i+1,j) - \boldsymbol{v}(i,j-1) - \boldsymbol{v}(i,j+1) = h^2 f_{i,j}, \quad i,j = 1,\ldots,m,$$

with homogenous boundary conditions also given in (10.4). Solving for $\boldsymbol{v}(i,j)$ we obtain the **fixed point form**

$$\boldsymbol{v}(i,j) = \big(\boldsymbol{v}(i-1,j) + \boldsymbol{v}(i+1,j) + \boldsymbol{v}(i,j-1) + v(i,j+1) + e_{i,j}\big)/4, \tag{12.6}$$

where $e_{i,j} := f_{i,j}/(m+1)^2$. The J, GS , and SOR methods take the form

$$J : \boldsymbol{v}_{k+1}(i,j) = \big(\boldsymbol{v}_k(i-1,j) + \boldsymbol{v}_k(i,j-1) + \boldsymbol{v}_k(i+1,j) + \boldsymbol{v}_k(i,j+1) \\ + \boldsymbol{e}(i,j)\big)/4$$

$$GS : \boldsymbol{v}_{k+1}(i,j) = \big(\boldsymbol{v}_{k+1}(i-1,j) + \boldsymbol{v}_{k+1}(i,j-1) + \boldsymbol{v}_k(i+1,j) + \boldsymbol{v}_k(i,j+1) \\ + \boldsymbol{e}(i,j)\big)/4$$

Table 12.1 The number of iterations k_n to solve the discrete Poisson problem with n unknowns using the methods of Jacobi, Gauss-Seidel, and SOR (see text) with a tolerance 10^{-8}

	k_{100}	k_{2500}	$k_{10\,000}$	$k_{40\,000}$	$k_{160\,000}$
J	385	8386			
GS	194	4194			
SOR	35	164	324	645	1286

$$SOR: \boldsymbol{v}_{k+1}(i,j) = \omega\big(\boldsymbol{v}_{k+1}(i-1,j) + \boldsymbol{v}_{k+1}(i,j-1) + \boldsymbol{v}_k(i+1,j) + \boldsymbol{v}_k(i,j+1) + \boldsymbol{e}(i,j)\big)/4 + (1-\omega)\boldsymbol{v}_k(i,j). \tag{12.7}$$

We note that for GS and SOR we have used the **natural ordering**, i.e., $(i_1, j_1) < (i_2, j_2)$ if and only if $j_1 \le j_2$ and $i_1 < i_2$ if $j_1 = j_2$. For the J method any ordering can be used.

In Algorithm 12.1 we give a MATLAB program to test the convergence of Jacobi's method on the discrete Poisson problem. We carry out Jacobi iterations on the linear system (12.6) with $\boldsymbol{F} = (f_{i,j}) \in \mathbb{R}^{m\times m}$, starting with $\boldsymbol{V}_0 = \boldsymbol{0} \in \mathbb{R}^{(m+2)\times(m+2)}$. The output is the number of iterations k, to obtain $\|\boldsymbol{V}^{(k)} - \boldsymbol{U}\|_M := \max_{i,j}|v_{ij} - u_{ij}| < tol$. Here $[u_{ij}] \in \mathbb{R}^{(m+2)\times(m+2)}$ is the "exact" solution of (12.6) computed using the fast Poisson solver in Algorithm 11.1. We set $k = K+1$ if convergence is not obtained in K iterations. In Table 12.1 we show the output $k = k_n$ from this algorithm using $\boldsymbol{F} = \texttt{ones}(m,m)$ for $m = 10, 50$, $K = 10^4$, and $tol = 10^{-8}$. We also show the number of iterations for Gauss-Seidel and SOR with a value of ω known as the optimal acceleration parameter $\omega^* := 2/\big(1+\sin(\frac{\pi}{m+1})\big)$. We will derive this value later.

```
function k=jdp(F,K,tol)
% k=jdp(F,K,tol)
m=length(F); U=fastpoisson(F);
V=zeros(m+2,m+2); E=F/(m+1)^2;
for k=1:K
  V(2:m+1,2:m+1)=(V(1:m,2:m+1)+V(3:m+2,2:m+1)...
       +V(2:m+1,1:m)+V(2:m+1,3:m+2)+E)/4;
  if max(max(abs(V-U)))<tol, return
  end
end
k=K+1;
end
```

Listing 12.1 jdp

For the GS and SOR methods we have used Algorithm 12.2. This is the analog of Algorithm 12.1 using SOR instead of J to solve the discrete Poisson problem. w is an acceleration parameter with $0 < w < 2$. For $w = 1$ we obtain Gauss-Seidel's method.

```
function k=sordp(F,K,w,tol)
% k=sordp(F,K,w,tol)
m=length(F); U=fastpoisson(F); V=zeros(m+2,m+2); E=F/(m+1)^2;
for k=1:K
  for j=2:m+1
    for i=2:m+1
      V(i,j)=w*(V(i-1,j)+V(i+1,j)+V(i,j-1)...
             +V(i,j+1)+E(i-1,j-1))/4+(1-w)*V(i,j);
    end
  end
  if max(max(abs(V-U)))<tol, return
  end
end
k=K+1;
end
```

Listing 12.2 sordp

We make several remarks about these programs and the results in Table 12.1.

1. The rate (speed) of convergence is quite different for the four methods. The J and GS methods converge, but rather slowly. The J method needs about twice as many iterations as the GS method. The improvement using the SOR method with optimal ω is spectacular.
2. We show in Sect. 12.3.4 that the number of iterations k_n for a size n problem is $k_n = O(n)$ for the J and GS method and $k_n = O(\sqrt{n})$ for SOR with optimal ω. The choice of tol will only influence the constants multiplying n or $\sqrt{n}$.
3. From (12.1.1) it follows that each iteration requires $O(n)$ arithmetic operations. Thus the number of arithmetic operations to achieve a given tolerance is $O(k_n \times n)$. Therefore the number of arithmetic operations for the J and GS method is $O(n^2)$, while it is only $O(n^{3/2})$ for the SOR method with optimal ω. Asymptotically, for J and GS this is the same as using banded Cholesky, while SOR competes with the fast method (without FFT).
4. We do not need to store the coefficient matrix so the storage requirements for these methods on the discrete Poisson problem is $O(n)$, asymptotically the same as for the fast methods.
5. Jacobi's method has the advantage that it can be easily parallelized.

12.2 Classical Iterative Methods; Matrix Form

To study convergence we need matrix formulations of the classical methods.

12.2.1 Fixed-Point Form

In general we can construct an iterative method by choosing a nonsingular matrix $\boldsymbol{M}$ and write $\boldsymbol{A}\boldsymbol{x} = \boldsymbol{b}$ in the equivalent form

$$\boldsymbol{M}\boldsymbol{x} = (\boldsymbol{M} - \boldsymbol{A})\boldsymbol{x} + \boldsymbol{b}. \tag{12.8}$$

The matrix $\boldsymbol{M}$ is known as a **splitting matrix**.

The corresponding iterative method is given by

$$\boldsymbol{M}\boldsymbol{x}_{k+1} = (\boldsymbol{M} - \boldsymbol{A})\boldsymbol{x}_k + \boldsymbol{b} \tag{12.9}$$

or

$$\boldsymbol{x}_{k+1} := \boldsymbol{G}\boldsymbol{x}_k + \boldsymbol{c}, \quad \boldsymbol{G} = \boldsymbol{I} - \boldsymbol{M}^{-1}\boldsymbol{A}, \quad , \boldsymbol{c} = \boldsymbol{M}^{-1}b. \tag{12.10}$$

This is known as a **fixed-point iteration**. Starting with $\boldsymbol{x}_0$ this defines a sequence $\{\boldsymbol{x}_k\}$ of vectors in $\mathbb{C}^n$. For a general $\boldsymbol{G} \in \mathbb{C}^{n\times n}$ and $\boldsymbol{c} \in \mathbb{C}^n$ a solution of $\boldsymbol{x} = \boldsymbol{G}\boldsymbol{x} + \boldsymbol{c}$ is called a **fixed-point**. The fixed-point is unique if $\boldsymbol{I} - \boldsymbol{G}$ is nonsingular.

If $\lim_{k\to\infty} \boldsymbol{x}_k = \boldsymbol{x}$ for some $\boldsymbol{x} \in \mathbb{C}^n$ then $\boldsymbol{x}$ is a fixed point since

$$\boldsymbol{x} = \lim_{k\to\infty} \boldsymbol{x}_{k+1} = \lim_{k\to\infty} (\boldsymbol{G}\boldsymbol{x}_k + \boldsymbol{c}) = \boldsymbol{G} \lim_{k\to\infty} \boldsymbol{x}_k + \boldsymbol{c} = \boldsymbol{G}\boldsymbol{x} + \boldsymbol{c}.$$

The matrix $\boldsymbol{M}$ can also be interpreted as a **preconditioning matrix**. We first write $\boldsymbol{A}\boldsymbol{x} = \boldsymbol{b}$ in the equivalent form $\boldsymbol{M}^{-1}\boldsymbol{A}\boldsymbol{x} = \boldsymbol{M}^{-1}\boldsymbol{b}$ or $\boldsymbol{x} = \boldsymbol{x} - \boldsymbol{M}^{-1}\boldsymbol{A}\boldsymbol{x} + \boldsymbol{M}^{-1}\boldsymbol{b}$. This again leads to the iterative method (12.10), and $\boldsymbol{M}$ is chosen to reduce the condition number of $\boldsymbol{A}$.

12.2.2 The Splitting Matrices for the Classical Methods

Different choices of $\boldsymbol{M}$ in (12.9) lead to different iterative methods. We now derive $\boldsymbol{M}$ for the classical methods. For GS and SOR it is convenient to write $\boldsymbol{A}$ as a sum of three matrices, $\boldsymbol{A} = \boldsymbol{D} - \boldsymbol{A}_L - \boldsymbol{A}_R$, where $-\boldsymbol{A}_L$, $\boldsymbol{D}$, and $-\boldsymbol{A}_R$ are the lower, diagonal, and upper part of $\boldsymbol{A}$, respectively. Thus $\boldsymbol{D} := \operatorname{diag}(a_{11}, \ldots, a_{nn})$,

$$\boldsymbol{A}_L := \begin{bmatrix} 0 & & & \\ -a_{2,1} & 0 & & \\ \vdots & \ddots & \ddots & \\ -a_{n,1} & \cdots & -a_{n,n-1} & 0 \end{bmatrix}, \quad \boldsymbol{A}_R := \begin{bmatrix} 0 & -a_{1,2} & \cdots & -a_{1,n} \\ & \ddots & \ddots & \vdots \\ & & 0 & -a_{n-1,n} \\ & & & 0 \end{bmatrix}. \tag{12.11}$$

Theorem 12.1 (Splitting Matrices for J, GS and SOR) *The splitting matrices* $\boldsymbol{M}_J$, $\boldsymbol{M}_1$ *and* $\boldsymbol{M}_\omega$ *for the J, GS and SOR methods are given by*

$$\boldsymbol{M}_J = \boldsymbol{D}, \quad \boldsymbol{M}_1 = \boldsymbol{D} - \boldsymbol{A}_L, \quad \boldsymbol{M}_\omega = \omega^{-1}\boldsymbol{D} - \boldsymbol{A}_L. \tag{12.12}$$

Proof To find $\boldsymbol{M}$ we write the methods in the form (12.9) where the coefficient of $\boldsymbol{b}$ is equal to one. Moving a_{ii} to the left hand side of the Jacobi iteration (12.2) we obtain the matrix form $\boldsymbol{D}\boldsymbol{x}_{k+1} = (\boldsymbol{D} - \boldsymbol{A})\boldsymbol{x}_k + \boldsymbol{b}$ showing that $\boldsymbol{M}_J = \boldsymbol{D}$.

For the SOR method a matrix form is

$$\boldsymbol{D}\boldsymbol{x}_{k+1} = \omega\big(\boldsymbol{A}_L\boldsymbol{x}_{k+1} + \boldsymbol{A}_R\boldsymbol{x}_k + \boldsymbol{b}\big) + (1-\omega)\boldsymbol{D}\boldsymbol{x}_k. \tag{12.13}$$

Dividing both sides by ω and moving $\boldsymbol{A}_L\boldsymbol{x}_{k+1}$ to the left hand side this takes the form $\big(\omega^{-1}\boldsymbol{D} - \boldsymbol{A}_L\big)\boldsymbol{x}_{k+1} = \boldsymbol{A}_R\boldsymbol{x}_k + \boldsymbol{b} + (\omega^{-1} - 1)\boldsymbol{D}\boldsymbol{x}_k$ showing that $\boldsymbol{M}_\omega = \omega^{-1}\boldsymbol{D} - \boldsymbol{A}_L$. We obtain $\boldsymbol{M}_1$ by letting $\omega = 1$ in $\boldsymbol{M}_\omega$. □

Example 12.2 (Splitting Matrices) For the system

$$\begin{bmatrix} 2 & -1 \\ -1 & 2 \end{bmatrix}\begin{bmatrix} x_1 \\ x_2 \end{bmatrix} = \begin{bmatrix} 1 \\ 1 \end{bmatrix}$$

we find

$$\boldsymbol{A}_L = \begin{bmatrix} 0 & 0 \\ 1 & 0 \end{bmatrix}, \quad \boldsymbol{D} = \begin{bmatrix} 2 & 0 \\ 0 & 2 \end{bmatrix}, \quad \boldsymbol{A}_R = \begin{bmatrix} 0 & 1 \\ 0 & 0 \end{bmatrix},$$

and

$$\boldsymbol{M}_J = \boldsymbol{D} = \begin{bmatrix} 2 & 0 \\ 0 & 2 \end{bmatrix}, \quad \boldsymbol{M}_\omega = \omega^{-1}\boldsymbol{D} - \boldsymbol{A}_L = \begin{bmatrix} 2\omega^{-1} & 0 \\ -1 & 2\omega^{-1} \end{bmatrix}.$$

The iteration matrix $\boldsymbol{G}_\omega = \boldsymbol{I} - \boldsymbol{M}_\omega^{-1}\boldsymbol{A}$ is given by

$$\boldsymbol{G}_\omega = \begin{bmatrix} 1 & 0 \\ 0 & 1 \end{bmatrix} - \begin{bmatrix} \omega/2 & 0 \\ \omega^2/4 & \omega/2 \end{bmatrix}\begin{bmatrix} 2 & -1 \\ -1 & 2 \end{bmatrix} = \begin{bmatrix} 1-\omega & \omega/2 \\ \omega(1-\omega)/2 & 1-\omega+\omega^2/4 \end{bmatrix}. \tag{12.14}$$

For the J and GS method we have

$$\boldsymbol{G}_J = \boldsymbol{I} - \boldsymbol{D}^{-1}\boldsymbol{A} = \begin{bmatrix} 0 & 1/2 \\ 1/2 & 0 \end{bmatrix}, \quad \boldsymbol{G}_1 = \begin{bmatrix} 0 & 1/2 \\ 0 & 1/4 \end{bmatrix}. \tag{12.15}$$

We could have derived these matrices directly from the component form of the iteration. For example, for the GS method we have the component form

$$\boldsymbol{x}_{k+1}(1) = \frac{1}{2}\boldsymbol{x}_k(2) + \frac{1}{2}, \quad \boldsymbol{x}_{k+1}(2) = \frac{1}{2}\boldsymbol{x}_{k+1}(1) + \frac{1}{2}.$$

Substituting the value of $\boldsymbol{x}_{k+1}(1)$ from the first equation into the second equation we find

$$\boldsymbol{x}_{k+1}(2) = \frac{1}{2}(\frac{1}{2}\boldsymbol{x}_k(2) + \frac{1}{2}) + \frac{1}{2} = \frac{1}{4}\boldsymbol{x}_k(2) + \frac{3}{4}.$$

Thus

$$\boldsymbol{x}_{k+1} = \begin{bmatrix} \boldsymbol{x}_{k+1}(1) \\ \boldsymbol{x}_{k+1}(2) \end{bmatrix} = \begin{bmatrix} 0 & 1/2 \\ 0 & 1/4 \end{bmatrix} \begin{bmatrix} \boldsymbol{x}_k(1) \\ \boldsymbol{x}_k(2) \end{bmatrix} + \begin{bmatrix} 1/2 \\ 3/4 \end{bmatrix} = \boldsymbol{G}_1\boldsymbol{x}_k + \boldsymbol{c}.$$

12.3 Convergence

For Newton's method the choice of starting value is important. This is not the case for methods of the form $\boldsymbol{x}_{k+1} := \boldsymbol{G}\boldsymbol{x}_k + \boldsymbol{c}$.

In the following we assume that $\boldsymbol{G} \in \mathbb{C}^{n\times n}$, $\boldsymbol{c} \in \mathbb{C}^n$ and $\boldsymbol{I} - \boldsymbol{G}$ is nonsingular. We let $\boldsymbol{x} \in \mathbb{C}^n$ be the unique fixed point satisfying $\boldsymbol{x} = \boldsymbol{G}\boldsymbol{x} + \boldsymbol{c}$.

Definition 12.1 (Convergence of $\boldsymbol{x}_{k+1} := \boldsymbol{G}\boldsymbol{x}_k + \boldsymbol{c}$) We say that the iterative method $\boldsymbol{x}_{k+1} := \boldsymbol{G}\boldsymbol{x}_k + \boldsymbol{c}$ **converges** if the sequence $\{\boldsymbol{x}_k\}$ converges for **any** starting vector $\boldsymbol{x}_0$.

We have the following necessary and sufficient condition for convergence:

Theorem 12.2 (Convergence of an Iterative Method) *The iterative method $\boldsymbol{x}_{k+1} := \boldsymbol{G}\boldsymbol{x}_k + \boldsymbol{c}$ converges if and only if* $\lim_{k\to\infty} \boldsymbol{G}^k = \boldsymbol{0}$.

Proof Suppose $\lim_{k\to\infty} \boldsymbol{G}^k = \boldsymbol{0}$, and let $\boldsymbol{x}$ be the unique fixed point. We subtract $\boldsymbol{x} = \boldsymbol{G}\boldsymbol{x} + \boldsymbol{c}$ from $\boldsymbol{x}_{k+1} = \boldsymbol{G}\boldsymbol{x}_k + \boldsymbol{c}$. The vector $\boldsymbol{c}$ cancels and we obtain $\boldsymbol{x}_{k+1} - \boldsymbol{x} = \boldsymbol{G}(\boldsymbol{x}_k - \boldsymbol{x})$. By induction on k

$$\boldsymbol{x}_k - \boldsymbol{x} = \boldsymbol{G}^k(\boldsymbol{x}_0 - \boldsymbol{x}), \quad k = 0, 1, 2, \ldots \tag{12.16}$$

Clearly $\boldsymbol{x}_k - \boldsymbol{x} \to \boldsymbol{0}$ if $\boldsymbol{G}^k \to \boldsymbol{0}$ and the method converges. For the converse we let $\boldsymbol{x}$ be as before and choose $\boldsymbol{x}_0 - \boldsymbol{x} = \boldsymbol{e}_j$, the jth unit vector for $j = 1, \ldots, n$. Since $\boldsymbol{x}_k - \boldsymbol{x} \to \boldsymbol{0}$ for any $\boldsymbol{x}_0$ we have $\boldsymbol{G}^k\boldsymbol{e}_j \to \boldsymbol{0}$ for $j = 1, \ldots, n$ which implies that $\lim_{k\to\infty} \boldsymbol{G}^k = \boldsymbol{0}$. □

Theorem 12.3 (Sufficient Condition for Convergence) *If* $\|\boldsymbol{G}\| < 1$ *then the iteration* $\boldsymbol{x}_{k+1} = \boldsymbol{G}\boldsymbol{x}_k + \boldsymbol{c}$ *converges.*

Proof We have

$$\|\boldsymbol{x}_k - \boldsymbol{x}\| = \|\boldsymbol{G}^k(\boldsymbol{x}_0 - \boldsymbol{x})\| \le \|\boldsymbol{G}^k\|\|\boldsymbol{x}_0 - \boldsymbol{x}\| \le \|\boldsymbol{G}\|^k\|\boldsymbol{x}_0 - \boldsymbol{x}\| \to \boldsymbol{0}, \quad k \to \infty.$$

□

A necessary and sufficient condition for convergence involves the eigenvalues of $\boldsymbol{G}$. We define the **spectral radius** of a matrix $\boldsymbol{A} \in \mathbb{C}^{n\times n}$ as the maximum absolute value of its eigenvalues.

$$\rho(\boldsymbol{A}) := \max_{\lambda\in\sigma(\boldsymbol{A})} |\lambda|. \tag{12.17}$$

Theorem 12.4 (When Does an Iterative Method Converge?) *Suppose* $\boldsymbol{G} \in \mathbb{C}^{n\times n}$ *with* $\boldsymbol{I} - \boldsymbol{G}$ *nonsingular and let* $\boldsymbol{c} \in \mathbb{C}^n$. *The iteration* $\boldsymbol{x}_{k+1} = \boldsymbol{G}\boldsymbol{x}_k + \boldsymbol{c}$ *converges if and only if* $\rho(\boldsymbol{G}) < 1$.

We will prove this theorem using Theorem 12.10 in Sect. 12.4.

12.3.1 Richardson's Method

The **Richardson's method (R method)** is defined by

$$\boldsymbol{x}_{k+1} = \boldsymbol{x}_k + \alpha\big(\boldsymbol{b} - \boldsymbol{A}\boldsymbol{x}_k\big). \tag{12.18}$$

Here we pick an acceleration parameter α and compute a new approximation by adding a multiple of the residual vector $\boldsymbol{r}_k := \boldsymbol{b} - \boldsymbol{A}\boldsymbol{x}_k$. Note that we do not need the assumption of nonzero diagonal elements. Richardson considered this method in 1910.

We will assume that α is real. If all eigenvalues of $\boldsymbol{A}$ have positive real parts then the R method converges provided α is positive and sufficiently small. We show this result for positive eigenvalues and leave the more general case to Exercise 12.2.

Theorem 12.5 (Convergence of Richardson's Method) *If* $\boldsymbol{A}$ *has positive eigenvalues* $\lambda_1 \ge \lambda_2 \ge \cdots \ge \lambda_n > 0$ *then the R method given by* $\boldsymbol{x}_{k+1} = \big(\boldsymbol{I} - \alpha\boldsymbol{A}\big)\boldsymbol{x}_k + \boldsymbol{b}$ *converges if and only if* $0 < \alpha < 2/\lambda_1$. *Moreover,*

$$\begin{aligned} \rho(\boldsymbol{I} - \alpha\boldsymbol{A}) &> \rho(\boldsymbol{I} - \alpha_o\boldsymbol{A}) = \frac{\kappa - 1}{\kappa + 1}, \quad \alpha \in \mathbb{R}\setminus\{\alpha_o\}, \\ \kappa &:= \frac{\lambda_1}{\lambda_n}, \quad \alpha_o := \frac{2}{\lambda_1 + \lambda_n}. \end{aligned} \tag{12.19}$$

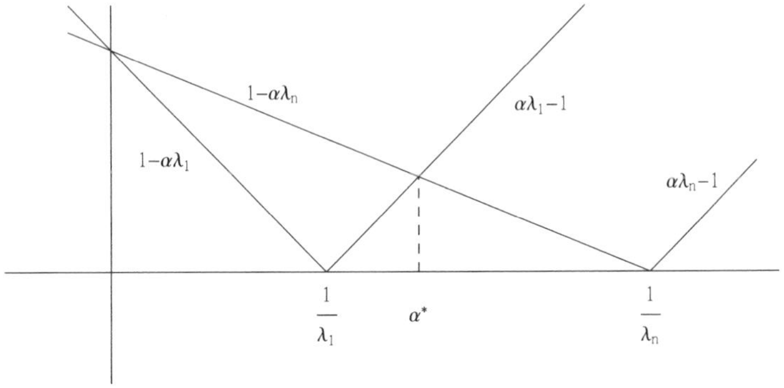

Fig. 12.1 The functions $\alpha \to |1 - \alpha\lambda_1|$ and $\alpha \to |1 - \alpha\lambda_n|$

Proof The eigenvalues of $\boldsymbol{I} - \alpha\boldsymbol{A}$ are $1 - \alpha\lambda_j, \quad j = 1, \ldots, n$. We have

$$\rho_\alpha := \rho(\boldsymbol{I} - \alpha\boldsymbol{A}) := \max_j |1 - \alpha\lambda_j| = \begin{cases} 1 - \alpha\lambda_1, & \text{if } \alpha \le 0 \\ 1 - \alpha\lambda_n, & \text{if } 0 < \alpha \le \alpha_o \\ \alpha\lambda_1 - 1, & \text{if } \alpha > \alpha_o, \end{cases}$$

see Fig. 12.1. Clearly $1 - \alpha\lambda_n = \alpha\lambda_1 - 1$ for $\alpha = \alpha_o$ and

$$\rho_{\alpha_o} = \alpha_o\lambda_1 - 1 = \frac{\lambda_1 - \lambda_n}{\lambda_1 + \lambda_n} = \frac{\kappa - 1}{\kappa + 1} < 1.$$

We have $\rho_\alpha < 1$ if and only if $\alpha > 0$ and $\alpha\lambda_1 - 1 < 1$ showing convergence if and only if $0 < \alpha < 2/\lambda_1$ and $\rho_\alpha > \rho_{\alpha_o}$ for $\alpha \le 0$ and $\alpha \ge 2/\lambda_1$. Finally, if $0 < \alpha < \alpha_o$ then $\rho_\alpha = 1 - \alpha\lambda_n > 1 - \alpha_o\lambda_n = \rho_{\alpha_o}$ and if $\alpha_o < \alpha < 2/\lambda_1$ then $\rho_\alpha = \alpha\lambda_1 - 1 > \alpha_o\lambda_1 - 1 = \rho_{\alpha_o}$. □

For a positive definite matrix we obtain

Corollary 12.1 (Rate of Convergence for the R Method) *Suppose* $\boldsymbol{A}$ *is positive definite with largest and smallest eigenvalue* λ_{max} *and* λ_{min}, *respectively. Richardson's method* $\boldsymbol{x}_{k+1} = \big(\boldsymbol{I} - \alpha\boldsymbol{A}\big)\boldsymbol{x}_k + \boldsymbol{b}$ *converges if and only if* $0 < \alpha < 2/\lambda_{max}$. *With* $\alpha = \alpha_o := \frac{2}{\lambda_{max} + \lambda_{min}}$ *we have the error estimate*

$$\|\boldsymbol{x}_k - \boldsymbol{x}\|_2 \le \Big(\frac{\kappa - 1}{\kappa + 1}\Big)^k \|\boldsymbol{x}_0 - \boldsymbol{x}\|_2, \quad k = 0, 1, 2, \ldots \tag{12.20}$$

where $\kappa := \lambda_{max}/\lambda_{min}$ *is the spectral condition number of* $\boldsymbol{A}$.

Proof The spectral norm $\|\ \|_2$ is consistent and therefore $\|\boldsymbol{x}_k - \boldsymbol{x}\|_2 \le \|\boldsymbol{I} - \alpha_o\boldsymbol{A}\|_2^k \|\boldsymbol{x}_0 - \boldsymbol{x}\|_2$. But for a positive definite matrix the spectral norm is equal to the spectral radius and the result follows form (12.19). □

12.3.2 Convergence of SOR

The condition $\omega \in (0, 2)$ is necessary for convergence of the SOR method.

Theorem 12.6 (Necessary Condition for Convergence of SOR) *Suppose $\boldsymbol{A} \in \mathbb{C}^{n\times n}$ is nonsingular with nonzero diagonal elements. If the SOR method applied to $\boldsymbol{A}$ converges then $\omega \in (0, 2)$.*

Proof We have (cf. (12.13)) $\boldsymbol{D}\boldsymbol{x}_{k+1} = \omega\big(\boldsymbol{A}_L\boldsymbol{x}_{k+1} + \boldsymbol{A}_R\boldsymbol{x}_k + \boldsymbol{b}\big) + (1-\omega)\boldsymbol{D}\boldsymbol{x}_k$ or $\boldsymbol{x}_{k+1} = \omega\big(\boldsymbol{L}\boldsymbol{x}_{k+1} + \boldsymbol{R}\boldsymbol{x}_k + \boldsymbol{D}^{-1}\boldsymbol{b}\big) + (1-\omega)\boldsymbol{x}_k$, where $\boldsymbol{L} := \boldsymbol{D}^{-1}\boldsymbol{A}_L$ and $\boldsymbol{R} := \boldsymbol{D}^{-1}\boldsymbol{A}_R$. Thus $(\boldsymbol{I} - \omega\boldsymbol{L})\boldsymbol{x}_{k+1} = (\omega\boldsymbol{R} + (1-\omega)\boldsymbol{I})\boldsymbol{x}_k + \omega\boldsymbol{D}^{-1}\boldsymbol{b}$ so the following form of the iteration matrix is obtained

$$\boldsymbol{G}_\omega = (\boldsymbol{I} - \omega\boldsymbol{L})^{-1}\big(\omega\boldsymbol{R} + (1-\omega)\boldsymbol{I}\big). \tag{12.21}$$

We next compute the determinant of $\boldsymbol{G}_\omega$. Since $\boldsymbol{I} - \omega\boldsymbol{L}$ is lower triangular with ones on the diagonal, the same holds for the inverse by Lemma 2.5, and therefore the determinant of this matrix is equal to one. The matrix $\omega\boldsymbol{R} + (1-\omega)\boldsymbol{I}$ is upper triangular with $1-\omega$ on the diagonal and therefore its determinant equals $(1-\omega)^n$. It follows that $\det(\boldsymbol{G}_\omega) = (1-\omega)^n$. Since the determinant of a matrix equals the product of its eigenvalues we must have $|\lambda| \geq |1-\omega|$ for at least one eigenvalue λ of $\boldsymbol{G}_\omega$ and we conclude that $\rho(\boldsymbol{G}_\omega) \geq |\omega - 1|$. But then $\rho(\boldsymbol{G}_\omega) \geq 1$ if ω is not in the interval $(0, 2)$ and by Theorem 12.4 SOR diverges. □

The SOR method always converges for a positive definite matrix.

Theorem 12.7 (SOR on Positive Definite Matrix) *SOR converges for a positive definite matrix $\boldsymbol{A} \in \mathbb{R}^{n\times n}$ if and only if $0 < \omega < 2$. In particular, Gauss-Seidel's method converges for a positive definite matrix.*

Proof By Theorem 12.6 convergence implies $0 < \omega < 2$. Suppose now $0 < \omega < 2$ and let $(\lambda, \boldsymbol{x})$ be an eigenpair for $\boldsymbol{G}_\omega$. Note that λ and $\boldsymbol{x}$ can be complex. We need to show that $|\lambda| < 1$. The following identity will be shown below:

$$\omega^{-1}(2-\omega)|1-\lambda|^2\boldsymbol{x}^*\boldsymbol{D}\boldsymbol{x} = (1-|\lambda|^2)\boldsymbol{x}^*\boldsymbol{A}\boldsymbol{x}, \tag{12.22}$$

where $\boldsymbol{D} := \mathrm{diag}(a_{11}, \ldots, a_{nn})$. Now $\boldsymbol{x}^*\boldsymbol{A}\boldsymbol{x}$ and $\boldsymbol{x}^*\boldsymbol{D}\boldsymbol{x}$ are positive for all nonzero $x \in \mathbb{C}^n$ since a positive definite matrix has positive diagonal elements $a_{ii} = \boldsymbol{e}_i^T\boldsymbol{A}\boldsymbol{e}_i > 0$. It follows that the left hand side of (12.22) is nonnegative and then the right hand side must be nonnegative as well. This implies $|\lambda| \leq 1$. If $|\lambda| = 1$ then (12.22) implies that $\lambda = 1$ and it remains to show that this is not possible. By (12.10) and (12.12) we have

$$\boldsymbol{G}_\omega\boldsymbol{x} = (\boldsymbol{I} - \boldsymbol{M}_\omega^{-1}\boldsymbol{A})\boldsymbol{x} = \boldsymbol{x} - (\omega^{-1}\boldsymbol{D} - \boldsymbol{A}_L)^{-1}\boldsymbol{A}\boldsymbol{x}$$

and the eigenpair equation $\boldsymbol{G}_\omega \boldsymbol{x} = \lambda \boldsymbol{x}$ can be written $\boldsymbol{x} - (\omega^{-1}\boldsymbol{D} - \boldsymbol{A}_L)^{-1}\boldsymbol{A}\boldsymbol{x} = \lambda \boldsymbol{x}$ or

$$\boldsymbol{A}\boldsymbol{x} = (\omega^{-1}\boldsymbol{D} - \boldsymbol{A}_L)\boldsymbol{y}, \quad \boldsymbol{y} := (1-\lambda)\boldsymbol{x}. \tag{12.23}$$

Now $\boldsymbol{A}\boldsymbol{x} \neq \boldsymbol{0}$ implies that $\lambda \neq 1$.

To prove equation (12.22) we first show that

$$\boldsymbol{E}\boldsymbol{y} = \lambda \boldsymbol{A}\boldsymbol{x}, \quad \boldsymbol{E} := \omega^{-1}\boldsymbol{D} + \boldsymbol{A}_R - \boldsymbol{D} = \omega^{-1}\boldsymbol{D} - \boldsymbol{A}_L - \boldsymbol{A}. \tag{12.24}$$

The second equality follows immediately from $\boldsymbol{A} = \boldsymbol{D} - \boldsymbol{A}_L - \boldsymbol{A}_R$. By (12.23) and (12.24) we have

$$\boldsymbol{E}\boldsymbol{y} = (\omega^{-1}\boldsymbol{D} - \boldsymbol{A}_L - \boldsymbol{A})\boldsymbol{y} = \boldsymbol{A}\boldsymbol{x} - \boldsymbol{A}\boldsymbol{y} = \boldsymbol{A}\boldsymbol{x} - (1-\lambda)\boldsymbol{A}\boldsymbol{x} = \lambda \boldsymbol{A}\boldsymbol{x}.$$

Again using (12.23), (12.24) and adding $(\boldsymbol{A}\boldsymbol{x})^*\boldsymbol{y} = \boldsymbol{y}^*(\omega^{-1}\boldsymbol{D} - \boldsymbol{A}_L)^*\boldsymbol{y} = \boldsymbol{y}^*(\omega^{-1}\boldsymbol{D} - \boldsymbol{A}_R)\boldsymbol{y}$ and $\boldsymbol{y}^*(\lambda \boldsymbol{A}\boldsymbol{x}) = \boldsymbol{y}^*\boldsymbol{E}\boldsymbol{y} = \boldsymbol{y}^*(\omega^{-1}\boldsymbol{D} + \boldsymbol{A}_R - \boldsymbol{D})\boldsymbol{y}$ we find

$$\begin{aligned}(\boldsymbol{A}\boldsymbol{x})^*\boldsymbol{y} + \boldsymbol{y}^*(\lambda \boldsymbol{A}\boldsymbol{x}) &= \boldsymbol{y}^*(\omega^{-1}\boldsymbol{D} - \boldsymbol{A}_R)\boldsymbol{y} + \boldsymbol{y}^*(\omega^{-1}\boldsymbol{D} + \boldsymbol{A}_R - \boldsymbol{D})\boldsymbol{y} \\ &= \boldsymbol{y}^*(2\omega^{-1} - 1)\boldsymbol{D}\boldsymbol{y} = \omega^{-1}(2-\omega)|1-\lambda|^2\boldsymbol{x}^*\boldsymbol{D}\boldsymbol{x}.\end{aligned}$$

Since $(\boldsymbol{A}\boldsymbol{x})^* = \boldsymbol{x}^*\boldsymbol{A}^* = \boldsymbol{x}^*\boldsymbol{A}$, $\boldsymbol{y} := (1-\lambda)\boldsymbol{x}$ and $\boldsymbol{y}^* = (1-\overline{\lambda})\boldsymbol{x}^*$ this also equals

$$(\boldsymbol{A}\boldsymbol{x})^*\boldsymbol{y} + \boldsymbol{y}^*(\lambda \boldsymbol{A}\boldsymbol{x}) = (1-\lambda)\boldsymbol{x}^*\boldsymbol{A}\boldsymbol{x} + \lambda(1-\overline{\lambda})\boldsymbol{x}^*\boldsymbol{A}\boldsymbol{x} = (1-|\lambda|^2)\boldsymbol{x}^*\boldsymbol{A}\boldsymbol{x},$$

and (12.22) follows. □

12.3.3 Convergence of the Classical Methods for the Discrete Poisson Matrix

We know the eigenvalues of the discrete Poisson matrix $\boldsymbol{A}$ given by (10.8) and we can use this to estimate the number of iterations necessary to achieve a given accuracy for the various methods.

Recall that by (10.15) the eigenvalues $\lambda_{j,k}$ of $\boldsymbol{A}$ are

$$\lambda_{j,k} = 4 - 2\cos(j\pi h) - 2\cos(k\pi h), \quad j,k = 1,\ldots,m,\ h = 1/(m+1).$$

Consider first the J method. The matrix $\boldsymbol{G}_J = \boldsymbol{I} - \boldsymbol{D}^{-1}\boldsymbol{A} = \boldsymbol{I} - \boldsymbol{A}/4$ has eigenvalues

$$\mu_{j,k} = 1 - \frac{1}{4}\lambda_{j,k} = \frac{1}{2}\cos(j\pi h) + \frac{1}{2}\cos(k\pi h), \quad j,k = 1,\ldots,m. \tag{12.25}$$

It follows that $\rho(\boldsymbol{G}_J) = \cos(\pi h) < 1$. Since $\boldsymbol{G}_J$ is symmetric it is normal, and the spectral norm is equal to the spectral radius (cf. Theorem 8.4). We obtain

$$\|\boldsymbol{x}_k - \boldsymbol{x}\|_2 \le \|\boldsymbol{G}_J\|_2^k \|\boldsymbol{x}_0 - \boldsymbol{x}\|_2 = \cos^k(\pi h)\|\boldsymbol{x}_0 - \boldsymbol{x}\|_2, \quad k = 0, 1, 2, \ldots \tag{12.26}$$

The R method given by $\boldsymbol{x}_{k+1} = \boldsymbol{x}_k + \alpha \boldsymbol{r}_k$ with $\alpha = 2/(\lambda_{max} + \lambda_{min}) = 1/4$ is the same as the J-method so (12.26) holds in this case as well. This also follows from Corollary 12.1 with κ given by (10.20).

For the SOR method it is possible to explicitly determine $\rho(\boldsymbol{G}_\omega)$ for any $\omega \in (0, 2)$. The following result will be shown in Sect. 12.5.

Theorem 12.8 (The Spectral Radius of SOR Matrix) *Consider the SOR iteration (12.1.1), with the natural ordering. The spectral radius of $\boldsymbol{G}_\omega$ is*

$$\rho(\boldsymbol{G}_\omega) = \begin{cases} \frac{1}{4}\left(\omega\beta + \sqrt{(\omega\beta)^2 - 4(\omega - 1)}\right)^2, & \text{for } 0 < \omega \le \omega^*, \\ \omega - 1, & \text{for } \omega^* < \omega < 2, \end{cases} \tag{12.27}$$

where $\beta := \rho(\boldsymbol{G}_J) = \cos(\pi h)$ and

$$\omega^* := \frac{2}{1 + \sqrt{1 - \beta^2}} > 1. \tag{12.28}$$

Moreover,

$$\rho(\boldsymbol{G}_\omega) > \rho(\boldsymbol{G}_{\omega^*}) \text{ for } \omega \in (0, 2) \setminus \{\omega^*\}. \tag{12.29}$$

A plot of $\rho(\boldsymbol{G}_\omega)$ as a function of $\omega \in (0, 2)$ is shown in Fig. 12.2 for $n = 100$ (lower curve) and $n = 2500$ (upper curve). As ω increases the spectral radius of $\boldsymbol{G}_\omega$ decreases monotonically to the minimum ω^*. Then it increases linearly to the value one for $\omega = 2$. We call ω^* the **optimal relaxation parameter**.

For the discrete Poisson problem we have $\beta = \cos(\pi h)$ and it follows from (12.27), (12.28) that

$$\omega^* = \frac{2}{1 + \sin(\pi h)}, \quad \rho(\boldsymbol{G}_{\omega^*}) = \omega^* - 1 = \frac{1 - \sin(\pi h)}{1 + \sin(\pi h)}, \quad h = \frac{1}{m+1}. \tag{12.30}$$

Letting $\omega = 1$ in (12.27) we find $\rho(\boldsymbol{G}_1) = \beta^2 = \rho(\boldsymbol{G}_J)^2 = \cos^2(\pi h)$ for the GS method. Thus, for the discrete Poisson problem the J method needs twice as many iterations as the GS method for a given accuracy.

The values of $\rho(\boldsymbol{G}_J)$, $\rho(\boldsymbol{G}_1)$, and $\rho(\boldsymbol{G}_{\omega^*}) = \omega^* - 1$ are shown in Table 12.2 for $n = 100$ and $n = 2500$. We also show the smallest integer k_n such that $\rho(\boldsymbol{G})^{k_n} \le 10^{-8}$. This is an estimate for the number of iteration needed to obtain an accuracy of 10^{-8}. These values are comparable to the exact values given in Table 12.1.

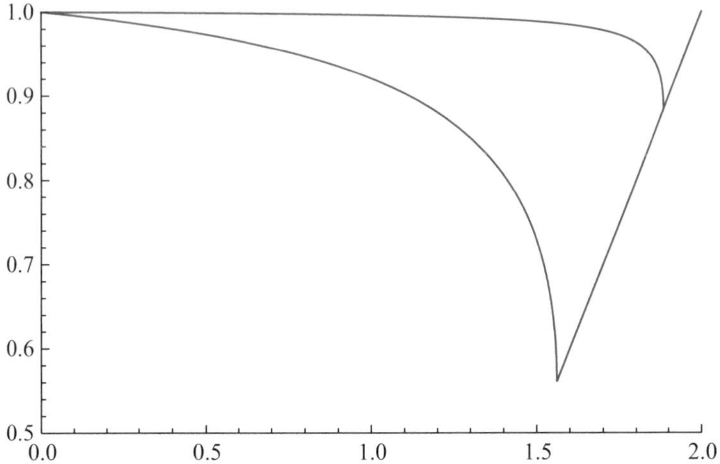

Fig. 12.2 $\rho(\boldsymbol{G}_\omega)$ with $\omega \in [0, 2]$ for $n = 100$, (lower curve) and $n = 2500$ (upper curve)

Table 12.2 Spectral radial for $\boldsymbol{G}_J$, $\boldsymbol{G}_1$, $\boldsymbol{G}_{\omega^*}$ and the smallest integer k_n such that $\rho(\boldsymbol{G})^{k_n} \le 10^{-8}$

	n=100	n = 2500	k_{100}	k_{2500}
J	0.959493	0.998103	446	9703
GS	0.920627	0.99621	223	4852
SOR	0.56039	0.88402	32	150

12.3.4 Number of Iterations

Consider next the **rate of convergence** of the iteration $\boldsymbol{x}_{k+1} = \boldsymbol{G}\boldsymbol{x}_k + \boldsymbol{c}$. We like to know how fast the iterative method converges. Recall that $\boldsymbol{x}_k - \boldsymbol{x} = \boldsymbol{G}^k(\boldsymbol{x}_0 - \boldsymbol{x})$. For k sufficiently large

$$\|\boldsymbol{x}_k - \boldsymbol{x}\| \le \|\boldsymbol{G}^k\|\|\boldsymbol{x}_0 - \boldsymbol{x}\| \approx \rho(\boldsymbol{G})^k\|\boldsymbol{x}_0 - \boldsymbol{x}\|.$$

For the last formula we apply Theorem 12.13 which says that $\lim_{k\to\infty}\|\boldsymbol{G}^k\|^{1/k} = \rho(\boldsymbol{G})$. For Jacobi's method and the spectral norm we have $\|\boldsymbol{G}_J^k\|_2 = \rho(\boldsymbol{G}_J)^k$ (cf. (12.26)).

For fast convergence we should use a $\boldsymbol{G}$ with small spectral radius.

Lemma 12.1 (Number of Iterations) *Suppose $\rho(\boldsymbol{G}) = 1 - \eta$ for some $0 < \eta < 1$ and let $s \in \mathbb{N}$. Then*

$$\tilde{k} := \frac{s\log(10)}{\eta} \tag{12.31}$$

is an estimate for the smallest number of iterations k so that $\rho(\boldsymbol{G})^k \le 10^{-s}$.

Proof The estimate $\tilde{k}$ is an approximate solution of the equation $\rho(\boldsymbol{G})^k = 10^{-s}$. Thus, since $-\log(1-\eta) \approx \eta$ when η is small

$$k = -\frac{s\log(10)}{\log(1-\eta)} \approx \frac{s\log(10)}{\eta} = \tilde{k}.$$

□

The following estimates are obtained. They agree with those we found numerically in Sect. 12.1.1.

- R and J: $\rho(\boldsymbol{G}_J) = \cos(\pi h) = 1-\eta$, $\eta = 1-\cos(\pi h) = \frac{1}{2}\pi^2h^2 + O(h^4) = \frac{\pi^2}{2}/n + O(n^{-2})$. Thus,

$$\tilde{k}_n = \frac{2\log(10)s}{\pi^2}n + O(n^{-1}) = O(n).$$

- GS: $\rho(\boldsymbol{G}_1) = \cos^2(\pi h) = 1-\eta$, $\eta = 1-\cos^2(\pi h) = \sin^2\pi h = \pi^2h^2 + O(h^4) = \pi^2/n + O(n^{-2})$. Thus,

$$\tilde{k}_n = \frac{\log(10)s}{\pi^2}n + O(n^{-1}) = O(n).$$

- SOR: $\rho(\boldsymbol{G}_{\omega^*}) = \frac{1-\sin(\pi h)}{1+\sin(\pi h)} = 1 - 2\pi h + O(h^2)$. Thus,

$$\tilde{k}_n = \frac{\log(10)s}{2\pi}\sqrt{n} + O(n^{-1/2}) = O(\sqrt{n}).$$

We note that

1. The convergence depends on the behavior of the powers $\boldsymbol{G}^k$ as k increases. The matrix $\boldsymbol{M}$ should be chosen so that all elements in $\boldsymbol{G}^k$ converge quickly to zero and such that the linear system (12.9) is easy to solve for $\boldsymbol{x}_{k+1}$. These are conflicting demands. $\boldsymbol{M}$ should be an approximation to $\boldsymbol{A}$ to obtain a $\boldsymbol{G}$ with small elements, but then (12.9) might not be easy to solve for $\boldsymbol{x}_{k+1}$.
2. The convergence $\lim_{k\to\infty}\|\boldsymbol{G}^k\|^{1/k} = \rho(\boldsymbol{G})$ can be quite slow (cf. Exercise 12.15).

12.3.5 Stopping the Iteration

In Algorithms 12.1 and 12.2 we had access to the exact solution and could stop the iteration when the error was sufficiently small in the infinity norm. The decision when to stop is obviously more complicated when the exact solution is not known. One possibility is to choose a vector norm, keep track of $\|\boldsymbol{x}_{k+1} - \boldsymbol{x}_k\|$, and stop when this number is sufficiently small. The following result indicates that $\|\boldsymbol{x}_k - \boldsymbol{x}\|$ can be quite large if $\|\boldsymbol{G}\|$ is close to one.

Lemma 12.2 (Be Careful When Stopping) *If* $\boldsymbol{x}_k = \boldsymbol{G}\boldsymbol{x}_{k-1} + \boldsymbol{c}$, $\boldsymbol{x} = \boldsymbol{G}\boldsymbol{x} + \boldsymbol{c}$ *and* $\|\boldsymbol{G}\| < 1$ *then*

$$\|\boldsymbol{x}_k - \boldsymbol{x}_{k-1}\| \geq \frac{1 - \|\boldsymbol{G}\|}{\|\boldsymbol{G}\|}\|\boldsymbol{x}_k - \boldsymbol{x}\|, \quad k \geq 1. \tag{12.32}$$

Proof We find

$$\begin{aligned}\|\boldsymbol{x}_k - \boldsymbol{x}\| &= \|\boldsymbol{G}(\boldsymbol{x}_{k-1} - \boldsymbol{x})\| \leq \|\boldsymbol{G}\|\|\boldsymbol{x}_{k-1} - \boldsymbol{x}\| \\ &= \|\boldsymbol{G}\|\|\boldsymbol{x}_{k-1} - \boldsymbol{x}_k + \boldsymbol{x}_k - \boldsymbol{x}\| \leq \|\boldsymbol{G}\|\big(\|\boldsymbol{x}_{k-1} - \boldsymbol{x}_k\| + \|\boldsymbol{x}_k - \boldsymbol{x}\|\big).\end{aligned}$$

Thus $(1 - \|\boldsymbol{G}\|)\|\boldsymbol{x}_k - \boldsymbol{x}\| \leq \|\boldsymbol{G}\|\|\boldsymbol{x}_{k-1} - \boldsymbol{x}_k\|$ which implies (12.32). □

Another possibility is to stop when the residual vector $\boldsymbol{r}_k := \boldsymbol{b} - \boldsymbol{A}\boldsymbol{x}_k$ is sufficiently small in some norm. To use the residual vector for stopping it is convenient to write the iterative method (12.10) in an alternative form. If $\boldsymbol{M}$ is the splitting matrix of the method then by (12.9) we have $\boldsymbol{M}\boldsymbol{x}_{k+1} = \boldsymbol{M}\boldsymbol{x}_k - \boldsymbol{A}\boldsymbol{x}_k + \boldsymbol{b}$. This leads to

$$\boldsymbol{x}_{k+1} = \boldsymbol{x}_k + \boldsymbol{M}^{-1}\boldsymbol{r}_k, \quad \boldsymbol{r}_k = \boldsymbol{b} - \boldsymbol{A}\boldsymbol{x}_k. \tag{12.33}$$

Testing on $\boldsymbol{r}_k$ works fine if $\boldsymbol{A}$ is well conditioned, but Theorem 8.8 shows that the relative error in the solution can be much larger than the relative error in $\boldsymbol{r}_k$ if $\boldsymbol{A}$ is ill-conditioned.

12.4 Powers of a Matrix

Let $\boldsymbol{A} \in \mathbb{C}^{n \times n}$ be a square matrix. In this section we consider the special matrix sequence $\{\boldsymbol{A}^k\}$ of powers of $\boldsymbol{A}$. We want to know when this sequence converges to the zero matrix. Such a sequence occurs in iterative methods (cf. (12.16)), in Markov processes in statistics, in the converge of geometric series of matrices (Neumann series cf. Sect. 12.4.2) and in many other applications.

12.4.1 The Spectral Radius

In this section we show the following important theorem.

Theorem 12.10 (When Is $\lim_{k\to\infty} A^k = 0$?) *For any $A \in \mathbb{C}^{n\times n}$ we have*

$$\lim_{k\to\infty} A^k = 0 \Longleftrightarrow \rho(A) < 1,$$

where $\rho(A)$ is the spectral radius of A given by (12.17).

Clearly $\rho(A) < 1$ is a necessary condition for $\lim_{k\to\infty} A^k = 0$. For if (λ, x) is an eigenpair of A with $|\lambda| \geq 1$ and $\|x\|_2 = 1$ then $A^k x = \lambda^k x$, and this implies $\|A^k\|_2 \geq \|A^k x\|_2 = \|\lambda^k x\|_2 = |\lambda|^k$, and it follows that A^k does not tend to zero.

The sufficiency condition is harder to show. We construct a consistent matrix norm on $\mathbb{C}^{n\times n}$ such that $\|A\| < 1$ and then use Theorems 12.2 and 12.3.

We start with

Theorem 12.11 (Any Consistent Norm Majorizes the Spectral Radius) *For any matrix norm $\|\cdot\|$ that is consistent on $\mathbb{C}^{n\times n}$ and any $A \in \mathbb{C}^{n\times n}$ we have $\rho(A) \leq \|A\|$.*

Proof Let (λ, x) be an eigenpair for A, $\|\ \|$ a consistent matrix norm on $\mathbb{C}^{n\times n}$ and define $X := [x, \ldots, x] \in \mathbb{C}^{n\times n}$. Then $\lambda X = AX$, which implies $|\lambda|\,\|X\| = \|\lambda X\| = \|AX\| \leq \|A\|\,\|X\|$. Since $\|X\| \neq 0$ we obtain $|\lambda| \leq \|A\|$. □

The next theorem shows that if $\rho(A) < 1$ then $\|A\| < 1$ for some consistent matrix norm on $\mathbb{C}^{n\times n}$, thus completing the proof of Theorem 12.10.

Theorem 12.12 (The Spectral Radius Can Be Approximated by a Norm) *Let $A \in \mathbb{C}^{n\times n}$ and $\epsilon > 0$ be given. There is a consistent matrix norm $\|\cdot\|$ on $\mathbb{C}^{n\times n}$ such that $\rho(A) \leq \|A\| \leq \rho(A) + \epsilon$.*

Proof Let A have eigenvalues $\lambda_1, \ldots, \lambda_n$. By the Schur Triangulation Theorem 6.5 there is a unitary matrix U and an upper triangular matrix $R = [r_{ij}]$ such that $U^*AU = R$. For $t > 0$ we define $D_t := \operatorname{diag}(t, t^2, \ldots, t^n) \in \mathbb{R}^{n\times n}$, and note that the (i, j) element in $D_t R D_t^{-1}$ is given by $t^{i-j} r_{ij}$ for all i, j. For $n = 3$

$$D_t R D_t^{-1} = \begin{bmatrix} \lambda_1 & t^{-1}r_{12} & t^{-2}r_{13} \\ 0 & \lambda_2 & t^{-1}r_{23} \\ 0 & 0 & \lambda_3 \end{bmatrix}.$$

For each $B \in \mathbb{C}^{n\times n}$ and $t > 0$ we use the one norm to define the matrix norm $\|B\|_t := \|D_t U^* B U D_t^{-1}\|_1$. We leave it as an exercise to show that $\|\ \|_t$ is a consistent matrix norm on $\mathbb{C}^{n\times n}$. We define $\|B\| := \|B\|_t$, where t is chosen so large that the sum of the absolute values of all off-diagonal elements in $D_t R D_t^{-1}$ is less than ϵ. Then

$$\|A\| = \|D_t U^* A U D_t^{-1}\|_1 = \|D_t R D_t^{-1}\|_1 = \max_{1\leq j\leq n} \sum_{i=1}^{n} \left|\left(D_t R D_t^{-1}\right)_{ij}\right|$$

$$\leq \max_{1\leq j\leq n} (|\lambda_j| + \epsilon) = \rho(A) + \epsilon.$$

□

A consistent matrix norm of a matrix can be much larger than the spectral radius. However the following result holds.

Theorem 12.13 (Spectral Radius Convergence) *For any consistent matrix norm* $\|\cdot\|$ *on* $\mathbb{C}^{n\times n}$ *and any* $\boldsymbol{A} \in \mathbb{C}^{n\times n}$ *we have*

$$\lim_{k\to\infty} \|\boldsymbol{A}^k\|^{1/k} = \rho(\boldsymbol{A}). \tag{12.34}$$

Proof Let $\|\ \|$ be a consistent matrix norm on $\mathbb{C}^{n\times n}$. If λ is an eigenvalue of $\boldsymbol{A}$ then λ^k is an eigenvalue of $\boldsymbol{A}^k$ for any $k \in \mathbb{N}$. By Theorem 12.11 we then obtain $\rho(\boldsymbol{A})^k = \rho(\boldsymbol{A}^k) \le \|\boldsymbol{A}^k\|$ for any $k \in \mathbb{N}$ so that $\rho(\boldsymbol{A}) \le \|\boldsymbol{A}^k\|^{1/k}$. Let $\epsilon > 0$ and consider the matrix $\boldsymbol{B} := (\rho(\boldsymbol{A}) + \epsilon)^{-1}\boldsymbol{A}$. Then $\rho(\boldsymbol{B}) = \rho(\boldsymbol{A})/(\rho(\boldsymbol{A}) + \epsilon) < 1$ and $\|\boldsymbol{B}^k\| \to 0$ by Theorem 12.10 as $k \to \infty$. Choose $N \in \mathbb{N}$ such that $\|\boldsymbol{B}^k\| < 1$ for all $k \ge N$. Then for $k \ge N$

$$\|\boldsymbol{A}^k\| = \|(\rho(\boldsymbol{A}) + \epsilon)^k \boldsymbol{B}^k\| = \big(\rho(\boldsymbol{A}) + \epsilon\big)^k \|\boldsymbol{B}^k\| < \big(\rho(\boldsymbol{A}) + \epsilon\big)^k.$$

We have shown that $\rho(\boldsymbol{A}) \le \|\boldsymbol{A}^k\|^{1/k} \le \rho(\boldsymbol{A}) + \epsilon$ for $k \ge N$. Since ϵ is arbitrary the result follows. □

12.4.2 Neumann Series

Let $\boldsymbol{B}$ be a square matrix. In this section we consider the **Neumann series**

$$\sum_{k=0}^{\infty} \boldsymbol{B}^k$$

which is a matrix analogue of a geometric series of numbers.

Consider an infinite series $\sum_{k=0}^{\infty} \boldsymbol{A}_k$ of matrices in $\mathbb{C}^{n\times n}$. We say that the series converges if the sequence of partial sums $\{\boldsymbol{S}_m\}$ given by $\boldsymbol{S}_m = \sum_{k=0}^{m} \boldsymbol{A}_k$ converges. The series converges if and only if $\{\boldsymbol{S}_m\}$ is a Cauchy sequence, i.e. to each $\epsilon > 0$ there exists an integer N so that $\|\boldsymbol{S}_l - \boldsymbol{S}_m\| < \epsilon$ for all $l > m \ge N$.

Theorem 12.14 (Neumann Series) *Suppose* $\boldsymbol{B} \in \mathbb{C}^{n\times n}$. *Then*

1. *The series* $\sum_{k=0}^{\infty} \boldsymbol{B}^k$ *converges if and only if* $\rho(\boldsymbol{B}) < 1$.
2. *If* $\rho(\boldsymbol{B}) < 1$ *then* $(\boldsymbol{I} - \boldsymbol{B})$ *is nonsingular and* $(\boldsymbol{I} - \boldsymbol{B})^{-1} = \sum_{k=0}^{\infty} \boldsymbol{B}^k$.
3. *If* $\|\boldsymbol{B}\| < 1$ *for some consistent matrix norm* $\|\cdot\|$ *on* $\mathbb{C}^{n\times n}$ *then*

$$\|(\boldsymbol{I} - \boldsymbol{B})^{-1}\| \le \frac{1}{1 - \|\boldsymbol{B}\|}. \tag{12.35}$$

Proof

1. Suppose $\rho(\boldsymbol{B}) < 1$. We show that $\boldsymbol{S}_m := \sum_{k=0}^{m} \boldsymbol{B}^k$ is a Cauchy sequence and hence convergent. Let $\epsilon > 0$. By Theorem 12.12 there is a consistent matrix norm $\|\cdot\|$ on $\mathbb{C}^{n\times n}$ such that $\|\boldsymbol{B}\| < 1$. Then for $l > m$

$$\|\boldsymbol{S}_l - \boldsymbol{S}_m\| = \|\sum_{k=m+1}^{l} \boldsymbol{B}^k\| \le \sum_{k=m+1}^{l} \|\boldsymbol{B}\|^k \le \|\boldsymbol{B}\|^{m+1} \sum_{k=0}^{\infty} \|\boldsymbol{B}\|^k = \frac{\|\boldsymbol{B}\|^{m+1}}{1 - \|\boldsymbol{B}\|}.$$

But then $\{\boldsymbol{S}_m\}$ is a Cauchy sequence provided N is such that $\frac{\|\boldsymbol{B}\|^{N+1}}{1-\|\boldsymbol{B}\|} < \epsilon$.
   Conversely, suppose $(\lambda, \boldsymbol{x})$ is an eigenpair for $\boldsymbol{B}$ with $|\lambda| \ge 1$. We find $\boldsymbol{S}_m\boldsymbol{x} = \sum_{k=0}^{m} \boldsymbol{B}^k \boldsymbol{x} = \big(\sum_{k=0}^{m} \lambda^k\big)\boldsymbol{x}$. Since λ^k does not tend to zero the series $\sum_{k=0}^{\infty} \lambda^k$ is not convergent and therefore $\{\boldsymbol{S}_m\boldsymbol{x}\}$ and hence $\{\boldsymbol{S}_m\}$ does not converge.
2. We have

$$\big(\sum_{k=0}^{m} \boldsymbol{B}^k\big)(\boldsymbol{I} - \boldsymbol{B}) = \boldsymbol{I} + \boldsymbol{B} + \cdots + \boldsymbol{B}^m - (\boldsymbol{B} + \cdots + \boldsymbol{B}^{m+1}) = \boldsymbol{I} - \boldsymbol{B}^{m+1}. \tag{12.36}$$

   Since $\rho(\boldsymbol{B}) < 1$ we conclude that $\boldsymbol{B}^{m+1} \to 0$ and hence taking limits in (12.36) we obtain $\big(\sum_{k=0}^{\infty} \boldsymbol{B}^k\big)(\boldsymbol{I} - \boldsymbol{B}) = \boldsymbol{I}$ which completes the proof of 2.
3. By 2: $\|(\boldsymbol{I} - \boldsymbol{B})^{-1}\| = \|\sum_{k=0}^{\infty} \boldsymbol{B}^k\| \le \sum_{k=0}^{\infty}\|\boldsymbol{B}\|^k = \frac{1}{1-\|\boldsymbol{B}\|}$.

□

12.5 The Optimal SOR Parameter ω

The following analysis is only carried out for the discrete Poisson matrix. It also holds for the averaging matrix given by (10.10). A more general theory is presented in [22]. We will compare the eigenpair equations for $\boldsymbol{G}_J$ and $\boldsymbol{G}_\omega$. It is convenient to write these equations using the matrix formulation $\boldsymbol{T}\boldsymbol{V} + \boldsymbol{V}\boldsymbol{T} = h^2\boldsymbol{F}$. If $\boldsymbol{G}_J\boldsymbol{v} = \mu\boldsymbol{v}$ is an eigenpair of $\boldsymbol{G}_J$ then

$$\frac{1}{4}(v_{i-1,j} + v_{i,j-1} + v_{i+1,j} + v_{i,j+1}) = \mu v_{i,j}, \quad i, j = 1, \ldots, m, \tag{12.37}$$

where $\boldsymbol{v} := \operatorname{vec}(\boldsymbol{V}) \in \mathbb{R}^{m^2}$ and $v_{i,j} = 0$ if $i \in \{0, m+1\}$ or $j \in \{0, m+1\}$.
   Suppose $(\lambda, \boldsymbol{w})$ is an eigenpair for $\boldsymbol{G}_\omega$. By (12.21) $(\boldsymbol{I} - \omega\boldsymbol{L})^{-1}\big(\omega\boldsymbol{R} + (1 - \omega)\boldsymbol{I}\big)\boldsymbol{w} = \lambda\boldsymbol{w}$ or

$$(\omega\boldsymbol{R} + \lambda\omega\boldsymbol{L})\boldsymbol{w} = (\lambda + \omega - 1)\boldsymbol{w}. \tag{12.38}$$

Let $\boldsymbol{w} = \text{vec}(\boldsymbol{W})$, where $\boldsymbol{W} \in \mathbb{C}^{m\times m}$. Then (12.38) can be written

$$\frac{\omega}{4}(\lambda w_{i-1,j} + \lambda w_{i,j-1} + w_{i+1,j} + w_{i,j+1}) = (\lambda + \omega - 1)w_{i,j}, \tag{12.39}$$

where $w_{i,j} = 0$ if $i \in \{0, m+1\}$ or $j \in \{0, m+1\}$.

Theorem 12.15 (The Optimal ω**)** *Consider the SOR method applied to the discrete Poisson matrix* (10.10)*, where we use the natural ordering. Moreover, assume* $\omega \in (0, 2)$.

1. *If* $\lambda \neq 0$ *is an eigenvalue of* $\boldsymbol{G}_\omega$ *then*

$$\mu := \frac{\lambda + \omega - 1}{\omega \lambda^{1/2}} \tag{12.40}$$

is an eigenvalue of $\boldsymbol{G}_J$.

2. *If* μ *is an eigenvalue of* $\boldsymbol{G}_J$ *and* λ *satisfies the equation*

$$\mu\omega\lambda^{1/2} = \lambda + \omega - 1 \tag{12.41}$$

then λ *is an eigenvalue of* $\boldsymbol{G}_\omega$.

Proof Suppose $(\lambda, \boldsymbol{w})$ is an eigenpair for $\boldsymbol{G}_\omega$. We claim that $(\mu, \boldsymbol{v})$ is an eigenpair for $\boldsymbol{G}_J$, where μ is given by (12.40) and $\boldsymbol{v} = (\boldsymbol{V})$ with $v_{i,j} := \lambda^{-(i+j)/2} w_{i,j}$. Indeed, replacing $w_{i,j}$ by $\lambda^{(i+j)/2} v_{i,j}$ in (12.39) and cancelling the common factor $\lambda^{(i+j)/2}$ we obtain

$$\frac{\omega}{4}(v_{i-1,j} + v_{i,j-1} + v_{i+1,j} + v_{i,j+1}) = \lambda^{-1/2}(\lambda + \omega - 1)v_{i,j}.$$

But then

$$\boldsymbol{G}_J \boldsymbol{v} = (\boldsymbol{L} + \boldsymbol{R})\boldsymbol{v} = \frac{\lambda + \omega - 1}{\omega \lambda^{1/2}} \boldsymbol{v} = \mu \boldsymbol{v}.$$

For the converse let $(\mu, \boldsymbol{v})$ be an eigenpair for $\boldsymbol{G}_J$ and let et λ be a solution of (12.41). We define as before $\boldsymbol{v} =: \text{vec}(\boldsymbol{V})$, $\boldsymbol{W} = \text{vec}(\boldsymbol{W})$ with $w_{i,j} := \lambda^{(i+j)/2} v_{i,j}$. Inserting this in (12.37) and canceling $\lambda^{-(i+j)/2}$ we obtain

$$\frac{1}{4}(\lambda^{1/2} w_{i-1,j} + \lambda^{1/2} w_{i,j-1} + \lambda^{-1/2} w_{i+1,j} + \lambda^{-1/2} w_{i,j+1}) = \mu w_{i,j}.$$

Multiplying by $\omega\lambda^{1/2}$ we obtain

$$\frac{\omega}{4}(\lambda w_{i-1,j} + \lambda w_{i,j-1} + w_{i+1,j} + w_{i,j+1}) = \omega\mu\lambda^{1/2} w_{i,j},$$

Thus, if $\omega\mu\lambda^{1/2} = \lambda + \omega - 1$ then by (12.39) $(\lambda, \boldsymbol{w})$ is an eigenpair for $\boldsymbol{G}_\omega$. □

Proof of Theorem 12.8 Combining statement 1 and 2 in Theorem 12.15 we see that $\rho(\boldsymbol{G}_\omega) = |\lambda(\mu)|$, where $\lambda(\mu)$ is an eigenvalue of $\boldsymbol{G}_\omega$ satisfying (12.41) for some eigenvalue μ of $\boldsymbol{G}_J$. The eigenvalues of $\boldsymbol{G}_J$ are $\frac{1}{2}\cos(j\pi h) + \frac{1}{2}\cos(k\pi h)$, $j, k = 1, \dots, m$, so μ is real and both μ and $-\mu$ are eigenvalues. Thus, to compute $\rho(\boldsymbol{G}_\omega)$ it is enough to consider (12.41) for a positive eigenvalue μ of $\boldsymbol{G}_J$. Solving (12.41) for $\lambda = \lambda(\mu)$ gives

$$\lambda(\mu) := \frac{1}{4}\Big(\omega\mu \pm \sqrt{(\omega\mu)^2 - 4(\omega - 1)}\Big)^2. \tag{12.42}$$

Both roots $\lambda(\mu)$ are eigenvalues of $\boldsymbol{G}_\omega$. The discriminant

$$d(\omega) := (\omega\mu)^2 - 4(\omega - 1).$$

is strictly decreasing on $(0, 2)$ since

$$d'(\omega) = 2(\omega\mu^2 - 2) < 2(\omega - 2) < 0.$$

Moreover $d(0) = 4 > 0$ and $d(2) = 4\mu^2 - 4 < 0$. As a function of ω, $\lambda(\mu)$ changes from real to complex when $d(\omega) = 0$. The root in $(0, 2)$ is

$$\omega = \tilde{\omega}(\mu) := 2\frac{1 - \sqrt{1 - \mu^2}}{\mu^2} = \frac{2}{1 + \sqrt{1 - \mu^2}}. \tag{12.43}$$

In the complex case we find

$$|\lambda(\mu)| = \frac{1}{4}\Big((\omega\mu)^2 + 4(\omega - 1) - (\omega\mu)^2\Big) = \omega - 1, \quad \tilde{\omega}(\mu) < \omega < 2.$$

In the real case both roots of (12.42) are positive and the larger one is

$$\lambda(\mu) = \frac{1}{4}\Big(\omega\mu + \sqrt{(\omega\mu)^2 - 4(\omega - 1)}\Big)^2, \quad 0 < \omega \le \tilde{\omega}(\mu). \tag{12.44}$$

Both $\lambda(\mu)$ and $\tilde{\omega}(\mu)$ are strictly increasing as functions of μ. It follows that $|\lambda(\mu)|$ is maximized for $\mu = \rho(\boldsymbol{G}_J) =: \beta$ and for this value of μ we obtain (12.27) for $0 < \omega \le \tilde{\omega}(\beta) = \omega^*$.

Evidently $\rho(\boldsymbol{G}_\omega) = \omega - 1$ is strictly increasing in $\omega^* < \omega < 2$. Equation (12.29) will follow if we can show that $\rho(\boldsymbol{G}_\omega)$ is strictly decreasing in $0 < \omega < \omega^*$. By differentiation

$$\frac{d}{d\omega}\Big(\omega\beta + \sqrt{(\omega\beta)^2 - 4(\omega - 1)}\Big) = \frac{\beta\sqrt{(\omega\beta)^2 - 4(\omega - 1)} + \omega\beta^2 - 2}{\sqrt{(\omega\beta)^2 - 4(\omega - 1)}}.$$

Since $\beta^2(\omega^2\beta^2 - 4\omega + 4) < (2 - \omega\beta^2)^2$ the numerator is negative and the strict decrease of $\rho(\boldsymbol{G}_\omega)$ in $0 < \omega < \omega^*$ follows. □

12.6 Exercises Chap. 12

12.6.1 Exercises Sect. 12.3

Exercise 12.1 (Richardson and Jacobi) Show that if $a_{ii} = d \neq 0$ for all i then Richardson's method with $\alpha := 1/d$ is the same as Jacobi's method.

Exercise 12.2 (R-Method When Eigenvalues Have Positive Real Part) Suppose all eigenvalues λ_j of $\boldsymbol{A}$ have positive real parts u_j for $j = 1, \ldots, n$ and that α is real. Show that the R method converges if and only if $0 < \alpha < \min_j(2u_j/|\lambda_j|^2)$.

Exercise 12.3 (Divergence Example for J and GS) Show that both Jacobi's method and Gauss-Seidel's method diverge for $\boldsymbol{A} = \begin{bmatrix} 1 & 2 \\ 3 & 4 \end{bmatrix}$.

Exercise 12.4 (2 by 2 Matrix) We want to show that Gauss-Seidel converges if and only if Jacobi converges for a 2 by 2 matrix $\boldsymbol{A} := \begin{bmatrix} a_{11} & a_{12} \\ a_{21} & a_{22} \end{bmatrix} \in \mathbb{R}^{2\times 2}$.

a) Show that the spectral radius for the Jacobi method is

$$\rho(\boldsymbol{G}_J) = \sqrt{|a_{21}a_{12}/a_{11}a_{22}|}.$$

b) Show that the spectral radius for the Gauss-Seidel method is

$$\rho(\boldsymbol{G}_1) = |a_{21}a_{12}/a_{11}a_{22}|.$$

c) Conclude that Gauss-Seidel converges if and only if Jacobi converges.

Exercise 12.5 (Example: GS Converges, J Diverges) Show (by finding its eigenvalues) that the matrix $\begin{bmatrix} 1 & a & a \\ a & 1 & a \\ a & a & 1 \end{bmatrix}$ is positive definite for $-1/2 < a < 1$. Thus, GS converges for these values of a. Show that the J method does not converge for $1/2 < a < 1$.

Exercise 12.6 (Example: GS Diverges, J Converges) Let $\boldsymbol{G}_J$ and $\boldsymbol{G}_1$ be the iteration matrices for the Jacobi and Gauss-Seidel methods applied to the matrix $\boldsymbol{A} := \begin{bmatrix} 1 & 0 & 1/2 \\ 1 & 1 & 0 \\ -1 & 1 & 1 \end{bmatrix}$.[1]

a) Show that $\boldsymbol{G}_1 := \begin{bmatrix} 0 & 0 & -1/2 \\ 0 & 0 & 1/2 \\ 0 & 0 & -1 \end{bmatrix}$ and conclude that GS diverges.
b) Show that $p(\lambda) := \det(\lambda \boldsymbol{I} - \boldsymbol{G}_J) = \lambda^3 + \frac{1}{2}\lambda + \frac{1}{2}$.
c) Show that if $|\lambda| \geq 1$ then $p(\lambda) \neq 0$. Conclude that J converges.

Exercise 12.7 (Strictly Diagonally Dominance; The J Method) Show that the J method converges if $|a_{ii}| > \sum_{j\neq i} |a_{ij}|$ for $i = 1, \ldots, n$.

[1] Stewart Venit, "The convergence of Jacobi and Gauss-Seidel iteration", Mathematics Magazine **48** (1975), 163–167.

Exercise 12.8 (Strictly Diagonally Dominance; The GS Method) Consider the GS method. Suppose $r := \max_i r_i < 1$, where $r_i = \sum_{j\neq i} \frac{|a_{ij}|}{|a_{ii}|}$. Show using induction on i that $|\boldsymbol{\epsilon}_{k+1}(j)| \le r\|\boldsymbol{\epsilon}_k\|_\infty$ for $j = 1, \ldots, i$. Conclude that Gauss-Seidel's method is convergent when $\boldsymbol{A}$ is strictly diagonally dominant.

Exercise 12.9 (Convergence Example for Fix Point Iteration) Consider for $a \in \mathbb{C}$

$$\boldsymbol{x} := \begin{bmatrix} x_1 \\ x_2 \end{bmatrix} = \begin{bmatrix} 0 & a \\ a & 0 \end{bmatrix} \begin{bmatrix} x_1 \\ x_2 \end{bmatrix} + \begin{bmatrix} 1-a \\ 1-a \end{bmatrix} =: \boldsymbol{G}\boldsymbol{x} + \boldsymbol{c}.$$

Starting with $\boldsymbol{x}_0 = \boldsymbol{0}$ show by induction

$$\boldsymbol{x}_k(1) = \boldsymbol{x}_k(2) = 1 - a^k, \quad k \ge 0,$$

and conclude that the iteration converges to the fixed-point $\boldsymbol{x} = [1, 1]^T$ for $|a| < 1$ and diverges for $|a| > 1$. Show that $\rho(\boldsymbol{G}) = 1 - \eta$ with $\eta = 1 - |a|$. Compute the estimate (12.31) for the rate of convergence for $a = 0.9$ and $s = 16$ and compare with the true number of iterations determined from $|a|^k \le 10^{-16}$.

Exercise 12.10 (Estimate in Lemma 12.1 Can Be Exact) Consider the iteration in Example 12.2. Show that $\rho(\boldsymbol{G}_J) = 1/2$. Then show that $\boldsymbol{x}_k(1) = \boldsymbol{x}_k(2) = 1 - 2^{-k}$ for $k \ge 0$. Thus the estimate in Lemma 12.1 is exact in this case.

Exercise 12.11 (Iterative Method (Exam Exercise 1991-3)) Let $\boldsymbol{A} \in \mathbb{R}^{n\times n}$ be a symmetric positive definite matrix with ones on the diagonal and let $\boldsymbol{b} \in \mathbb{R}^n$. We will consider an iterative method for the solution of $\boldsymbol{A}\boldsymbol{x} = \boldsymbol{b}$. Observe that $\boldsymbol{A}$ may be written $\boldsymbol{A} = \boldsymbol{I} - \boldsymbol{L} - \boldsymbol{L}^T$, where $\boldsymbol{L}$ is lower triangular with zero's on the diagonal, $l_{i,j} = 0$, when $j >= i$. The method is defined by

$$\boldsymbol{M}\boldsymbol{x}_{k+1} = \boldsymbol{N}\boldsymbol{x}_k + \boldsymbol{b}, \tag{12.45}$$

where $\boldsymbol{M}$ and $\boldsymbol{N}$ are given by the splitting

$$\boldsymbol{A} = \boldsymbol{M} - \boldsymbol{N}, \quad \boldsymbol{M} = (\boldsymbol{I} - \boldsymbol{L})(\boldsymbol{I} - \boldsymbol{L}^T), \quad \boldsymbol{N} = \boldsymbol{L}\boldsymbol{L}^T. \tag{12.46}$$

a) Let $\boldsymbol{x} \neq \boldsymbol{0}$ be an eigenvector of $\boldsymbol{M}^{-1}\boldsymbol{N}$ with eigenvalue λ. Show that

$$\lambda = \frac{\boldsymbol{x}^T\boldsymbol{N}\boldsymbol{x}}{\boldsymbol{x}^T\boldsymbol{A}\boldsymbol{x} + \boldsymbol{x}^T\boldsymbol{N}\boldsymbol{x}}. \tag{12.47}$$

b) Show that the sequence $\{\boldsymbol{x}_k\}$ generated by (12.45) converges to the solution $\boldsymbol{x}$ of $\boldsymbol{A}\boldsymbol{x} = \boldsymbol{b}$ for any starting vector $\boldsymbol{x}_0$.

c) Consider the following algorithm

$$\boxed{\begin{array}{l} \text{1. Choose } \boldsymbol{x} = [x(1), x(2), \ldots, x(n)]^T. \\ \text{2. for } k = 1, 2, 3, \ldots \\ \quad \text{for } i = 1, 2, \ldots, n-1, n, n, n-1, n-2, \ldots, 1 \\ \quad x(i) = b(i) - \sum_{j \neq i} a(i, j) x(j) \end{array}} \tag{12.48}$$

Is there a connection between this algorithm and the method of Gauss-Seidel? Show that the algorithm (12.48) leads up to the splitting (12.46).

Exercise 12.12 (Gauss-Seidel Method (Exam Exercise 2008-1)) Consider the linear system $\boldsymbol{Ax} = \boldsymbol{b}$ in which

$$\boldsymbol{A} := \begin{bmatrix} 3 & 0 & 1 \\ 0 & 7 & 2 \\ 1 & 2 & 4 \end{bmatrix}$$

and $\boldsymbol{b} := [1, 9, -2]^T$.

a) With $\boldsymbol{x}_0 = [1, 1, 1]^t$, carry out one iteration of the Gauss-Seidel method to find $\boldsymbol{x}_1 \in \mathbb{R}^3$.
b) If we continue the iteration, will the method converge? Why?
c) Write a MATLAB program for the Gauss-Seidel method applied to a matrix $\boldsymbol{A} \in \mathbb{R}^{n \times n}$ and right-hand side $\boldsymbol{b} \in \mathbb{R}^n$. Use the ratio of the current residual to the initial residual as the stopping criterion, as well as a maximum number of iterations.[2]

12.6.2 Exercises Sect. 12.4

Exercise 12.13 (A Special Norm) Show that $\|\boldsymbol{B}\|_t := \|\boldsymbol{D}_t \boldsymbol{U}^* \boldsymbol{B} \boldsymbol{U} \boldsymbol{D}_t^{-1}\|_1$ defined in the proof of Theorem 12.12 is a consistent matrix norm on $\mathbb{C}^{n \times n}$.

Exercise 12.14 (Is $A + E$ Nonsingular?) Suppose $\boldsymbol{A} \in \mathbb{C}^{n \times n}$ is nonsingular and $\boldsymbol{E} \in \mathbb{C}^{n \times n}$. Show that $\boldsymbol{A} + \boldsymbol{E}$ is nonsingular if $\rho(\boldsymbol{A}^{-1}\boldsymbol{E}) < 1$.

[2]Hint: The function C=tril(A) extracts the lower part of A into a lower triangular matrix C.

Exercise 12.15 (Slow Spectral Radius Convergence) The convergence $\lim_{k\to\infty}\|A^k\|^{1/k} = \rho(A)$ can be quite slow. Consider

$$A := \begin{bmatrix} \lambda & a & 0 & \cdots & 0 & 0 \\ 0 & \lambda & a & \cdots & 0 & 0 \\ 0 & 0 & \lambda & \cdots & 0 & 0 \\ \vdots & & & & & \vdots \\ 0 & 0 & 0 & \cdots & \lambda & a \\ 0 & 0 & 0 & \cdots & 0 & \lambda \end{bmatrix} \in \mathbb{R}^{n\times n}.$$

If $|\lambda| = \rho(A) < 1$ then $\lim_{k\to\infty} A^k = \mathbf{0}$ for any $a \in \mathbb{R}$. We show below that the $(1, n)$ element of A^k is given by $f(k) := \binom{k}{n-1} a^{n-1}\lambda^{k-n+1}$ for $k \ge n-1$.

a) Pick an n, e.g. $n = 5$, and make a plot of $f(k)$ for $\lambda = 0.9$, $a = 10$, and $n-1 \le k \le 200$. Your program should also compute $\max_k f(k)$. Use your program to determine how large k must be before $f(k) < 10^{-8}$.

b) We can determine the elements of A^k explicitly for any k. Let $E := (A - \lambda I)/a$. Show by induction that $E^k = \begin{bmatrix} \mathbf{0} & I_{n-k} \\ \mathbf{0} & \mathbf{0} \end{bmatrix}$ for $1 \le k \le n-1$ and that $E^n = \mathbf{0}$.

c) We have $A^k = (aE + \lambda I)^k = \sum_{j=0}^{\min\{k,n-1\}} \binom{k}{j} a^j \lambda^{k-j} E^j$ and conclude that the $(1, n)$ element is given by $f(k)$ for $k \ge n-1$.

12.7 Review Questions

12.7.1 Consider a matrix $A \in \mathbb{C}^{n\times n}$ with nonzero diagonal elements.

- Define the J and GS method in component form,
- Do they always converge?
- Give a necessary and sufficient condition that $A^n \to \mathbf{0}$.
- Is there a matrix norm $\|\ \|$ consistent on $\mathbb{C}^{n\times n}$ such that $\|A\| < \rho(A)$?

12.7.2 What is a Neumann series? when does it converge?

12.7.3 How do we define convergence of a fixed point iteration $x_{k+1} = Gx_k + c$? When does it converge?

12.7.4 Define Richardson's method.

Chapter 13
The Conjugate Gradient Method

The **conjugate gradient method** was published by Hestenes and Stiefel in 1952, [6] as a direct method for solving linear systems. Today its main use is as an iterative method for solving large sparse linear systems. On a test problem we show that it performs as well as the SOR method with optimal acceleration parameter, and we do not have to estimate any such parameter. However the conjugate gradient method is restricted to positive definite systems. We also consider a mathematical formulation of the **preconditioned conjugate gradient method**. It is used to speed up convergence of the conjugate gradient method. We only give one example of a possible preconditioner. See [1] for a more complete treatment of iterative methods and preconditioning.

The conjugate gradient method can also be used for minimization and is related to a method known as **steepest descent**. This method and the conjugate gradient method are both minimization methods, and iterative methods, for solving equations.

Throughout this chapter $\boldsymbol{A} \in \mathbb{R}^{n\times n}$ will be a symmetric and positive definite matrix. We recall that $\boldsymbol{A}$ has positive eigenvalues and that the spectral (2-norm) condition number of $\boldsymbol{A}$ is given by $\kappa := \frac{\lambda_{max}}{\lambda_{min}}$, where λ_{max} and λ_{min} are the largest and smallest eigenvalue of $\boldsymbol{A}$.

The analysis of the methods in this chapter is in terms of two inner products on $\mathbb{R}^n$, the usual inner product $\langle \boldsymbol{x}, \boldsymbol{y}\rangle = \boldsymbol{x}^T\boldsymbol{y}$ with the associated Euclidian norm $\|\boldsymbol{x}\|_2 = \sqrt{\boldsymbol{x}^T\boldsymbol{x}}$, and the **$\boldsymbol{A}$-inner product** and the corresponding **$\boldsymbol{A}$-norm** given by

$$\langle \boldsymbol{x}, \boldsymbol{y}\rangle_A := \boldsymbol{x}^T\boldsymbol{A}\boldsymbol{y}, \quad \|\boldsymbol{y}\|_A := \sqrt{\boldsymbol{y}^T\boldsymbol{A}\boldsymbol{y}}, \quad \boldsymbol{x}, \boldsymbol{y} \in \mathbb{R}^n. \tag{13.1}$$

T. Lyche, *Numerical Linear Algebra and Matrix Factorizations*,
Texts in Computational Science and Engineering 22,
https://doi.org/10.1007/978-3-030-36468-7_13

We note that the $\boldsymbol{A}$-inner product is an inner product on $\mathbb{R}^n$. Indeed, for any $\boldsymbol{x}, \boldsymbol{y}, \boldsymbol{z} \in \mathbb{R}^n$

1. $\langle \boldsymbol{x}, \boldsymbol{x} \rangle_A = \boldsymbol{x}^T \boldsymbol{A}\boldsymbol{x} \geq 0$ and $\langle \boldsymbol{x}, \boldsymbol{x} \rangle_A = 0$ if and only if $\boldsymbol{x} = \boldsymbol{0}$, since $\boldsymbol{A}$ is positive definite,
2. $\langle \boldsymbol{x}, \boldsymbol{y} \rangle_A := \boldsymbol{x}^T \boldsymbol{A}\boldsymbol{y} = (\boldsymbol{x}^T \boldsymbol{A}\boldsymbol{y})^T = \boldsymbol{y}^T \boldsymbol{A}^T \boldsymbol{x} = \boldsymbol{y}^T \boldsymbol{A}\boldsymbol{x} = \langle \boldsymbol{y}, \boldsymbol{x} \rangle_A$ by symmetry of $\boldsymbol{A}$,
3. $\langle \boldsymbol{x} + \boldsymbol{y}, \boldsymbol{z} \rangle_A := \boldsymbol{x}^T \boldsymbol{A}\boldsymbol{z} + \boldsymbol{y}^T \boldsymbol{A}\boldsymbol{z} = \langle \boldsymbol{x}, \boldsymbol{z} \rangle_A + \langle \boldsymbol{y}, \boldsymbol{z} \rangle_A$, true for any $\boldsymbol{A}$.

By Theorem 5.2 the $\boldsymbol{A}$-norm is a vector norm on $\mathbb{R}^n$ since it is an inner product norm, and the Cauchy-Schwarz inequality holds

$$|\boldsymbol{x}^T \boldsymbol{A}\boldsymbol{y}|^2 \leq (\boldsymbol{x}^T \boldsymbol{A}\boldsymbol{x})(\boldsymbol{y}^T \boldsymbol{A}\boldsymbol{y}), \quad \boldsymbol{x}, \boldsymbol{y} \in \mathbb{R}^n. \tag{13.2}$$

13.1 Quadratic Minimization and Steepest Descent

We start by discussing some aspect of quadratic minimization and its relation to solving linear systems.

Consider for a positive definite $\boldsymbol{A} \in \mathbb{R}^{n \times n}$, $\boldsymbol{b} \in \mathbb{R}^n$ and $c \in \mathbb{R}$ the quadratic function $Q : \mathbb{R}^n \to \mathbb{R}$ given by

$$Q(\boldsymbol{y}) := \frac{1}{2}\boldsymbol{y}^T \boldsymbol{A}\boldsymbol{y} - \boldsymbol{b}^T \boldsymbol{y} + c. \tag{13.3}$$

As an example, some level curves of

$$Q(x, y) := \frac{1}{2}\begin{bmatrix} x & y \end{bmatrix}\begin{bmatrix} 2 & -1 \\ -1 & 2 \end{bmatrix}\begin{bmatrix} x \\ y \end{bmatrix} = x^2 - xy + y^2 \tag{13.4}$$

are shown in Fig. 13.1. The level curves are ellipses and the graph of Q is a paraboloid (cf. Exercise 13.2).

The following expansion will be used repeatedly. For $\boldsymbol{y}, \boldsymbol{h} \in \mathbb{R}^n$ and $\varepsilon \in \mathbb{R}$

$$Q(\boldsymbol{y} + \varepsilon\boldsymbol{h}) = Q(\boldsymbol{y}) - \varepsilon\boldsymbol{h}^T r(\boldsymbol{y}) + \frac{1}{2}\varepsilon^2 \boldsymbol{h}^T \boldsymbol{A}\boldsymbol{h}, \text{ where } r(\boldsymbol{y}) := \boldsymbol{b} - \boldsymbol{A}\boldsymbol{y}. \tag{13.5}$$

Minimizing a quadratic function is equivalent to solving a linear system.

Lemma 13.1 (Quadratic Function) *A vector $\boldsymbol{x} \in \mathbb{R}^n$ minimizes Q given by* (13.3) *if and only if $\boldsymbol{A}\boldsymbol{x} = \boldsymbol{b}$. Moreover, the residual $\boldsymbol{r}(\boldsymbol{y}) := \boldsymbol{b} - \boldsymbol{A}\boldsymbol{y}$ for any $\boldsymbol{y} \in \mathbb{R}^n$ is equal to the negative gradient, i.e., $\boldsymbol{r}(\boldsymbol{y}) = -\nabla Q(\boldsymbol{y})$, where $\nabla := \left[\frac{\partial}{\partial y_1}, \ldots, \frac{\partial}{\partial y_n}\right]^T$.*

Proof If $\boldsymbol{y} = \boldsymbol{x}$, $\varepsilon = 1$, and $\boldsymbol{A}\boldsymbol{x} = \boldsymbol{b}$, then (13.5) simplifies to $Q(\boldsymbol{x} + \boldsymbol{h}) = Q(\boldsymbol{x}) + \frac{1}{2}\boldsymbol{h}^T \boldsymbol{A}\boldsymbol{h}$, and since $\boldsymbol{A}$ is positive definite $Q(\boldsymbol{x} + \boldsymbol{h}) > Q(\boldsymbol{x})$ for all nonzero $\boldsymbol{h} \in \mathbb{R}^n$. It follows that $\boldsymbol{x}$ is the unique minimum of Q. Conversely, if $\boldsymbol{A}\boldsymbol{x} \neq \boldsymbol{b}$ and

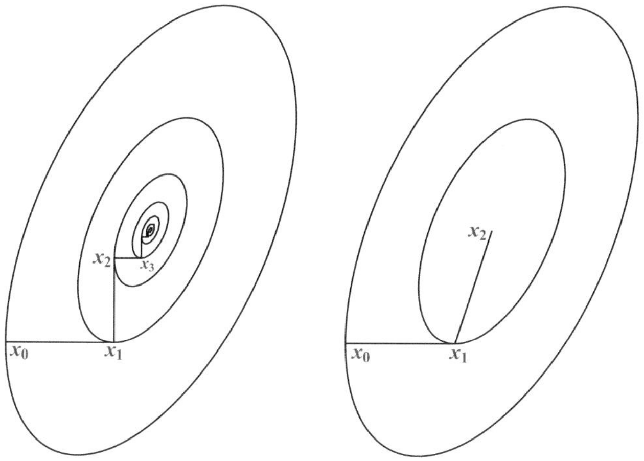

Fig. 13.1 Level curves for $Q(x, y)$ given by (13.4). Also shown is a steepest descent iteration (left) and a conjugate gradient iteration (right) to find the minimum of Q (cf Examples 13.1,13.2)

$\boldsymbol{h} := \boldsymbol{r}(\boldsymbol{x})$, then by (13.5), $Q(\boldsymbol{x} + \varepsilon\boldsymbol{h}) - Q(\boldsymbol{x}) = -\varepsilon(\boldsymbol{h}^T\boldsymbol{r}(x) - \frac{1}{2}\varepsilon\boldsymbol{h}^T\boldsymbol{A}\boldsymbol{h}) < 0$ for $\varepsilon > 0$ sufficiently small. Thus $\boldsymbol{x}$ does not minimize Q. By (13.5) for $\boldsymbol{y} \in \mathbb{R}^n$

$$\begin{aligned}\frac{\partial}{\partial y_i}Q(\boldsymbol{y}) &:= \lim_{\varepsilon\to 0}\frac{1}{\varepsilon}\left(Q(\boldsymbol{y} + \varepsilon\boldsymbol{e}_i) - Q(\boldsymbol{y})\right)\\ &= \lim_{\varepsilon\to 0}\frac{1}{\varepsilon}\left(-\varepsilon\boldsymbol{e}_i^T\boldsymbol{r}(\boldsymbol{y})) + \frac{1}{2}\varepsilon^2\boldsymbol{e}_i^T\boldsymbol{A}\boldsymbol{e}_i\right) = -\boldsymbol{e}_i^T\boldsymbol{r}(\boldsymbol{y}), \quad i = 1, \ldots, n,\end{aligned}$$

showing that $\boldsymbol{r}(\boldsymbol{y}) = -\nabla Q(\boldsymbol{y})$. □

A general class of minimization algorithms for Q and solution algorithms for a linear system is given as follows:

1. Choose $\boldsymbol{x}_0 \in \mathbb{R}^n$.
2. For $k = 0, 1, 2, \ldots$

$$\boxed{\begin{array}{l}\text{Choose a “search direction” } \boldsymbol{p}_k,\\ \text{Choose a “step length” } \alpha_k,\\ \text{Compute } \boldsymbol{x}_{k+1} = \boldsymbol{x}_k + \alpha_k\boldsymbol{p}_k.\end{array}} \tag{13.6}$$

We would like to generate a sequence $\{\boldsymbol{x}_k\}$ that converges quickly to the minimum $\boldsymbol{x}$ of Q.

For a fixed direction $\boldsymbol{p}_k$ we say that α_k is **optimal** if $Q(\boldsymbol{x}_{k+1})$ is as small as possible, i.e.

$$Q(\boldsymbol{x}_{k+1}) = Q(\boldsymbol{x}_k + \alpha_k \boldsymbol{p}_k) = \min_{\alpha \in \mathbb{R}} Q(\boldsymbol{x}_k + \alpha \boldsymbol{p}_k).$$

By (13.5) we have $Q(\boldsymbol{x}_k + \alpha \boldsymbol{p}_k) = Q(\boldsymbol{x}_k) - \alpha \boldsymbol{p}_k^T \boldsymbol{r}_k + \frac{1}{2}\alpha^2 \boldsymbol{p}_k^T \boldsymbol{A} \boldsymbol{p}_k$, where $\boldsymbol{r}_k := \boldsymbol{b} - \boldsymbol{A}\boldsymbol{x}_k$. Since $\boldsymbol{p}_k^T \boldsymbol{A} \boldsymbol{p}_k \geq 0$ we find a minimum α_k by solving $\frac{\partial}{\partial \alpha} Q(\boldsymbol{x}_k + \alpha \boldsymbol{p}_k) = 0$. It follows that the optimal α_k is uniquely given by

$$\alpha_k := \frac{\boldsymbol{p}_k^T \boldsymbol{r}_k}{\boldsymbol{p}_k^T \boldsymbol{A} \boldsymbol{p}_k}. \tag{13.7}$$

In the method of **steepest descent**, also known as the **gradient method**, we choose $\boldsymbol{p}_k = \boldsymbol{r}_k$ the negative gradient, and the optimal α_k. Starting from $\boldsymbol{x}_0$ we compute for $k = 0, 1, 2 \ldots$

$$\boldsymbol{x}_{k+1} = \boldsymbol{x}_k + \Big(\frac{\boldsymbol{r}_k^T \boldsymbol{r}_k}{\boldsymbol{r}_k^T \boldsymbol{A} \boldsymbol{r}_k}\Big) \boldsymbol{r}_k. \tag{13.8}$$

This is similar to Richardson's method (12.18), but in that method we used a constant step length. Computationally, a step in the steepest descent iteration can be organized as follows

$$\boxed{\begin{aligned} \boldsymbol{p}_k &= \boldsymbol{r}_k, \ \boldsymbol{t}_k = \boldsymbol{A}\boldsymbol{p}_k, \\ \alpha_k &= (\boldsymbol{p}_k^T \boldsymbol{r}_k)/(\boldsymbol{p}_k^T \boldsymbol{t}_k), \\ \boldsymbol{x}_{k+1} &= \boldsymbol{x}_k + \alpha_k \boldsymbol{p}_k, \\ \boldsymbol{r}_{k+1} &= \boldsymbol{r}_k - \alpha_k \boldsymbol{t}_k. \end{aligned}} \tag{13.9}$$

Here, and in general, the following update of the residual is used:

$$\boldsymbol{r}_{k+1} = \boldsymbol{b} - \boldsymbol{A}\boldsymbol{x}_{k+1} = \boldsymbol{b} - \boldsymbol{A}(\boldsymbol{x}_k + \alpha_k \boldsymbol{p}_k) = \boldsymbol{r}_k - \alpha_k \boldsymbol{A} \boldsymbol{p}_k. \tag{13.10}$$

In the steepest descent method the choice $\boldsymbol{p}_k = \boldsymbol{r}_k$ implies that the last two gradients are orthogonal. Indeed, by (13.10), $\boldsymbol{r}_{k+1}^T \boldsymbol{r}_k = (\boldsymbol{r}_k - \alpha_k \boldsymbol{A}\boldsymbol{r}_k)^T \boldsymbol{r}_k = 0$ since $\alpha_k = \frac{\boldsymbol{r}_k^T \boldsymbol{r}_k}{\boldsymbol{r}_k^T \boldsymbol{A} \boldsymbol{r}_k}$ and $\boldsymbol{A}$ is symmetric.

Example 13.1 (Steepest Descent Iteration) Suppose $Q(x, y)$ is given by (13.4). Starting with $\boldsymbol{x}_0 = [-1, -1/2]^T$ and $\boldsymbol{r}_0 = -\boldsymbol{A}\boldsymbol{x}_0 = [3/2, 0]^T$ we find

$$\boldsymbol{t}_0 = 3\begin{bmatrix} 1 \\ -1/2 \end{bmatrix}, \quad \alpha_0 = \frac{1}{2}, \quad \boldsymbol{x}_1 = -4^{-1}\begin{bmatrix} 1 \\ 2 \end{bmatrix}, \quad \boldsymbol{r}_1 = 3 * 4^{-1}\begin{bmatrix} 0 \\ 1 \end{bmatrix}$$

$$\boldsymbol{t}_1 = 3 * 4^{-1}\begin{bmatrix} -1 \\ 2 \end{bmatrix}, \quad \alpha_1 = \frac{1}{2}, \quad \boldsymbol{x}_2 = -4^{-1}\begin{bmatrix} 1 \\ 1/2 \end{bmatrix}, \quad \boldsymbol{r}_2 = 3 * 4^{-1}\begin{bmatrix} 1/2 \\ 0 \end{bmatrix},$$

and in general for $k \geq 1$

$$t_{2k-2} = 3 * 4^{1-k} \begin{bmatrix} 1 \\ -1/2 \end{bmatrix}, \quad x_{2k-1} = -4^{-k} \begin{bmatrix} 1 \\ 2 \end{bmatrix}, \quad r_{2k-1} = 3 * 4^{-k} \begin{bmatrix} 0 \\ 1 \end{bmatrix}$$

$$t_{2k-1} = 3 * 4^{-k} \begin{bmatrix} -1 \\ 2 \end{bmatrix}, \quad x_{2k} = -4^{-k} \begin{bmatrix} 1 \\ 1/2 \end{bmatrix}, \quad r_{2k} = 3 * 4^{-k} \begin{bmatrix} 1/2 \\ 0 \end{bmatrix}.$$

Since $\alpha_k = 1/2$ is constant for all k the methods of Richardson, Jacobi and steepest descent are the same on this simple problem. See the left part of Fig. 13.1. The rate of convergence is determined from $\|x_{j+1}\|_2/\|x_j\| = \|r_{j+1}\|_2/\|r_j\|_2 = 1/2$ for all j.

13.2 The Conjugate Gradient Method

In the steepest descent method the last two gradients are orthogonal. In the conjugate gradient method all gradients are orthogonal.[1] We achieve this by using **A-orthogonal search directions** i.e., $p_i^T A p_j = 0$ for all $i \neq j$.

13.2.1 Derivation of the Method

As in the steepest descent method we choose a starting vector $x_0 \in \mathbb{R}^n$. If $r_0 = b - Ax_0 = 0$ then x_0 is the exact solution and we are finished, otherwise we initially make a steepest descent step. It follows that $r_1^T r_0 = 0$ and $p_0 := r_0$.

For the general case we define for $j \geq 0$

$$p_j := r_j - \sum_{i=0}^{j-1} \left(\frac{r_j^T A p_i}{p_i^T A p_i}\right) p_i, \tag{13.11}$$

$$x_{j+1} := x_j + \alpha_j p_j \quad \alpha_j := \frac{r_j^T r_j}{p_j^T A p_j}, \tag{13.12}$$

$$r_{j+1} = r_j - \alpha_j A p_j. \tag{13.13}$$

We note that

1. p_j is computed by the Gram-Schmidt orthogonalization process applied to the residuals $r_0, \ldots, r_j$ using the A-inner product. The search directions are therefore A-orthogonal and nonzero as long as the residuals are linearly independent.
2. Equation (13.13) follows from (13.10).
3. It can be shown that the step length α_j is optimal for all j (cf. Exercise 13.7).

[1] It is this property that has given the method its name.

Lemma 13.2 (The Residuals Are Orthogonal) *Suppose that for some $k \geq 0$ that $\boldsymbol{x}_j$ is well defined, $\boldsymbol{r}_j \neq 0$, and $\boldsymbol{r}_i^T \boldsymbol{r}_j = 0$ for $i, j = 0, 1, \ldots, k$, $i \neq j$. Then $\boldsymbol{x}_{k+1}$ is well defined and $\boldsymbol{r}_{k+1}^T \boldsymbol{r}_j = 0$ for $j = 0, 1, \ldots, k$.*

Proof Since the residuals $\boldsymbol{r}_j$ are orthogonal and nonzero for $j \leq k$, they are linearly independent, and it follows form the Gram-Schmidt Theorem 5.4 that $\boldsymbol{p}_k$ is nonzero and $\boldsymbol{p}_k^T \boldsymbol{A} \boldsymbol{p}_i = 0$ for $i < k$. But then $\boldsymbol{x}_{k+1}$ and $\boldsymbol{r}_{k+1}$ are well defined. Now

$$\begin{aligned}
\boldsymbol{r}_{k+1}^T \boldsymbol{r}_j &\overset{(13.13)}{=} (\boldsymbol{r}_k - \alpha_k \boldsymbol{A} \boldsymbol{p}_k)^T \boldsymbol{r}_j \\
&\overset{(13.11)}{=} \boldsymbol{r}_k^T \boldsymbol{r}_j - \alpha_k \boldsymbol{p}_k^T \boldsymbol{A}\big(\boldsymbol{p}_j + \sum_{i=0}^{j-1} (\frac{\boldsymbol{r}_j^T \boldsymbol{A} \boldsymbol{p}_i}{\boldsymbol{p}_i^T \boldsymbol{A} \boldsymbol{p}_i}) \boldsymbol{p}_i\big) \\
&\overset{\boldsymbol{p}_k^T \boldsymbol{A} \boldsymbol{p}_i = 0}{=} \boldsymbol{r}_k^T \boldsymbol{r}_j - \alpha_k \boldsymbol{p}_k^T \boldsymbol{A} \boldsymbol{p}_j = 0, \quad j = 0, 1, \ldots, k.
\end{aligned}$$

That the final expression is equal to zero follows by orthogonality and $\boldsymbol{A}$-orthogonality for $j < k$ and by the definition of α_k for $j = k$. This completes the proof. □

The conjugate gradient method is also a direct method. The residuals are orthogonal and therefore linearly independent if they are nonzero. Since $\dim \mathbb{R}^n = n$ the $n + 1$ residuals $\boldsymbol{r}_0, \ldots, \boldsymbol{r}_n$ cannot all be nonzero and we must have $\boldsymbol{r}_k = 0$ for some $k \leq n$. Thus we find the exact solution in at most n iterations.

The expression (13.11) for $\boldsymbol{p}_k$ can be greatly simplified. All terms except the last one vanish, since by orthogonality of the residuals

$$\boldsymbol{r}_j^T \boldsymbol{A} \boldsymbol{p}_i \overset{(13.13)}{=} \boldsymbol{r}_j^T \Big(\frac{\boldsymbol{r}_i - \boldsymbol{r}_{i+1}}{\alpha_i}\Big) = 0, \quad i = 0, 1, \ldots, j-2.$$

With $j = k + 1$ (13.11) therefore takes the simple form $\boldsymbol{p}_{k+1} = \boldsymbol{r}_{k+1} + \beta_k \boldsymbol{p}_k$ and we find

$$\beta_k := -\frac{\boldsymbol{r}_{k+1}^T \boldsymbol{A} \boldsymbol{p}_k}{\boldsymbol{p}_k^T \boldsymbol{A} \boldsymbol{p}_k} \overset{(13.13)}{=} \frac{\boldsymbol{r}_{k+1}^T (\boldsymbol{r}_{k+1} - \boldsymbol{r}_k)}{\alpha_k \boldsymbol{p}_k^T \boldsymbol{A} \boldsymbol{p}_k} \overset{(13.12)}{=} \frac{\boldsymbol{r}_{k+1}^T \boldsymbol{r}_{k+1}}{\boldsymbol{r}_k^T \boldsymbol{r}_k}. \tag{13.14}$$

To summarize, in the **conjugate gradient method** we start with $\boldsymbol{x}_0$, $\boldsymbol{p}_0 = \boldsymbol{r}_0 = \boldsymbol{b} - \boldsymbol{A}\boldsymbol{x}_0$ and then generate a sequence of vectors $\{\boldsymbol{x}_k\}$ as follows:

For $k = 0, 1, 2, \ldots$

$$\boldsymbol{x}_{k+1} := \boldsymbol{x}_k + \alpha_k \boldsymbol{p}_k, \quad \alpha_k := \frac{\boldsymbol{r}_k^T \boldsymbol{r}_k}{\boldsymbol{p}_k^T \boldsymbol{A} \boldsymbol{p}_k}, \tag{13.15}$$

$$\boldsymbol{r}_{k+1} := \boldsymbol{r}_k - \alpha_k \boldsymbol{A} \boldsymbol{p}_k, \tag{13.16}$$

$$p_{k+1} := r_{k+1} + \beta_k p_k, \quad \beta_k := \frac{r_{k+1}^T r_{k+1}}{r_k^T r_k}. \tag{13.17}$$

The residuals and search directions are orthogonal and A-orthogonal, respectively.

For computation we organize the iterations as follows for $k = 0, 1, 2, \ldots$

$$\boxed{\begin{aligned} t_k &= A p_k, \\ \alpha_k &= (r_k^T r_k)/(p_k^T t_k), \\ x_{k+1} &= x_k + \alpha_k p_k, \\ r_{k+1} &= r_k - \alpha_k t_k, \\ \beta_k &= (r_{k+1}^T r_{k+1})/(r_k^T r_k), \\ p_{k+1} &:= r_{k+1} + \beta_k p_k. \end{aligned}} \tag{13.18}$$

Note that (13.18) differs from (13.9) only in the computation of the search direction.

Example 13.2 (Conjugate Gradient Iteration) Consider (13.18) applied to the positive definite linear system $\begin{bmatrix} 2 & -1 \\ -1 & 2 \end{bmatrix}\begin{bmatrix} x_1 \\ x_2 \end{bmatrix} = \begin{bmatrix} 0 \\ 0 \end{bmatrix}$. Starting as in Example 13.1 with $x_0 = \begin{bmatrix} -1 \\ -1/2 \end{bmatrix}$ we find $p_0 = r_0 = \begin{bmatrix} 3/2 \\ 0 \end{bmatrix}$ and then

$$t_0 = \begin{bmatrix} 3 \\ -3/2 \end{bmatrix}, \quad \alpha_0 = 1/2, \quad x_1 = \begin{bmatrix} -1/4 \\ -1/2 \end{bmatrix}, \quad r_1 = \begin{bmatrix} 0 \\ 3/4 \end{bmatrix}, \quad \beta_0 = 1/4,$$

$$p_1 = \begin{bmatrix} 3/8 \\ 3/4 \end{bmatrix}, \quad t_1 = \begin{bmatrix} 0 \\ 9/8 \end{bmatrix}, \quad \alpha_1 = 2/3, \quad x_2 = \mathbf{0}, \quad r_2 = \mathbf{0}.$$

Thus x_2 is the exact solution as illustrated in the right part of Fig. 13.1.

13.2.2 The Conjugate Gradient Algorithm

In this section we give numerical examples and discuss implementation.

The formulas in (13.18) form a basis for the following algorithm, which solves the positive definite linear system $Ax = b$ by the conjugate gradient method. x is a starting vector for the iteration. The iteration is stopped when $||r_k||_2/||b||_2 \le \text{tol}$ or

$k >$ itmax. K is the number of iterations used:

```
function [x,K]=cg(A,b,x,tol,itmax)
% [x,K]=cg(A,b,x,tol,itmax)
r=b-A*x; p=r; rho0=b'*b; rho=r'*r;
for k=0:itmax
    if sqrt(rho/rho0)<= tol
       K=k; return
    end
    t=A*p; a=rho/(p'*t);
    x=x+a*p; r=r-a*t;
    rhos=rho; rho=r'*r;
    p=r+(rho/rhos)*p;
end
K=itmax+1;
end
```

Listing 13.1 cg

The work involved in each iteration is

1. one matrix times vector ($\boldsymbol{t} = \boldsymbol{A}\boldsymbol{p}$),
2. two inner products (($\boldsymbol{p}^T\boldsymbol{t}$ and $\boldsymbol{r}^T\boldsymbol{r}$),
3. three vector-plus-scalar-times-vector ($\boldsymbol{x} = \boldsymbol{x} + a\boldsymbol{p}$, $\boldsymbol{r} = \boldsymbol{r} - a\boldsymbol{t}$ and $\boldsymbol{p} = \boldsymbol{r} + (rho/rhos)\boldsymbol{p}$),

The dominating part is the computation of $\boldsymbol{t} = \boldsymbol{A}\boldsymbol{p}$.

13.2.3 Numerical Example

We test the conjugate gradient method on two examples. For a similar test for the steepest descent method see Exercise 13.9. Consider the matrix given by the Kronecker sum $\boldsymbol{T}_2 := \boldsymbol{T}_1 \otimes \boldsymbol{I} + \boldsymbol{I} \otimes \boldsymbol{T}_1$, where $\boldsymbol{T}_1 := \mathrm{tridiag}_m(a, d, a) \in \mathbb{R}^{m\times m}$ and $a, d \in \mathbb{R}$. We recall that this matrix is positive definite if $d > 0$ and $d \geq 2|a|$ (cf. Theorem 10.2). We set $h = 1/(m+1)$ and $\boldsymbol{f} = [1, \dots, 1]^T \in \mathbb{R}^n$.

We consider two problems.

1. $a = 1/9$, $d = 5/18$, the Averaging matrix.
2. $a = -1$, $d = 2$, the Poisson matrix.

13.2.4 Implementation Issues

Note that for our test problems $\boldsymbol{T}_2$ only has $O(5n)$ nonzero elements. Therefore, taking advantage of the sparseness of $\boldsymbol{T}_2$ we can compute $\boldsymbol{t}$ in Algorithm 13.1

in $O(n)$ arithmetic operations. With such an implementation the total number of arithmetic operations in one iteration is $O(n)$. We also note that it is not necessary to store the matrix $\boldsymbol{T}_2$.

To use the conjugate gradient algorithm on the test matrix for large n it is advantageous to use a matrix equation formulation. We define matrices $\boldsymbol{V}, \boldsymbol{R}, \boldsymbol{P}, \boldsymbol{B}, \boldsymbol{T} \in \mathbb{R}^{m\times m}$ by $\boldsymbol{x} = \text{vec}(\boldsymbol{V})$, $\boldsymbol{r} = \text{vec}(\boldsymbol{R})$, $\boldsymbol{p} = \text{vec}(\boldsymbol{P})$, $\boldsymbol{t} = \text{vec}(\boldsymbol{T})$, and $h^2\boldsymbol{f} = \text{vec}(\boldsymbol{B})$. Then $\boldsymbol{T}_2\boldsymbol{x} = h^2\boldsymbol{f} \iff \boldsymbol{T}_1\boldsymbol{V} + \boldsymbol{V}\boldsymbol{T}_1 = \boldsymbol{B}$, and $\boldsymbol{t} = \boldsymbol{T}_2\boldsymbol{p} \iff \boldsymbol{T} = \boldsymbol{T}_1\boldsymbol{P} + \boldsymbol{P}\boldsymbol{T}_1$.

This leads to the following algorithm for testing the conjugate gradient algorithm on the matrix

$$\boldsymbol{A} = \text{tridiag}_m(a, d, a) \otimes \boldsymbol{I}_m + \boldsymbol{I}_m \otimes \text{tridiag}_m(a, d, a) \in \mathbb{R}^{(m^2)\times(m^2)}.$$

```
function [V,K]=cgtest(m,a,d,tol,itmax)
% [V,K]=cgtest(m,a,d,tol,itmax)
R=ones(m)/(m+1)^2; rho=sum(sum(R.*R)); rho0=rho; P=R;
V=zeros(m,m); T1=sparse(tridiagonal(a,d,a,m));
for k=1:itmax
    if sqrt(rho/rho0)<= tol
        K=k; return
    end
    T=T1*P+P*T1;
    a=rho/sum(sum(P.*T)); V=V+a*P; R=R-a*T;
    rhos=rho; rho=sum(sum(R.*R)); P=R+(rho/rhos)*P;
end
K=itmax+1;
end
```

Listing 13.2 cgtest

For both the averaging- and Poison matrix we use $tol = 10^{-8}$.

For the averaging matrix we obtain the values in Table 13.1.

The convergence is quite rapid. It appears that the number of iterations can be bounded independently of n, and therefore we solve the problem in $O(n)$ operations. This is the best we can do for a problem with n unknowns.

Consider next the Poisson problem. In Table 13.2 we list K, the required number of iterations, and $K/\sqrt{n}$.

Table 13.1 The number of iterations K for the averaging problem on a $\sqrt{n} \times \sqrt{n}$ grid for various n

n	2500	10000	40000	1000000	4000000
K	19	18	18	16	15

Table 13.2 The number of iterations K for the Poisson problem on a $\sqrt{n} \times \sqrt{n}$ grid for various n

n	2500	10000	40000	160000
K	94	188	370	735
$K/\sqrt{n}$	1.88	1.88	1.85	1.84

The results show that K is much smaller than n and appears to be proportional to $\sqrt{n}$. This is the same speed as for SOR and we don't have to estimate any acceleration parameter.

13.3 Convergence

13.3.1 The Main Theorem

Recall that the $\boldsymbol{A}$-norm of a vector $\boldsymbol{x} \in \mathbb{R}^n$ is given by $\|\boldsymbol{x}\|_A := \sqrt{\boldsymbol{x}^T\boldsymbol{A}\boldsymbol{x}}$. The following theorem gives upper bounds for the $\boldsymbol{A}$-norm of the error in both steepest descent and conjugate gradients.

Theorem 13.3 (Error Bound for Steepest Descent and Conjugate Gradients) *Suppose $\boldsymbol{A}$ is positive definite. For the $\boldsymbol{A}$-norms of the errors in the steepest descent method* (13.8) *the following upper bounds hold*

$$\frac{||\boldsymbol{x}-\boldsymbol{x}_k||_A}{||\boldsymbol{x}-\boldsymbol{x}_0||_A} \le \left(\frac{\kappa-1}{\kappa+1}\right)^k < e^{-\frac{2}{\kappa}k}, \quad , k > 0, \tag{13.19}$$

while for the conjugate gradient method we have

$$\frac{||\boldsymbol{x}-\boldsymbol{x}_k||_A}{||\boldsymbol{x}-\boldsymbol{x}_0||_A} \le 2\left(\frac{\sqrt{\kappa}-1}{\sqrt{\kappa}+1}\right)^k < 2e^{-\frac{2}{\sqrt{\kappa}}k}, \quad k \ge 0. \tag{13.20}$$

Here $\kappa = cond_2(\boldsymbol{A}) := \lambda_{max}/\lambda_{min}$ is the spectral condition number of $\boldsymbol{A}$, where λ_{max} and λ_{min} are the largest and smallest eigenvalue of $\boldsymbol{A}$, respectively.

Theorem 13.3 implies

1. Since $\frac{\kappa-1}{\kappa+1} < 1$ the steepest descent method always converges for a positive definite matrix. The convergence can be slow when $\frac{\kappa-1}{\kappa+1}$ is close to one, and this happens even for a moderately ill-conditioned $\boldsymbol{A}$.
2. The rate of convergence for the conjugate gradient method appears to be determined by the square root of the spectral condition number. This is much better than the estimate for the steepest descent method. Especially for problems with large condition numbers.
3. The proofs of the estimates in (13.19) and (13.20) are quite different. This is in spite of their similar appearance.

13.3.2 The Number of Iterations for the Model Problems

Consider the test matrix

$$\boldsymbol{T}_2 := \text{tridiag}_m(a, d, a) \otimes \boldsymbol{I}_m + \boldsymbol{I}_m \otimes \text{tridiag}_m(a, d, a) \in \mathbb{R}^{(m^2)\times(m^2)}.$$

The eigenvalues were given in (10.15) as

$$\lambda_{j,k} = 2d + 2a\cos(j\pi h) + 2a\cos(k\pi h), \quad j, k = 1, \ldots, m. \tag{13.21}$$

For the averaging problem given by $d = 5/18$, $a = 1/9$, the largest and smallest eigenvalue of $\boldsymbol{T}_2$ are given by $\lambda_{max} = \frac{5}{9} + \frac{4}{9}\cos(\pi h)$ and $\lambda_{min} = \frac{5}{9} - \frac{4}{9}\cos(\pi h)$. Thus

$$\kappa_A = \frac{5 + 4\cos(\pi h)}{5 - 4\cos(\pi h)} \leq 9,$$

and the condition number is bounded independently of n. It follows from (13.20) that the number of iterations can be bounded independently of the size n of the problem, and this is in agreement with what we observed in Table 13.1.

For the Poisson problem we have by (10.20) the condition number

$$\kappa_P = \frac{\lambda_{max}}{\lambda_{min}} = \frac{\cos^2(\pi h/2)}{\sin^2(\pi h/2)} \text{ and } \sqrt{\kappa_P} = \frac{\cos(\pi h/2)}{\sin(\pi h/2)} \approx \frac{2}{\pi h} \approx \frac{2}{\pi}\sqrt{n}.$$

Thus, (see also Exercise 8.19) we solve the discrete Poisson problem in $O(n^{3/2})$ arithmetic operations using the conjugate gradient method. This is the same as for the SOR method and for the fast method without the FFT. In comparison the Cholesky Algorithm requires $O(n^2)$ arithmetic operations both for the averaging and the Poisson problem.

13.3.3 Krylov Spaces and the Best Approximation Property

For the convergence analysis of the conjugate gradient method certain subspaces of $\mathbb{R}^n$ called **Krylov spaces** play a central role. In fact the iterates in the conjugate gradient method are best approximation of the solution from these subspaces using the $\boldsymbol{A}$-norm to measure the error.

The Krylov spaces are defined by $\mathbb{W}_0 = \{\boldsymbol{0}\}$ and

$$\mathbb{W}_k = \text{span}(\boldsymbol{r}_0, \boldsymbol{A}\boldsymbol{r}_0, \boldsymbol{A}^2\boldsymbol{r}_0, \ldots, \boldsymbol{A}^{k-1}\boldsymbol{r}_0), \quad k = 1, 2, 3, \cdots.$$

They are nested subspaces

$$\mathbb{W}_0 \subset \mathbb{W}_1 \subset \mathbb{W}_2 \subset \cdots \subset \mathbb{W}_n \subset \mathbb{R}^n$$

with $\dim(\mathbb{W}_k) \leq k$ for all $k \geq 0$. Moreover, If $\boldsymbol{v} \in \mathbb{W}_k$ then $\boldsymbol{A}\boldsymbol{v} \in \mathbb{W}_{k+1}$.

Lemma 13.3 (Krylov Space) *For the iterates in the conjugate gradient method we have*

$$\boldsymbol{x}_k - \boldsymbol{x}_0 \in \mathbb{W}_k, \quad \boldsymbol{r}_k, \boldsymbol{p}_k \in \mathbb{W}_{k+1}, \quad k = 0, 1, \ldots, \tag{13.22}$$

and

$$\boldsymbol{r}_k^T \boldsymbol{w} = \boldsymbol{p}_k^T \boldsymbol{A}\boldsymbol{w} = 0, \quad \boldsymbol{w} \in \mathbb{W}_k. \tag{13.23}$$

Proof Equation (13.22) clearly holds for $k = 0$ since $\boldsymbol{p}_0 = \boldsymbol{r}_0$. Suppose it holds for some $k \geq 0$. Then $\boldsymbol{r}_{k+1} = \boldsymbol{r}_k - \alpha_k \boldsymbol{A}\boldsymbol{p}_k \in \mathbb{W}_{k+2}$, $\boldsymbol{p}_{k+1} = \boldsymbol{r}_{k+1} + \beta_k \boldsymbol{p}_k \in \mathbb{W}_{k+2}$ and $\boldsymbol{x}_{k+1} - \boldsymbol{x}_0 \overset{(13.12)}{=} \boldsymbol{x}_k - \boldsymbol{x}_0 + \alpha_k \boldsymbol{p}_k \in \mathbb{W}_{k+1}$. Thus (13.22) follows by induction. The equation (13.23) follows since any $\boldsymbol{w} \in \mathbb{W}_k$ is a linear combination of $\{\boldsymbol{r}_0, \boldsymbol{r}_1, \ldots, \boldsymbol{r}_{k-1}\}$ and also $\{\boldsymbol{p}_0, \boldsymbol{p}_1, \ldots, \boldsymbol{p}_{k-1}\}$. □

Theorem 13.4 (Best Approximation Property) *Suppose* $\boldsymbol{A}\boldsymbol{x} = \boldsymbol{b}$, *where* $\boldsymbol{A} \in \mathbb{R}^{n\times n}$ *is positive definite and* $\{\boldsymbol{x}_k\}$ *is generated by the conjugate gradient method (cf.* (13.15)*). Then*

$$\|\boldsymbol{x} - \boldsymbol{x}_k\|_A = \min_{\boldsymbol{w} \in \mathbb{W}_k} \|\boldsymbol{x} - \boldsymbol{x}_0 - \boldsymbol{w}\|_A. \tag{13.24}$$

Proof Fix k, let $\boldsymbol{w} \in \mathbb{W}_k$ and $\boldsymbol{u} := \boldsymbol{x}_k - \boldsymbol{x}_0 - \boldsymbol{w}$. By (13.22) $\boldsymbol{u} \in \mathbb{W}_k$ and then (13.23) implies that $\langle \boldsymbol{x} - \boldsymbol{x}_k, \boldsymbol{u}\rangle = \boldsymbol{r}_k^T \boldsymbol{u} = 0$. Using Corollary 5.2 we obtain

$$\|\boldsymbol{x} - \boldsymbol{x}_0 - \boldsymbol{w}\|_A = \|\boldsymbol{x} - \boldsymbol{x}_k + \boldsymbol{u}\|_A \geq \|\boldsymbol{x} - \boldsymbol{x}_k\|_A,$$

with equality for $\boldsymbol{u} = \boldsymbol{0}$. □

If $\boldsymbol{x}_0 = \boldsymbol{0}$ then (13.24) says that $\boldsymbol{x}_k$ is the element in $\mathbb{W}_k$ that is closest to the solution $\boldsymbol{x}$ in the $\boldsymbol{A}$-norm. More generally, if $\boldsymbol{x}_0 \neq \boldsymbol{0}$ then $\boldsymbol{x} - \boldsymbol{x}_k = (\boldsymbol{x} - \boldsymbol{x}_0) - (\boldsymbol{x}_k - \boldsymbol{x}_0)$ and $\boldsymbol{x}_k - \boldsymbol{x}_0$ is the element in $\mathbb{W}_k$ that is closest to $\boldsymbol{x} - \boldsymbol{x}_0$ in the $\boldsymbol{A}$-norm. This is the orthogonal projection of $\boldsymbol{x} - \boldsymbol{x}_0$ into $\mathbb{W}_k$, see Fig. 13.2.

Recall that to each polynomial $p(t) := \sum_{j=0}^{m} a_j t^m$ there corresponds a matrix polynomial $p(\boldsymbol{A}) := a_0\boldsymbol{I} + a_1\boldsymbol{A} + \cdots + a_m\boldsymbol{A}^m$. Moreover, if $(\lambda_j, \boldsymbol{u}_j)$ are eigenpairs of $\boldsymbol{A}$ then $(p(\lambda_j), \boldsymbol{u}_j)$ are eigenpairs of $p(\boldsymbol{A})$ for $j = 1, \ldots, n$.

Lemma 13.4 (Krylov Space and Polynomials) *Suppose* $\boldsymbol{A}\boldsymbol{x} = \boldsymbol{b}$ *where* $\boldsymbol{A} \in \mathbb{R}^{n\times n}$ *is positive definite with orthonormal eigenpairs* $(\lambda_j, \boldsymbol{u}_j)$, $j = 1, 2, \ldots, n$, *and let* $\boldsymbol{r}_0 := \boldsymbol{b} - \boldsymbol{A}\boldsymbol{x}_0$ *for some* $\boldsymbol{x}_0 \in \mathbb{R}^n$. *To each* $\boldsymbol{w} \in \mathbb{W}_k$ *there corresponds*

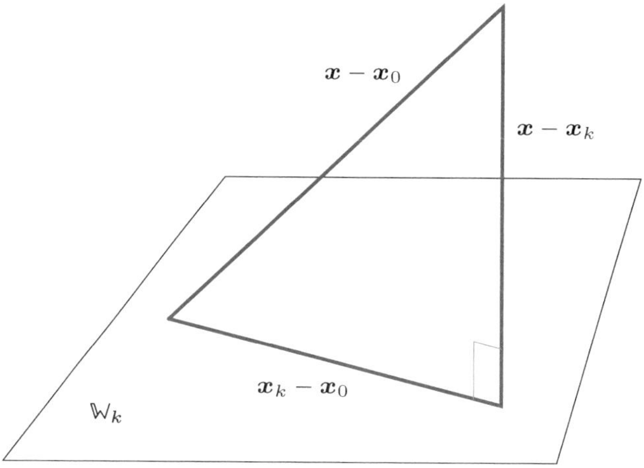

Fig. 13.2 The orthogonal projection of $\boldsymbol{x} - \boldsymbol{x}_0$ into $\mathbb{W}_k$

a polynomial $P(t) := \sum_{j=0}^{k-1} a_j t^{k-1}$ *such that* $\boldsymbol{w} = P(\boldsymbol{A})\boldsymbol{r}_0$. *Moreover, if* $\boldsymbol{r}_0 = \sum_{j=1}^{n} \sigma_j \boldsymbol{u}_j$ *then*

$$||\boldsymbol{x} - \boldsymbol{x}_0 - \boldsymbol{w}||_{\boldsymbol{A}}^2 = \sum_{j=1}^{n} \frac{\sigma_j^2}{\lambda_j} Q(\lambda_j)^2, \quad Q(t) := 1 - tP(t). \tag{13.25}$$

Proof If $\boldsymbol{w} \in \mathbb{W}_k$ then $\boldsymbol{w} = a_0\boldsymbol{r}_0 + a_1\boldsymbol{A}\boldsymbol{r}_0 + \cdots + a_{k-1}\boldsymbol{A}^{k-1}\boldsymbol{r}_0$ for some scalars $a_0, \ldots, a_{k-1}$. But then $\boldsymbol{w} = P(\boldsymbol{A})\boldsymbol{r}_0$. We find $\boldsymbol{x} - \boldsymbol{x}_0 - P(\boldsymbol{A})\boldsymbol{r}_0 = \boldsymbol{A}^{-1}(\boldsymbol{r}_0 - \boldsymbol{A}P(\boldsymbol{A}))\boldsymbol{r}_0 = \boldsymbol{A}^{-1}Q(\boldsymbol{A})\boldsymbol{r}_0$ and $\boldsymbol{A}(\boldsymbol{x} - \boldsymbol{x}_0 - P(\boldsymbol{A})\boldsymbol{r}_0) = Q(\boldsymbol{A})\boldsymbol{r}_0$. Therefore,

$$\|\boldsymbol{x} - \boldsymbol{x}_0 - P(\boldsymbol{A})\boldsymbol{r}_0\|_{\boldsymbol{A}}^2 = \boldsymbol{c}^T\boldsymbol{A}^{-1}\boldsymbol{c} \text{ where } \boldsymbol{c} = (\boldsymbol{I} - \boldsymbol{A}P(\boldsymbol{A}))\boldsymbol{r}_0 = Q(\boldsymbol{A})\boldsymbol{r}_0. \tag{13.26}$$

Using the eigenvector expansion for $\boldsymbol{r}_0$ we obtain

$$\boldsymbol{c} = \sum_{j=1}^{n} \sigma_j Q(\lambda_j)\boldsymbol{u}_j, \quad \boldsymbol{A}^{-1}\boldsymbol{c} = \sum_{i=1}^{n} \sigma_i \frac{Q(\lambda_i)}{\lambda_i}\boldsymbol{u}_i. \tag{13.27}$$

Now (13.25) follows by the orthonormality of the eigenvectors. □

We will use the following theorem to estimate the rate of convergence.

Theorem 13.5 (cg and Best Polynomial Approximation) *Suppose $[a, b]$ with $0 < a < b$ is an interval containing all the eigenvalues of $\boldsymbol{A}$. Then in the conjugate gradient method*

$$\frac{||\boldsymbol{x} - \boldsymbol{x}_k||_{\boldsymbol{A}}}{||\boldsymbol{x} - \boldsymbol{x}_0||_{\boldsymbol{A}}} \leq \min_{\substack{Q \in \Pi_k \\ Q(0)=1}} \max_{a \leq x \leq b} |Q(x)|, \tag{13.28}$$

where Π_k denotes the class of univariate polynomials of degree $\leq k$ with real coefficients.

Proof By (13.25) with $Q(t) = 1$ (corresponding to $P(\boldsymbol{A}) = \boldsymbol{0}$) we find $||\boldsymbol{x} - \boldsymbol{x}_0||_{\boldsymbol{A}}^2 = \sum_{j=1}^n \frac{\sigma_j^2}{\lambda_j}$. Therefore, by the best approximation property Theorem 13.4 and (13.25), for any $\boldsymbol{w} \in \mathbb{W}_k$

$$||\boldsymbol{x}-\boldsymbol{x}_k||_{\boldsymbol{A}}^2 \leq ||\boldsymbol{x}-\boldsymbol{x}_0-\boldsymbol{w}||_{\boldsymbol{A}}^2 \leq \max_{a \leq x \leq b} |Q(x)|^2 \sum_{j=1}^n \frac{\sigma_j^2}{\lambda_j} = \max_{a \leq x \leq b} |Q(x)|^2 ||\boldsymbol{x}-\boldsymbol{x}_0||_{\boldsymbol{A}}^2,$$

where $Q \in \Pi_k$ and $Q(0) = 1$. Minimizing over such polynomials Q and taking square roots the result follows. □

In the next section we use properties of the Chebyshev polynomials to show that

$$\frac{||\boldsymbol{x} - \boldsymbol{x}_k||_{\boldsymbol{A}}}{||\boldsymbol{x} - \boldsymbol{x}_0||_{\boldsymbol{A}}} \leq \min_{\substack{Q \in \Pi_k \\ Q(0)=1}} \max_{\lambda_{min} \leq x \leq \lambda_{max}} |Q(x)| = \frac{2}{a^{-k} + a^k}, \quad a := \frac{\sqrt{\kappa} - 1}{\sqrt{\kappa} + 1}, \tag{13.29}$$

where $\kappa = \lambda_{max}/\lambda_{min}$ is the spectral condition number of $\boldsymbol{A}$. Ignoring the second term in the denominator this implies the first inequality in (13.20).

Consider the second inequality in (13.20). The inequality

$$\frac{x - 1}{x + 1} < e^{-2/x} \quad \text{for} \quad x > 1 \tag{13.30}$$

follows from the familiar series expansion of the exponential function. Indeed, with $y = 1/x$, using $2^k/k! = 2$, $k = 1, 2$, and $2^k/k! < 2$ for $k > 2$, we find

$$e^{2/x} = e^{2y} = \sum_{k=0}^{\infty} \frac{(2y)^k}{k!} < -1 + 2\sum_{k=0}^{\infty} y^k = \frac{1 + y}{1 - y} = \frac{x + 1}{x - 1}$$

and (13.30) follows.

13.4 Proof of the Convergence Estimates

13.4.1 Chebyshev Polynomials

The proof of the estimate (13.29) for the error in the conjugate gradient method is based on an extremal property of the Chebyshev polynomials. Suppose $a < b$, $c \notin [a, b]$ and $k \in \mathbb{N}$. Consider the set $\mathcal{S}_k$ of all polynomials Q of degree $\leq k$ such that $Q(c) = 1$. For any continuous function f on $[a, b]$ we define

$$\|f\|_\infty = \max_{a \leq x \leq b} |f(x)|.$$

We want to find a polynomial $Q^* \in \mathcal{S}_k$ such that

$$\|Q^*\|_\infty = \min_{Q \in \mathcal{S}_k} \|Q\|_\infty.$$

We will show that Q^* is uniquely given as a suitably shifted and normalized version of the **Chebyshev polynomial**. The Chebyshev polynomial T_n of degree n can be defined recursively by

$$T_{n+1}(t) = 2tT_n(t) - T_{n-1}(t), \quad n \geq 1, \quad t \in \mathbb{R},$$

starting with $T_0(t) = 1$ and $T_1(t) = t$. Thus $T_2(t) = 2t^2 - 1$, $T_3(t) = 4t^3 - 3t$ etc. In general T_n is a polynomial of degree n.

There are some convenient closed form expressions for T_n.

Lemma 13.5 (Closed Forms of Chebyshev Polynomials) *For $n \geq 0$*

1. $T_n(t) = \cos(n \arccos t)$ *for* $t \in [-1, 1]$,
2. $T_n(t) = \frac{1}{2}\left[\left(t + \sqrt{t^2 - 1}\right)^n + \left(t + \sqrt{t^2 - 1}\right)^{-n}\right]$ *for* $|t| \geq 1$.

Proof

1. With $P_n(t) = \cos(n \arccos t)$ we have $P_n(t) = \cos n\phi$, where $t = \cos\phi$. Therefore,

$$P_{n+1}(t) + P_{n-1}(t) = \cos(n+1)\phi + \cos(n-1)\phi = 2\cos\phi\cos n\phi = 2tP_n(t),$$

   and it follows that P_n satisfies the same recurrence relation as T_n. Since $P_0 = T_0$ and $P_1 = T_1$ we have $P_n = T_n$ for all $n \geq 0$.
2. Fix t with $|t| \geq 1$ and let $x_n := T_n(t)$ for $n \geq 0$. The recurrence relation for the Chebyshev polynomials can then be written

$$x_{n+1} - 2tx_n + x_{n-1} = 0 \text{ for } n \geq 1, \text{ with } x_0 = 1, x_1 = t. \tag{13.31}$$

To solve this difference equation we insert $x_n = z^n$ into (13.31) and obtain $z^{n+1} - 2tz^n + z^{n-1} = 0$ or $z^2 - 2tz + 1 = 0$. The roots of this equation are

$$z_1 = t + \sqrt{t^2 - 1}, \quad z_2 = t - \sqrt{t^2 - 1} = \left(t + \sqrt{t^2 - 1}\right)^{-1}.$$

Now z_1^n, z_2^n and more generally $c_1 z_1^n + c_2 z_2^n$ are solutions of (13.31) for any constants c_1 and c_2. We find these constants from the initial conditions $x_0 = c_1 + c_2 = 1$ and $x_1 = c_1 z_1 + c_2 z_2 = t$. Since $z_1 + z_2 = 2t$ the solution is $c_1 = c_2 = \frac{1}{2}$.

□

We show that the unique solution to our minimization problem is

$$Q^*(x) = \frac{T_k(u(x))}{T_k(u(c))}, \quad u(x) = \frac{b + a - 2x}{b - a}. \tag{13.32}$$

Clearly $Q^* \in \mathcal{S}_k$.

Theorem 13.6 (A Minimal Norm Problem) *Suppose $a < b$, $c \notin [a, b]$ and $k \in \mathbb{N}$. If $Q \in \mathcal{S}_k$ and $Q \neq Q^*$ then $\|Q\|_\infty > \|Q^*\|_\infty$.*

Proof Recall that a nonzero polynomial p of degree k can have at most k zeros. If $p(z) = p'(z) = 0$, we say that p has a double zero at z. Counting such a zero as two zeros it is still true that a nonzero polynomial of degree k has at most k zeros.

$|Q^*|$ takes on its maximum $1/|T_k(u(c))|$ at the $k + 1$ points $\mu_0, \ldots, \mu_k$ in $[a, b]$ such that $u(\mu_i) = \cos(i\pi/k)$ for $i = 0, 1, \ldots, k$. Suppose $Q \in \mathcal{S}_k$ and that $\|Q\|_\infty \leq \|Q^*\|_\infty$. We have to show that $Q \equiv Q^*$. Let $f \equiv Q - Q^*$. We show that f has at least k zeros in $[a, b]$. Since f is a polynomial of degree $\leq k$ and $f(c) = 0$, this means that $f \equiv 0$ or equivalently $Q \equiv Q^*$.

Consider $I_j = [\mu_{j-1}, \mu_j]$ for a fixed j. Let

$$\sigma_j = f(\mu_{j-1}) f(\mu_j).$$

We have $\sigma_j \leq 0$. For if say $Q^*(\mu_j) > 0$ then

$$Q(\mu_j) \leq \|Q\|_\infty \leq \|Q^*\|_\infty = Q^*(\mu_j)$$

so that $f(\mu_j) \leq 0$. Moreover,

$$-Q(\mu_{j-1}) \leq \|Q\|_\infty \leq \|Q^*\|_\infty = -Q^*(\mu_{j-1}).$$

Thus $f(\mu_{j-1}) \geq 0$ and it follows that $\sigma_j \leq 0$. Similarly, $\sigma_j \leq 0$ if $Q^*(\mu_j) < 0$.

If $\sigma_j < 0$, f must have a zero in I_j since it is continuous. Suppose $\sigma_j = 0$. Then $f(\mu_{j-1}) = 0$ or $f(\mu_j) = 0$. If $f(\mu_j) = 0$ then $Q(\mu_j) = Q^*(\mu_j)$. But then μ_j is a maximum or minimum both for Q and Q^*. If $\mu_j \in (a, b)$ then $Q'(\mu_j) =$

$Q^{*\prime}(\mu_j) = 0$. Thus $f(\mu_j) = f'(\mu_j) = 0$, and f has a double zero at μ_j. We can count this as one zero for I_j and one for I_{j+1}. If $\mu_j = b$, we still have a zero in I_j. Similarly, if $f(\mu_{j-1}) = 0$, a double zero of f at μ_{j-1} appears if $\mu_{j-1} \in (a, b)$. We count this as one zero for I_{j-1} and one for I_j.

In this way we associate one zero of f for each of the k intervals I_j, $j = 1, 2, \ldots, k$. We conclude that f has at least k zeros in $[a, b]$. □

Theorem 13.6 with a, and b, the smallest and largest eigenvalue of $\boldsymbol{A}$, and $c = 0$ implies that the minimizing polynomial in (13.29) is given by

$$Q^*(x) = T_k\left(\frac{b+a-2x}{b-a}\right) / T_k\left(\frac{b+a}{b-a}\right). \tag{13.33}$$

By Lemma 13.5

$$\max_{a \le x \le b} \left| T_k\left(\frac{b+a-2x}{b-a}\right) \right| = \max_{-1 \le t \le 1} \left| T_k(t) \right| = 1. \tag{13.34}$$

Moreover with $t = (b+a)/(b-a)$ we have

$$t + \sqrt{t^2 - 1} = \frac{\sqrt{\kappa}+1}{\sqrt{\kappa}-1}, \quad \kappa = b/a.$$

Thus again by Lemma 13.5 we find

$$T_k\left(\frac{b+a}{b-a}\right) = T_k\left(\frac{\kappa+1}{\kappa-1}\right) = \frac{1}{2}\left[\left(\frac{\sqrt{\kappa}+1}{\sqrt{\kappa}-1}\right)^k + \left(\frac{\sqrt{\kappa}-1}{\sqrt{\kappa}+1}\right)^k\right] \tag{13.35}$$

and (13.29) follows (Fig. 13.3).

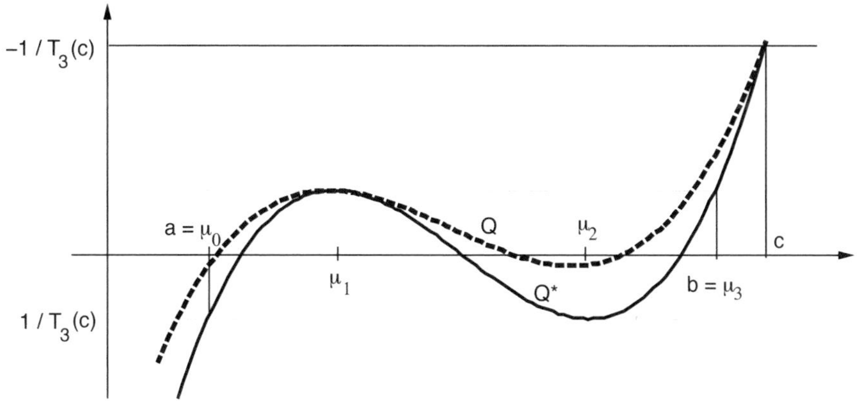

Fig. 13.3 This is an illustration of the proof of Theorem 13.6 for $k = 3$. $f \equiv Q - Q^*$ has a double zero at μ_1 and one zero between μ_2 and μ_3

13.4.2 Convergence Proof for Steepest Descent

For the proof of (13.19) the following inequality will be used.

Theorem 13.7 (Kantorovich Inequality) *For any positive definite matrix* $\boldsymbol{A} \in \mathbb{R}^{n\times n}$

$$1 \le \frac{(\boldsymbol{y}^T\boldsymbol{A}\boldsymbol{y})(\boldsymbol{y}^T\boldsymbol{A}^{-1}\boldsymbol{y})}{(\boldsymbol{y}^T\boldsymbol{y})^2} \le \frac{(M+m)^2}{4Mm} \quad \boldsymbol{y} \ne \boldsymbol{0},\ \boldsymbol{y} \in \mathbb{R}^n, \tag{13.36}$$

where $M := \lambda_{max}$ *and* $m := \lambda_{min}$ *are the largest and smallest eigenvalue of* $\boldsymbol{A}$*, respectively.*

Proof If $(\lambda_j, \boldsymbol{u}_j)$ are orthonormal eigenpairs of $\boldsymbol{A}$ then $(\lambda_j^{-1}, \boldsymbol{u}_j)$ are eigenpairs for $\boldsymbol{A}^{-1}$, $j = 1, \ldots, n$. Let $\boldsymbol{y} = \sum_{j=1}^n c_j \boldsymbol{u}_j$ be the corresponding eigenvector expansion of a vector $\boldsymbol{y} \in \mathbb{R}^n$. By orthonormality, (cf. (6.9))

$$a := \frac{\boldsymbol{y}^T\boldsymbol{A}\boldsymbol{y}}{\boldsymbol{y}^T\boldsymbol{y}} = \sum_{i=1}^n t_i\lambda_i, \quad b := \frac{\boldsymbol{y}^T\boldsymbol{A}^{-1}\boldsymbol{y}}{\boldsymbol{y}^T\boldsymbol{y}} = \sum_{i=1}^n \frac{t_i}{\lambda_i}, \tag{13.37}$$

where

$$t_i = \frac{c_i^2}{\sum_{j=1}^n c_j^2} \ge 0, \quad i = 1, \ldots, n \text{ and } \sum_{i=1}^n t_i = 1. \tag{13.38}$$

Thus a and b are **convex combinations** of the eigenvalues of $\boldsymbol{A}$ and $\boldsymbol{A}^{-1}$, respectively. Let c be a positive constant to be chosen later. By the geometric/arithmetic mean inequality (8.33) and (13.37)

$$\sqrt{ab} = \sqrt{(ac)(b/c)} \le (ac + b/c)/2 = \frac{1}{2}\sum_{i=1}^n t_i\big(\lambda_i c + 1/(\lambda_i c)\big) = \frac{1}{2}\sum_{i=1}^n t_i f(\lambda_i c),$$

where $f : [mc, Mc] \to \mathbb{R}$ is given by $f(x) := x + 1/x$. By (13.38)

$$\sqrt{ab} \le \frac{1}{2} \max_{mc \le x \le Mc} f(x).$$

Since $f \in C^2$ and f'' is positive it follows from Lemma 8.2 that f is a convex function. But a convex function takes it maximum at one of the endpoints of the range (cf. Exercise 13.16) and we obtain

$$\sqrt{ab} \le \frac{1}{2}\max\{f(mc), f(Mc)\}. \tag{13.39}$$

Choosing $c := 1/\sqrt{mM}$ we find $f(mc) = f(Mc) = \sqrt{\frac{M}{m}} + \sqrt{\frac{m}{M}} = \frac{M+m}{\sqrt{mM}}$. By (13.39) we obtain

$$\frac{(\boldsymbol{y}^T\boldsymbol{A}\boldsymbol{y})(\boldsymbol{y}^T\boldsymbol{A}^{-1}\boldsymbol{y})}{(\boldsymbol{y}^T\boldsymbol{y})^2} = ab \le \frac{(M+m)^2}{4Mm},$$

the upper bound in (13.36). For the lower bound we use the Cauchy-Schwarz inequality as follows

$$1 = \Big(\sum_{i=1}^n t_i\Big)^2 = \left(\sum_{i=1}^n (t_i\lambda_i)^{1/2}(t_i/\lambda_i)^{1/2}\right)^2 \le \Big(\sum_{i=1}^n t_i\lambda_i\Big)\Big(\sum_{i=1}^n t_i/\lambda_i\Big) = ab.$$

□

Proof of (13.19) Let $\boldsymbol{\epsilon}_j := \boldsymbol{x} - \boldsymbol{x}_j$, $j = 0, 1, \ldots$, where $\boldsymbol{A}\boldsymbol{x} = \boldsymbol{b}$. It is enough to show that

$$\frac{\|\boldsymbol{\epsilon}_{k+1}\|_{\boldsymbol{A}}^2}{\|\boldsymbol{\epsilon}_k\|_{\boldsymbol{A}}^2} \le \left(\frac{\kappa-1}{\kappa+1}\right)^2, \quad k = 0, 1, 2, \ldots, \tag{13.40}$$

for then $\|\boldsymbol{\epsilon}_k\|_{\boldsymbol{A}} \le \left(\frac{\kappa-1}{\kappa+1}\right)\|\boldsymbol{\epsilon}_{k-1}\| \le \cdots \le \left(\frac{\kappa-1}{\kappa+1}\right)^k\|\boldsymbol{\epsilon}_0\|$. It follows from (13.8) that

$$\boldsymbol{\epsilon}_{k+1} = \boldsymbol{\epsilon}_k - \alpha_k\boldsymbol{r}_k, \quad \alpha_k := \frac{\boldsymbol{r}_k^T\boldsymbol{r}_k}{\boldsymbol{r}_k^T\boldsymbol{A}\boldsymbol{r}_k}.$$

We find

$$\begin{aligned}
\|\boldsymbol{\epsilon}_k\|_{\boldsymbol{A}}^2 &= \boldsymbol{\epsilon}_k^T\boldsymbol{A}\boldsymbol{\epsilon}_k = \boldsymbol{r}_k^T\boldsymbol{A}^{-1}\boldsymbol{r}_k,\\
\|\boldsymbol{\epsilon}_{k+1}\|_{\boldsymbol{A}}^2 &= (\boldsymbol{\epsilon}_k - \alpha_k\boldsymbol{r}_k)^T\boldsymbol{A}(\boldsymbol{\epsilon}_k - \alpha_k\boldsymbol{r}_k)\\
&= \boldsymbol{\epsilon}_k^T\boldsymbol{A}\boldsymbol{\epsilon}_k - 2\alpha_k\boldsymbol{r}_k^T\boldsymbol{A}\boldsymbol{\epsilon}_k + \alpha_k^2\boldsymbol{r}_k^T\boldsymbol{A}\boldsymbol{r}_k = \|\boldsymbol{\epsilon}_k\|_{\boldsymbol{A}}^2 - \frac{(\boldsymbol{r}_k^T\boldsymbol{r}_k)^2}{\boldsymbol{r}_k^T\boldsymbol{A}\boldsymbol{r}_k}.
\end{aligned}$$

Combining these and using Kantorovich inequality

$$\frac{\|\boldsymbol{\epsilon}_{k+1}\|_{\boldsymbol{A}}^2}{\|\boldsymbol{\epsilon}_k\|_{\boldsymbol{A}}^2} = 1 - \frac{(\boldsymbol{r}_k^T\boldsymbol{r}_k)^2}{(\boldsymbol{r}_k^T\boldsymbol{A}\boldsymbol{r}_k)(\boldsymbol{r}_k^T\boldsymbol{A}^{-1}\boldsymbol{r}_k)} \le 1 - \frac{4\lambda_{min}\lambda_{max}}{(\lambda_{min}+\lambda_{max})^2} = \left(\frac{\kappa-1}{\kappa+1}\right)^2$$

and (13.40) is proved.

□

13.4.3 Monotonicity of the Error

The error analysis for the conjugate gradient method is based on the $\boldsymbol{A}$-norm. We end this chapter by considering the Euclidian norm of the error, and show that it is strictly decreasing.

Theorem 13.8 (The Error in cg Is Strictly Decreasing) *Let in the conjugate gradient method m be the smallest integer such that $\boldsymbol{r}_{m+1} = \boldsymbol{0}$. For $k \le m$ we have $\|\boldsymbol{\epsilon}_{k+1}\|_2 < \|\boldsymbol{\epsilon}_k\|_2$. More precisely,*

$$\|\boldsymbol{\epsilon}_k\|_2^2 - \|\boldsymbol{\epsilon}_{k+1}\|_2^2 = \frac{\|\boldsymbol{p}_k\|_2^2}{\|\boldsymbol{p}_k\|_{\boldsymbol{A}}^2}(\|\boldsymbol{\epsilon}_k\|_{\boldsymbol{A}}^2 + \|\boldsymbol{\epsilon}_{k+1}\|_{\boldsymbol{A}}^2)$$

where $\boldsymbol{\epsilon}_j = \boldsymbol{x} - \boldsymbol{x}_j$ and $\boldsymbol{A}\boldsymbol{x} = \boldsymbol{b}$.

Proof For $j \le m$

$$\boldsymbol{\epsilon}_j = \boldsymbol{x}_{m+1} - \boldsymbol{x}_j = \boldsymbol{x}_m - \boldsymbol{x}_j + \alpha_m \boldsymbol{p}_m = \boldsymbol{x}_{m-1} - \boldsymbol{x}_j + \alpha_{m-1}\boldsymbol{p}_{m-1} + \alpha_m \boldsymbol{p}_m = \ldots$$

so that

$$\boldsymbol{\epsilon}_j = \sum_{i=j}^{m} \alpha_i \boldsymbol{p}_i, \quad \alpha_i = \frac{\boldsymbol{r}_i^T \boldsymbol{r}_i}{\boldsymbol{p}_i^T \boldsymbol{A}\boldsymbol{p}_i}. \tag{13.41}$$

By (13.41) and $\boldsymbol{A}$-orthogonality

$$\|\boldsymbol{\epsilon}_j\|_{\boldsymbol{A}}^2 = \boldsymbol{\epsilon}_j \boldsymbol{A}\boldsymbol{\epsilon}_j = \sum_{i=j}^{m} \alpha_i^2 \boldsymbol{p}_i^T \boldsymbol{A}\boldsymbol{p}_i = \sum_{i=j}^{m} \frac{(\boldsymbol{r}_i^T \boldsymbol{r}_i)^2}{\boldsymbol{p}_i^T \boldsymbol{A}\boldsymbol{p}_i}. \tag{13.42}$$

By (13.17) and Lemma 13.3

$$\boldsymbol{p}_i^T \boldsymbol{p}_k = (\boldsymbol{r}_i + \beta_{i-1}\boldsymbol{p}_{i-1})^T \boldsymbol{p}_k = \beta_{i-1}\boldsymbol{p}_{i-1}^T \boldsymbol{p}_k = \cdots = \beta_{i-1}\cdots\beta_k(\boldsymbol{p}_k^T \boldsymbol{p}_k),$$

and since $\beta_{i-1}\cdots\beta_k = (\boldsymbol{r}_i^T \boldsymbol{r}_i)/(\boldsymbol{r}_k^T \boldsymbol{r}_k)$ we find

$$\boldsymbol{p}_i^T \boldsymbol{p}_k = \frac{\boldsymbol{r}_i^T \boldsymbol{r}_i}{\boldsymbol{r}_k^T \boldsymbol{r}_k} \boldsymbol{p}_k^T \boldsymbol{p}_k, \quad i \ge k. \tag{13.43}$$

Since

$$\|\boldsymbol{\epsilon}_k\|_2^2 = \|\boldsymbol{\epsilon}_{k+1} + \boldsymbol{x}_{k+1} - \boldsymbol{x}_k\|_2^2 = \|\boldsymbol{\epsilon}_{k+1} + \alpha_k \boldsymbol{p}_k\|_2^2,$$

we obtain

$$\begin{aligned}\|\boldsymbol{\epsilon}_k\|_2^2 - \|\boldsymbol{\epsilon}_{k+1}\|_2^2 &= \alpha_k\big(2\boldsymbol{p}_k^T\boldsymbol{\epsilon}_{k+1} + \alpha_k\boldsymbol{p}_k^T\boldsymbol{p}_k\big)\\ &\overset{(13.41)}{=} \alpha_k\big(2\sum_{i=k+1}^{m}\alpha_i\boldsymbol{p}_i^T\boldsymbol{p}_k + \alpha_k\boldsymbol{p}_k^T\boldsymbol{p}_k\big) = \big(\sum_{i=k}^{m} + \sum_{i=k+1}^{m}\big)\alpha_k\alpha_i\boldsymbol{p}_i^T\boldsymbol{p}_k\\ &\overset{(13.43)}{=} \big(\sum_{i=k}^{m} + \sum_{i=k+1}^{m}\big)\frac{\boldsymbol{r}_k^T\boldsymbol{r}_k}{\boldsymbol{p}_k^T\boldsymbol{A}\boldsymbol{p}_k}\frac{\boldsymbol{r}_i^T\boldsymbol{r}_i}{\boldsymbol{p}_i^T\boldsymbol{A}\boldsymbol{p}_i}\frac{\boldsymbol{r}_i^T\boldsymbol{r}_i}{\boldsymbol{r}_k^T\boldsymbol{r}_k}\boldsymbol{p}_k^T\boldsymbol{p}_k\\ &\overset{(13.42)}{=} \frac{\|\boldsymbol{p}_k\|_2^2}{\|\boldsymbol{p}_k\|_{\boldsymbol{A}}^2}\big(\|\boldsymbol{\epsilon}_k\|_{\boldsymbol{A}}^2 + \|\boldsymbol{\epsilon}_{k+1}\|_{\boldsymbol{A}}^2\big).\end{aligned}$$

and the Theorem is proved. □

13.5 Preconditioning

For problems $\boldsymbol{A}\boldsymbol{x} = \boldsymbol{b}$ of size n, where both n and $\mathrm{cond}_2(\boldsymbol{A})$ are large, it is often possible to improve the performance of the conjugate gradient method by using a technique known as **preconditioning**. Instead of $\boldsymbol{A}\boldsymbol{x} = \boldsymbol{b}$ we consider an equivalent system $\boldsymbol{B}\boldsymbol{A}\boldsymbol{x} = \boldsymbol{B}\boldsymbol{b}$, where $\boldsymbol{B}$ is nonsingular and $\mathrm{cond}_2(\boldsymbol{B}\boldsymbol{A})$ is smaller than $\mathrm{cond}_2(\boldsymbol{A})$. The matrix $\boldsymbol{B}$ will in many cases be the inverse of another matrix, $\boldsymbol{B} = \boldsymbol{M}^{-1}$. We cannot use CG on $\boldsymbol{B}\boldsymbol{A}\boldsymbol{x} = \boldsymbol{B}\boldsymbol{b}$ directly since $\boldsymbol{B}\boldsymbol{A}$ in general is not symmetric even if both $\boldsymbol{A}$ and $\boldsymbol{B}$ are. But if $\boldsymbol{B}$ (and hence $\boldsymbol{M}$) is positive definite then we can apply CG to a symmetrized system and then transform the recurrence formulas to an iterative method for the original system $\boldsymbol{A}\boldsymbol{x} = \boldsymbol{b}$. This iterative method is known as the **preconditioned conjugate gradient method**. We shall see that the convergence properties of this method is determined by the eigenvalues of $\boldsymbol{B}\boldsymbol{A}$.

Suppose $\boldsymbol{B}$ is positive definite. By Theorem 4.4 there is a nonsingular matrix $\boldsymbol{C}$ such that $\boldsymbol{B} = \boldsymbol{C}^T\boldsymbol{C}$. ($\boldsymbol{C}$ is only needed for the derivation and will not appear in the final formulas). Now

$$\boldsymbol{B}\boldsymbol{A}\boldsymbol{x} = \boldsymbol{B}\boldsymbol{b} \Leftrightarrow \boldsymbol{C}^T(\boldsymbol{C}\boldsymbol{A}\boldsymbol{C}^T)\boldsymbol{C}^{-T}\boldsymbol{x} = \boldsymbol{C}^T\boldsymbol{C}\boldsymbol{b} \Leftrightarrow (\boldsymbol{C}\boldsymbol{A}\boldsymbol{C}^T)\boldsymbol{y} = \boldsymbol{C}\boldsymbol{b},$$

where $\boldsymbol{y} := \boldsymbol{C}^{-T}\boldsymbol{x}$. We have 3 linear systems

$$\boldsymbol{A}\boldsymbol{x} = \boldsymbol{b} \tag{13.44}$$

$$\boldsymbol{B}\boldsymbol{A}\boldsymbol{x} = \boldsymbol{B}\boldsymbol{b} \tag{13.45}$$

$$(\boldsymbol{C}\boldsymbol{A}\boldsymbol{C}^T)\boldsymbol{y} = \boldsymbol{C}\boldsymbol{b}, \ \& \ \boldsymbol{x} = \boldsymbol{C}^T\boldsymbol{y}. \tag{13.46}$$

Note that (13.44) and (13.46) are positive definite linear systems. In addition to being positive definite the matrix $\boldsymbol{C}\boldsymbol{A}\boldsymbol{C}^T$ is similar to $\boldsymbol{B}\boldsymbol{A}$. Indeed,

$$\boldsymbol{C}^T(\boldsymbol{C}\boldsymbol{A}\boldsymbol{C}^T)\boldsymbol{C}^{-T} = \boldsymbol{B}\boldsymbol{A}.$$

Thus $\boldsymbol{C}\boldsymbol{A}\boldsymbol{C}^T$ and $\boldsymbol{B}\boldsymbol{A}$ have the same eigenvalues. Therefore, if we apply the conjugate gradient method to (13.46) then the rate of convergence will be determined by the eigenvalues of $\boldsymbol{B}\boldsymbol{A}$.

We apply the conjugate gradient method to $(\boldsymbol{C}\boldsymbol{A}\boldsymbol{C}^T)\boldsymbol{y} = \boldsymbol{C}\boldsymbol{b}$. Denoting the search direction by $\boldsymbol{q}_k$ and the residual by $\boldsymbol{z}_k := \boldsymbol{C}\boldsymbol{b} - \boldsymbol{C}\boldsymbol{A}\boldsymbol{C}^T\boldsymbol{y}_k$ we obtain the following from (13.15), (13.16), and (13.17).

$$\begin{aligned}
\boldsymbol{y}_{k+1} &= \boldsymbol{y}_k + \alpha_k \boldsymbol{q}_k, \quad \alpha_k = \boldsymbol{z}_k^T \boldsymbol{z}_k / \boldsymbol{q}_k^T(\boldsymbol{C}\boldsymbol{A}\boldsymbol{C}^T)\boldsymbol{q}_k,\\
\boldsymbol{z}_{k+1} &= \boldsymbol{z}_k - \alpha_k(\boldsymbol{C}\boldsymbol{A}\boldsymbol{C}^T)\boldsymbol{q}_k,\\
\boldsymbol{q}_{k+1} &= \boldsymbol{z}_{k+1} + \beta_k \boldsymbol{q}_k, \quad \beta_k = \boldsymbol{z}_{k+1}^T \boldsymbol{z}_{k+1} / \boldsymbol{z}_k^T \boldsymbol{z}_k.
\end{aligned}$$

With

$$\boldsymbol{x}_k := \boldsymbol{C}^T\boldsymbol{y}_k, \quad \boldsymbol{p}_k := \boldsymbol{C}^T\boldsymbol{q}_k, \quad \boldsymbol{s}_k := \boldsymbol{C}^T\boldsymbol{z}_k, \quad \boldsymbol{r}_k := \boldsymbol{C}^{-1}\boldsymbol{z}_k \tag{13.47}$$

this can be transformed into

$$\boldsymbol{x}_{k+1} = \boldsymbol{x}_k + \alpha_k \boldsymbol{p}_k, \quad \alpha_k = \frac{\boldsymbol{s}_k^T \boldsymbol{r}_k}{\boldsymbol{p}_k^T \boldsymbol{A} \boldsymbol{p}_k}, \tag{13.48}$$

$$\boldsymbol{r}_{k+1} = \boldsymbol{r}_k - \alpha_k \boldsymbol{A}\boldsymbol{p}_k, \tag{13.49}$$

$$\boldsymbol{s}_{k+1} = \boldsymbol{s}_k - \alpha_k \boldsymbol{B}\boldsymbol{A}\boldsymbol{p}_k, \tag{13.50}$$

$$\boldsymbol{p}_{k+1} = \boldsymbol{s}_{k+1} + \beta_k \boldsymbol{p}_k, \quad \beta_k = \frac{\boldsymbol{s}_{k+1}^T \boldsymbol{r}_{k+1}}{\boldsymbol{s}_k^T \boldsymbol{r}_k}. \tag{13.51}$$

Here $\boldsymbol{x}_k$ will be an approximation to the solution $\boldsymbol{x}$ of $\boldsymbol{A}\boldsymbol{x} = \boldsymbol{b}$, $\boldsymbol{r}_k = \boldsymbol{b} - \boldsymbol{A}\boldsymbol{x}_k$ is the residual in the original system, and $\boldsymbol{s}_k = \boldsymbol{C}^T\boldsymbol{z}_k = \boldsymbol{C}^T(\boldsymbol{C} - \boldsymbol{C}\boldsymbol{A}\boldsymbol{C}^T)\boldsymbol{y}_k = \boldsymbol{B}\boldsymbol{b} - \boldsymbol{B}\boldsymbol{A}\boldsymbol{x}_k$ is the residual in the preconditioned system. If we set $\boldsymbol{r}_0 = \boldsymbol{b} - \boldsymbol{A}\boldsymbol{x}_0$, $\boldsymbol{p}_0 = \boldsymbol{s}_0 = \boldsymbol{B}\boldsymbol{r}_0$, we obtain the following preconditioned conjugate gradient algorithm for determining approximations $\boldsymbol{x}_k$ to the solution of a positive definite system $\boldsymbol{A}\boldsymbol{x} = \boldsymbol{b}$, by considering the system $\boldsymbol{B}\boldsymbol{A}\boldsymbol{x} = \boldsymbol{B}\boldsymbol{b}$, with $\boldsymbol{B}$ positive definite. The iteration is stopped when $||\boldsymbol{r}_k||_2/||\boldsymbol{b}||_2 \le \text{tol}$ or $k > \text{itmax}$. K is the number of iterations used, and x$(= \boldsymbol{x}_0)$ is the starting iteration.

```
function [x,K]=pcg(A,B,b,x,tol,itmax)
% [x,K]=pcg(A,B,b,x,tol,itmax)
r=b-A*x; p=B*r; s=p; rho=s'*r; rho0=b'*b;
for k=0:itmax
    if sqrt(rho/rho0)<= tol
        K=k; return
    end
    t=A*p; a=rho/(p'*t);
    x=x+a*p; r=r-a*t;
    w=B*t; s=s-a*w;
    rhos=rho; rho=s'*r;
    p=r+(rho/rhos)*p;
end
K=itmax+1;
end
```

Listing 13.3 pcg

Apart from the calculation of ρ this algorithm is quite similar to Algorithm 13.1. The main additional work is contained in $w = B * t$. We'll discuss this further in connection with an example. There the inverse of $\boldsymbol{B}$ is known and we have to solve a linear system to find $\boldsymbol{w}$.

We have the following convergence result for this algorithm.

Theorem 13.9 (Error Bound Preconditioned cg) *Suppose we apply a positive definite preconditioner* $\boldsymbol{B}$ *to the positive definite system* $\boldsymbol{A}\boldsymbol{x} = \boldsymbol{b}$. *Then the quantities* $\boldsymbol{x}_k$ *computed in Algorithm 13.3 satisfy the following bound:*

$$\frac{||\boldsymbol{x} - \boldsymbol{x}_k||_A}{||\boldsymbol{x} - \boldsymbol{x}_0||_A} \le 2\left(\frac{\sqrt{\kappa}-1}{\sqrt{\kappa}+1}\right)^k, \quad \text{for} \quad k \ge 0,$$

where $\kappa = \lambda_{max}/\lambda_{min}$ *is the ratio of the largest and smallest eigenvalue of* $\boldsymbol{B}\boldsymbol{A}$.

Proof Since Algorithm 13.3 is equivalent to solving (13.46) by the conjugate gradient method Theorem 13.3 implies that

$$\frac{||\boldsymbol{y} - \boldsymbol{y}_k||_{\boldsymbol{C}\boldsymbol{A}\boldsymbol{C}^T}}{||\boldsymbol{y} - \boldsymbol{y}_0||_{\boldsymbol{C}\boldsymbol{A}\boldsymbol{C}^T}} \le 2\left(\frac{\sqrt{\kappa}-1}{\sqrt{\kappa}+1}\right)^k, \quad \text{for} \quad k \ge 0,$$

where $\boldsymbol{y}_k$ is the conjugate gradient approximation to the solution $\boldsymbol{y}$ of (13.46) and κ is the ratio of the largest and smallest eigenvalue of $\boldsymbol{C}\boldsymbol{A}\boldsymbol{C}^T$. Since $\boldsymbol{B}\boldsymbol{A}$ and $\boldsymbol{C}\boldsymbol{A}\boldsymbol{C}^T$ are similar this is the same as the κ in the theorem. By (13.47) we have

$$\begin{aligned}\|\boldsymbol{y} - \boldsymbol{y}_k\|^2_{\boldsymbol{C}\boldsymbol{A}\boldsymbol{C}^T} &= (\boldsymbol{y} - \boldsymbol{y}_k)^T(\boldsymbol{C}\boldsymbol{A}\boldsymbol{C}^T)(\boldsymbol{y} - \boldsymbol{y}_k)\\ &= (\boldsymbol{C}^T(\boldsymbol{y} - \boldsymbol{y}_k))^T\boldsymbol{A}(\boldsymbol{C}^T(\boldsymbol{y} - \boldsymbol{y}_k)) = \|\boldsymbol{x} - \boldsymbol{x}_k\|^2_{\boldsymbol{A}},\end{aligned}$$

and the proof is complete. □

We conclude that $\boldsymbol{B}$ should satisfy the following requirements for a problem of size n:

1. The eigenvalues of $\boldsymbol{BA}$ should be located in a narrow interval. Preferably one should be able to bound the length of the interval independently of n.
2. The evaluation of $\boldsymbol{Bx}$ for a given vector $\boldsymbol{x}$ should not be expensive in storage and arithmetic operations, ideally $O(n)$ for both.

In this book we only consider one example of a preconditioner. For a comprehensive treatment of preconditioners see [1].

13.6 Preconditioning Example

13.6.1 A Variable Coefficient Problem

Consider the problem

$$\begin{aligned} -\tfrac{\partial}{\partial x}\Big(c(x,y)\tfrac{\partial u}{\partial x}\Big) - \tfrac{\partial}{\partial y}\Big(c(x,y)\tfrac{\partial u}{\partial y}\Big) &= f(x,y), \qquad (x,y) \in \Omega = (0,1)^2, \\ u(x,y) &= 0, \qquad (x,y) \in \partial\Omega. \end{aligned} \tag{13.52}$$

Here Ω is the open unit square while $\partial\Omega$ is the boundary of Ω. The functions f and c are given and we seek a function $u = u(x,y)$ such that (13.52) holds. We assume that c and f are defined and continuous on Ω and that $c(x,y) > 0$ for all $(x,y) \in \Omega$. The problem (13.52) reduces to the Poisson problem (10.1) in the special case where $c(x,y) = 1$ for $(x,y) \in \Omega$.

To solve (13.52) numerically, we choose $m \in \mathbb{N}$, set $h := 1/(m+1)$, and define index sets

$$\begin{aligned} I_m &:= \{(j,k) : 1 \le j,k \le m\}, \\ \overline{I}_m &:= \{(j,k) : 0 \le j,k \le m+1\}, \\ \partial I_m &:= \overline{I}_m \setminus I_m. \end{aligned}$$

We compute approximations $v_{j,k} \approx u(x_j, y_k)$ on a grid of points

$$\{(x_j, y_k) = (jh, kh) : (j,k) \in \overline{I}_m\}$$

using a finite difference method. For univariate functions f, g we approximate derivatives by using the central difference approximations

$$\frac{d}{dt}\Big(f(t)\frac{d}{dt}g(t)\Big) \approx \Big(f(t+\frac{h}{2})\frac{d}{dt}g(t+h/2) - f(t-\frac{h}{2})\frac{d}{dt}g(t-\frac{h}{2})\Big)/h$$
$$\approx \Big(f(t+\frac{h}{2})\big(g(t+h)-g(t)\big) - f(t-\frac{h}{2})\big(g(t)-g(t-h)\big)\Big)/h^2$$

to obtain

$$(L_h v)_{j,k} := \frac{(dv)_{j,k}}{h^2} = f_{j,k}, \qquad (j,k) \in I_m, \tag{13.53}$$

$$v_{j,k} = 0, \qquad (j,k) \in \partial I_m, \tag{13.54}$$

where $f_{j,k} := f(x_j, y_k)$, $(dv)_{j,k} := (d_1 v)_{j,k} + (d_2 v)_{j,k}$,

$$\begin{aligned}(d_1 v)_{j,k} &:= c_{j-\frac{1}{2},k}(v_{j,k} - v_{j-1,k}) - c_{j+\frac{1}{2},k}(v_{j+1,k} - v_{j,k}) \approx -h^2 \frac{\partial}{\partial x}\big(c\frac{\partial u}{\partial x}\big)_{j,k},\\ (d_2 v)_{j,k} &:= c_{j,k-\frac{1}{2}}(v_{j,k} - v_{j,k-1}) - c_{j,k+\frac{1}{2}}(v_{j,k+1} - v_{j,k}) \approx -h^2 \frac{\partial}{\partial y}\big(c\frac{\partial u}{\partial y}\big)_{j,k},\end{aligned} \tag{13.55}$$

and where $c_{p,q} = c(ph, qh)$ for $p, q \in \mathbb{R}$. The equation (13.53) can be written in matrix form as

$$L_h v = \boldsymbol{F}, \quad L_h v := \frac{1}{h^2}\begin{bmatrix} (dv)_{1,1} & \dots & (dv)_{1,m} \\ \vdots & & \vdots \\ (dv)_{m,1} & \dots & (dv)_{m,m} \end{bmatrix}, \quad \boldsymbol{F} := \begin{bmatrix} f_{1,1} & \dots & f_{1,m} \\ \vdots & & \vdots \\ f_{m,1} & \dots & f_{m,m} \end{bmatrix}. \tag{13.56}$$

This is a linear system with the elements of

$$\boldsymbol{V} := \begin{bmatrix} v_{1,1} & \dots & v_{1,m} \\ \vdots & & \vdots \\ v_{m,1} & \dots & v_{m,m} \end{bmatrix}$$

as unknowns. The system $h^2 L_h v = h^2 \boldsymbol{F}$ can be written in standard form $\boldsymbol{A}\boldsymbol{x} = \boldsymbol{b}$ where $\boldsymbol{x} = \text{vec}(\boldsymbol{V})$, $\boldsymbol{b} = h^2\text{vec}(\boldsymbol{F})$, and the coefficient matrix $\boldsymbol{A} \in \mathbb{R}^{n\times n}$ is defined as follows

$$\boldsymbol{A}\boldsymbol{x} = \boldsymbol{A}\text{vec}(\boldsymbol{V}) := h^2\text{vec}(L_h v). \tag{13.57}$$

If $c(x, y) = 1$ for all $(x, y) \in \Omega$ we recover the Poisson matrix (10.8). In general we can show that $\boldsymbol{A}$ is positive definite for all $m \in \mathbb{N}$ provided $c(x, y) > 0$ for all $(x, y) \in \Omega$. For this we do not need the explicit form of $\boldsymbol{A}$.

To start we define for $m \in \mathbb{N}$ a discrete inner product on the space of matrices $\mathbb{R}^{m\times m}$

$$\langle \boldsymbol{V}, \boldsymbol{W}\rangle := h^2 \sum_{j,k=1}^{m} v_{j,k} w_{j,k}, \tag{13.58}$$

We then have the following lemma.

Lemma 13.6 (Discrete Inner Product) *If* $\boldsymbol{V}, \boldsymbol{W} \in \mathbb{R}^{m\times m}$ *and* $v_{j,k} = w_{j,k} = 0$ *for* $(j, k) \in \partial I_m$, *then*

$$\begin{aligned}\langle L_h v, \boldsymbol{W}\rangle &= \sum_{j=1}^{m}\sum_{k=0}^{m} c_{j,k+\frac{1}{2}}\big(v_{j,k+1} - v_{j,k}\big)\big(w_{j,k+1} - w_{j,k}\big)\\ &+ \sum_{j=0}^{m}\sum_{k=1}^{m} c_{j+\frac{1}{2},k}\big(v_{j+1,k} - v_{j,k}\big)\big(w_{j+1,k} - w_{j,k}\big).\end{aligned} \tag{13.59}$$

Proof If $m \in \mathbb{N}$, $a_i, b_i, c_i \in \mathbb{R}$ for $i = 0, \ldots, m$ and $b_0 = c_0 = b_{m+1} = c_{m+1} = 0$ then

$$\sum_{i=1}^{m} \big(a_{i-1}(b_i - b_{i-1}) - a_i(b_{i+1} - b_i)\big)c_i = \sum_{i=0}^{m} a_i(b_{i+1} - b_i)(c_{i+1} - c_i). \tag{13.60}$$

Indeed, the left hand side can be written

$$\sum_{i=0}^{m} a_i(b_{i+1} - b_i)c_{i+1} - \sum_{i=0}^{m} a_i(b_{i+1} - b_i)c_i,$$

and the right hand side of (13.60) follows. We apply (13.60) to $(d_1 v)_{j,k} w_{j,k}$ and $(d_2 v)_{j,k} w_{j,k}$ given by (13.55) and (13.59) follows. □

Theorem 13.10 (Positive Definite Matrix) *If* $c(x, y) > 0$ *for* $(x, y) \in \Omega$ *then the matrix* $\boldsymbol{A}$ *defined by* (13.57) *via the linear system* (13.56) *is positive definite.*

Proof By (13.59) $\langle L_h v, \boldsymbol{W}\rangle = \langle \boldsymbol{W}, L_h v\rangle$ and symmetry follows. We take $\boldsymbol{W} = \boldsymbol{V}$ and obtain quadratic factors in (13.59). Since $c_{j+\frac{1}{2},k}$ and $c_{j,k+\frac{1}{2}}$ correspond to values of c in Ω for the values of j, k in the sums, it follows that they are positive and $\langle L_h v, \boldsymbol{V}\rangle \geq 0$ for all $\boldsymbol{V} \in \mathbb{R}^{m\times m}$. If $\langle L_h v, \boldsymbol{V}\rangle = 0$ then all the quadratic factors must be zero, and $v_{j,k+1} = v_{j,k}$ for $k = 0, 1, \ldots, m$ and $j = 1, \ldots, m$. Now $v_{j,0} = v_{j,m+1} = 0$ implies that $\boldsymbol{V} = \boldsymbol{0}$. It follows that the linear system (13.56) is positive definite. □

13.6.2 Applying Preconditioning

Consider solving $\boldsymbol{Ax} = \boldsymbol{b}$, where $\boldsymbol{A}$ is given by (13.57) and $b \in \mathbb{R}^n$. Since $\boldsymbol{A}$ is positive definite it is nonsingular and the system has a unique solution $x \in \mathbb{R}^n$. Moreover we can use either Cholesky factorization or the block tridiagonal solver to find $\boldsymbol{x}$. Since the bandwidth of $\boldsymbol{A}$ is $m = \sqrt{n}$ both of these methods require $O(n^2)$ arithmetic operations for large n.

If we choose $c(x, y) \equiv 1$ in (13.52), we get the Poisson problem. With this in mind, we may think of the coefficient matrix $\boldsymbol{A}_p$ arising from the discretization of the Poisson problem as an approximation to the matrix (13.57). This suggests using $\boldsymbol{B} = \boldsymbol{A}_p^{-1}$, the inverse of the discrete Poisson matrix as a preconditioner for the system (13.53).

Consider Algorithm 13.3. With this preconditioner the calculation $\boldsymbol{w} = \boldsymbol{Bt}$ takes the form $\boldsymbol{A}_p\boldsymbol{w}_k = \boldsymbol{t}_k$.

In Sect. 11.2 we developed a Simple fast Poisson Solver, Cf. Algorithm 11.1. This method can be utilized to solve $\boldsymbol{A}_p\boldsymbol{w} = \boldsymbol{t}$.

Consider the specific problem where

$$c(x, y) = e^{-x+y} \text{ and } f(x, y) = 1.$$

We have used Algorithm 13.1 (conjugate gradient without preconditioning), and Algorithm 13.3 (conjugate gradient with preconditioning) to solve the problem (13.52). We used $\boldsymbol{x}_0 = 0$ and $\epsilon = 10^{-8}$. The results are shown in Table 13.3.

Without preconditioning the number of iterations still seems to be more or less proportional to $\sqrt{n}$ although the convergence is slower than for the constant coefficient problem. Using preconditioning speeds up the convergence considerably. The number of iterations appears to be bounded independently of n.

Using a preconditioner increases the work in each iteration. For the present example the number of arithmetic operations in each iteration changes from $O(n)$ without preconditioning to $O(n^{3/2})$ or $O(n \log_2 n)$ with preconditioning. This is not a large increase and both the number of iterations and the computing time is reduced significantly.

Let us finally show that the number $\kappa = \lambda_{max}/\lambda_{min}$ which determines the rate of convergence for the preconditioned conjugate gradient method applied to (13.52) can be bounded independently of n.

Table 13.3 The number of iterations K (no preconditioning) and K_{pre} (with preconditioning) for the problem (13.52) using the discrete Poisson problem as a preconditioner

n	2500	10000	22500	40000	62500
K	222	472	728	986	1246
$K/\sqrt{n}$	4.44	4.72	4.85	4.93	4.98
K_{pre}	22	23	23	23	23

Theorem 13.12 (Eigenvalues of Preconditioned Matrix) *Suppose* $0 < c_0 \le c(x, y) \le c_1$ *for all* $(x, y) \in [0, 1]^2$. *For the eigenvalues of the matrix* $\boldsymbol{B}\boldsymbol{A} = \boldsymbol{A}_p^{-1}\boldsymbol{A}$ *just described we have*

$$\kappa = \frac{\lambda_{max}}{\lambda_{min}} \le \frac{c_1}{c_0}.$$

Proof Suppose $\boldsymbol{A}_p^{-1}\boldsymbol{A}\boldsymbol{x} = \lambda x$ for some $\boldsymbol{x} \in \mathbb{R}^n \setminus \{0\}$. Then $\boldsymbol{A}\boldsymbol{x} = \lambda \boldsymbol{A}_p x$. Multiplying this by $\boldsymbol{x}^T$ and solving for λ we find

$$\lambda = \frac{x^T \boldsymbol{A}\boldsymbol{x}}{x^T \boldsymbol{A}_p x}.$$

We computed $\boldsymbol{x}^T\boldsymbol{A}\boldsymbol{x}$ in (13.59) and we obtain $\boldsymbol{x}^T\boldsymbol{A}_p\boldsymbol{x}$ by setting all the c's there equal to one

$$\boldsymbol{x}^T\boldsymbol{A}_p x = \sum_{i=1}^{m}\sum_{j=0}^{m}\left(v_{i,j+1} - v_{i,j}\right)^2 + \sum_{j=1}^{m}\sum_{i=0}^{m}\left(v_{i+1,j} - v_{i,j}\right)^2.$$

Thus $\boldsymbol{x}^T\boldsymbol{A}_p x > 0$ and bounding all the c's in (13.59) from below by c_0 and above by c_1 we find

$$c_0(x^T\boldsymbol{A}_p x) \le x^T\boldsymbol{A}\boldsymbol{x} \le c_1(x^T\boldsymbol{A}_p x)$$

which implies that $c_0 \le \lambda \le c_1$ for all eigenvalues λ of $\boldsymbol{B}\boldsymbol{A} = \boldsymbol{A}_p^{-1}\boldsymbol{A}$. □

Using $c(x, y) = e^{-x+y}$ as above, we find $c_0 = e^{-2}$ and $c_1 = 1$. Thus $\kappa \le e^2 \approx 7.4$, a quite acceptable matrix condition number which explains the convergence results from our numerical experiment.

13.7 Exercises Chap. 13

13.7.1 Exercises Sect. 13.1

Exercise 13.1 ($\boldsymbol{A}$-Norm) One can show that the $\boldsymbol{A}$-norm is a vector norm on $\mathbb{R}^n$ without using the fact that it is an inner product norm. Show this with the help of the Cholesky factorization of $\boldsymbol{A}$.

Exercise 13.2 (Paraboloid) Let $\boldsymbol{A} = \boldsymbol{U}\boldsymbol{D}\boldsymbol{U}^T$ be the spectral decomposition of $\boldsymbol{A}$, i.e., $\boldsymbol{U}$ is orthogonal and $\boldsymbol{D} = \mathrm{diag}(\lambda_1, \ldots, \lambda_n)$ is diagonal. Define new variables

$\boldsymbol{v} = [v_1, \ldots, v_n]^T := \boldsymbol{U}^T \boldsymbol{y}$, and set $\boldsymbol{c} := \boldsymbol{U}^T \boldsymbol{b} = [c_1, \ldots, c_n]^T$. Show that

$$Q(\boldsymbol{y}) = \frac{1}{2}\sum_{j=1}^{n} \lambda_j v_j^2 - \sum_{j=1}^{n} c_j v_j.$$

Exercise 13.3 (Steepest Descent Iteration) Verify the numbers in Example 13.1.

Exercise 13.4 (Steepest Descent (Exam Exercise 2011-1)) The method of steepest descent can be used to solve a linear system $\boldsymbol{A}\boldsymbol{x} = \boldsymbol{b}$ for $\boldsymbol{x} \in \mathbb{R}^n$, where $\boldsymbol{A} \in \mathbb{R}^{n,n}$ is symmetric and positive definite, and $\boldsymbol{b} \in \mathbb{R}^n$. With $\boldsymbol{x}_0 \in \mathbb{R}^n$ an initial guess, the iteration is $\boldsymbol{x}_{k+1} = \boldsymbol{x}_k + \alpha_k \boldsymbol{r}_k$, where $\boldsymbol{r}_k$ is the residual, $\boldsymbol{r}_k = \boldsymbol{b} - \boldsymbol{A}\boldsymbol{x}_k$, and $\alpha_k = \frac{\boldsymbol{r}_k^T \boldsymbol{r}_k}{\boldsymbol{r}_k^T \boldsymbol{A} \boldsymbol{r}_k}$.

a) Compute $\boldsymbol{x}_1$ if $\boldsymbol{A} = \begin{bmatrix} 2 & -1 \\ -1 & 2 \end{bmatrix}$, $\boldsymbol{b} = [1\ 1]^T$ and $\boldsymbol{x}_0 = \boldsymbol{0}$.

b) If the k-th error, $\boldsymbol{e}_k = \boldsymbol{x}_k - \boldsymbol{x}$, is an eigenvector of $\boldsymbol{A}$, what can you say about $\boldsymbol{x}_{k+1}$?

13.7.2 Exercises Sect. 13.2

Exercise 13.5 (Conjugate Gradient Iteration, II) Do one iteration with the conjugate gradient method when $\boldsymbol{x}_0 = \boldsymbol{0}$. (Answer: $\boldsymbol{x}_1 = \left(\frac{\boldsymbol{b}^T \boldsymbol{b}}{\boldsymbol{b}^T \boldsymbol{A} \boldsymbol{b}}\right) \boldsymbol{b}$.)

Exercise 13.6 (Conjugate Gradient Iteration, III) Do two conjugate gradient iterations for the system

$$\begin{bmatrix} 2 & -1 \\ -1 & 2 \end{bmatrix} \begin{bmatrix} x_1 \\ x_2 \end{bmatrix} = \begin{bmatrix} 0 \\ 3 \end{bmatrix}$$

starting with $\boldsymbol{x}_0 = \boldsymbol{0}$.

Exercise 13.7 (The cg Step Length Is Optimal) Show that the step length α_k in the conjugate gradient method is optimal.[2]

Exercise 13.8 (Starting Value in cg) Show that the conjugate gradient method (13.18) for $\boldsymbol{A}\boldsymbol{x} = \boldsymbol{b}$ starting with $\boldsymbol{x}_0$ is the same as applying the method to the system $\boldsymbol{A}\boldsymbol{y} = \boldsymbol{r}_0 := \boldsymbol{b} - \boldsymbol{A}\boldsymbol{x}_0$ starting with $\boldsymbol{y}_0 = \boldsymbol{0}$.[3]

---

[2]Hint: use induction on k to show that $\boldsymbol{p}_k = \boldsymbol{r}_k + \sum_{j=0}^{k-1} a_{k,j} \boldsymbol{r}_j$ for some constants $a_{k,j}$.

[3]Hint: The conjugate gradient method for $\boldsymbol{A}\boldsymbol{y} = \boldsymbol{r}_0$ can be written $\boldsymbol{y}_{k+1} := \boldsymbol{y}_k + \gamma_k \boldsymbol{q}_k$, $\gamma_k := \frac{\boldsymbol{s}_k^T \boldsymbol{s}_k}{\boldsymbol{q}_k^T \boldsymbol{A} \boldsymbol{q}_k}$, $\boldsymbol{s}_{k+1} := \boldsymbol{s}_k - \gamma_k \boldsymbol{A} \boldsymbol{q}_k$, $\boldsymbol{q}_{k+1} := \boldsymbol{s}_{k+1} + \delta_k \boldsymbol{q}_k$, $\delta_k := \frac{\boldsymbol{s}_{k+1}^T \boldsymbol{s}_{k+1}}{\boldsymbol{s}_k^T \boldsymbol{s}_k}$. Show that $\boldsymbol{y}_k = \boldsymbol{x}_k - \boldsymbol{x}_0$, $\boldsymbol{s}_k = \boldsymbol{r}_k$, and $\boldsymbol{q}_k = \boldsymbol{p}_k$, for $k = 0, 1, 2 \ldots$.

Exercise 13.9 (Program Code for Testing Steepest Descent) Write a function `K=sdtest(m,a,d,tol,itmax)` to test the steepest descent method on the matrix $\boldsymbol{T}_2$. Make the analogues of Tables 13.1 and 13.2. For Table 13.2 it is enough to test for say $n = 100, 400, 1600, 2500$, and tabulate K/n instead of $K/\sqrt{n}$ in the last row. Conclude that the upper bound (13.19) is realistic. Compare also with the number of iterations for the J and GS method in Table 12.1.

Exercise 13.10 (Using cg to Solve Normal Equations) Consider solving the linear system $\boldsymbol{A}^T\boldsymbol{A}\boldsymbol{x} = \boldsymbol{A}^T\boldsymbol{b}$ by using the conjugate gradient method. Here $\boldsymbol{A} \in \mathbb{R}^{m,n}$, $\boldsymbol{b} \in \mathbb{R}^m$ and $\boldsymbol{A}^T\boldsymbol{A}$ is positive definite.[4] Explain why only the following modifications in Algorithm 13.1 are necessary

1. r=A'(b-A*x); p=r;
2. a=rho/(t'*t);
3. r=r-a*A'*t;

Note that the condition number of the normal equations is $\text{cond}_2(\boldsymbol{A})^2$, the square of the condition number of $\boldsymbol{A}$.

Exercise 13.11 ($\boldsymbol{A}^T\boldsymbol{A}$ Inner Product (Exam Exercise 2018-3)) In this problem we consider linear systems of the form $\boldsymbol{A}\boldsymbol{x} = \boldsymbol{b}$, where $\boldsymbol{A} \in \mathbb{R}^{n\times n}$ and $\boldsymbol{b} \in \mathbb{R}^n$ are given, and $\boldsymbol{x} \in \mathbb{R}^n$ is the unknown vector. We assume throughout that $\boldsymbol{A}$ is nonsingular.

a) Let $\{\boldsymbol{v}_i\}_{i=1}^k$ be a set of linearly independent vectors in $\mathbb{R}^n$, and let $\langle\cdot,\cdot\rangle$ be an inner product in $\mathbb{R}^n$. Explain that the $k \times k$-matrix $\boldsymbol{N}$ with entries $n_{ij} = \langle \boldsymbol{v}_i, \boldsymbol{v}_j\rangle$ is symmetric positive definite.

b) Let $\mathbb{W} \subset \mathbb{R}^n$ be any linear subspace. Show that there is one and only one vector $\hat{\boldsymbol{x}} \in \mathbb{W}$ so that

$$\boldsymbol{w}^T\boldsymbol{A}^T\boldsymbol{A}\hat{\boldsymbol{x}} = \boldsymbol{w}^T\boldsymbol{A}^T\boldsymbol{b}, \qquad \text{for all } \boldsymbol{w} \in \mathbb{W},$$

and that $\hat{\boldsymbol{x}}$ satisfies

$$\|\boldsymbol{b} - \boldsymbol{A}\hat{\boldsymbol{x}}\|_2 \le \|\boldsymbol{b} - \boldsymbol{A}\boldsymbol{w}\|_2, \qquad \text{for all } \boldsymbol{w} \in \mathbb{W}.$$

c) In the rest of this problem we consider the situation above, but where the vector space $\mathbb{W}$ is taken to be the Krylov space

$$\mathbb{W}_k := \text{span}(\boldsymbol{b}, \boldsymbol{A}\boldsymbol{b}, \ldots, \boldsymbol{A}^{k-1}\boldsymbol{b}).$$

We use the inner product in $\mathbb{R}^n$ given by

$$\langle \boldsymbol{v}, \boldsymbol{w}\rangle_A := \boldsymbol{v}^T\boldsymbol{A}^T\boldsymbol{A}\boldsymbol{w}, \qquad \boldsymbol{v}, \boldsymbol{w} \in \mathbb{R}^n.$$

---

[4] This system known as the **normal equations** appears in linear least squares problems and was considered in this context in Chap. 9.

The associated approximations of $\boldsymbol{x}$, corresponding to $\hat{\boldsymbol{x}}$ in $\mathbb{W}_k$, are then denoted $\boldsymbol{x}_k$. Assume that $\boldsymbol{x}_k \in \mathbb{W}_k$ is already determined. In addition, assume that we already have computed a "search direction" $\boldsymbol{p}_k \in \mathbb{W}_{k+1}$ such that $\|\boldsymbol{A}\boldsymbol{p}_k\|_2 = \|\boldsymbol{p}_k\|_A = 1$, and such that

$$\langle \boldsymbol{p}_k, \boldsymbol{w}\rangle_A = 0, \quad \text{for all } \boldsymbol{w} \in \mathbb{W}_k.$$

Show that $\boldsymbol{x}_{k+1} = \boldsymbol{x}_k + \alpha_k \boldsymbol{p}_k$ for a suitable $\alpha_k \in \mathbb{R}$, and express α_k in terms of the residual $\boldsymbol{r}_k := \boldsymbol{b} - \boldsymbol{A}\boldsymbol{x}_k$, and $\boldsymbol{p}_k$.

d) Assume that $\boldsymbol{A}$ is symmetric, but not necessarily positive definite. Assume further that the vectors $\boldsymbol{p}_{k-2}$, $\boldsymbol{p}_{k-1}$, and $\boldsymbol{p}_k$ are already known with properties as above. Show that

$$\boldsymbol{A}\boldsymbol{p}_{k-1} \in \operatorname{span}(\boldsymbol{p}_{k-2}, \boldsymbol{p}_{k-1}, \boldsymbol{p}_k).$$

Use this to suggest how the search vectors $\boldsymbol{p}_k$ can be computed recursively.

13.7.3 Exercises Sect. 13.3

Exercise 13.12 (Krylov Space and cg Iterations) Consider the linear system $\boldsymbol{A}\boldsymbol{x} = \boldsymbol{b}$ where

$$\boldsymbol{A} = \begin{bmatrix} 2 & -1 & 0 \\ -1 & 2 & -1 \\ 0 & -1 & 2 \end{bmatrix}, \quad \text{and} \quad \boldsymbol{b} = \begin{bmatrix} 4 \\ 0 \\ 0 \end{bmatrix}.$$

a) Determine the vectors defining the Krylov spaces for $k \le 3$ taking as initial approximation $\boldsymbol{x} = \boldsymbol{0}$. Answer: $[\boldsymbol{b}, \boldsymbol{A}\boldsymbol{b}, \boldsymbol{A}^2\boldsymbol{b}] = \begin{bmatrix} 4 & 8 & 20 \\ 0 & -4 & -16 \\ 0 & 0 & 4 \end{bmatrix}$.

b) Carry out three CG-iterations on $\boldsymbol{A}\boldsymbol{x} = \boldsymbol{b}$. Answer:

$$[\boldsymbol{x}_0, \boldsymbol{x}_1, \boldsymbol{x}_2, \boldsymbol{x}_3] = \begin{bmatrix} 0 & 2 & 8/3 & 3 \\ 0 & 0 & 4/3 & 2 \\ 0 & 0 & 0 & 1 \end{bmatrix},$$

$$[\boldsymbol{r}_0, \boldsymbol{r}_1, \boldsymbol{r}_2, \boldsymbol{r}_3] = \begin{bmatrix} 4 & 0 & 0 & 0 \\ 0 & 2 & 0 & 0 \\ 0 & 0 & 4/3 & 0 \end{bmatrix},$$

$$[\boldsymbol{A}\boldsymbol{p}_0, \boldsymbol{A}\boldsymbol{p}_1, \boldsymbol{A}\boldsymbol{p}_2] = \begin{bmatrix} 8 & 0 & 0 \\ -4 & 3 & 0 \\ 0 & -2 & 16/9 \end{bmatrix},$$

$$[\boldsymbol{p}_0, \boldsymbol{p}_1, \boldsymbol{p}_2, \boldsymbol{p}_3] = \begin{bmatrix} 4 & 1 & 4/9 & 0 \\ 0 & 2 & 8/9 & 0 \\ 0 & 0 & 12/9 & 0 \end{bmatrix},$$

c) Verify that

- $\dim(\mathbb{W}_k) = k$ for $k = 0, 1, 2, 3$.
- $\boldsymbol{x}_3$ is the exact solution of $\boldsymbol{A}\boldsymbol{x} = \boldsymbol{b}$.
- $\boldsymbol{r}_0, \ldots, \boldsymbol{r}_{k-1}$ is an orthogonal basis for $\mathbb{W}_k$ for $k = 1, 2, 3$.
- $\boldsymbol{p}_0, \ldots, \boldsymbol{p}_{k-1}$ is an $\boldsymbol{A}$-orthogonal basis for $\mathbb{W}_k$ for $k = 1, 2, 3$.
- $\{\|\boldsymbol{r}_k\|$ is monotonically decreasing.
- $\{\|\boldsymbol{x}_k - \boldsymbol{x}\|$ is monotonically decreasing.

Exercise 13.13 (Antisymmetric System (Exam Exercise 1983-3)) In this and the next exercise $\langle \boldsymbol{x}, \boldsymbol{y}\rangle = \boldsymbol{x}^T\boldsymbol{y}$ is the usual inner product in $\mathbb{R}^n$. We note that

$$\langle \boldsymbol{x}, \boldsymbol{y}\rangle = \langle \boldsymbol{y}, \boldsymbol{x}\rangle, \quad \boldsymbol{x}, \boldsymbol{y} \in \mathbb{R}^n, \tag{13.61}$$

$$\langle \boldsymbol{C}\boldsymbol{x}, \boldsymbol{y}\rangle = \langle \boldsymbol{y}, \boldsymbol{C}^T\boldsymbol{x}\rangle \overset{(13.61)}{=} \langle \boldsymbol{C}^T\boldsymbol{x}, \boldsymbol{y}\rangle, \quad \boldsymbol{x}, \boldsymbol{y} \in \mathbb{R}^n, \boldsymbol{C} \in \mathbb{R}^{n\times n}. \tag{13.62}$$

Let $\boldsymbol{B} \in \mathbb{R}^{n\times n}$ be an antisymmetric matrix, i.e., $\boldsymbol{B}^T = -\boldsymbol{B}$, and let $\boldsymbol{A} := \boldsymbol{I} - \boldsymbol{B}$, where $\boldsymbol{I}$ is the unit matrix in $\mathbb{R}^n$.

a) Show that

$$\langle \boldsymbol{B}\boldsymbol{x}, \boldsymbol{x}\rangle = 0, \quad \boldsymbol{x} \in \mathbb{R}^n, \tag{13.63}$$

$$\langle \boldsymbol{A}\boldsymbol{x}, \boldsymbol{x}\rangle = \langle \boldsymbol{x}, \boldsymbol{x}\rangle = \|\boldsymbol{x}\|_2^2. \tag{13.64}$$

b) Show that $\|\boldsymbol{A}\boldsymbol{x}\|_2^2 = \|\boldsymbol{x}\|_2^2 + \|\boldsymbol{B}\boldsymbol{x}\|_2^2$ and that $\|\boldsymbol{A}\|_2 = \sqrt{1 + \|\boldsymbol{B}\|_2^2}$.

c) Show that $\boldsymbol{A}$ is nonsingular,

$$\|\boldsymbol{A}^{-1}\|_2 = \max_{\boldsymbol{x}\neq \boldsymbol{0}} \frac{\|\boldsymbol{x}\|_2}{\|\boldsymbol{A}\boldsymbol{x}\|_2},$$

and $\|\boldsymbol{A}\|_2 \le 1$.

d) Let $1 \le k \le n$, $\mathcal{W} = \operatorname{span}(\boldsymbol{w}_1, \ldots, \boldsymbol{w}_k)$ a k-dimensional subspace of $\mathbb{R}^n$ and $\boldsymbol{b} \in \mathbb{R}^n$. Show that if $\boldsymbol{x} \in \mathcal{W}$ is such that

$$\langle \boldsymbol{A}\boldsymbol{x}, \boldsymbol{w}\rangle = \langle \boldsymbol{b}, \boldsymbol{w}\rangle \text{ for all } \boldsymbol{w} \in \mathcal{W}, \tag{13.65}$$

then $\|\boldsymbol{x}\|_2 \le \|\boldsymbol{b}\|_2$.

With $\boldsymbol{x} := \sum_{j=1}^{k} x_j \boldsymbol{w}_j$ the problem (13.65) is equivalent to finding real numbers $x_1, \ldots, x_k$ solving the linear system

$$\sum_{j=1}^{k} x_j \langle \boldsymbol{A}\boldsymbol{w}_j, \boldsymbol{w}_i \rangle = \langle \boldsymbol{b}, \boldsymbol{w}_i \rangle, \quad i = 1, \ldots, k. \tag{13.66}$$

Show that (13.65) has a unique solution $\boldsymbol{x} \in \mathcal{W}$.

e) Let $\boldsymbol{x}^* := \boldsymbol{A}^{-1}\boldsymbol{b}$. Show that

$$\|\boldsymbol{x}^* - \boldsymbol{x}\|_2 \leq \|\boldsymbol{A}\|_2 \min_{\boldsymbol{w}\in\mathcal{W}} \|\boldsymbol{x}^* - \boldsymbol{w}\|_2. \tag{13.67}$$

Exercise 13.14 (cg Antisymmetric System (Exam Exercise 1983-4)) (It is recommended to study Exercise 13.13 before starting this exercise.) As in Exercise 13.13 let $\boldsymbol{B} \in \mathbb{R}^{n\times n}$ be an antisymmetric matrix, i.e., $\boldsymbol{B}^T = -\boldsymbol{B}$, let $\langle \boldsymbol{x}, \boldsymbol{y} \rangle = \boldsymbol{x}^T\boldsymbol{y}$ be the usual inner product in $\mathbb{R}^n$, let $\boldsymbol{A} := \boldsymbol{I} - \boldsymbol{B}$, where $\boldsymbol{I}$ is the unit matrix in $\mathbb{R}^n$ and $\boldsymbol{b} \in \mathbb{R}^n$. The purpose of this exercise is to develop an iterative algorithm for the linear system $\boldsymbol{A}\boldsymbol{x} = \boldsymbol{b}$. The algorithm is partly built on the same idea as for the conjugate gradient method for positive definite systems.

Let $\boldsymbol{x}_0 = 0$ be the initial approximation to the exact solution $\boldsymbol{x}^* := \boldsymbol{A}^{-1}\boldsymbol{b}$. For $k = 1, 2, \ldots, n$ we let

$$\mathcal{W}_k := \operatorname{span}(\boldsymbol{b}, \boldsymbol{B}\boldsymbol{b}, \ldots, \boldsymbol{B}^{k-1}\boldsymbol{b}).$$

For $k = 1, 2, \ldots, n$ we define $\boldsymbol{x}_k \in \mathcal{W}_k$ by

$$\langle \boldsymbol{A}\boldsymbol{x}_k, \boldsymbol{w} \rangle = \langle \boldsymbol{b}, \boldsymbol{w} \rangle, \text{ for all } \boldsymbol{w} \in \mathcal{W}_k.$$

The vector $\boldsymbol{x}_k$ is uniquely determined as shown in Exercise 13.13d) and that it is a "good" approximation to $\boldsymbol{x}*$ follows from (13.67). In this exercise we will derive a recursive algorithm to determine $\boldsymbol{x}_k$.

For $k = 0, \ldots, n$ we set

$$\boldsymbol{r}_k := \boldsymbol{b} - \boldsymbol{A}\boldsymbol{x}_k, \text{ and } \rho_k := \|\boldsymbol{r}_k\|_2^2.$$

Let $m \in \mathbb{N}$ be such that

$$\rho_k \neq 0, \quad k = 0, \ldots, m.$$

Let $\omega_0, \omega_1, \ldots, \omega_m$ be real numbers defined recursively for $k = 1, 2, \ldots, m$ by

$$\omega_k := \begin{cases} 1, & \text{if } k = 0 \\ (1 + \omega_{k-1}^{-1}\rho_k/\rho_{k-1})^{-1}, & \text{otherwise.} \end{cases} \tag{13.68}$$

We will show below that $\boldsymbol{x}_k$ and $\boldsymbol{r}_k$ satisfy the following recurrence relations for $k = 0, 1, \ldots, m-1$

$$\boldsymbol{x}_{k+1} = (1 - \omega_k)\boldsymbol{x}_{k-1} + \omega_k(\boldsymbol{x}_k + \boldsymbol{r}_k), \tag{13.69}$$

$$\boldsymbol{r}_{k+1} = (1 - \omega_k)\boldsymbol{r}_{k-1} + \omega_k \boldsymbol{B}\boldsymbol{r}_k, \tag{13.70}$$

starting with $\boldsymbol{x}_0 = \boldsymbol{x}_{-1} = 0$ and $\boldsymbol{r}_0 = \boldsymbol{r}_{-1} = \boldsymbol{b}$.

a) Show that $0 < \omega_k < 1$ for $k = 1, 2, \ldots, m$.
b) Explain briefly how to define an iterative algorithm for determining $\boldsymbol{x}_k$ using the formulas (13.68), (13.69), (13.70) and estimate the number of arithmetic operations in each iteration.
c) Show that $\langle \boldsymbol{r}_k, \boldsymbol{r}_j \rangle = 0$ for $j = 0, 1, \ldots, k-1$.
d) Show that if $k \le m+1$ then $\mathcal{W}_k = \text{span}(\boldsymbol{r}_0, \boldsymbol{r}_1, \ldots, \boldsymbol{r}_{k-1})$ and $\dim \mathcal{W}_k = k$.
e) Show that if $1 \le k \le m-1$ then

$$\boldsymbol{B}\boldsymbol{r}_k = \alpha_k \boldsymbol{r}_{k+1} + \beta_k \boldsymbol{r}_{k-1}, \tag{13.71}$$

where $\alpha_k := \langle \boldsymbol{B}\boldsymbol{r}_k, \boldsymbol{r}_{k+1} \rangle / \rho_{k+1}$ and $\beta_k := \langle \boldsymbol{B}\boldsymbol{r}_k, \boldsymbol{r}_{k-1} \rangle / \rho_{k-1}$.
f) Define $\alpha_0 := \langle \boldsymbol{B}\boldsymbol{r}_0, \boldsymbol{r}_1 \rangle / \rho_1$ and show that $\alpha_0 = 1$.
g) Show that if $1 \le k \le m-1$ then $\beta_k = -\alpha_{k-1}\rho_k / \rho_{k-1}$.
h) Show that[5]

$$\langle \boldsymbol{r}_{k+1}, \boldsymbol{A}^{-1}\boldsymbol{r}_{k+1} \rangle = \langle \boldsymbol{r}_{k+1}, \boldsymbol{A}^{-1}\boldsymbol{r}_j \rangle, \quad j = 0, 1, \ldots, k. \tag{13.72}$$

i) Use (13.71) and (13.72) to show that $\alpha_k + \beta_k = 1$ for $k = 1, 2, \ldots, m-1$.
j) Show that $\alpha_k \ge 1$ for $k = 1, 2, \ldots, m-1$.
k) Show that $\boldsymbol{x}_k$, $\boldsymbol{r}_k$ and $\boldsymbol{\omega}_k$ satisfy the recurrence relations (13.68), (13.69) and (13.70).

13.7.4 Exercises Sect. 13.4

Exercise 13.15 (Another Explicit Formula for the Chebyshev Polynomial) Show that

$$T_n(t) = \cosh(n \operatorname{arccosh} t) \text{ for } t \ge 1,$$

where arccosh is the inverse function of $\cosh x := (e^x + e^{-x})/2$.

[5] Hint: Show that $\boldsymbol{A}^{-1}(\boldsymbol{r}_{k+1} - \boldsymbol{r}_j) \in \mathcal{W}_{k+1}$.

Exercise 13.16 (Maximum of a Convex Function) Show that if $f : [a, b] \to \mathbb{R}$ is convex then $\max_{a \le x \le b} f(x) \le \max\{f(a), f(b)\}$.

13.7.5 Exercises Sect. 13.5

Exercise 13.17 (Variable Coefficient) For $m = 2$, show that (13.57) takes the form

$$\boldsymbol{A}\boldsymbol{x} = \begin{bmatrix} a_{1,1} & -c_{\frac{3}{2},1} & -c_{1,\frac{3}{2}} & 0 \\ -c_{\frac{3}{2},1} & a_{2,2} & 0 & -c_{2,\frac{3}{2}} \\ -c_{1,\frac{3}{2}} & 0 & a_{3,3} & -c_{\frac{3}{2},2} \\ 0 & -c_{2,\frac{3}{2}} & -c_{\frac{3}{2},2} & a_{4,4} \end{bmatrix} \begin{bmatrix} v_{1,1} \\ v_{2,1} \\ v_{1,2} \\ v_{2,2} \end{bmatrix} = \begin{bmatrix} (dv)_{1,1} \\ (dv)_{2,1} \\ (dv)_{1,2} \\ (dv)_{2,2} \end{bmatrix},$$

where

$$\begin{bmatrix} a_{1,1} \\ a_{2,2} \\ a_{3,3} \\ a_{4,4} \end{bmatrix} = \begin{bmatrix} c_{\frac{1}{2},1} + c_{1,\frac{1}{2}} + c_{1,\frac{3}{2}} + c_{\frac{3}{2},1} \\ c_{\frac{3}{2},1} + c_{2,\frac{1}{2}} + c_{2,\frac{3}{2}} + c_{\frac{5}{2},1} \\ c_{\frac{1}{2},2} + c_{1,\frac{3}{2}} + c_{1,\frac{5}{2}} + c_{\frac{3}{2},2} \\ c_{\frac{3}{2},2} + c_{2,\frac{3}{2}} + c_{2,\frac{5}{2}} + c_{\frac{5}{2},2} \end{bmatrix}.$$

Show that the matrix $\boldsymbol{A}$ is symmetric, and if $c(x, y) > 0$ for all $(x, y) \in \Omega$ then it is strictly diagonally dominant.

13.8 Review Questions

13.8.1 Does the steepest descent and conjugate gradient method always converge?
13.8.2 What kind of orthogonalities occur in the conjugate gradient method?
13.8.3 What is a Krylov space?
13.8.4 What is a convex function?
13.8.5 How do SOR and conjugate gradient compare?

Part VI
Eigenvalues and Eigenvectors

In this and the next chapter we briefly give some numerical methods for finding one or more eigenvalues and eigenvectors of a matrix. Both Hermitian and non hermitian matrices are considered.

But first we consider a location result for eigenvalues and then give a useful upper bound for how much an eigenvalue can change when the elements of the matrix is perturbed.

Chapter 14
Numerical Eigenvalue Problems

14.1 Eigenpairs

Consider the eigenpair problem for some classes of matrices $\boldsymbol{A} \in \mathbb{C}^{n \times n}$.

Diagonal Matrices The eigenpairs are easily determined. Since $\boldsymbol{A}\boldsymbol{e}_i = a_{ii}\boldsymbol{e}_i$ the eigenpairs are $(\lambda_i, \boldsymbol{e}_i)$, where $\lambda_i = a_{ii}$ for $i = 1, \ldots, n$. Moreover, the eigenvectors of $\boldsymbol{A}$ are linearly independent.

Triangular Matrices Suppose $\boldsymbol{A}$ is upper or lower triangular. Consider finding the eigenvalues Since $\det(\boldsymbol{A} - \lambda \boldsymbol{I}) = \prod_{i=1}^{n}(a_{ii} - \lambda)$ the eigenvalues are $\lambda_i = a_{ii}$ for $i = 1, \ldots, n$, the diagonal elements of $\boldsymbol{A}$. To determine the eigenvectors can be more challenging since $\boldsymbol{A}$ can be defective, i.e., the eigenvectors are not necessarily linearly independent, cf. Chap. 6.

Block Diagonal Matrices Suppose

$$\boldsymbol{A} = \operatorname{diag}(\boldsymbol{A}_1, \boldsymbol{A}_2, \ldots, \boldsymbol{A}_r), \quad \boldsymbol{A}_i \in \mathbb{C}^{m_i \times m_i}.$$

Here the eigenpair problem reduces to r smaller problems. Let $\boldsymbol{A}_i\boldsymbol{X}_i = \boldsymbol{X}_i\boldsymbol{D}_i$ define the eigenpairs of $\boldsymbol{A}_i$ for $i = 1, \ldots, r$ and let $\boldsymbol{X} := \operatorname{diag}(\boldsymbol{X}_1, \ldots, \boldsymbol{X}_r)$, $\boldsymbol{D} := \operatorname{diag}(\boldsymbol{D}_1, \ldots, \boldsymbol{D}_r)$. Then the eigenpairs for $\boldsymbol{A}$ are given by

$$\begin{aligned}\boldsymbol{A}\boldsymbol{D} &= \operatorname{diag}(\boldsymbol{A}_1, \ldots, \boldsymbol{A}_r)\operatorname{diag}(\boldsymbol{X}_1, \ldots, \boldsymbol{X}_r) = \operatorname{diag}(\boldsymbol{A}_1\boldsymbol{X}_1, \ldots, \boldsymbol{A}_r\boldsymbol{X}_r) \\ &= \operatorname{diag}(\boldsymbol{X}_1\boldsymbol{D}_1, \ldots, \boldsymbol{X}_r\boldsymbol{D}_r) = \boldsymbol{X}\boldsymbol{D}.\end{aligned}$$

T. Lyche, *Numerical Linear Algebra and Matrix Factorizations*,
Texts in Computational Science and Engineering 22,
https://doi.org/10.1007/978-3-030-36468-7_14

Block Triangular matrices Matrices Let $\boldsymbol{A}_{11}, \boldsymbol{A}_{22}, \ldots, \boldsymbol{A}_{rr}$ be the diagonal blocks of $\boldsymbol{A}$. By Property 8. of determinants

$$\det(\boldsymbol{A} - \lambda \boldsymbol{I}) = \prod_{i=1}^{r} \det(\boldsymbol{A}_{ii} - \lambda \boldsymbol{I})$$

and the eigenvalues are found from the eigenvalues of the diagonal blocks.

In this and the next chapter we consider some numerical methods for finding one or more of the eigenvalues and eigenvectors of a matrix $\boldsymbol{A} \in \mathbb{C}^{n\times n}$. Maybe the first method which comes to mind is to form the characteristic polynomial $\pi_{\boldsymbol{A}}$ of $\boldsymbol{A}$, and then use a polynomial root finder, like Newton's method to determine one or several of the eigenvalues.

It turns out that this is not suitable as an all purpose method. One reason is that a small change in one of the coefficients of $\pi_{\boldsymbol{A}}(\lambda)$ can lead to a large change in the roots of the polynomial. For example, if $\pi_{\boldsymbol{A}}(\lambda :) = \lambda^{16}$ and $q(\lambda) = \lambda^{16} - 10^{-16}$ then the roots of $\pi_{\boldsymbol{A}}$ are all equal to zero, while the roots of q are $\lambda_j = 10^{-1} e^{2\pi i j/16}$, $j = 1, \ldots, 16$. The roots of q have absolute value 0.1 and a perturbation in one of the polynomial coefficients of magnitude 10^{-16} has led to an error in the roots of approximately 0.1. The situation can be somewhat remedied by representing the polynomials using a different basis.

In this text we will only consider methods which work directly with the matrix. But before that, in Sect. 14.3 we consider how much the eigenvalues change when the elements in the matrix are perturbed. We start with a simple but useful result for locating the eigenvalues.

14.2 Gershgorin's Theorem

The following theorem is useful for locating eigenvalues of an arbitrary square matrix.

Theorem 14.1 (Gershgorin's Circle Theorem) *Suppose* $\boldsymbol{A} \in \mathbb{C}^{n\times n}$. *Define for* $i = 1, 2, \ldots, n$

$$R_i = \{z \in \mathbb{C} : |z - a_{ii}| \le r_i\}, \quad r_i := \sum_{\substack{j=1 \\ j\neq i}}^{n} |a_{ij}|,$$

$$C_j = \{z \in \mathbb{C} : |z - a_{jj}| \le c_j\}, \quad c_j := \sum_{\substack{i=1 \\ i\neq j}}^{n} |a_{ij}|.$$

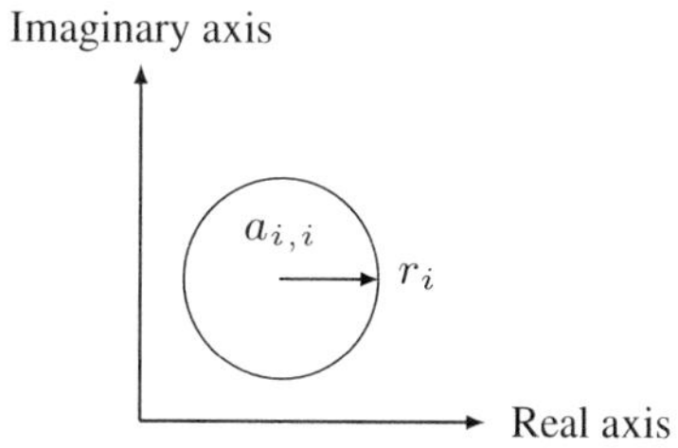

Fig. 14.1 The Gershgorin disk R_i

Then any eigenvalue of $\boldsymbol{A}$ *lies in* $R \cap C$ *where* $R = R_1 \cup R_2 \cup \cdots \cup R_n$ *and* $C = C_1 \cup C_2 \cup \cdots \cup C_n$.

Proof Suppose $(\lambda, \boldsymbol{x})$ is an eigenpair for $\boldsymbol{A}$. We claim that $\lambda \in R_i$, where i is such that $|x_i| = \|\boldsymbol{x}\|_\infty$. Indeed, $\boldsymbol{A}\boldsymbol{x} = \lambda\boldsymbol{x}$ implies that $\sum_j a_{ij}x_j = \lambda x_i$ or $(\lambda - a_{ii})x_i = \sum_{j \neq i} a_{ij}x_j$. Dividing by x_i and taking absolute values we find

$$|\lambda - a_{ii}| = |\sum_{j \neq i} a_{ij}x_j/x_i| \leq \sum_{j \neq i} |a_{ij}||x_j/x_i| \leq r_i$$

since $|x_j/x_i| \leq 1$ for all j. Thus $\lambda \in R_i$.

Since λ is also an eigenvalue of $\boldsymbol{A}^T$, it must be in one of the row disks of $\boldsymbol{A}^T$. But these are the column disks C_j of $\boldsymbol{A}$. Hence $\lambda \in C_j$ for some j. □

The set R_i is a subset of the complex plane consisting of all points inside a circle with center at a_{ii} and radius r_i, c.f. Fig. 14.1. R_i is called a (Gerschgorin) row disk.

An eigenvalue λ lies in the union of the row disks $R_1, \ldots, R_n$ and also in the union of the column disks $C_1, \ldots, C_n$. If $\boldsymbol{A}$ is Hermitian then $R_i = C_i$ for $i = 1, 2, \ldots, n$. Moreover, in this case the eigenvalues of $\boldsymbol{A}$ are real, and the Gerschgorin disks can be taken to be intervals on the real line.

Example 14.1 (Gershgorin) Let $\boldsymbol{T} = \text{tridiag}(-1, 2, -1) \in \mathbb{R}^{m \times m}$ be the second derivative matrix. Since $\boldsymbol{A}$ is Hermitian we have $R_i = C_i$ for all i and the eigenvalues are real. We find $R_1 = R_m = \{z \in \mathbb{R} : |z - 2| \leq 1\}$ and

$$R_i = \{z \in \mathbb{R} : |z - 2| \leq 2\}, \quad i = 2, 3, \ldots, m - 1.$$

We conclude that $\lambda \in [0, 4]$ for any eigenvalue λ of $\boldsymbol{T}$. To check this, we recall that by Lemma 2.2 the eigenvalues of $\boldsymbol{T}$ are given by

$$\lambda_j = 4\left[\sin \frac{j\pi}{2(m+1)}\right]^2, \quad j = 1, 2, \ldots, m.$$

When m is large the smallest eigenvalue $4\left[\sin\frac{\pi}{2(m+1)}\right]^2$ is very close to zero and the largest eigenvalue $4\left[\sin\frac{m\pi}{2(m+1)}\right]^2$ is very close to 4. Thus Gerschgorin's theorem gives a remarkably good estimate for large m.

Sometimes some of the Gerschgorin disks are distinct and we have

Corollary 14.1 (Disjoint Gershgorin Disks) *If p of the Gershgorin row disks are disjoint from the others, the union of these disks contains precisely p eigenvalues. The same result holds for the column disks.*

Proof Consider a family of matrices

$$\boldsymbol{A}(t) := \boldsymbol{D} + t(\boldsymbol{A} - \boldsymbol{D}), \quad \boldsymbol{D} := \operatorname{diag}(a_{11}, \ldots, a_{nn}), \quad t \in [0, 1].$$

We have $\boldsymbol{A}(0) = \boldsymbol{D}$ and $\boldsymbol{A}(1) = \boldsymbol{A}$. As a function of t, every eigenvalue of $\boldsymbol{A}(t)$ is a continuous function of t. This follows from Theorem 14.2, see Exercise 14.5. The row disks $R_i(t)$ of $\boldsymbol{A}(t)$ have radius proportional to t, indeed

$$R_i(t) = \{z \in \mathbb{C} : |z - a_{ii}| \le t r_i\}, \quad r_i := \sum_{\substack{j=1\\ j\neq i}}^{n} |a_{ij}|.$$

Clearly $0 \le t_1 < t_2 \le 1$ implies $R_i(t_1) \subset R_i(t_2)$ and $R_i(1)$ is a row disk of $\boldsymbol{A}$ for all i. Suppose $\bigcup_{k=1}^{p} R_{i_k}(1)$ are disjoint from the other disks of $\boldsymbol{A}$ and set $R^p(t) := \bigcup_{k=1}^{p} R_{i_k}(t)$ for $t \in [0, 1]$. Now $R^p(0)$ contains only the p eigenvalues $a_{i_1,i_1}, \ldots, a_{i_p,i_p}$ of $\boldsymbol{A}(0) = \boldsymbol{D}$. As t increases from zero to one the set $R^p(t)$ is disjoint from the other row disks of $\boldsymbol{A}$ and by the continuity of the eigenvalues cannot loose or gain eigenvalues. It follows that $R^p(1)$ must contain p eigenvalues of $\boldsymbol{A}$. □

Example 14.2 Consider the matrix $\boldsymbol{A} = \begin{bmatrix} 1 & \epsilon_1 & \epsilon_2 \\ \epsilon_3 & 2 & \epsilon_4 \\ \epsilon_5 & \epsilon_6 & 3 \end{bmatrix}$, where $|\epsilon_i| \le 10^{-15}$ all i. By Corollary 14.1 the eigenvalues $\lambda_1, \lambda_2, \lambda_3$ of $\boldsymbol{A}$ are distinct and satisfy $|\lambda_j - j| \le 2 \times 10^{-15}$ for $j = 1, 2, 3$.

14.3 Perturbation of Eigenvalues

In this section we study the following problem. Given matrices $\boldsymbol{A}, \boldsymbol{E} \in \mathbb{C}^{n\times n}$, where we think of $\boldsymbol{E}$ as a perturbation of $\boldsymbol{A}$. By how much do the eigenvalues of $\boldsymbol{A}$ and $\boldsymbol{A}+\boldsymbol{E}$ differ? Not surprisingly this problem is more complicated than the corresponding problem for linear systems.

We illustrate this by considering two examples. Suppose $\boldsymbol{A}_0 := \boldsymbol{0}$ is the zero matrix. If $\lambda \in \sigma(\boldsymbol{A}_0 + \boldsymbol{E}) = \sigma(\boldsymbol{E})$, then $|\lambda| \le \|\boldsymbol{E}\|_\infty$ by Theorem 12.11, and any zero eigenvalue of A_0 is perturbed by at most $\|E\|_\infty$. On the other hand consider for $\epsilon > 0$ the matrices

$$A_1 := \begin{bmatrix} 0 & 1 & 0 & \cdots & 0 & 0 \\ 0 & 0 & 1 & \cdots & 0 & 0 \\ \vdots & \vdots & \vdots & & \vdots & \vdots \\ 0 & 0 & 0 & \cdots & 0 & 1 \\ 0 & 0 & 0 & \cdots & 0 & 0 \end{bmatrix}, \quad E := \begin{bmatrix} 0 & 0 & 0 & \cdots & 0 & 0 \\ 0 & 0 & 0 & \cdots & 0 & 0 \\ \vdots & \vdots & \vdots & & \vdots & \vdots \\ 0 & 0 & 0 & \cdots & 0 & 0 \\ \epsilon & 0 & 0 & \cdots & 0 & 0 \end{bmatrix} = \epsilon e_n e_1^T.$$

The characteristic polynomial of $A_1 + E$ is $\pi(\lambda) := (-1)^n(\lambda^n - \epsilon)$, and the zero eigenvalues of A_1 are perturbed by the amount $|\lambda| = \|E\|_\infty^{1/n}$. Thus, for $n = 16$, a perturbation of say $\epsilon = 10^{-16}$ gives a change in eigenvalue of 0.1.

The following theorem shows that a dependence $\|E\|_\infty^{1/n}$ is the worst that can happen.

Theorem 14.2 (Elsner's Theorem (1985)) *Suppose $A, E \in \mathbb{C}^{n\times n}$. To every $\mu \in \sigma(A+E)$ there is a $\lambda \in \sigma(A)$ such that*

$$|\mu - \lambda| \le K\|E\|_2^{1/n}, \quad K = \big(\|A\|_2 + \|A+E\|_2\big)^{1-1/n}. \tag{14.1}$$

Proof Suppose A has eigenvalues $\lambda_1, \ldots, \lambda_n$ and let λ_1 be one which is closest to μ. Let u_1 with $\|u_1\|_2 = 1$ be an eigenvector corresponding to μ, and extend u_1 to an orthonormal basis $\{u_1, \ldots, u_n\}$ of $\mathbb{C}^n$. Note that

$$\|(\mu I - A)u_1\|_2 = \|(A+E)u_1 - Au_1\|_2 = \|Eu_1\|_2 \le \|E\|_2,$$

$$\prod_{j=2}^{n} \|(\mu I - A)u_j\|_2 \le \prod_{j=2}^{n} (|\mu| + \|Au_j\|_2) \le \big(\|(A+E)\|_2 + \|A\|_2\big)^{n-1}.$$

Using this and Hadamard's inequality (5.21) we find

$$|\mu - \lambda_1|^n \le \prod_{j=1}^{n} |\mu - \lambda_j| = |\det(\mu I - A)| = \big|\det\big((\mu I - A)[u_1, \ldots, u_n]\big)\big|$$

$$\le \|(\mu I - A)u_1\|_2 \prod_{j=2}^{n} \|(\mu I - A)u_j\|_2 \le \|E\|_2\big(\|(A+E)\|_2 + \|A\|_2\big)^{n-1}.$$

The result follows by taking nth roots in this inequality. □

It follows from this theorem that the eigenvalues depend continuously on the elements of the matrix. The factor $\|E\|_2^{1/n}$ shows that this dependence is almost, but not quite, differentiable. As an example, the eigenvalues of the matrix $\left[\begin{smallmatrix} 1 & 1 \\ \epsilon & 1 \end{smallmatrix}\right]$ are $1 \pm \sqrt{\epsilon}$ and these expressions are not differentiable at $\epsilon = 0$.

14.3.1 Nondefective Matrices

Recall that a matrix is nondefective if the eigenvectors form a basis for $\mathbb{C}^n$. For nondefective matrices we can get rid of the annoying exponent $1/n$ in $\|\boldsymbol{E}\|_2$ in (14.1). For a more general discussion than the one in the following theorem see [19].

Theorem 14.3 (Absolute Errors) *Suppose $\boldsymbol{A} \in \mathbb{C}^{n\times n}$ has linearly independent eigenvectors $\{\boldsymbol{x}_1, \ldots, \boldsymbol{x}_n\}$ and let $\boldsymbol{X} = [\boldsymbol{x}_1, \ldots, \boldsymbol{x}_n]$ be the eigenvector matrix. To any $\mu \in \mathbb{C}$ and $\boldsymbol{x} \in \mathbb{C}^n$ with $\|\boldsymbol{x}\|_p = 1$ we can find an eigenvalue λ of $\boldsymbol{A}$ such that*

$$|\lambda - \mu| \leq K_p(\boldsymbol{X})\|\boldsymbol{r}\|_p, \quad 1 \leq p \leq \infty, \tag{14.2}$$

where $\boldsymbol{r} := \boldsymbol{A}\boldsymbol{x} - \mu\boldsymbol{x}$ and $K_p(\boldsymbol{X}) := \|\boldsymbol{X}\|_p\, \|\boldsymbol{X}^{-1}\|_p$. If for some $\boldsymbol{E} \in \mathbb{C}^{n\times n}$ it holds that $(\mu, \boldsymbol{x})$ is an eigenpair for $\boldsymbol{A} + \boldsymbol{E}$, then we can find an eigenvalue λ of $\boldsymbol{A}$ such that

$$|\lambda - \mu| \leq K_p(\boldsymbol{X})\|\boldsymbol{E}\|_p, \quad 1 \leq p \leq \infty, \tag{14.3}$$

Proof If $\mu \in \sigma(\boldsymbol{A})$ then we can take $\lambda = \mu$ and (14.2), (14.3) hold trivially. So assume $\mu \notin \sigma(\boldsymbol{A})$. Since $\boldsymbol{A}$ is nondefective it can be diagonalized, we have $\boldsymbol{A} = \boldsymbol{X}\boldsymbol{D}\boldsymbol{X}^{-1}$, where $\boldsymbol{D} = \operatorname{diag}(\lambda_1, \ldots, \lambda_n)$ and $(\lambda_j, \boldsymbol{x}_j)$ are the eigenpairs of $\boldsymbol{A}$ for $j = 1, \ldots, n$. Define $\boldsymbol{D}_1 := \boldsymbol{D} - \mu\boldsymbol{I}$. Then $\boldsymbol{D}_1^{-1} = \operatorname{diag}\big((\lambda_1 - \mu)^{-1}, \ldots, (\lambda_n - \mu)^{-1}\big)$ exists and

$$\boldsymbol{X}\boldsymbol{D}_1^{-1}\boldsymbol{X}^{-1}\boldsymbol{r} = \big(\boldsymbol{X}(\boldsymbol{D} - \mu\boldsymbol{I})\boldsymbol{X}^{-1}\big)^{-1}\boldsymbol{r} = (\boldsymbol{A} - \mu\boldsymbol{I})^{-1}(\boldsymbol{A} - \mu\boldsymbol{I})\boldsymbol{x} = \boldsymbol{x}.$$

Using this and Lemma 14.1 below we obtain

$$1 = \|\boldsymbol{x}\|_p = \|\boldsymbol{X}\boldsymbol{D}_1^{-1}\boldsymbol{X}^{-1}\boldsymbol{r}\|_p \leq \|\boldsymbol{D}_1^{-1}\|_p K_p(\boldsymbol{X})\|\boldsymbol{r}\|_p = \frac{K_p(\boldsymbol{X})\|\boldsymbol{r}\|_p}{\min_j|\lambda_j - \mu|}.$$

But then (14.2) follows. If $(\boldsymbol{A} + \boldsymbol{E})\boldsymbol{x} = \mu\boldsymbol{x}$ then $\boldsymbol{0} = \boldsymbol{A}\boldsymbol{x} - \mu\boldsymbol{x} + \boldsymbol{E}\boldsymbol{x} = \boldsymbol{r} + \boldsymbol{E}\boldsymbol{x}$. But then $\|\boldsymbol{r}\|_p = \|-\boldsymbol{E}\boldsymbol{x}\|_p \leq \|\boldsymbol{E}\|_p$. Inserting this in (14.2) proves (14.3). □

The equation (14.3) shows that for a nondefective matrix the absolute error can be magnified by at most $K_p(\boldsymbol{X})$, the condition number of the eigenvector matrix with respect to inversion. If $K_p(\boldsymbol{X})$ is small then a small perturbation changes the eigenvalues by small amounts.

Even if we get rid of the exponent $1/n$, the equation (14.3) illustrates that it can be difficult or sometimes impossible to compute accurate eigenvalues and eigenvectors of matrices with almost linearly dependent eigenvectors. On the other hand the eigenvalue problem for normal matrices is better conditioned. Indeed, if $\boldsymbol{A}$ is normal then it has a set of orthonormal eigenvectors and the eigenvector matrix is unitary. If we restrict attention to the 2-norm then $K_2(\boldsymbol{X}) = 1$ and (14.3) implies the following result.

Theorem 14.4 (Perturbations, Normal Matrix) *Suppose $\boldsymbol{A} \in \mathbb{C}^{n\times n}$ is normal and let μ be an eigenvalue of $\boldsymbol{A}+\boldsymbol{E}$ for some $\boldsymbol{E} \in \mathbb{C}^{n\times n}$. Then we can find an eigenvalue λ of $\boldsymbol{A}$ such that $|\lambda-\mu| \le \|\boldsymbol{E}\|_2$.*

For an even stronger result for Hermitian matrices see Corollary 6.13. We conclude that the situation for the absolute error in an eigenvalue of a Hermitian matrix is quite satisfactory. Small perturbations in the elements are not magnified in the eigenvalues.

In the proof of Theorem 14.3 we used that the p-norm of a diagonal matrix is equal to its spectral radius.

Lemma 14.1 (p-Norm of a Diagonal Matrix) *If $\boldsymbol{A} = \operatorname{diag}(\lambda_1,\dots,\lambda_n)$ is a diagonal matrix then $\|\boldsymbol{A}\|_p = \rho(\boldsymbol{A})$ for $1 \le p \le \infty$.*

Proof For $p=\infty$ the proof is left as an exercise. For any $\boldsymbol{x} \in \mathbb{C}^n$ and $p<\infty$ we have

$$\|\boldsymbol{A}\boldsymbol{x}\|_p = \|[\lambda_1 x_1,\dots,\lambda_n x_n]^T\|_p = \Big(\sum_{j=1}^n |\lambda_j|^p |x_j|^p\Big)^{1/p} \le \rho(\boldsymbol{A})\|\boldsymbol{x}\|_p.$$

Thus $\|\boldsymbol{A}\|_p = \max_{\boldsymbol{x}\ne\boldsymbol{0}} \frac{\|\boldsymbol{A}\boldsymbol{x}\|_p}{\|\boldsymbol{x}\|_p} \le \rho(\boldsymbol{A})$. But from Theorem 12.11 we have $\rho(\boldsymbol{A}) \le \|\boldsymbol{A}\|_p$ and the proof is complete. □

For the accuracy of an eigenvalue of small magnitude we are interested in the size of the relative error.

Theorem 14.5 (Relative Errors) *Suppose in Theorem 14.3 that $\boldsymbol{A} \in \mathbb{C}^{n\times n}$ is nonsingular. To any $\mu \in \mathbb{C}$ and $\boldsymbol{x} \in \mathbb{C}^n$ with $\|\boldsymbol{x}\|_p = 1$, we can find an eigenvalue λ of $\boldsymbol{A}$ such that*

$$\frac{|\lambda-\mu|}{|\lambda|} \le K_p(\boldsymbol{X})K_p(\boldsymbol{A})\frac{\|\boldsymbol{r}\|_p}{\|\boldsymbol{A}\|_p}, \quad 1 \le p \le \infty, \tag{14.4}$$

where $\boldsymbol{r} := \boldsymbol{A}\boldsymbol{x} - \mu\boldsymbol{x}$. If for some $\boldsymbol{E} \in \mathbb{C}^{n\times n}$ it holds that $(\mu,\boldsymbol{x})$ is an eigenpair for $\boldsymbol{A}+\boldsymbol{E}$, then we can find an eigenvalue λ of $\boldsymbol{A}$ such that

$$\frac{|\lambda-\mu|}{|\lambda|} \le K_p(\boldsymbol{X})\|\boldsymbol{A}^{-1}\boldsymbol{E}\|_p \le K_p(\boldsymbol{X})K_p(\boldsymbol{A})\frac{\|\boldsymbol{E}\|_p}{\|\boldsymbol{A}\|_p}, \quad 1 \le p \le \infty, \tag{14.5}$$

Proof Applying Theorem 12.11 to $\boldsymbol{A}^{-1}$ we have for any $\lambda \in \sigma(\boldsymbol{A})$

$$\frac{1}{\lambda} \le \|\boldsymbol{A}^{-1}\|_p = \frac{K_p(\boldsymbol{A})}{\|\boldsymbol{A}\|_p}$$

and (14.4) follows from (14.2). To prove (14.5) we define the matrices $\boldsymbol{B} := \mu \boldsymbol{A}^{-1}$ and $\boldsymbol{F} := -\boldsymbol{A}^{-1}\boldsymbol{E}$. If $(\lambda_j, \boldsymbol{x})$ are the eigenpairs for $\boldsymbol{A}$ then $\left(\frac{\mu}{\lambda_j}, \boldsymbol{x}\right)$ are the eigenpairs for $\boldsymbol{B}$ for $j = 1, \dots, n$. Since $(\mu, \boldsymbol{x})$ is an eigenpair for $\boldsymbol{A} + \boldsymbol{E}$ we find

$$(\boldsymbol{B} + \boldsymbol{F} - \boldsymbol{I})\boldsymbol{x} = (\mu \boldsymbol{A}^{-1} - \boldsymbol{A}^{-1}\boldsymbol{E} - \boldsymbol{I})\boldsymbol{x} = \boldsymbol{A}^{-1}\big(\mu \boldsymbol{I} - (\boldsymbol{E} + \boldsymbol{A})\big)\boldsymbol{x} = \boldsymbol{0}.$$

Thus $(1, \boldsymbol{x})$ is an eigenpair for $\boldsymbol{B} + \boldsymbol{F}$. Applying Theorem 14.3 to this eigenvalue we can find $\lambda \in \sigma(\boldsymbol{A})$ such that $|\frac{\mu}{\lambda} - 1| \le K_p(\boldsymbol{X})\|\boldsymbol{F}\|_p = K_p(\boldsymbol{X})\|\boldsymbol{A}^{-1}\boldsymbol{E}\|_p$ which proves the first estimate in (14.5). The second inequality in (14.5) follows from the submultiplicativity of the p-norm. □

14.4 Unitary Similarity Transformation of a Matrix into Upper Hessenberg Form

Before attempting to find eigenvalues and eigenvectors of a matrix (exceptions are made for certain sparse matrices), it is often advantageous to reduce it by similarity transformations to a simpler form. Orthogonal or unitary similarity transformations are particularly important since they are insensitive to round-off errors in the elements of the matrix. In this section we show how this reduction can be carried out.

Recall that a matrix $\boldsymbol{A} \in \mathbb{C}^{n \times n}$ is upper Hessenberg if $a_{i,j} = 0$ for $j = 1, 2, \dots, i-2, i = 3, 4, \dots, n$. We will reduce $\boldsymbol{A} \in \mathbb{C}^{n \times n}$ to upper Hessenberg form by unitary similarity transformations. Let $\boldsymbol{A}_1 = \boldsymbol{A}$ and define $\boldsymbol{A}_{k+1} = \boldsymbol{H}_k \boldsymbol{A}_k \boldsymbol{H}_k$ for $k = 1, 2, \dots, n-2$. Here $\boldsymbol{H}_k$ is a Householder transformation chosen to introduce zeros in the elements of column k of $\boldsymbol{A}_k$ under the subdiagonal. The final matrix $\boldsymbol{A}_{n-1}$ will be upper Hessenberg. Householder transformations were used in Chap. 5 to reduce a matrix to upper triangular form. To preserve eigenvalues similarity transformations are needed and then the final matrix in the reduction cannot in general be upper triangular.

If $\boldsymbol{A}_1 = \boldsymbol{A}$ is Hermitian, the matrix $\boldsymbol{A}_{n-1}$ will be Hermitian and tridiagonal. For if $\boldsymbol{A}_k^* = \boldsymbol{A}_k$ then

$$\boldsymbol{A}_{k+1}^* = (\boldsymbol{H}_k \boldsymbol{A}_k \boldsymbol{H}_k)^* = \boldsymbol{H}_k \boldsymbol{A}_k^* \boldsymbol{H}_k = \boldsymbol{A}_{k+1}.$$

Since $\boldsymbol{A}_{n-1}$ is upper Hessenberg and Hermitian, it must be tridiagonal.

To describe the reduction to upper Hessenberg or tridiagonal form in more detail we partition $\boldsymbol{A}_k$ as follows

$$\boldsymbol{A}_k = \begin{bmatrix} \boldsymbol{B}_k & \boldsymbol{C}_k \\ \boldsymbol{D}_k & \boldsymbol{E}_k \end{bmatrix}.$$

Suppose $\boldsymbol{B}_k \in \mathbb{C}^{k,k}$ is upper Hessenberg, and the first $k-1$ columns of $\boldsymbol{D}_k \in \mathbb{C}^{n-k,k}$ are zero, i.e. $\boldsymbol{D}_k = [\boldsymbol{0}, \boldsymbol{0}, \ldots, \boldsymbol{0}, \boldsymbol{d}_k]$. Let $\boldsymbol{V}_k = \boldsymbol{I} - \boldsymbol{v}_k \boldsymbol{v}_k^* \in \mathbb{C}^{n-k,n-k}$ be a Householder transformation such that $\boldsymbol{V}_k \boldsymbol{d}_k = \alpha_k \boldsymbol{e}_1$. Define

$$\boldsymbol{H}_k = \begin{bmatrix} \boldsymbol{I}_k & \boldsymbol{0} \\ \boldsymbol{0} & \boldsymbol{V}_k \end{bmatrix} \in \mathbb{C}^{n\times n}.$$

The matrix $\boldsymbol{H}_k$ is a Householder transformation, and we find

$$\begin{aligned} \boldsymbol{A}_{k+1} = \boldsymbol{H}_k \boldsymbol{A}_k \boldsymbol{H}_k &= \begin{bmatrix} \boldsymbol{I}_k & \boldsymbol{0} \\ \boldsymbol{0} & \boldsymbol{V}_k \end{bmatrix} \begin{bmatrix} \boldsymbol{B}_k & \boldsymbol{C}_k \\ \boldsymbol{D}_k & \boldsymbol{E}_k \end{bmatrix} \begin{bmatrix} \boldsymbol{I}_k & \boldsymbol{0} \\ \boldsymbol{0} & \boldsymbol{V}_k \end{bmatrix} \\ &= \begin{bmatrix} \boldsymbol{B}_k & \boldsymbol{C}_k \boldsymbol{V}_k \\ \boldsymbol{V}_k \boldsymbol{D}_k & \boldsymbol{V}_k \boldsymbol{E}_k \boldsymbol{V}_k \end{bmatrix}. \end{aligned}$$

Now $\boldsymbol{V}_k \boldsymbol{D}_k = [\boldsymbol{V}_k \boldsymbol{0}, \ldots, \boldsymbol{V}_k \boldsymbol{0}, \boldsymbol{V}_k \boldsymbol{d}_k] = (\boldsymbol{0}, \ldots, \boldsymbol{0}, \alpha_k \boldsymbol{e}_1)$. Moreover, the matrix $\boldsymbol{B}_k$ is not affected by the $\boldsymbol{H}_k$ transformation. Therefore the upper left $(k+1)\times(k+1)$ corner of $\boldsymbol{A}_{k+1}$ is upper Hessenberg and the reduction is carried one step further. The reduction stops with $\boldsymbol{A}_{n-1}$ which is upper Hessenberg.

To find $\boldsymbol{A}_{k+1}$ we use Algorithm 5.1 to find $\boldsymbol{v}_k$ and α_k. We store $\boldsymbol{v}_k$ in the kth column of a matrix $\boldsymbol{L}$ as $\boldsymbol{L}(k+1:n, k) = \boldsymbol{v}_k$. This leads to the following algorithm for reducing a matrix $\boldsymbol{A} \in \mathbb{C}^{n\times n}$ to upper Hessenberg form using Householder transformations. The algorithm returns the reduced matrix $\boldsymbol{B}$. $\boldsymbol{B}$ is tridiagonal if $\boldsymbol{A}$ is symmetric. Details of the transformations are stored in a lower triangular matrix $\boldsymbol{L}$, also returned by the algorithm. The elements of $\boldsymbol{L}$ can be used to assemble a unitary matrix $\boldsymbol{Q}$ such that $\boldsymbol{B} = \boldsymbol{Q}^* \boldsymbol{A} \boldsymbol{Q}$. Algorithm 5.1 is used in each step of the reduction:

```
function [L,B] = hesshousegen(A)
n=length(A); L=zeros(n,n); B=A;
for k=1:n-2
    [v,B(k+1,k)]=housegen(B(k+1:n,k));
    L((k+1):n,k)=v; B((k+2):n,k)=zeros(n-k-1,1);
    C=B((k+1):n,(k+1):n);
    B((k+1):n,(k+1):n)=C-v*(v'*C);
    C=B(1:n,(k+1):n); B(1:n,(k+1):n)=C-(C*v)*v';
end
end
```

Listing 14.1 hesshousegen

14.4.1 Assembling Householder Transformations

We can use the output of Algorithm 14.1 to assemble the matrix $\boldsymbol{Q} \in \mathbb{R}^{n\times n}$ such that $\boldsymbol{Q}$ is orthogonal and $\boldsymbol{Q}^*\boldsymbol{A}\boldsymbol{Q}$ is upper Hessenberg. We need to compute the product $\boldsymbol{Q} = \boldsymbol{H}_1\boldsymbol{H}_2\cdots\boldsymbol{H}_{n-2}$, where $\boldsymbol{H}_k = \begin{bmatrix} \boldsymbol{I} & \boldsymbol{0} \\ \boldsymbol{0} & \boldsymbol{I}-\boldsymbol{v}_k\boldsymbol{v}_k^T \end{bmatrix}$ and $\boldsymbol{v}_k \in \mathbb{R}^{n-k}$. Since $\boldsymbol{v}_1 \in \mathbb{R}^{n-1}$ and $\boldsymbol{v}_{n-2} \in \mathbb{R}^2$ it is most economical to assemble the product from right to left. We compute

$$\boldsymbol{Q}_{n-1} = \boldsymbol{I} \text{ and } \boldsymbol{Q}_k = \boldsymbol{H}_k\boldsymbol{Q}_{k+1} \text{ for } k = n-2, n-3, \ldots, 1.$$

Suppose $\boldsymbol{Q}_{k+1}$ has the form $\begin{bmatrix} \boldsymbol{I}_k & \boldsymbol{0} \\ \boldsymbol{0} & \boldsymbol{U}_k \end{bmatrix}$, where $\boldsymbol{U}_k \in \mathbb{R}^{n-k,n-k}$. Then

$$\boldsymbol{Q}_k = \begin{bmatrix} \boldsymbol{I}_k & \boldsymbol{0} \\ \boldsymbol{0} & \boldsymbol{I}-\boldsymbol{v}_k\boldsymbol{v}_k^T \end{bmatrix} * \begin{bmatrix} \boldsymbol{I}_k & \boldsymbol{0} \\ \boldsymbol{0} & \boldsymbol{U}_k \end{bmatrix} = \begin{bmatrix} \boldsymbol{I}_k & \boldsymbol{0} \\ \boldsymbol{0} & \boldsymbol{U}_k - \boldsymbol{v}_k(\boldsymbol{v}_k^T\boldsymbol{U}_k) \end{bmatrix}.$$

This leads to the following algorithm for assembling Householder transformations. The algorithm assumes that `L` is output from Algorithm 14.1, and assembles an orthogonal matrix $\boldsymbol{Q}$ from the columns of $\boldsymbol{L}$ so that $\boldsymbol{Q}^*\boldsymbol{A}\boldsymbol{Q}$ is upper Hessenberg.

```
function Q = accumulateQ(L)
n=length(L); Q=eye(n);
for k=n-2:-1:1
    v=L((k+1):n,k); C=Q((k+1):n,(k+1):n);
    Q((k+1):n,(k+1):n)=C-v*(v'*C);
end
```

Listing 14.2 accumulateQ

14.5 Computing a Selected Eigenvalue of a Symmetric Matrix

Let $\boldsymbol{A} \in \mathbb{R}^{n\times n}$ be symmetric with eigenvalues $\lambda_1 \ge \lambda_2 \ge \cdots \ge \lambda_n$. In this section we consider a method to compute an approximation to the mth eigenvalue λ_m for some $1 \le m \le n$. Using Householder similarity transformations as outlined in the previous section we can assume that $\boldsymbol{A}$ is symmetric and tridiagonal.

$$\boldsymbol{A} = \begin{bmatrix} d_1 & c_1 & & & \\ c_1 & d_2 & c_2 & & \\ & \ddots & \ddots & \ddots & \\ & & c_{n-2} & d_{n-1} & c_{n-1} \\ & & & c_{n-1} & d_n \end{bmatrix}. \tag{14.6}$$

Suppose one of the off-diagonal elements is equal to zero, say $c_i = 0$. We then have $\boldsymbol{A} = \begin{bmatrix} \boldsymbol{A}_1 & \boldsymbol{0} \\ \boldsymbol{0} & \boldsymbol{A}_2 \end{bmatrix}$, where

$$\boldsymbol{A}_1 = \begin{bmatrix} d_1 & c_1 & & & \\ c_1 & d_2 & c_2 & & \\ & \ddots & \ddots & \ddots & \\ & & c_{i-2} & d_{i-1} & c_{i-1} \\ & & & c_{i-1} & d_i \end{bmatrix} \text{ and } \boldsymbol{A}_2 = \begin{bmatrix} d_{i+1} & c_{i+1} & & & \\ c_{i+1} & d_{i+2} & c_{i+2} & & \\ & \ddots & \ddots & \ddots & \\ & & c_{n-2} & d_{n-1} & c_{n-1} \\ & & & c_{n-1} & d_n \end{bmatrix}.$$

Thus $\boldsymbol{A}$ is block diagonal and we can split the eigenvalue problem into two smaller problems involving $\boldsymbol{A}_1$ and $\boldsymbol{A}_2$. We assume that this reduction has been carried out so that $\boldsymbol{A}$ is irreducible, i.e., $c_i \neq 0$ for $i = 1, \ldots, n-1$.

We first show that irreducibility implies that the eigenvalues are distinct.

Lemma 14.2 (Distinct Eigenvalues of a Tridiagonal Matrix) *An irreducible, tridiagonal and symmetric matrix $\boldsymbol{A} \in \mathbb{R}^{n \times n}$ has n real and distinct eigenvalues.*

Proof Let $\boldsymbol{A}$ be given by (14.6). By Theorem 6.10 the eigenvalues are real. Define for $x \in \mathbb{R}$ the polynomial $p_k(x) := \det(x\boldsymbol{I}_k - \boldsymbol{A}_k)$ for $k = 1, \ldots, n$, where $\boldsymbol{A}_k$ is the upper left $k \times k$ corner of $\boldsymbol{A}$ (the leading principal submatrix of order k). The eigenvalues of $\boldsymbol{A}$ are the roots of the polynomial p_n. Using the last column to expand for $k \geq 2$ the determinant $p_{k+1}(x)$ we find

$$p_{k+1}(x) = (x - d_{k+1})p_k(x) - c_k^2 p_{k-1}(x). \tag{14.7}$$

Since $p_1(x) = x - d_1$ and $p_2(x) = (x - d_2)(x - d_1) - c_1^2$ this also holds for $k = 0, 1$ if we define $p_{-1}(x) = 0$ and $p_0(x) = 1$. For M sufficiently large we have

$$p_2(-M) > 0, \quad p_2(d_1) < 0, \quad p_2(+M) > 0.$$

Since p_2 is continuous there are $y_1 \in (-M, d_1)$ and $y_2 \in (d_1, M)$ such that $p_2(y_1) = p_2(y_2) = 0$. It follows that the root d_1 of p_1 separates the roots of p_2, so y_1 and y_2 must be distinct. Consider next

$$p_3(x) = (x - d_3)p_2(x) - c_2^2 p_1(x) = (x - d_3)(x - y_1)(x - y_2) - c_2^2(x - d_1).$$

Since $y_1 < d_1 < y_2$ we have for M sufficiently large

$$p_3(-M) < 0, \quad p_3(y_1) > 0, \quad p_3(y_2) < 0, \quad p_3(+M) > 0.$$

Thus the roots x_1, x_2, x_3 of p_3 are separated by the roots y_1, y_2 of p_2. In the general case suppose for $k \geq 2$ that the roots $z_1, \ldots, z_{k-1}$ of p_{k-1} separate the roots $y_1, \ldots, y_k$ of p_k. Choose M so that $y_0 := -M < y_1$, $y_{k+1} := M > y_k$. Then

$$y_0 < y_1 < z_1 < y_2 < z_2 \cdots < z_{k-1} < y_k < y_{k+1}.$$

We claim that for M sufficiently large

$$p_{k+1}(y_j) = (-1)^{k+1-j}|p_{k+1}(y_j)| \neq 0, \text{ for } j = 0, 1, \ldots, k+1.$$

This holds for $j = 0, k+1$, and for $j = 1, \ldots, k$ since

$$p_{k+1}(y_j) = -c_k^2 p_{k-1}(y_j) = -c_k^2(y_j - z_1)\cdots(y_j - z_{k-1}).$$

It follows that the roots $x_1, \ldots, x_{k+1}$ are separated by the roots $y_1, \ldots, y_k$ of p_k and by induction the roots of p_n (the eigenvalues of $\boldsymbol{A}$) are distinct. □

14.5.1 The Inertia Theorem

We say that two matrices $\boldsymbol{A}, \boldsymbol{B} \in \mathbb{C}^{n\times n}$ are **congruent** if $\boldsymbol{A} = \boldsymbol{E}^*\boldsymbol{B}\boldsymbol{E}$ for some nonsingular matrix $\boldsymbol{E} \in \mathbb{C}^{n\times n}$. By Theorem 6.7 a Hermitian matrix $\boldsymbol{A}$ is both congruent and similar to a diagonal matrix $\boldsymbol{D}$, $\boldsymbol{U}^*\boldsymbol{A}\boldsymbol{U} = \boldsymbol{D}$ where $\boldsymbol{U}$ is unitary. The eigenvalues of $\boldsymbol{A}$ are the diagonal elements of $\boldsymbol{D}$. Let $\pi(\boldsymbol{A})$, $\zeta(\boldsymbol{A})$ and $\upsilon(\boldsymbol{A})$ denote the number of positive, zero and negative eigenvalues of $\boldsymbol{A}$. If $\boldsymbol{A}$ is Hermitian then all eigenvalues are real and $\pi(\boldsymbol{A}) + \zeta(\boldsymbol{A}) + \upsilon(\boldsymbol{A}) = n$.

Theorem 14.6 (Sylvester's Inertia Theorem) *If $\boldsymbol{A}, \boldsymbol{B} \in \mathbb{C}^{n\times n}$ are Hermitian and congruent then $\pi(\boldsymbol{A}) = \pi(\boldsymbol{B})$, $\zeta(\boldsymbol{A}) = \zeta(\boldsymbol{B})$ and $\upsilon(\boldsymbol{A}) = \upsilon(\boldsymbol{B})$.*

Proof Suppose $\boldsymbol{A} = \boldsymbol{E}^*\boldsymbol{B}\boldsymbol{E}$, where $\boldsymbol{E}$ is nonsingular. Assume first that $\boldsymbol{A}$ and $\boldsymbol{B}$ are diagonal matrices. Suppose $\pi(\boldsymbol{A}) = k$ and $\pi(\boldsymbol{B}) = m < k$. We shall show that this leads to a contradiction. Let $\boldsymbol{E}_1$ be the upper left $m \times k$ corner of $\boldsymbol{E}$. Since $m < k$, we can find a nonzero $\boldsymbol{x}$ such that $\boldsymbol{E}_1\boldsymbol{x} = \boldsymbol{0}$ (cf. Lemma 1.3). Let $\boldsymbol{y}^T = [\boldsymbol{x}^T, \boldsymbol{0}^T] \in \mathbb{C}^n$, and $\boldsymbol{z} = [z_1, \ldots, z_n]^T = \boldsymbol{E}\boldsymbol{y}$. Then $z_i = 0$ for $i = 1, 2, \ldots, m$. If $\boldsymbol{A}$ has positive eigenvalues $\lambda_1, \ldots, \lambda_k$ and $\boldsymbol{B}$ has eigenvalues $\mu_1, \ldots, \mu_n$, where $\mu_i \leq 0$ for $i \geq m+1$ then

$$\boldsymbol{y}^*\boldsymbol{A}\boldsymbol{y} = \sum_{i=1}^{n} \lambda_i |y_i|^2 = \sum_{i=1}^{k} \lambda_i |x_i|^2 > 0.$$

But

$$y^* A y = y^* E^* B E y = z^* B z = \sum_{i=m+1}^{n} \mu_i |z_i|^2 \le 0,$$

a contradiction.

We conclude that $\pi(A) = \pi(B)$ if A and B are diagonal. Moreover, $\upsilon(A) = \pi(-A) = \pi(-B) = \upsilon(B)$ and $\zeta(A) = n - \pi(A) - \upsilon(A) = n - \pi(B) - \upsilon(B) = \zeta(B)$. This completes the proof for diagonal matrices.

Let in the general case U_1 and U_2 be unitary matrices such that $U_1^* A U_1 = D_1$ and $U_2^* B U_2 = D_2$ where D_1 and D_2 are diagonal matrices. Since $A = E^* B E$, we find $D_1 = F^* D_2 F$ where $F = U_2^* E U_1$ is nonsingular. Thus D_1 and D_2 are congruent diagonal matrices. But since A and D_1, B and D_2 have the same eigenvalues, we find $\pi(A) = \pi(D_1) = \pi(D_2) = \pi(B)$. Similar results hold for ζ and υ. □

Corollary 14.2 (Counting Eigenvalues Using the LDL* Factorization) *Suppose $A = \operatorname{tridiag}(c_i, d_i, c_i) \in \mathbb{R}^{n\times n}$ is symmetric and that $\alpha \in \mathbb{R}$ is such that $A - \alpha I$ has an symmetric LU factorization, i.e. $A - \alpha I = LDL^T$ where L is unit lower triangular and D is diagonal. Then the number of eigenvalues of A strictly less than α equals the number of negative diagonal elements in D. The diagonal elements $d_1(\alpha), \ldots, d_n(\alpha)$ in D can be computed recursively as follows*

$$d_1(\alpha) = d_1 - \alpha, \; d_k(\alpha) = d_k - \alpha - c_{k-1}^2 / d_{k-1}(\alpha), \; k = 2, 3, \ldots, n. \tag{14.8}$$

Proof Since the diagonal elements in L in an LU factorization equal the diagonal elements in D in an LDL^T factorization we see that the formulas in (14.8) follows immediately from (2.16). Since L is nonsingular, $A - \alpha I$ and D are congruent. By the previous theorem $\upsilon(A - \alpha I) = \upsilon(D)$, the number of negative diagonal elements in D. If $Ax = \lambda x$ then $(A - \alpha I)x = (\lambda - \alpha)x$, and $\lambda - \alpha$ is an eigenvalue of $A - \alpha I$. But then $\upsilon(A - \alpha I)$ equals the number of eigenvalues of A which are less than α. □

14.5.2 Approximating λ_m

Corollary 14.2 can be used to determine the mth eigenvalue of A, where $\lambda_1 \ge \lambda_2 \ge \cdots \ge \lambda_n$. Using Gerschgorin's theorem we first find an interval $[a, b]$, such that (a, b) contains the eigenvalues of A. Let for $x \in [a, b]$

$$\rho(x) := \#\{k : d_k(x) > 0 \text{ for } k = 1, \ldots, n\}$$

be the number of eigenvalues of A which are strictly greater than x. Clearly $\rho(a) = n$, $\rho(b) = 0$. Choosing a tolerance ϵ and using bisection we proceed as follows:

$$
\begin{array}{|l|}
\hline
h = b - a; \\
for\ j = 1 : itmax \\
\quad c = (a + b)/2; \\
\quad if\ b - a < eps * h \\
\qquad \lambda = (a + b)/2;\ \text{return} \\
\quad end \\
\quad k = \rho(c); \\
\quad if\ k \geq m\ a = c\ else\ b = c; \\
\text{end} \\
\hline
\end{array}
\tag{14.9}
$$

We generate a sequence $\{[a_j, b_j]\}$ of intervals, each containing λ_m and $b_j - a_j = 2^{-j}(b - a)$.

As it stands this method will fail if in (14.8) one of the $d_k(\alpha)$ is zero. One possibility is to replace such a $d_k(\alpha)$ by a suitable small number, say $\delta_k = c_k\epsilon_M$, where ϵ_M is the Machine epsilon, typically 2×10^{-16} for MATLAB. This replacement is done if $|d_k(\alpha)| < |\delta_k|$.

14.6 Exercises Chap. 14

14.6.1 Exercises Sect. 14.1

Exercise 14.1 (Yes or No (Exam Exercise 2006-1)) Answer simply yes or no to the following questions:

a) Every matrix $A \in \mathbb{C}^{m \times n}$ has a singular value decomposition?
b) The algebraic multiplicity of an eigenvalue is always less than or equal to the geometric multiplicity?
c) The QR factorization of a matrix $A \in \mathbb{R}^{n \times n}$ can be determined by Householder transformations in $O(n^2)$ arithmetic operations?
d) Let $\rho(A)$ be the spectral radius of $A \in \mathbb{C}^{n \times n}$. Then $\lim_{k\to\infty} A^k = \mathbf{0}$ if and only if $\rho(A) < 1$?

14.6.2 Exercises Sect. 14.2

Exercise 14.2 (Nonsingularity Using Gershgorin) Consider the matrix

$$A = \begin{pmatrix} 4 & 1 & 0 & 0 \\ 1 & 4 & 1 & 0 \\ 0 & 1 & 4 & 1 \\ 0 & 0 & 1 & 4 \end{pmatrix}.$$

Show using Gershgorin's theorem that $\boldsymbol{A}$ is nonsingular.

Exercise 14.3 (Gershgorin, Strictly Diagonally Dominant Matrix) Show using Gershgorin's circle theorem that a strictly diagonally dominant matrix $\boldsymbol{A}$ ($|a_{i,i}| > \sum_{j \neq i} |a_{i,j}|$ for all i) is nonsingular.

Exercise 14.4 (Gershgorin Disks (Exam Exercise 2009-2)) The eigenvalues of $\boldsymbol{A} \in \mathbb{R}^{n,n}$ lie inside $R \cap C$, where $R := R_1 \cup \cdots \cup R_n$ is the union of the row disks R_i of $\boldsymbol{A}$, and $C = C_1 \cup \cdots \cup C_n$ is the union of the column disks C_j. You do not need to prove this. Write a MATLAB function `[s,r,c]=gershgorin(A)` that computes the centres $\boldsymbol{s} = [s_1, \ldots, s_n] \in \mathbb{R}^n$ of the row and column disks, and their radii $\boldsymbol{r} = [r_1, \ldots, r_n] \in \mathbb{R}^n$ and $\boldsymbol{c} = [c_1, \ldots, c_n] \in \mathbb{R}^n$, respectively.

14.6.3 Exercises Sect. 14.3

Exercise 14.5 (Continuity of Eigenvalues) Suppose

$$\boldsymbol{A}(t) := \boldsymbol{D} + t(\boldsymbol{A} - \boldsymbol{D}), \quad \boldsymbol{D} := \operatorname{diag}(a_{11}, \ldots, a_{nn}), \quad t \in \mathbb{R}.$$

$0 \leq t_1 < t_2 \leq 1$ and that μ is an eigenvalue of $\boldsymbol{A}(t_2)$. Show, using Theorem 14.2 with $\boldsymbol{A} = \boldsymbol{A}(t_1)$ and $\boldsymbol{E} = \boldsymbol{A}(t_2) - \boldsymbol{A}(t_1)$, that $\boldsymbol{A}(t_1)$ has an eigenvalue λ such that

$$|\lambda - \mu| \leq C(t_2 - t_1)^{1/n}, \text{ where } C \leq 2\big(\|\boldsymbol{D}\|_2 + \|\boldsymbol{A} - \boldsymbol{D}\|_2\big).$$

Thus, as a function of t, every eigenvalue of $\boldsymbol{A}(t)$ is a continuous function of t.

Exercise 14.6 (∞-Norm of a Diagonal Matrix) Give a direct proof that $\|\boldsymbol{A}\|_\infty = \rho(\boldsymbol{A})$ if $\boldsymbol{A}$ is diagonal.

Exercise 14.7 (Eigenvalue Perturbations (Exam Exercise 2010-2)) Let $\boldsymbol{A} = [a_{kj}]$, $\boldsymbol{E} = [e_{kj}]$, and $\boldsymbol{B} = [b_{kj}]$ be matrices in $\mathbb{R}^{n,n}$ with

$$a_{kj} = \begin{cases} 1, & j = k+1, \\ 0, & \text{otherwise}, \end{cases} \qquad e_{kj} = \begin{cases} \epsilon, & k = n, j = 1, \\ 0, & \text{otherwise}, \end{cases} \tag{14.10}$$

and $\boldsymbol{B} = \boldsymbol{A} + \boldsymbol{E}$, where $0 < \epsilon < 1$. Thus for $n = 4$,

$$\boldsymbol{A} := \begin{bmatrix} 0 & 1 & 0 & 0 \\ 0 & 0 & 1 & 0 \\ 0 & 0 & 0 & 1 \\ 0 & 0 & 0 & 0 \end{bmatrix}, \quad \boldsymbol{E} := \begin{bmatrix} 0 & 0 & 0 & 0 \\ 0 & 0 & 0 & 0 \\ 0 & 0 & 0 & 0 \\ \epsilon & 0 & 0 & 0 \end{bmatrix}, \quad \boldsymbol{B} := \begin{bmatrix} 0 & 1 & 0 & 0 \\ 0 & 0 & 1 & 0 \\ 0 & 0 & 0 & 1 \\ \epsilon & 0 & 0 & 0 \end{bmatrix}.$$

a) Find the eigenvalues of $\boldsymbol{A}$ and $\boldsymbol{B}$.
b) Show that $\|\boldsymbol{A}\|_2 = \|\boldsymbol{B}\|_2 = 1$ for arbitrary $n \in \mathbb{N}$.
c) Recall Elsner's Theorem (Theorem 14.2). Let $\boldsymbol{A}, \boldsymbol{E}, \boldsymbol{B}$ be given by (14.10). What upper bound does (14.1) in Elsner's theorem give for the eigenvalue $\mu = \epsilon^{1/n}$ of $\boldsymbol{B}$? How sharp is this upper bound?

14.6.4 Exercises Sect. 14.4

Exercise 14.8 (Number of Arithmetic Operations, Hessenberg Reduction) Show that the number of arithmetic operations for Algorithm 14.1 is $\frac{10}{3}n^3 = 5G_n$.

Exercise 14.9 (Assemble Householder Transformations) Show that the number of arithmetic operations required by Algorithm 14.2 is $\frac{4}{3}n^3 = 2G_n$.

Exercise 14.10 (Tridiagonalize a Symmetric Matrix) If $\boldsymbol{A}$ is real and symmetric we can modify Algorithm 14.1 as follows. To find $\boldsymbol{A}_{k+1}$ from $\boldsymbol{A}_k$ we have to compute $\boldsymbol{V}_k\boldsymbol{E}_k\boldsymbol{V}_k$ where $\boldsymbol{E}_k$ is symmetric. Dropping subscripts we have to compute a product of the form $\boldsymbol{G} = (\boldsymbol{I} - \boldsymbol{v}\boldsymbol{v}^T)\boldsymbol{E}(\boldsymbol{I} - \boldsymbol{v}\boldsymbol{v}^T)$. Let $\boldsymbol{w} := \boldsymbol{E}\boldsymbol{v}$, $\beta := \frac{1}{2}\boldsymbol{v}^T\boldsymbol{w}$ and $\boldsymbol{z} := \boldsymbol{w} - \beta\boldsymbol{v}$. Show that $\boldsymbol{G} = \boldsymbol{E} - \boldsymbol{v}\boldsymbol{z}^T - \boldsymbol{z}\boldsymbol{v}^T$. Since $\boldsymbol{G}$ is symmetric, only the sub- or superdiagonal elements of $\boldsymbol{G}$ need to be computed. Computing $\boldsymbol{G}$ in this way, it can be shown that we need $O(4n^3/3)$ operations to tridiagonalize a symmetric matrix by orthonormal similarity transformations. This is less than half the work to reduce a nonsymmetric matrix to upper Hessenberg form. We refer to [18] for a detailed algorithm.

14.6.5 Exercises Sect. 14.5

Exercise 14.11 (Counting Eigenvalues) Consider the matrix in Exercise 14.2. Determine the number of eigenvalues greater than 4.5.

Exercise 14.12 (Overflow in LDL* Factorization) Let for $n \in \mathbb{N}$

$$A_n = \begin{bmatrix} 10 & 1 & 0 & \cdots & 0 \\ 1 & 10 & 1 & \ddots & \vdots \\ 0 & \ddots & \ddots & \ddots & 0 \\ \vdots & \ddots & 1 & 10 & 1 \\ 0 & \cdots & 0 & 1 & 10 \end{bmatrix} \in \mathbb{R}^{n \times n}.$$

a) Let d_k be the diagonal elements of D in an LDL* factorization of A_n. Show that $5+\sqrt{24} < d_k \leq 10$, $k = 1, 2, \ldots, n$.
b) Show that $D_n := \det(A_n) > (5+\sqrt{24})^n$. Give $n_0 \in \mathbb{N}$ such that your computer gives an overflow when D_{n_0} is computed in floating point arithmetic.

Exercise 14.13 (Simultaneous Diagonalization) (Simultaneous diagonalization of two symmetric matrices by a congruence transformation). Let $A, B \in \mathbb{R}^{n \times n}$ where $A^T = A$ and B is symmetric positive definite. Then $B = U^T D U$ where U is orthonormal and $D = \operatorname{diag}(d_1, \ldots, d_n)$ has positive diagonal elements. Let $\hat{A} = D^{-1/2} U A U^T D^{-1/2}$ where

$$D^{-1/2} := \operatorname{diag}\left(d_1^{-1/2}, \ldots, d_n^{-1/2}\right).$$

a) Show that $\hat{A}$ is symmetric.

Let $\hat{A} = \hat{U}^T \hat{D} \hat{U}$ where $\hat{U}$ is orthonormal and $\hat{D}$ is diagonal. Set $E = U^T D^{-1/2} \hat{U}^T$.

b) Show that E is nonsingular and that $E^T A E = \hat{D}$, $E^T B E = I$.

For a more general result see Theorem 10.1 in [11].

Exercise 14.14 (Program Code for One Eigenvalue) Suppose $A = \operatorname{tridiag}(c, d, c)$ is symmetric and tridiagonal with elements $d_1, \ldots, d_n$ on the diagonal and $c_1, \ldots, c_{n-1}$ on the neighboring subdiagonals. Let $\lambda_1 \geq \lambda_2 \geq \cdots \geq \lambda_n$ be the eigenvalues of A. We shall write a program to compute one eigenvalue λ_m for a given m using bisection and the method outlined in (14.9).

a) Write a `function k=counting(c,d,x)` which for given x counts the number of eigenvalues of A strictly greater than x. Use the replacement described above if one of the $d_j(x)$ is close to zero.
b) Write a `function lambda=findeigv(c,d,m)` which first estimates an interval $(a, b]$ containing all eigenvalues of A and then generates a sequence $\{(a_j, b_j]\}$ of intervals each containing λ_m. Iterate until $b_j - a_j \leq (b-a)\epsilon_M$, where ϵ_M is MATLAB's machine epsilon `eps`. Typically $\epsilon_M \approx 2.22 \times 10^{-16}$.
c) Test the program on $T := \operatorname{tridiag}(-1, 2, -1)$ of size 100. Compare the exact value of λ_5 with your result and the result obtained by using MATLAB's built-in function `eig`.

Exercise 14.15 (Determinant of Upper Hessenberg Matrix) Suppose $\boldsymbol{A} \in \mathbb{C}^{n\times n}$ is upper Hessenberg and $x \in \mathbb{C}$. We will study two algorithms to compute $f(x) = \det(\boldsymbol{A} - x\boldsymbol{I})$.

a) Show that Gaussian elimination without pivoting requires $O(n^2)$ arithmetic operations.
b) Show that the number of arithmetic operations is the same if partial pivoting is used.
c) Estimate the number of arithmetic operations if Given's rotations are used.
d) Compare the two methods discussing advantages and disadvantages.

14.7 Review Questions

14.7.1 Suppose $\boldsymbol{A}, \boldsymbol{E} \in \mathbb{C}^{n\times n}$. To every $\mu \in \sigma(\boldsymbol{A}+\boldsymbol{E})$ there is a $\lambda \in \sigma(\boldsymbol{A})$ which is in some sense close to μ.

- What is the general result (Elsner's theorem)?
- what if $\boldsymbol{A}$ is non defective?
- what if $\boldsymbol{A}$ is normal?
- what if $\boldsymbol{A}$ is Hermitian?

14.7.2 Can Gerschgorin's theorem be used to check if a matrix is nonsingular?
14.7.3 How many arithmetic operation does it take to reduce a matrix by similarity transformations to upper Hessenberg form by Householder transformations?
14.7.4 Give a condition ensuring that a tridiagonal symmetric matrix has real and distinct eigenvalues:
14.7.5 What is the content of Sylvester's inertia theorem?
14.7.6 Give an application of this theorem.

Chapter 15
The QR Algorithm

The QR algorithm is a method to find all eigenvalues and eigenvectors of a matrix. In this chapter we give a brief informal introduction to this important algorithm. For a more complete treatment see [18].

The QR algorithm is related to a simpler method called the power method and we start studying this method and its variants.

15.1 The Power Method and Its Variants

These methods can be used to compute a single eigenpair of a matrix. They also play a role when studying properties of the QR algorithm.

15.1.1 The Power Method

The **power method** in its basic form is a technique to compute an approximation to the eigenvector corresponding to the largest (in absolute value) eigenvalue of a matrix $\boldsymbol{A} \in \mathbb{C}^{n\times n}$. As a by product we can also find an approximation to the corresponding eigenvalue. We define a sequence $\{z_k\}$ of vectors in $\mathbb{C}^n$ by

$$z_k := \boldsymbol{A}^k z_0 = \boldsymbol{A} z_{k-1}, \quad k = 1, 2, \ldots. \tag{15.1}$$

Example 15.1 (Power Method) Let

$$\boldsymbol{A} = \begin{bmatrix} 2 & -1 \\ -1 & 2 \end{bmatrix}, \quad z_0 := \begin{bmatrix} 1 \\ 0 \end{bmatrix}.$$

T. Lyche, *Numerical Linear Algebra and Matrix Factorizations*,
Texts in Computational Science and Engineering 22,
https://doi.org/10.1007/978-3-030-36468-7_15

We find

$$z_1 = Az_0 = \begin{bmatrix} 2 \\ -1 \end{bmatrix}, \quad z_2 = Az_1 = \begin{bmatrix} 5 \\ -4 \end{bmatrix}, \cdots, z_k = \frac{1}{2}\begin{bmatrix} 1+3^k \\ 1-3^k \end{bmatrix}, \cdots.$$

It follows that $2z_k/3^k$ converges to $[1, -1]$, an eigenvector corresponding to the dominant eigenvalue $\lambda = 3$. The sequence of Rayleigh quotients $\{z_k^T Az_k/z_k^T z_k\}$ will converge to the dominant eigenvalue $\lambda = 3$.

To understand better what happens we expand z_0 in terms of the eigenvectors

$$z_0 = \frac{1}{2}\begin{bmatrix} 1 \\ -1 \end{bmatrix} + \frac{1}{2}\begin{bmatrix} 1 \\ 1 \end{bmatrix} = c_1 v_1 + c_2 v_2.$$

Since A^k has eigenpairs (λ_j^k, v_j), $j = 1, 2$ we find

$$z_k = c_1\lambda_1^k v_1 + c_2\lambda_2^k v_2 = c_1 3^k v_1 + c_2 1^k v_2.$$

Thus $3^{-k}z_k = c_1 v_1 + 3^{-k}c_2 v_2 \to c_1 v_1$. Since $c_1 \neq 0$ we obtain convergence to the dominant eigenvector.

Let $A \in \mathbb{C}^{n\times n}$ have eigenpairs (λ_j, v_j), $j = 1, \ldots, n$ with $|\lambda_1| > |\lambda_2| \geq \cdots \geq |\lambda_n|$.

Given $z_0 \in \mathbb{C}^n$ we assume that

$$\begin{aligned} &(i) && |\lambda_1| > |\lambda_2| \geq |\lambda_3| \geq \cdots \geq |\lambda_n|, \\ &(ii) && z_0^T v_1 \neq 0 \\ &(iii) && A \text{ has linearly independent eigenvectors.} \end{aligned} \tag{15.2}$$

The first assumption means that A has a dominant eigenvalue λ_1 of algebraic multiplicity one. The second assumption says that z_0 has a component in the direction v_1. The third assumption is not necessary, but is included in order to simplify the analysis.

To see what happens let $z_0 = c_1 v_1 + c_2 v_2 + \cdots + c_n v_n$, where by assumption (ii) of (15.2) we have $c_1 \neq 0$. Since $A^k v_j = \lambda_j^k v_j$ for all j we see that

$$z_k = c_1\lambda_1^k v_1 + c_2\lambda_2^k v_2 + \cdots + c_n\lambda_n^k v_n, \quad k = 0, 1, 2, \ldots. \tag{15.3}$$

Dividing by λ_1^k we find

$$\frac{z_k}{\lambda_1^k} = c_1 v_1 + c_2\Big(\frac{\lambda_2}{\lambda_1}\Big)^k v_2 + \cdots + c_n\Big(\frac{\lambda_n}{\lambda_1}\Big)^k v_n, \quad k = 0, 1, 2, \ldots. \tag{15.4}$$

Assumption (i) of (15.2) implies that $(\lambda_j/\lambda_1)^k \to 0$ as $k \to \infty$ for all $j \geq 2$ and we obtain

$$\lim_{k\to\infty} \frac{z_k}{\lambda_1^k} = c_1 \boldsymbol{v}_1, \tag{15.5}$$

the dominant eigenvector of $\boldsymbol{A}$. It can be shown that this also holds for defective matrices as long as (i) and (ii) of (15.2) hold, see for example page 58 of [18].

In practice we need to scale the iterates z_k somehow and we normally do not know λ_1. Instead we choose a norm on $\mathbb{C}^n$, set $\boldsymbol{x}_0 = z_0/\|z_0\|$ and generate for $k = 1, 2, \ldots$ unit vectors as follows:

$$\begin{aligned} (i) &\quad \boldsymbol{y}_k = \boldsymbol{A}\boldsymbol{x}_{k-1} \\ (ii) &\quad \boldsymbol{x}_k = \boldsymbol{y}_k/\|\boldsymbol{y}_k\|. \end{aligned} \tag{15.6}$$

Lemma 15.1 (Convergence of the Power Method) *Suppose* (15.2) *holds. Then*

$$\lim_{k\to\infty} \left(\frac{|\lambda_1|}{\lambda_1}\right)^k \boldsymbol{x}_k = \frac{c_1}{|c_1|} \frac{\boldsymbol{v}_1}{\|\boldsymbol{v}_1\|}.$$

In particular, if $\lambda_1 > 0$ and $c_1 > 0$ then the sequence $\{\boldsymbol{x}_k\}$ will converge to the eigenvector $\boldsymbol{u}_1 := \boldsymbol{v}_1/\|\boldsymbol{v}_1\|$ of unit length.

Proof By induction on k it follows that $\boldsymbol{x}_k = z_k/\|z_k\|$ for all $k \geq 0$, where $z_k = \boldsymbol{A}^k z_0$. Indeed, this holds for $k = 1$, and if it holds for $k - 1$ then $\boldsymbol{y}_k = \boldsymbol{A}\boldsymbol{x}_{k-1} = \boldsymbol{A}z_{k-1}/\|z_{k-1}\| = z_k/\|z_{k-1}\|$ and $\boldsymbol{x}_k = (z_k/\|z_{k-1}\|)(\|z_{k-1}\|/\|z_k\|) = z_k/\|z_k\|$. But then

$$\boldsymbol{x}_k = \frac{z_k}{\|z_k\|} = \frac{c_1\lambda_1^k}{|c_1\lambda_1^k|} \frac{\boldsymbol{v}_1 + \frac{c_2}{c_1}\left(\frac{\lambda_2}{\lambda_1}\right)^k \boldsymbol{v}_2 + \cdots + \frac{c_n}{c_1}\left(\frac{\lambda_n}{\lambda_1}\right)^k \boldsymbol{v}_n}{\|\boldsymbol{v}_1 + \frac{c_2}{c_1}\left(\frac{\lambda_2}{\lambda_1}\right)^k \boldsymbol{v}_2 + \cdots + \frac{c_n}{c_1}\left(\frac{\lambda_n}{\lambda_1}\right)^k \boldsymbol{v}_n\|}, \qquad k = 0, 1, 2, \ldots,$$

and this implies the lemma. □

Suppose we know an approximate eigenvector $\boldsymbol{u}$ of $\boldsymbol{A}$, but not the corresponding eigenvalue μ. One way of estimating μ is to minimize the Euclidian norm of the residual $r(\lambda) := \boldsymbol{A}\boldsymbol{u} - \lambda\boldsymbol{u}$.

Theorem 15.1 (The Rayleigh Quotient Minimizes the Residual) *Let $\boldsymbol{A} \in \mathbb{C}^{n\times n}$, $\boldsymbol{u} \in \mathbb{C}^n \setminus \{\boldsymbol{0}\}$, and let $\rho : \mathbb{C} \to \mathbb{R}$ be given by $\rho(\lambda) = \|\boldsymbol{A}\boldsymbol{u} - \lambda\boldsymbol{u}\|_2$. Then ρ is minimized when $\lambda := \frac{\boldsymbol{u}^*\boldsymbol{A}\boldsymbol{u}}{\boldsymbol{u}^*\boldsymbol{u}}$, the Rayleigh quotient for $\boldsymbol{A}$.*

Proof Assume $\boldsymbol{u}^*\boldsymbol{u} = 1$ and extend $\boldsymbol{u}$ to an orthonormal basis $\{\boldsymbol{u}, \boldsymbol{U}\}$ for $\mathbb{C}^n$. Then $\boldsymbol{U}^*\boldsymbol{u} = \boldsymbol{0}$ and

$$\begin{bmatrix} \boldsymbol{u}^* \\ \boldsymbol{U}^* \end{bmatrix} (\boldsymbol{A}\boldsymbol{u} - \lambda\boldsymbol{u}) = \begin{bmatrix} \boldsymbol{u}^*\boldsymbol{A}\boldsymbol{u} - \lambda\boldsymbol{u}^*\boldsymbol{u} \\ \boldsymbol{U}^*\boldsymbol{A}\boldsymbol{u} - \lambda\boldsymbol{U}^*\boldsymbol{u} \end{bmatrix} = \begin{bmatrix} \boldsymbol{u}^*\boldsymbol{A}\boldsymbol{u} - \lambda \\ \boldsymbol{U}^*\boldsymbol{A}\boldsymbol{u} \end{bmatrix}.$$

By unitary invariance of the Euclidian norm

$$\rho(\lambda)^2 = |\boldsymbol{u}^*\boldsymbol{A}\boldsymbol{u} - \lambda|^2 + \|\boldsymbol{U}^*\boldsymbol{A}\boldsymbol{u}\|_2^2,$$

and ρ has a global minimum at $\lambda = \boldsymbol{u}^*\boldsymbol{A}\boldsymbol{u}$. □

Using Rayleigh quotients we can incorporate the calculation of the eigenvalue into the power iteration. We can then compute the residual and stop the iteration when the residual is sufficiently small. But what does it mean to be sufficiently small? Recall that if $\boldsymbol{A}$ is nonsingular with a nonsingular eigenvector matrix $\boldsymbol{X}$ and $(\mu, \boldsymbol{u})$ is an approximate eigenpair with $\|\boldsymbol{u}\|_2 = 1$, then by (14.4) we can find an eigenvalue λ of $\boldsymbol{A}$ such that

$$\frac{|\lambda - \mu|}{|\lambda|} \le K_2(\boldsymbol{X})K_2(\boldsymbol{A})\frac{\|\boldsymbol{A}\boldsymbol{u} - \mu\boldsymbol{u}\|_2}{\|\boldsymbol{A}\|_2}.$$

Thus if the relative residual is small and both $\boldsymbol{A}$ and $\boldsymbol{X}$ are well conditioned then the relative error in the eigenvalue will be small.

This discussion leads to the power method with Rayleigh quotient computation. Given $\boldsymbol{A} \in \mathbb{C}^{n \times n}$, a starting vector $\boldsymbol{z} \in \mathbb{C}^n$, a maximum number K of iterations, and a convergence tolerance tol. The power method combined with a Rayleigh quotient estimate for the eigenvalue is used to compute a dominant eigenpair $(l, \boldsymbol{x})$ of $\boldsymbol{A}$ with $\|\boldsymbol{x}\|_2 = 1$. The integer it returns the number of iterations needed in order for $\|\boldsymbol{A}\boldsymbol{x} - l\boldsymbol{x}\|_2/\|\boldsymbol{A}\|_F < tol$. If no such eigenpair is found in K iterations the value $it = K + 1$ is returned.

```
function [l,x,it]=powerit(A,z,K,tol)
% [l,x,it]=powerit(A,z,K,tol)
af=norm(A,'fro'); x=z/norm(z);
for k=1:K
   y=A*x; l=x'*y;
   if norm(y-l*x)/af<tol
      it=k; x=y/norm(y); return
   end
   x=y/norm(y);
end
it=K+1;
end
```

Listing 15.1 powerit

Example 15.2 (Power Method) We try `powerit` on the three matrices

$$A_1 := \begin{bmatrix} 1 & 2 \\ 3 & 4 \end{bmatrix}, \quad A_2 := \begin{bmatrix} 1.7 & -0.4 \\ 0.15 & 2.2 \end{bmatrix}, \text{ and } A_3 = \begin{bmatrix} 1 & 2 \\ -3 & 4 \end{bmatrix}.$$

In each case we start with the random vector $z = [0.6602, 0.3420]$ and $tol = 10^{-6}$. For A_1 we get convergence in 7 iterations, for A_2 it takes 174 iterations, and for A_3 we do not get convergence.

The matrix A_3 does not have a dominant eigenvalue since the two eigenvalues are complex conjugate of each other. Thus the basic condition (i) of (15.2) is not satisfied and the power method diverges. The enormous difference in the rate of convergence for A_1 and A_2 can be explained by looking at (15.4). The rate of convergence depends on the ratio $\frac{|\lambda_2|}{|\lambda_1|}$. If this ratio is small then the convergence is fast, while it can be quite slow if the ratio is close to one. The eigenvalues of A_1 are $\lambda_1 = 5.3723$ and $\lambda_2 = -0.3723$ giving a quite small ratio of 0.07 and the convergence is fast. On the other hand the eigenvalues of A_2 are $\lambda_1 = 2$ and $\lambda_2 = 1.9$ and the corresponding ratio is 0.95 resulting in slow convergence.

A variant of the power method is the **shifted power method**. In this method we choose a number s and apply the power method to the matrix $A - sI$. The number s is called a shift since it shifts an eigenvalue λ of A to $\lambda - s$ of $A - sI$. Sometimes the convergence can be faster if the shift is chosen intelligently. For example, if we apply the shifted power method to A_2 in Example 15.2 with shift 1.8, then with the same starting vector and tol as above, we get convergence in 17 iterations instead of 174 for the unshifted algorithm.

15.1.2 *The Inverse Power Method*

Another variant of the power method with Rayleigh quotient is the **inverse power method**. This method can be used to determine any eigenpair (λ, x) of A as long as λ has algebraic multiplicity one. In the inverse power method we apply the power method to the inverse matrix $(A - sI)^{-1}$, where s is a shift. If A has eigenvalues $\lambda_1, \dots, \lambda_n$ in no particular order then $(A - sI)^{-1}$ has eigenvalues

$$\mu_1(s) = (\lambda_1 - s)^{-1}, \mu_2(s) = (\lambda_2 - s)^{-1}, \dots, \mu_n(s) = (\lambda_n - s)^{-1}.$$

Suppose λ_1 is a simple eigenvalue of A. Then $\lim_{s \to \lambda_1} |\mu_1(s)| = \infty$, while $\lim_{s \to \lambda_1} \mu_j(s) = (\lambda_j - \lambda_1)^{-1} < \infty$ for $j = 2, \dots, n$. Hence, by choosing s sufficiently close to λ_1 the inverse power method will converge to that eigenvalue.

For the inverse power method (15.6) is replaced by

$$\begin{aligned} &(i) \quad (\boldsymbol{A} - s\boldsymbol{I})\boldsymbol{y}_k = \boldsymbol{x}_{k-1} \\ &(ii) \quad \boldsymbol{x}_k = \boldsymbol{y}_k / \|\boldsymbol{y}_k\|. \end{aligned} \tag{15.7}$$

Note that we solve the linear system rather than computing the inverse matrix. Normally the PLU factorization of $\boldsymbol{A} - s\boldsymbol{I}$ is precomputed in order to speed up the computation.

15.1.3 Rayleigh Quotient Iteration

A variant of the inverse power method is known simply as **Rayleigh quotient iteration**. In this method we change the shift from iteration to iteration, using the previous Rayleigh quotient s_{k-1} as the current shift. In each iteration we need to compute the following quantities

$$\begin{aligned} &(i) \quad (\boldsymbol{A} - s_{k-1}\boldsymbol{I})\boldsymbol{y}_k = \boldsymbol{x}_{k-1}, \\ &(ii) \quad \boldsymbol{x}_k = \boldsymbol{y}_k / \|\boldsymbol{y}_k\|, \\ &(iii) \quad s_k = \boldsymbol{x}_k^* \boldsymbol{A} \boldsymbol{x}_k, \\ &(iv) \quad \boldsymbol{r}_k = \boldsymbol{A}\boldsymbol{x}_k - s_k \boldsymbol{x}_k. \end{aligned}$$

We can avoid the calculation of $\boldsymbol{A}\boldsymbol{x}_k$ in (iii) and (iv). Let

$$\rho_k := \frac{\boldsymbol{y}_k^* \boldsymbol{x}_{k-1}}{\boldsymbol{y}_k^* \boldsymbol{y}_k}, \quad \boldsymbol{w}_k := \frac{\boldsymbol{x}_{k-1}}{\|\boldsymbol{y}_k\|_2}.$$

Then

$$\begin{aligned} s_k &= \frac{\boldsymbol{y}_k^* \boldsymbol{A} \boldsymbol{y}_k}{\boldsymbol{y}_k^* \boldsymbol{y}_k} = s_{k-1} + \frac{\boldsymbol{y}_k^* (\boldsymbol{A} - s_{k-1}\boldsymbol{I}) \boldsymbol{y}_k}{\boldsymbol{y}_k^* \boldsymbol{y}_k} = s_{k-1} + \frac{\boldsymbol{y}_k^* \boldsymbol{x}_{k-1}}{\boldsymbol{y}_k^* \boldsymbol{y}_k} = s_{k-1} + \rho_k, \\ \boldsymbol{r}_k &= \boldsymbol{A}\boldsymbol{x}_k - s_k \boldsymbol{x}_k = \frac{\boldsymbol{A}\boldsymbol{y}_k - (s_{k-1} + \rho_k)\boldsymbol{y}_k}{\|\boldsymbol{y}_k\|_2} = \frac{\boldsymbol{x}_{k-1} - \rho_k \boldsymbol{y}_k}{\|\boldsymbol{y}_k\|_2} = \boldsymbol{w}_k - \rho_k \boldsymbol{x}_k. \end{aligned}$$

Another problem is that the linear system in (i) becomes closer and closer to singular as s_k converges to the eigenvalue. Thus the system becomes more and more ill-conditioned and we can expect large errors in the computed $\boldsymbol{y}_k$. This is indeed true, but we are lucky. Most of the error occurs in the direction of the eigenvector and this error disappears when we normalize $\boldsymbol{y}_k$ in (ii). Miraculously, the normalized eigenvector will be quite accurate.

Given an approximation $(s, \boldsymbol{x})$ to an eigenpair $(\lambda, \boldsymbol{v})$ of a matrix $\boldsymbol{A} \in \mathbb{C}^{n\times n}$. The following algorithm computes a hopefully better approximation to $(\lambda, \boldsymbol{v})$ by doing one Rayleigh quotient iteration. The length nr of the new residual is also returned.

```
function [x,s,nr]=rayleighit(A,x,s)
% [x,s,nr]=rayleighit(A,x,s)
n=length(x);
y=(A-s*eye(n,n))\x;
yn=norm(y);
w=x/yn;
x=y/yn;
rho=x'*w;
s=s+rho;
nr=norm(w-rho*x);
end
```

Listing 15.2 rayleighit

Since the shift changes from iteration to iteration the computation of $\boldsymbol{y}$ in `rayleighit` will require $O(n^3)$ arithmetic operations for a full matrix. For such a matrix it might be useful to reduce it to an upper Hessenberg form, or tridiagonal form, before starting the iteration. However, if we have a good approximation to an eigenpair then only a few iterations are necessary to obtain close to machine accuracy.

If Rayleigh quotient iteration converges the convergence will be quadratic and sometimes even cubic. We illustrate this with an example.

Example 15.3 (Rayleigh Quotient Iteration) The smallest eigenvalue of the matrix $\boldsymbol{A} = \begin{bmatrix} 1 & 2 \\ 3 & 4 \end{bmatrix}$ is $\lambda_1 = (5 - \sqrt{33})/2 \approx -0.37$. Starting with $\boldsymbol{x} = [1, 1]^T$ and $s = 0$ `rayleighit` converges to this eigenvalue and corresponding eigenvector. In Table 15.1 we show the rate of convergence by iterating `rayleighit` 5 times. The errors are approximately squared in each iteration indicating quadratic convergence.

Table 15.1 Quadratic convergence of Rayleigh quotient iteration

k	1	2	3	4	5
$\|\boldsymbol{r}\|_2$	1.0e+000	7.7e−002	1.6e−004	8.2e−010	2.0e−020
$\|s - \lambda_1\|$	3.7e−001	−1.2e−002	−2.9e−005	−1.4e−010	−2.2e−016

15.2 The Basic QR Algorithm

The QR algorithm is an iterative method to compute all eigenvalues and eigenvectors of a matrix $\boldsymbol{A} \in \mathbb{C}^{n\times n}$. The matrix is reduced to triangular form by a sequence of unitary similarity transformations computed from the QR factorization of $\boldsymbol{A}$. Recall that for a square matrix the QR factorization and the QR decomposition are the same. If $\boldsymbol{A} = \boldsymbol{Q}\boldsymbol{R}$ is a QR factorization then $\boldsymbol{Q} \in \mathbb{C}^{n\times n}$ is unitary, $\boldsymbol{Q}^*\boldsymbol{Q} = \boldsymbol{I}$ and $\boldsymbol{R} \in \mathbb{C}^{n\times n}$ is upper triangular.

The basic QR algorithm takes the following form:

$$\begin{array}{|l|}\hline \boldsymbol{A}_1 = \boldsymbol{A} \\ \text{for } k = 1, 2, \ldots \\ \quad \boldsymbol{Q}_k\boldsymbol{R}_k = \boldsymbol{A}_k \qquad \text{(QR factorization of } \boldsymbol{A}_k) \\ \quad \boldsymbol{A}_{k+1} = \boldsymbol{R}_k\boldsymbol{Q}_k. \\ \text{end} \\ \hline \end{array} \tag{15.8}$$

The determination of the QR factorization of $\boldsymbol{A}_k$ and the computation of $\boldsymbol{R}_k\boldsymbol{Q}_k$ is called a QR step. It is not at all clear that a QR step does anything useful. At this point, since $\boldsymbol{R}_k = \boldsymbol{Q}_k^*\boldsymbol{A}_k$ we find

$$\boldsymbol{A}_{k+1} = \boldsymbol{R}_k\boldsymbol{Q}_k = \boldsymbol{Q}_k^*\boldsymbol{A}_k\boldsymbol{Q}_k, \tag{15.9}$$

so $\boldsymbol{A}_{k+1}$ is unitary similar to $\boldsymbol{A}_k$. By induction $\boldsymbol{A}_{k+1}$ is unitary similar to $\boldsymbol{A}$. Thus, each $\boldsymbol{A}_k$ has the same eigenvalues as $\boldsymbol{A}$. We shall see that the basic QR algorithm is related to the power method.

Here are two examples to illustrate what happens.

Example 15.4 (QR Iteration; Real Eigenvalues) We start with

$$\boldsymbol{A}_1 = \boldsymbol{A} = \begin{bmatrix} 2 & 1 \\ 1 & 2 \end{bmatrix} = \Big(\frac{1}{\sqrt{5}}\begin{bmatrix} -2 & -1 \\ -1 & 2 \end{bmatrix}\Big) * \Big(\frac{1}{\sqrt{5}}\begin{bmatrix} -5 & -4 \\ 0 & 3 \end{bmatrix}\Big) = \boldsymbol{Q}_1\boldsymbol{R}_1$$

and obtain

$$\boldsymbol{A}_2 = \boldsymbol{R}_1\boldsymbol{Q}_1 = \frac{1}{5}\begin{bmatrix} -5 & -4 \\ 0 & 3 \end{bmatrix} * \begin{bmatrix} -2 & -1 \\ -1 & 2 \end{bmatrix} = \frac{1}{5}\begin{bmatrix} 14 & -3 \\ -3 & 6 \end{bmatrix} = \begin{bmatrix} 2.8 & -0.6 \\ -0.6 & 1.2 \end{bmatrix}.$$

Continuing we find

$$\boldsymbol{A}_4 \approx \begin{bmatrix} 2.997 & -0.074 \\ -0.074 & 1.0027 \end{bmatrix}, \quad \boldsymbol{A}_{10} \approx \begin{bmatrix} 3.0000 & -0.0001 \\ -0.0001 & 1.0000 \end{bmatrix}$$

$\boldsymbol{A}_{10}$ is almost diagonal and contains approximations to the eigenvalues $\lambda_1 = 3$ and $\lambda_2 = 1$ on the diagonal.

Example 15.5 (QR Iteration; Complex Eigenvalues) Applying the QR iteration (15.8) to the matrix

$$A_1 = A = \begin{bmatrix} 0.9501 & 0.8913 & 0.8214 & 0.9218 \\ 0.2311 & 0.7621 & 0.4447 & 0.7382 \\ 0.6068 & 0.4565 & 0.6154 & 0.1763 \\ 0.4860 & 0.0185 & 0.7919 & 0.4057 \end{bmatrix}$$

we obtain

$$A_{14} = \left[\begin{array}{c|cc|c} 2.323 & 0.047223 & -0.39232 & -0.65056 \\ \hline -2.1e-10 & 0.13029 & 0.36125 & 0.15946 \\ -4.1e-10 & -0.58622 & 0.052576 & -0.25774 \\ \hline 1.2e-14 & 3.3e-05 & -1.1e-05 & 0.22746 \end{array}\right].$$

This matrix is almost quasi-triangular and estimates for the eigenvalues $\lambda_1, \ldots, \lambda_4$ of $\boldsymbol{A}$ can now easily be determined from the diagonal blocks of $\boldsymbol{A}_{14}$. The 1×1 blocks give us two real eigenvalues $\lambda_1 \approx 2.323$ and $\lambda_4 \approx 0.2275$. The middle 2×2 block has complex eigenvalues resulting in $\lambda_2 \approx 0.0914 + 0.4586i$ and $\lambda_3 \approx 0.0914 - 0.4586i$. From Gerschgorin's circle Theorem 14.1 and Corollary 14.1 it follows that the approximations to the real eigenvalues are quite accurate. We would also expect the complex eigenvalues to have small absolute errors.

These two examples illustrate what most often happens in general. The sequence $(\boldsymbol{A}_k)_k$ converges to the triangular Schur form (Cf. Theorem 6.5) if all the eigenvalues are real or the quasi-triangular Schur form (Cf. Definition 6.5) if some of the eigenvalues are complex.

15.2.1 Relation to the Power Method

Let us show that the basic QR algorithm is related to the power method. We obtain the QR factorization of the powers $\boldsymbol{A}^k$ as follows:

Theorem 15.3 (QR and Power) *For $k = 1, 2, 3, \ldots$, the QR factorization of $\boldsymbol{A}^k$ is $\boldsymbol{A}^k = \tilde{\boldsymbol{Q}}_k \tilde{\boldsymbol{R}}_k$, where*

$$\tilde{\boldsymbol{Q}}_k := \boldsymbol{Q}_1 \cdots \boldsymbol{Q}_k \text{ and } \tilde{\boldsymbol{R}}_k := \boldsymbol{R}_k \cdots \boldsymbol{R}_1, \tag{15.10}$$

and $\boldsymbol{Q}_1, \ldots, \boldsymbol{Q}_k$, $\boldsymbol{R}_1, \ldots, \boldsymbol{R}_k$ are the matrices generated by the basic QR algorithm (15.8).

$$A = \begin{bmatrix} x & x & x & x \\ 0 & x & x & x \\ 0 & 0 & x & x \\ 0 & 0 & 0 & x \end{bmatrix} \overset{P_{12}^*}{\rightarrow} \begin{bmatrix} x & x & x & x \\ \mathbf{x} & x & x & x \\ 0 & 0 & x & x \\ 0 & 0 & 0 & x \end{bmatrix} \overset{P_{23}^*}{\rightarrow} \begin{bmatrix} x & x & x & x \\ x & x & x & x \\ 0 & \mathbf{x} & x & x \\ 0 & 0 & 0 & x \end{bmatrix} \overset{P_{34}^*}{\rightarrow} \begin{bmatrix} x & x & x & x \\ x & x & x & x \\ 0 & x & x & x \\ 0 & 0 & \mathbf{x} & x \end{bmatrix}.$$

Fig. 15.1 Post multiplication in a QR step

Proof By (15.9)

$$A_k = Q_{k-1}^* A_{k-1} Q_{k-1} = Q_{k-1}^* Q_{k-2}^* A_{k-2} Q_{k-2} Q_{k-1} = \cdots = \tilde{Q}_{k-1}^* A \tilde{Q}_{k-1}. \tag{15.11}$$

The proof is by induction on k. Clearly $\tilde{Q}_1 \tilde{R}_1 = Q_1 R_1 = A_1$. Suppose $\tilde{Q}_{k-1} \tilde{R}_{k-1} = A^{k-1}$ for some $k \geq 2$. Since $Q_k R_k = A_k$ and using (15.11)

$$\tilde{Q}_k \tilde{R}_k = \tilde{Q}_{k-1}(Q_k R_k)\tilde{R}_{k-1} = \tilde{Q}_{k-1} A_k \tilde{R}_{k-1} = (\tilde{Q}_{k-1} \tilde{Q}_{k-1}^*) A \tilde{Q}_{k-1} \tilde{R}_{k-1} = A^k.$$

□

Since $\tilde{R}_k$ is upper triangular, its first column is a multiple of e_1 so that

$$A^k e_1 = \tilde{Q}_k \tilde{R}_k e_1 = \tilde{r}_{11}^{(k)} \tilde{Q}_k e_1 \text{ or } \tilde{q}_1^{(k)} := \tilde{Q}_k e_1 = \frac{1}{\tilde{r}_{11}^{(k)}} A^k e_1.$$

Since $\|\tilde{q}_1^{(k)}\|_2 = 1$ the first column of $\tilde{Q}_k$ is the result of applying the normalized power iteration (15.6) to the starting vector $x_0 = e_1$. If this iteration converges we conclude that the first column of $\tilde{Q}_k$ must converge to a dominant eigenvector of A. It can be shown that the first column of A_k must then converge to $\lambda_1 e_1$, where λ_1 is a dominant eigenvalue of A. This is clearly what happens in Examples 15.4 and 15.5. Indeed, what is observed in practice is that the sequence $(\tilde{Q}_k^* A \tilde{Q}_k)_k$ converges to a (quasi-triangular) Schur form of A.

15.2.2 Invariance of the Hessenberg Form

One QR step requires $O(n^3)$ arithmetic operations for a matrix A of order n. By an initial reduction of A to upper Hessenberg form H_1 using Algorithm 14.1, the cost of a QR step can be reduced to $O(n^2)$. Consider a QR step on H_1. We first determine plane rotations $P_{i,i+1}, i = 1, \ldots, n-1$ so that $P_{n-1,n} \cdots P_{1,2} H_1 = R_1$ is upper triangular. The details were described in Sect. 5.6. Thus $H_1 = Q_1 R_1$, where $Q_1 = P_{1,2}^* \cdots P_{n-1,n}^*$ is a QR factorization of H_1. To finish the QR step we compute $R_1 Q_1 = R_1 P_{1,2}^* \cdots P_{n-1,n}^*$. This postmultiplication step is illustrated by the Wilkinson diagram in Fig. 15.1.

The postmultiplication by $P_{i,i+1}$ introduces a nonzero in position $(i+1, i)$ leaving the other elements marked by a zero in Fig. 15.1 unchanged. Thus the final matrix $\boldsymbol{R}\boldsymbol{P}_{1,2}^* \cdots \boldsymbol{P}_{n-1,n}^*$ is upper Hessenberg and a QR step leaves the Hessenberg form invariant.

In conclusion, to compute $\boldsymbol{A}_{k+1}$ from $\boldsymbol{A}_k$ requires $O(n^2)$ arithmetic operations if $\boldsymbol{A}_k$ is upper Hessenberg and $O(n)$ arithmetic operations if $\boldsymbol{A}_k$ is tridiagonal.

15.2.3 *Deflation*

If a subdiagonal element $a_{i+1,i}$ of an upper Hessenberg matrix $\boldsymbol{A}$ is equal to zero, then the eigenvalues of $\boldsymbol{A}$ are the union of the eigenvalues of the two smaller matrices $A(1:i, 1:i)$ and $A(i+1:n, i+1:n)$. Thus, if during the iteration the $(i+1, i)$ element of $\boldsymbol{A}_k$ is sufficiently small then we can continue the iteration on the two smaller submatrices separately.

To see what effect this can have on the eigenvalues of $\boldsymbol{A}$ suppose $|a_{i+1,i}^{(k)}| \le \epsilon$. Let $\hat{\boldsymbol{A}}_k := \boldsymbol{A}_k - a_{i+1,i}^{(k)} \boldsymbol{e}_{i+1} \boldsymbol{e}_i^T$ be the matrix obtained from $\boldsymbol{A}_k$ by setting the $(i+1, i)$ element equal to zero. Since $\boldsymbol{A}_k = \tilde{\boldsymbol{Q}}_{k-1}^* \boldsymbol{A} \tilde{\boldsymbol{Q}}_{k-1}$ we have

$$\hat{\boldsymbol{A}}_k = \tilde{\boldsymbol{Q}}_{k-1}^* (\boldsymbol{A} + \boldsymbol{E}) \tilde{\boldsymbol{Q}}_{k-1}, \quad \boldsymbol{E} = \tilde{\boldsymbol{Q}}_{k-1} (a_{i+1,i}^{(k)} \boldsymbol{e}_{i+1} \boldsymbol{e}_i^T) \tilde{\boldsymbol{Q}}_{k-1}^*.$$

Since $\tilde{\boldsymbol{Q}}_{k-1}$ is unitary, $\|\boldsymbol{E}\|_F = \|a_{i+1,i}^{(k)} \boldsymbol{e}_{i+1} \boldsymbol{e}_i^T\|_F = |a_{i+1,i}^{(k)}| \le \epsilon$ and setting $a_{i+1,i}^{(k)} = 0$ amounts to a perturbation in the original $\boldsymbol{A}$ of at most ϵ. For how to chose ϵ see the discussion on page 94-95 in [18].

This deflation occurs often in practice and can with a proper implementation reduce the computation time considerably. It should be noted that to find the eigenvectors of the original matrix one has to continue with some care, see [18].

15.3 The Shifted QR Algorithms

Like in the inverse power method it is possible to speed up the convergence by introducing shifts. The **explicitly shifted QR algorithm** works as follows:

$\boldsymbol{A}_1 = \boldsymbol{A}$

for $k = 1, 2, \ldots$

  Choose a shift s_k

  $\boldsymbol{Q}_k \boldsymbol{R}_k = \boldsymbol{A}_k - s_k \boldsymbol{I}$ (QR factorization of $\boldsymbol{A}_k - s\boldsymbol{I}$)

  $\boldsymbol{A}_{k+1} = \boldsymbol{R}_k \boldsymbol{Q}_k + s_k \boldsymbol{I}$.

end

Since $\boldsymbol{R}_k = \boldsymbol{Q}_k^*(\boldsymbol{A}_k - s_k\boldsymbol{I})$ we find

$$\boldsymbol{A}_{k+1} = \boldsymbol{Q}_k^*(\boldsymbol{A}_k - s_k\boldsymbol{I})\boldsymbol{Q}_k + s_k\boldsymbol{I} = \boldsymbol{Q}_k^*\boldsymbol{A}_k\boldsymbol{Q}_k$$

and $\boldsymbol{A}_{k+1}$ and $\boldsymbol{A}_k$ are unitary similar.

The shifted QR algorithm is related to the power method with shift, cf. Theorem 15.3 and also the inverse power method. In fact the last column of $\boldsymbol{Q}_k$ is the result of one iteration of the inverse power method to $\boldsymbol{A}^*$ with shift s_k. Indeed, since $\boldsymbol{A} - s_k\boldsymbol{I} = \boldsymbol{Q}_k\boldsymbol{R}_k$ we have $(\boldsymbol{A} - s_k\boldsymbol{I})^* = \boldsymbol{R}_k^*\boldsymbol{Q}_k^*$ and $(\boldsymbol{A} - s_k\boldsymbol{I})^*\boldsymbol{Q}_k = \boldsymbol{R}_k^*$. Thus, since $\boldsymbol{R}_k^*$ is lower triangular with n, n element $\overline{r}_{nn}^{(k)}$ we find $(\boldsymbol{A} - s_k\boldsymbol{I})^*\boldsymbol{Q}_k\boldsymbol{e}_n = \boldsymbol{R}_k^*\boldsymbol{e}_n = \overline{r}_{nn}^{(k)}\boldsymbol{e}_n$ from which the conclusion follows.

The shift $s_k := \boldsymbol{e}_n^T\boldsymbol{A}_k\boldsymbol{e}_n$ is called the **Rayleigh quotient shift**, while the eigenvalue of the lower right 2×2 corner of $\boldsymbol{A}_k$ closest to the n, n element of $\boldsymbol{A}_k$ is called the**Wilkinson shift**. This shift can be used to find complex eigenvalues of a real matrix. The convergence is very fast and at least quadratic both for the Rayleigh quotient shift and the Wilkinson shift.

By doing two QR iterations at a time it is possible to find both real and complex eigenvalues of a real matrix without using complex arithmetic. The corresponding algorithm is called the **implicitly shifted QR algorithm**.

After having computed the eigenvalues we can compute the eigenvectors in steps. First we find the eigenvectors of the triangular or quasi-triangular matrix. We then compute the eigenvectors of the upper Hessenberg matrix and finally we get the eigenvectors of $\boldsymbol{A}$.

The QR Algorithm without shifts does not always converge. A simple example is given by $\boldsymbol{A} :=? \begin{bmatrix} 0 & 1 \\ 1 & 0 \end{bmatrix}$. We obtain $\boldsymbol{A}_k = \boldsymbol{A}$ for all $k \in \mathbb{N}$. For convergence of the shifted QR algorithm for unitary upper Hessenberg matrices see [20].

Practical experience indicates that only $O(n)$ iterations are needed to find all eigenvalues of $\boldsymbol{A}$. Thus both the explicit- and implicit shift QR algorithms are normally $O(n^3)$ algorithms.

For further remarks and detailed algorithms see [18].

15.4 Exercises Chap. 15

15.4.1 Exercises Sect. 15.1

Exercise 15.1 (Orthogonal Vectors) Show that $\boldsymbol{u}$ and $\boldsymbol{A}\boldsymbol{u} - \lambda\boldsymbol{u}$ are orthogonal when $\lambda = \frac{\boldsymbol{u}^*\boldsymbol{A}\boldsymbol{u}}{\boldsymbol{u}^*\boldsymbol{u}}$.

15.5 Review Questions

15.4.1 What is the main use of the power method?
15.4.2 Can the QR method be used to find all eigenvectors of a matrix?
15.4.3 Can the power method be used to find an eigenvalue?
15.4.4 Do the power method converge to an eigenvector corresponding to a complex eigenvalue?
15.4.5 What is the inverse power method?
15.4.6 Give a relation between the QR algorithm and the power method.
15.4.7 How can we make the basic QR algorithm converge faster?

Part VII
Appendix

Chapter 16
Differentiation of Vector Functions

We give a short introduction to differentiation of vector functions.

For any sufficiently differentiable $f : \mathbb{R}^n \to \mathbb{R}$ we recall that the partial derivative with respect to the ith variable of f is defined by

$$D_i f(\boldsymbol{x}) := \frac{\partial f(\boldsymbol{x})}{\partial x_i} := \lim_{h \to \boldsymbol{0}} \frac{f(\boldsymbol{x} + h\boldsymbol{e}_i) - f(\boldsymbol{x})}{h}, \quad \boldsymbol{x} \in \mathbb{R}^n,$$

where $\boldsymbol{e}_i$ is the ith unit vector in $\mathbb{R}^n$. For each $\boldsymbol{x} \in \mathbb{R}^n$ we define the **gradient** $\nabla f(\boldsymbol{x}) \in \mathbb{R}^n$, and the **hessian** $\boldsymbol{H}f = \nabla\nabla^T f(\boldsymbol{x}) \in \mathbb{R}^{n \times n}$ of f by

$$\nabla f := \begin{bmatrix} D_1 f \\ \vdots \\ D_n f \end{bmatrix}, \quad \boldsymbol{H}f := \nabla\nabla^T f := \begin{bmatrix} D_1 D_1 f & \cdots & D_1 D_n f \\ \vdots & & \vdots \\ D_n D_1 & \cdots & D_n D_n f \end{bmatrix}, \tag{16.1}$$

where $\nabla^T f := (\nabla f)^T$ is the row vector gradient. The operators $\nabla\nabla^T$ and $\nabla^T\nabla$ are quite different. Indeed, $\nabla^T\nabla f = D_1^2 f + \cdots + D_n^2 f =: \nabla^2$ the **Laplacian** of f, while $\nabla\nabla^T$ can be thought of as an outer product resulting in a matrix.

Lemma 16.1 (Product Rules) *For $f, g : \mathbb{R}^n \to \mathbb{R}$ we have the product rules*

1. $\nabla(fg) = f\nabla g + g\nabla f, \quad \nabla^T(fg) = f\nabla^T g + g\nabla^T f,$
2. $\nabla\nabla^T(fg) = \nabla f \nabla^T g + \nabla g \nabla^T f + f\nabla\nabla^T g + g\nabla\nabla^T f.$
3. $\nabla^2(fg) = 2\nabla^T f \nabla g + f\nabla^2 g + g\nabla^2 f.$

T. Lyche, *Numerical Linear Algebra and Matrix Factorizations*,
Texts in Computational Science and Engineering 22,
https://doi.org/10.1007/978-3-030-36468-7_16

We define the **Jacobian** of a vector function $\boldsymbol{f} = [f_1, \ldots f_m]^T : \mathbb{R}^n \to \mathbb{R}^m$ as the m, n matrix

$$\nabla^T \boldsymbol{f} := \begin{bmatrix} D_1 f_1 & \cdots & D_n f_1 \\ \vdots & & \vdots \\ D_1 f_m & \cdots & D_n f_m \end{bmatrix}.$$

As an example, if $f(\boldsymbol{x}) = f(x, y) = x^2 - xy + y^2$ and $\boldsymbol{g}(x, y) := [f(x, y),\ x - y]^T$ then

$$\nabla f(x, y) = \begin{bmatrix} 2x - y \\ -x + 2y \end{bmatrix}, \quad \nabla^T \boldsymbol{g}(x, y) = \begin{bmatrix} 2x - y & -x + 2y \\ 1 & -1 \end{bmatrix},$$

$$\boldsymbol{H} f(x, y) = \begin{bmatrix} \frac{\partial^2 f}{\partial x^2} & \frac{\partial^2 f}{\partial x \partial y} \\ \frac{\partial^2 f}{\partial y \partial x} & \frac{\partial^2 f}{\partial y^2} \end{bmatrix} = \begin{bmatrix} 2 & -1 \\ -1 & 2 \end{bmatrix}.$$

The second order Taylor expansion in n variables can be expressed in terms of the gradient and the hessian.

Lemma 16.2 (Second Order Taylor Expansion) *Suppose $f \in C^2(\Omega)$, where $\Omega \in \mathbb{R}^n$ contains two points $\boldsymbol{x}, \boldsymbol{x}+\boldsymbol{h} \in \Omega$, such that the line segment $L := \{\boldsymbol{x}+t\boldsymbol{h} : t \in (0, 1)\} \subset \Omega$. Then*

$$f(\boldsymbol{x} + \boldsymbol{h}) = f(\boldsymbol{x}) + \boldsymbol{h}^T \nabla f(\boldsymbol{x}) + \frac{1}{2} \boldsymbol{h}^T \nabla \nabla^T f(\boldsymbol{c}) \boldsymbol{h}, \text{ for some } \boldsymbol{c} \in L. \tag{16.2}$$

Proof Let $g : [0, 1] \to \mathbb{R}$ be defined by $g(t) := f(\boldsymbol{x} + t\boldsymbol{h})$. Then $g \in C^2[0, 1]$ and by the chain rule

$$g(0) = f(\boldsymbol{x}) \quad g(1) = f(\boldsymbol{x} + \boldsymbol{h}),$$

$$g'(t) = \sum_{i=1}^{n} h_i \frac{\partial f(\boldsymbol{x} + t\boldsymbol{h})}{\partial x_i} = \boldsymbol{h}^T \nabla f(\boldsymbol{x} + t\boldsymbol{h}),$$

$$g''(t) = \sum_{i=1}^{n} \sum_{j=1}^{n} h_i h_j \frac{\partial^2 f(\boldsymbol{x} + t\boldsymbol{h})}{\partial x_i \partial x_j} = \boldsymbol{h}^T \nabla \nabla^T f(\boldsymbol{x} + t\boldsymbol{h}) \boldsymbol{h}.$$

Inserting these expressions in the second order Taylor expansion

$$g(1) = g(0) + g'(0) + \frac{1}{2} g''(u), \text{ for some } u \in (0, 1),$$

we obtain (16.2) with $\boldsymbol{c} = \boldsymbol{x} + u\boldsymbol{h}$. □

The gradient and hessian of some functions involving matrices can be found from the following lemma.

Lemma 16.3 (Functions Involving Matrices) *For any $m, n \in \mathbb{N}$, $\boldsymbol{B} \in \mathbb{R}^{n\times n}$, $\boldsymbol{C} \in \mathbb{R}^{m\times n}$, and $\boldsymbol{x} \in \mathbb{R}^n$, $\boldsymbol{y} \in \mathbb{R}^m$ we have*

1. $\nabla(\boldsymbol{y}^T\boldsymbol{C}) = \nabla^T(\boldsymbol{C}\boldsymbol{x}) = \boldsymbol{C}$,
2. $\nabla(\boldsymbol{x}^T\boldsymbol{B}\boldsymbol{x}) = (\boldsymbol{B} + \boldsymbol{B}^T)\boldsymbol{x}, \quad \nabla^T(\boldsymbol{x}^T\boldsymbol{B}\boldsymbol{x}) = \boldsymbol{x}^T(\boldsymbol{B} + \boldsymbol{B}^T)$,
3. $\nabla\nabla^T(\boldsymbol{x}^T\boldsymbol{B}\boldsymbol{x}) = \boldsymbol{B} + \boldsymbol{B}^T$.

Proof

1. We find $D_i(\boldsymbol{y}^T\boldsymbol{C}) = \lim_{h\to 0}\frac{1}{h}\big((\boldsymbol{y} + h\boldsymbol{e}_i)^T\boldsymbol{C} - \boldsymbol{y}^T\boldsymbol{C}\big) = \boldsymbol{e}_i^T\boldsymbol{C}$ and $D_i(\boldsymbol{C}\boldsymbol{x}) = \lim_{h\to 0}\frac{1}{h}(\boldsymbol{C}(\boldsymbol{x} + h\boldsymbol{e}_i) - \boldsymbol{C}\boldsymbol{x}) = \boldsymbol{C}\boldsymbol{e}_i$ and 1. follows.
2. Here we find

$$
\begin{aligned}
D_i(\boldsymbol{x}^T\boldsymbol{B}\boldsymbol{x}) &= \lim_{h\to 0}\frac{1}{h}\big((\boldsymbol{x} + h\boldsymbol{e}_i)^T\boldsymbol{B}(\boldsymbol{x} + h\boldsymbol{e}_i) - \boldsymbol{x}^T\boldsymbol{B}\boldsymbol{x}\big)\\
&= \lim_{h\to 0}\big(\boldsymbol{e}_i^T\boldsymbol{B}\boldsymbol{x} + \boldsymbol{x}^T\boldsymbol{B}\boldsymbol{e}_i + h\boldsymbol{e}_i^T\boldsymbol{e}_i\big) = \boldsymbol{e}_i^T(\boldsymbol{B} + \boldsymbol{B}^T)\boldsymbol{x},
\end{aligned}
$$

   and the first part of 2. follows. Taking transpose we obtain the second part.
3. Combining 1. and 2. we obtain 3.

□

References

1. O. Axelsson, *Iterative Solution Methods* (Cambridge University Press, 1994)
2. Å. Björck, *Numerical Methods in Matrix Computations* (Springer, 2015)
3. G.H. Golub, C.F. Van Loan, *Matrix Computations, 4th Edition* (The John Hopkins University Press, Baltimore, MD, 2013)
4. J.F. Grcar, Mathematicians of gaussian elimination. Not. AMS **58**, 782–792 (2011)
5. W. Hackbush, *Iterative Solution of Large Sparse Systems of Equations, Second edition* (Springer International Publishing, Berlin, 2016)
6. M. Hestenes, E. Stiefel, Methods of conjugate gradients for solving linear systems. J. Res. Natl. Bur. Stand. **29**, 409–439 (1952)
7. N.J. Higham, *Accuracy and Stability of Numerical Algorithms, Second Edition* (SIAM, Philadelphia, 2002)
8. M.W. Hirsch, S. Smale, *Differential Equations, Dynamical Systems, and Linear Algebra* (Academic Press, San Diego, 1974)
9. R.A. Horn, C.R. Johnson, *Topics in Matrix Analysis* (Cambridge University Press, Cambridge, UK, 1991)
10. R.A. Horn, C.R. Johnson, *Matrix Analysis, Second edition* (Cambridge University Press, Cambridge, UK, 2013)
11. P. Lancaster, L. Rodman, Canonical forms for hermitian matrix pairs under strict equivalence and congruence. SIAM Rev. **47**, 407–443 (2005)
12. C.L. Lawson, R.J. Hanson, *Solving Least Squares Problems* (Prentice-Hall, Englewood Cliffs, NJ, 1974)
13. D.C. Lay, S.R. Lay, J.J. McDonald, *Linear Algebra and Its Applications (Fifth edition)* (Pearson Education Limited, 2016)
14. C. Van Loan, *Computational Frameworks for the Fast Fourier Transform* (SIAM, Philadelphia, 1992)
15. C.D. Meyer, *Matrix Analysis and Applied Linear Algebra.* (SIAM, Philadelphia, 2000)
16. G.W. Stewart, *Introduction to Matrix Computations* (Academic press, New York, 1973)
17. G.W. Stewart, *Matrix Algorithms Volume I: Basic Decompositions* (SIAM, Philadelphia, 1998)
18. G.W. Stewart, *Matrix Algorithms Volume II: Eigensystems* (SIAM Philadelphia, 2001)
19. G.W. Stewart, J. Sun, *Matrix Perturbation Theory* (Academic Press, San Diego, 1990)
20. T.L. Wang, W.B. Gragg, Convergence of the shifted QR algorithm for unitary Hessenberg matrices. Math. Comput. **71**, 1473–1496 (2002)
21. J.H. Wilkinson, *The Algebraic Eigenvalue Problem* (Clarendon Press, Oxford, 1965)
22. D.M. Young, *Iterative Solution of Large Linear Systems* (Academic Press, New York, 1971)

T. Lyche, *Numerical Linear Algebra and Matrix Factorizations*,
Texts in Computational Science and Engineering 22,
https://doi.org/10.1007/978-3-030-36468-7

Index

T. Lyche, *Numerical Linear Algebra and Matrix Factorizations*,
Texts in Computational Science and Engineering 22,
https://doi.org/10.1007/978-3-030-36468-7

Editorial Policy

1. Textbooks on topics in the field of computational science and engineering will be considered. They should be written for courses in CSE education. Both graduate and undergraduate textbooks will be published in TCSE. Multidisciplinary topics and multidisciplinary teams of authors are especially welcome.

2. Format: Only works in English will be considered. For evaluation purposes, manuscripts may be submitted in print or electronic form, in the latter case, preferably as pdf- or zipped ps-files. Authors are requested to use the LaTeX style files available from Springer at: http://www.springer.com/gp/authors-editors/book-authors-editors/resources-guidelines/rights-permissions-licensing/manuscript-preparation/5636#c3324 (Layout & templates – LaTeX template – contributed books).
   Electronic material can be included if appropriate. Please contact the publisher.

3. Those considering a book which might be suitable for the series are strongly advised to contact the publisher or the series editors at an early stage.

General Remarks

Careful preparation of manuscripts will help keep production time short and ensure a satisfactory appearance of the finished book.

The following terms and conditions hold:

Regarding free copies and royalties, the standard terms for Springer mathematics textbooks hold. Please write to martin.peters@springer.com for details.

Authors are entitled to purchase further copies of their book and other Springer books for their personal use, at a discount of 33.3% directly from Springer-Verlag.

Series Editors

Timothy J. Barth
NASA Ames Research Center
NAS Division
Moffett Field, CA 94035, USA
barth@nas.nasa.gov

Michael Griebel
Institut für Numerische Simulation
der Universität Bonn
Wegelerstr. 6
53115 Bonn, Germany
griebel@ins.uni-bonn.de

David E. Keyes
Mathematical and Computer Sciences
and Engineering
King Abdullah University of Science
and Technology
P.O. Box 55455
Jeddah 21534, Saudi Arabia
david.keyes@kaust.edu.sa

and

Department of Applied Physics
and Applied Mathematics
Columbia University
500 W. 120 th Street
New York, NY 10027, USA
kd2112@columbia.edu

Risto M. Nieminen
Department of Applied Physics
Aalto University School of Science
and Technology
00076 Aalto, Finland
risto.nieminen@tkk.fi

Dirk Roose
Department of Computer Science
Katholieke Universiteit Leuven
Celestijnenlaan 200A
3001 Leuven-Heverlee, Belgium
dirk.roose@cs.kuleuven.be

Tamar Schlick
Department of Chemistry
and Courant Institute
of Mathematical Sciences
New York University
251 Mercer Street
New York, NY 10012, USA
schlick@nyu.edu

Editor for Computational Science
and Engineering at Springer:
Martin Peters
Springer-Verlag
Mathematics Editorial
Tiergartenstrasse 17
69121 Heidelberg, Germany
martin.peters@springer.com

Texts in Computational Science and Engineering

1. H. P. Langtangen, *Computational Partial Differential Equations.* Numerical Methods and Diffpack Programming. 2nd Edition
2. A. Quarteroni, F. Saleri, P. Gervasio, *Scientific Computing with MATLAB and Octave.* 4th Edition
3. H. P. Langtangen, *Python Scripting for Computational Science.* 3rd Edition
4. H. Gardner, G. Manduchi, *Design Patterns for e-Science.*
5. M. Griebel, S. Knapek, G. Zumbusch, *Numerical Simulation in Molecular Dynamics.*
6. H. P. Langtangen, *A Primer on Scientific Programming with Python.* 5th Edition
7. A. Tveito, H. P. Langtangen, B. F. Nielsen, X. Cai, *Elements of Scientific Computing.*
8. B. Gustafsson, *Fundamentals of Scientific Computing.*
9. M. Bader, *Space-Filling Curves.*
10. M. Larson, F. Bengzon, *The Finite Element Method: Theory, Implementation and Applications.*
11. W. Gander, M. Gander, F. Kwok, *Scientific Computing: An Introduction using Maple and MATLAB.*
12. P. Deuflhard, S. Röblitz, *A Guide to Numerical Modelling in Systems Biology.*
13. M. H. Holmes, *Introduction to Scientific Computing and Data Analysis.*
14. S. Linge, H. P. Langtangen, *Programming for Computations* - A Gentle Introduction to Numerical Simulations with MATLAB/Octave.
15. S. Linge, H. P. Langtangen, *Programming for Computations* - A Gentle Introduction to Numerical Simulations with Python.
16. H.P. Langtangen, S. Linge, *Finite Difference Computing with PDEs* - A Modern Software Approach.
17. B. Gustafsson, *Scientific Computing from a Historical Perspective*
18. J.A. Trangenstein, *Scientific Computing* - Vol. I. – Linear and Nonlinear Equations
19. J.A. Trangenstein, *Scientific Computing* - Vol. II. – Eigenvalues and Optimization
20. J.A. Trangenstein, *Scientific Computing* - Vol. III. – Approximation and Integration
21. H.P. Langtangen, K.-A. Mardal, *Introduction to Numerical Methods for Variational Problems.*
22. T. Lyche, *Numerical Linear Algebra and Matrix Factorizations.*

For further information on these books please have a look at our mathematics catalogue at the following URL: `www.springer.com/series/5151`

Monographs in Computational Science and Engineering

1. J. Sundnes, G.T. Lines, X. Cai, B.F. Nielsen, K.-A. Mardal, A. Tveito, *Computing the Electrical Activity in the Heart.*

For further information on this book, please have a look at our mathematics catalogue at the following URL: www.springer.com/series/7417

Lecture Notes in Computational Science and Engineering

1. D. Funaro, *Spectral Elements for Transport-Dominated Equations.*
2. H.P. Langtangen, *Computational Partial Differential Equations.* Numerical Methods and Diffpack Programming.
3. W. Hackbusch, G. Wittum (eds.), *Multigrid Methods V.*
4. P. Deuflhard, J. Hermans, B. Leimkuhler, A.E. Mark, S. Reich, R.D. Skeel (eds.), *Computational Molecular Dynamics: Challenges, Methods, Ideas.*
5. D. Kröner, M. Ohlberger, C. Rohde (eds.), *An Introduction to Recent Developments in Theory and Numerics for Conservation Laws.*
6. S. Turek, *Efficient Solvers for Incompressible Flow Problems.* An Algorithmic and Computational Approach.
7. R. von Schwerin, ***Multi Body System SIM**ulation.* Numerical Methods, Algorithms, and Software.
8. H.-J. Bungartz, F. Durst, C. Zenger (eds.), *High Performance Scientific and Engineering Computing.*
9. T.J. Barth, H. Deconinck (eds.), *High-Order Methods for Computational Physics.*
10. H.P. Langtangen, A.M. Bruaset, E. Quak (eds.), *Advances in Software Tools for Scientific Computing.*
11. B. Cockburn, G.E. Karniadakis, C.-W. Shu (eds.), *Discontinuous Galerkin Methods.* Theory, Computation and Applications.
12. U. van Rienen, *Numerical Methods in Computational Electrodynamics.* Linear Systems in Practical Applications.
13. B. Engquist, L. Johnsson, M. Hammill, F. Short (eds.), *Simulation and Visualization on the Grid.*
14. E. Dick, K. Riemslagh, J. Vierendeels (eds.), *Multigrid Methods VI.*
15. A. Frommer, T. Lippert, B. Medeke, K. Schilling (eds.), *Numerical Challenges in Lattice Quantum Chromodynamics.*
16. J. Lang, *Adaptive Multilevel Solution of Nonlinear Parabolic PDE Systems.* Theory, Algorithm, and Applications.

17. B.I. Wohlmuth, *Discretization Methods and Iterative Solvers Based on Domain Decomposition.*

18. U. van Rienen, M. Günther, D. Hecht (eds.), *Scientific Computing in Electrical Engineering.*

19. I. Babuška, P.G. Ciarlet, T. Miyoshi (eds.), *Mathematical Modeling and Numerical Simulation in Continuum Mechanics.*

20. T.J. Barth, T. Chan, R. Haimes (eds.), *Multiscale and Multiresolution Methods.* Theory and Applications.

21. M. Breuer, F. Durst, C. Zenger (eds.), *High Performance Scientific and Engineering Computing.*

22. K. Urban, *Wavelets in Numerical Simulation.* Problem Adapted Construction and Applications.

23. L.F. Pavarino, A. Toselli (eds.), *Recent Developments in Domain Decomposition Methods.*

24. T. Schlick, H.H. Gan (eds.), *Computational Methods for Macromolecules: Challenges and Applications.*

25. T.J. Barth, H. Deconinck (eds.), *Error Estimation and Adaptive Discretization Methods in Computational Fluid Dynamics.*

26. M. Griebel, M.A. Schweitzer (eds.), *Meshfree Methods for Partial Differential Equations.*

27. S. Müller, *Adaptive Multiscale Schemes for Conservation Laws.*

28. C. Carstensen, S. Funken, W. Hackbusch, R.H.W. Hoppe, P. Monk (eds.), *Computational Electromagnetics.*

29. M.A. Schweitzer, *A Parallel Multilevel Partition of Unity Method for Elliptic Partial Differential Equations.*

30. T. Biegler, O. Ghattas, M. Heinkenschloss, B. van Bloemen Waanders (eds.), *Large-Scale PDE-Constrained Optimization.*

31. M. Ainsworth, P. Davies, D. Duncan, P. Martin, B. Rynne (eds.), *Topics in Computational Wave Propagation.* Direct and Inverse Problems.

32. H. Emmerich, B. Nestler, M. Schreckenberg (eds.), *Interface and Transport Dynamics.* Computational Modelling.

33. H.P. Langtangen, A. Tveito (eds.), *Advanced Topics in Computational Partial Differential Equations.* Numerical Methods and Diffpack Programming.

34. V. John, *Large Eddy Simulation of Turbulent Incompressible Flows.* Analytical and Numerical Results for a Class of LES Models.

35. E. Bänsch (ed.), *Challenges in Scientific Computing - CISC 2002.*

36. B.N. Khoromskij, G. Wittum, *Numerical Solution of Elliptic Differential Equations by Reduction to the Interface.*

37. A. Iske, *Multiresolution Methods in Scattered Data Modelling.*

38. S.-I. Niculescu, K. Gu (eds.), *Advances in Time-Delay Systems.*

39. S. Attinger, P. Koumoutsakos (eds.), *Multiscale Modelling and Simulation.*

40. R. Kornhuber, R. Hoppe, J. Périaux, O. Pironneau, O. Wildlund, J. Xu (eds.), *Domain Decomposition Methods in Science and Engineering.*

41. T. Plewa, T. Linde, V.G. Weirs (eds.), *Adaptive Mesh Refinement – Theory and Applications.*

42. A. Schmidt, K.G. Siebert, *Design of Adaptive Finite Element Software.* The Finite Element Toolbox ALBERTA.

43. M. Griebel, M.A. Schweitzer (eds.), *Meshfree Methods for Partial Differential Equations II.*

44. B. Engquist, P. Lötstedt, O. Runborg (eds.), *Multiscale Methods in Science and Engineering.*

45. P. Benner, V. Mehrmann, D.C. Sorensen (eds.), *Dimension Reduction of Large-Scale Systems.*

46. D. Kressner, *Numerical Methods for General and Structured Eigenvalue Problems.*

47. A. Boriçi, A. Frommer, B. Joó, A. Kennedy, B. Pendleton (eds.), *QCD and Numerical Analysis III.*

48. F. Graziani (ed.), *Computational Methods in Transport.*

49. B. Leimkuhler, C. Chipot, R. Elber, A. Laaksonen, A. Mark, T. Schlick, C. Schütte, R. Skeel (eds.), *New Algorithms for Macromolecular Simulation.*

50. M. Bücker, G. Corliss, P. Hovland, U. Naumann, B. Norris (eds.), *Automatic Differentiation: Applications, Theory, and Implementations.*

51. A.M. Bruaset, A. Tveito (eds.), *Numerical Solution of Partial Differential Equations on Parallel Computers.*

52. K.H. Hoffmann, A. Meyer (eds.), *Parallel Algorithms and Cluster Computing.*

53. H.-J. Bungartz, M. Schäfer (eds.), *Fluid-Structure Interaction.*

54. J. Behrens, *Adaptive Atmospheric Modeling.*

55. O. Widlund, D. Keyes (eds.), *Domain Decomposition Methods in Science and Engineering XVI.*

56. S. Kassinos, C. Langer, G. Iaccarino, P. Moin (eds.), *Complex Effects in Large Eddy Simulations.*

57. M. Griebel, M.A Schweitzer (eds.), *Meshfree Methods for Partial Differential Equations III.*

58. A.N. Gorban, B. Kégl, D.C. Wunsch, A. Zinovyev (eds.), *Principal Manifolds for Data Visualization and Dimension Reduction.*

59. H. Ammari (ed.), *Modeling and Computations in Electromagnetics: A Volume Dedicated to Jean-Claude Nédélec.*

60. U. Langer, M. Discacciati, D. Keyes, O. Widlund, W. Zulehner (eds.), *Domain Decomposition Methods in Science and Engineering XVII.*

61. T. Mathew, *Domain Decomposition Methods for the Numerical Solution of Partial Differential Equations.*

62. F. Graziani (ed.), *Computational Methods in Transport: Verification and Validation.*

63. M. Bebendorf, *Hierarchical Matrices.* A Means to Efficiently Solve Elliptic Boundary Value Problems.

64. C.H. Bischof, H.M. Bücker, P. Hovland, U. Naumann, J. Utke (eds.), *Advances in Automatic Differentiation.*

65. M. Griebel, M.A. Schweitzer (eds.), *Meshfree Methods for Partial Differential Equations IV.*

66. B. Engquist, P. Lötstedt, O. Runborg (eds.), *Multiscale Modeling and Simulation in Science.*

67. I.H. Tuncer, Ü. Gülcat, D.R. Emerson, K. Matsuno (eds.), *Parallel Computational Fluid Dynamics 2007.*

68. S. Yip, T. Diaz de la Rubia (eds.), *Scientific Modeling and Simulations.*

69. A. Hegarty, N. Kopteva, E. O'Riordan, M. Stynes (eds.), *BAIL* 2008 – *Boundary and Interior Layers.*

70. M. Bercovier, M.J. Gander, R. Kornhuber, O. Widlund (eds.), *Domain Decomposition Methods in Science and Engineering XVIII.*

71. B. Koren, C. Vuik (eds.), *Advanced Computational Methods in Science and Engineering.*

72. M. Peters (ed.), *Computational Fluid Dynamics for Sport Simulation.*

73. H.-J. Bungartz, M. Mehl, M. Schäfer (eds.), *Fluid Structure Interaction II - Modelling, Simulation, Optimization.*

74. D. Tromeur-Dervout, G. Brenner, D.R. Emerson, J. Erhel (eds.), *Parallel Computational Fluid Dynamics 2008.*

75. A.N. Gorban, D. Roose (eds.), *Coping with Complexity: Model Reduction and Data Analysis.*

76. J.S. Hesthaven, E.M. Rønquist (eds.), *Spectral and High Order Methods for Partial Differential Equations.*

77. M. Holtz, *Sparse Grid Quadrature in High Dimensions with Applications in Finance and Insurance.*

78. Y. Huang, R. Kornhuber, O.Widlund, J. Xu (eds.), *Domain Decomposition Methods in Science and Engineering XIX.*

79. M. Griebel, M.A. Schweitzer (eds.), *Meshfree Methods for Partial Differential Equations V.*

80. P.H. Lauritzen, C. Jablonowski, M.A. Taylor, R.D. Nair (eds.), *Numerical Techniques for Global Atmospheric Models.*

81. C. Clavero, J.L. Gracia, F.J. Lisbona (eds.), *BAIL 2010 – Boundary and Interior Layers, Computational and Asymptotic Methods.*

82. B. Engquist, O. Runborg, Y.R. Tsai (eds.), *Numerical Analysis and Multiscale Computations.*

83. I.G. Graham, T.Y. Hou, O. Lakkis, R. Scheichl (eds.), *Numerical Analysis of Multiscale Problems.*

84. A. Logg, K.-A. Mardal, G. Wells (eds.), *Automated Solution of Differential Equations by the Finite Element Method.*

85. J. Blowey, M. Jensen (eds.), *Frontiers in Numerical Analysis - Durham 2010.*

86. O. Kolditz, U.-J. Gorke, H. Shao, W. Wang (eds.), *Thermo-Hydro-Mechanical-Chemical Processes in Fractured Porous Media - Benchmarks and Examples.*

87. S. Forth, P. Hovland, E. Phipps, J. Utke, A. Walther (eds.), *Recent Advances in Algorithmic Differentiation.*

88. J. Garcke, M. Griebel (eds.), *Sparse Grids and Applications.*

89. M. Griebel, M.A. Schweitzer (eds.), *Meshfree Methods for Partial Differential Equations VI.*

90. C. Pechstein, *Finite and Boundary Element Tearing and Interconnecting Solvers for Multiscale Problems.*

91. R. Bank, M. Holst, O. Widlund, J. Xu (eds.), *Domain Decomposition Methods in Science and Engineering XX.*

92. H. Bijl, D. Lucor, S. Mishra, C. Schwab (eds.), *Uncertainty Quantification in Computational Fluid Dynamics.*

93. M. Bader, H.-J. Bungartz, T. Weinzierl (eds.), *Advanced Computing.*

94. M. Ehrhardt, T. Koprucki (eds.), *Advanced Mathematical Models and Numerical Techniques for Multi-Band Effective Mass Approximations.*

95. M. Azaïez, H. El Fekih, J.S. Hesthaven (eds.), *Spectral and High Order Methods for Partial Differential Equations ICOSAHOM 2012.*

96. F. Graziani, M.P. Desjarlais, R. Redmer, S.B. Trickey (eds.), *Frontiers and Challenges in Warm Dense Matter.*

97. J. Garcke, D. Pflüger (eds.), *Sparse Grids and Applications – Munich 2012.*

98. J. Erhel, M. Gander, L. Halpern, G. Pichot, T. Sassi, O. Widlund (eds.), *Domain Decomposition Methods in Science and Engineering XXI.*

99. R. Abgrall, H. Beaugendre, P.M. Congedo, C. Dobrzynski, V. Perrier, M. Ricchiuto (eds.), *High Order Nonlinear Numerical Methods for Evolutionary PDEs - HONOM 2013.*

100. M. Griebel, M.A. Schweitzer (eds.), *Meshfree Methods for Partial Differential Equations VII.*

101. R. Hoppe (ed.), *Optimization with PDE Constraints - OPTPDE 2014.*

102. S. Dahlke, W. Dahmen, M. Griebel, W. Hackbusch, K. Ritter, R. Schneider, C. Schwab, H. Yserentant (eds.), *Extraction of Quantifiable Information from Complex Systems.*

103. A. Abdulle, S. Deparis, D. Kressner, F. Nobile, M. Picasso (eds.), *Numerical Mathematics and Advanced Applications - ENUMATH 2013.*

104. T. Dickopf, M.J. Gander, L. Halpern, R. Krause, L.F. Pavarino (eds.), *Domain Decomposition Methods in Science and Engineering XXII.*

105. M. Mehl, M. Bischoff, M. Schäfer (eds.), *Recent Trends in Computational Engineering - CE2014.* Optimization, Uncertainty, Parallel Algorithms, Coupled and Complex Problems.

106. R.M. Kirby, M. Berzins, J.S. Hesthaven (eds.), *Spectral and High Order Methods for Partial Differential Equations - ICOSAHOM'14.*

107. B. Jüttler, B. Simeon (eds.), *Isogeometric Analysis and Applications 2014.*

108. P. Knobloch (ed.), *Boundary and Interior Layers, Computational and Asymptotic Methods – BAIL 2014.*

109. J. Garcke, D. Pflüger (eds.), *Sparse Grids and Applications – Stuttgart 2014.*

110. H. P. Langtangen, *Finite Difference Computing with Exponential Decay Models.*

111. A. Tveito, G.T. Lines, *Computing Characterizations of Drugs for Ion Channels and Receptors Using Markov Models.*

112. B. Karazösen, M. Manguoğlu, M. Tezer-Sezgin, S. Göktepe, Ö. Uğur (eds.), *Numerical Mathematics and Advanced Applications - ENUMATH 2015.*

113. H.-J. Bungartz, P. Neumann, W.E. Nagel (eds.), *Software for Exascale Computing - SPPEXA 2013-2015.*

114. G.R. Barrenechea, F. Brezzi, A. Cangiani, E.H. Georgoulis (eds.), *Building Bridges: Connections and Challenges in Modern Approaches to Numerical Partial Differential Equations.*

115. M. Griebel, M.A. Schweitzer (eds.), *Meshfree Methods for Partial Differential Equations VIII.*

116. C.-O. Lee, X.-C. Cai, D.E. Keyes, H.H. Kim, A. Klawonn, E.-J. Park, O.B. Widlund (eds.), *Domain Decomposition Methods in Science and Engineering XXIII.*

117. T. Sakurai, S. Zhang, T. Imamura, Y. Yusaku, K. Yoshinobu, H. Takeo (eds.), *Eigenvalue Problems: Algorithms, Software and Applications, in Petascale Computing.* EPASA 2015, Tsukuba, Japan, September 2015.

118. T. Richter (ed.), *Fluid-structure Interactions.* Models, Analysis and Finite Elements.

119. M.L. Bittencourt, N.A. Dumont, J.S. Hesthaven (eds.), *Spectral and High Order Methods for Partial Differential Equations ICOSAHOM 2016.*

120. Z. Huang, M. Stynes, Z. Zhang (eds.), *Boundary and Interior Layers, Computational and Asymptotic Methods BAIL 2016.*

121. S.P.A. Bordas, E.N. Burman, M.G. Larson, M.A. Olshanskii (eds.), *Geometrically Unfitted Finite Element Methods and Applications*. Proceedings of the UCL Workshop 2016.

122. A. Gerisch, R. Penta, J. Lang (eds.), *Multiscale Models in Mechano and Tumor Biology*. Modeling, Homogenization, and Applications.

123. J. Garcke, D. Pflüger, C.G. Webster, G. Zhang (eds.), *Sparse Grids and Applications - Miami 2016.*

124. M. Schäfer, M. Behr, M. Mehl, B. Wohlmuth (eds.), *Recent Advances in Computational Engineering*. Proceedings of the 4th International Conference on Computational Engineering (ICCE 2017) in Darmstadt.

125. P.E. Bjørstad, S.C. Brenner, L. Halpern, R. Kornhuber, H.H. Kim, T. Rahman, O.B. Widlund (eds.), *Domain Decomposition Methods in Science and Engineering XXIV*. 24th International Conference on Domain Decomposition Methods, Svalbard, Norway, February 6–10, 2017.

126. F.A. Radu, K. Kumar, I. Berre, J.M. Nordbotten, I.S. Pop (eds.), *Numerical Mathematics and Advanced Applications – ENUMATH 2017.*

127. X. Roca, A. Loseille (eds.), *27th International Meshing Roundtable.*

128. Th. Apel, U. Langer, A. Meyer, O. Steinbach (eds.), *Advanced Finite Element Methods with Applications*. Selected Papers from the 30th Chemnitz Finite Element Symposium 2017.

129. M. Griebel, M. A. Schweitzer (eds.), *Meshfree Methods for Partial Differencial Equations IX.*

130. S. Weißer, BEM-based Finite Element *Approaches on Polytopal Meshes.*

131. V. A. Garanzha, L. Kamenski, H. Si (eds.), *Numerical Geometry, Grid Generation and Scientific Computing*. Proceedings of the 9th International Conference, NUMGRID 2018/Voronoi 150, Celebrating the 150th Anniversary of G. F. Voronoi, Moscow, Russia, December 2018.

132. E. H. van Brummelen, A. Corsini, S. Perotto, G. Rozza (eds.), *Numerical Methods for Flows.*

For further information on these books please have a look at our mathematics catalogue at the following URL: `www.springer.com/series/3527`

Printed in the United States
By Bookmasters